Lymphokines and interferons

a practical approach

TITLES PUBLISHED IN THE PRACTICAL APPROACH SERIES

Affinity chromatography
Animal cell culture
Biochemical toxicology
Biological membranes
Carbohydrate analysis
Centrifugation (2nd Edition)
DNA cloning
Drosophila
Electron microscopy in molecular biology
Gel electrophoresis of nucleic acids
Gel electrophoresis of proteins
H.p.l.c. of small molecules
Human cytogenetics
Human genetic diseases
Immobilised cells and enzymes
Iodinated density-gradient media
Microcomputers in biology
Mutagenicity testing
Neurochemistry
Nucleic acid and protein sequence analysis
Nucleic acid hybridisation
Oligonucleotide synthesis
Photosynthesis: energy transduction
Plant cell culture
Plasmids
Prostaglandins and related substances
Spectrophotometry and spectrofluorimetry
Steroid hormones
Teratocarcinomas and embryonic stem cells
Transcription and translation
Virology

Lymphokines and interferons

a practical approach

Edited by

M J Clemens

Department of Biochemistry, St. George's Hospital Medical School, Cranmer Terrace, London SW17 0RE, UK

A G Morris

Department of Biological Sciences, University of Warwick, Coventry CV4 7AL, UK

A J H Gearing

National Institute for Biological Standards and Control, Blanche Lane, South Mimms, Potters Bar, Herts EN6 3QG, UK

IRL Press Limited
PO Box 1,
Eynsham,
Oxford OX8 1JJ,
England

©1987 IRL Press Limited

All rights reserved by the publisher. No part of this book may be reproduced or transmitted in any form by any means, electronic or mechanical, including photocopying, recording or any information storage and retrieval system, without permission in writing from the publisher.

British Library Cataloguing in Publication Data

Lymphokines and interferons : a practical approach.—(Practical
approach series)
1. Interferon 2. Lymphokines
I. Clemens, M.J. II. Morris, A.G.
III. Gearing, A.J.H.
574.2'95 QR187.5

ISBN 1-85221-036-2 (Hardbound)
ISBN 0-85221-035-4 (Softbound)

Printed by Information Printing Ltd, Oxford, England

Preface

Interferons and interleukins comprise a group of polypeptides, now collectively known as cytokines, which are central to the establishment and regulation of inflammation and immune responses. Prior to 1979 researchers working with crude mixtures of poorly defined activities were divided into those studying interferon and others studying growth factors, colony stimulating factors and factors affecting cell migration. The development of specific bioassays, cloned recombinant cytokines and monoclonal antibodies for affinity purification has revealed, by 1987, at least 16 different human cytokines. Those that have been cloned are described in Appendix II; other factors will undoubtedly be added to this list in the future. It has become obvious that these cytokines have many overlapping functions and are interlinked in their production, action and regulation. This realisation has brought together what had been the distinct worlds of interferons and interleukins. We have attempted to reflect this union by producing a volume which combines both interferon and interleukin techniques in sufficient detail to enable research workers to set up cytokine laboratories. Production and assay protocols, sources of reagents, antibodies and cell lines are all described.

Where procedures are applicable to all cytokines we have illustrated their use with interferons. For example, production of antibodies and development of immunoassays (Chapter 8), measurement of mRNA levels (Chapter 7) and production of recombinant molecules in bacteria or mammalian cells (Chapters 3 and 5).

Production of murine and human interferons is dealt with in Chapters 1, 2 and 4 and measurement of interferon effects in biological assays is covered in Chapters 6, 9 and 10. Cell growth inhibition and cytotoxicity assays for interferons and/or tumour necrosis factors are described in Chapters 11 and 12. Macrophage activation is a significant property of interferon γ and other cytokines and two assays for the measurement of macrophage activation are covered in Chapters 13 and 14. The remaining chapters cover sources and assays for interleukin 1, interleukin 2, interleukin 3, other colony stimulating factors, eosinophil differentiation factor and finally B cell growth and differentiation factors. We hope that this book is helpful not only to immunologists wishing to extend their research into the cytokine field but also to scientists in other areas such as endocrinology, neuroscience and clinical medicine in which cytokines have been shown to be involved.

M J Clemens
A G Morris
A J H Gearing

Corresponding authors

F.R.Balkwill
Imperial Cancer Research Fund Laboratories, PO Box 123, Lincoln's Inn Fields, London WC2A 3PX, UK

R.E.Callard
Institute of Child Health, 30 Guilford Street, London WCIN IEH, UK

G.W.Duff
Rheumatic Diseases Unit, Department of Medicine, Northern General Hospital, Ferry Road, Edinburgh EH5 2DQ, UK

J.Garland
Department of Immunology, University of Manchester, Stopford Building, Oxford Road, Manchester M13 9PT, UK

A.J.H.Gearing
National Institute for Biological Standards and Control, Blanche Lane, South Mimms, Potters Bar, Herts EN6 3QG, UK

P.A.Kiener
Bristol-Myers Company, Pharmaceutical Research and Development Division, 5 Research Parkway, PO Box 5100, Wallingford, CT 06492-7660, USA

J.A.Lewis
Department of Anatomy and Cell Biology, SUNY Health Sciences Center at Brooklyn, 450 Clarkson Avenue, Brooklyn, NY 11203, USA

D.B.Lowrie
Department of Bacteriology, Royal Postgraduate Medical School, London W12 0HS, UK

N.Matthews
Department of Medical Microbiology, University of Wales College of Medicine, Cardiff, UK

A.Meager
National Institute for Biological Standards and Control, Blanche Lane, South Mimms, Potters Bar, Herts EN6 3QG, UK

A.G.Morris
Department of Biological Sciences, University of Warwick, Coventry CV4 7AL, UK

K.Nagata
Shionogi Research Laboratories, Shionogi & Co. Ltd, Fukushima-ku, Osaka 553, Japan

A.O'Garra
National Institute for Medical Research, The Ridgeway, Mill Hill, London NW7 1AA, UK

J.Shuttleworth
Department of Biological Sciences, University of Warwick, Coventry CV4 7AL, UK

K.W.Siggens
Leicester Biocentre, University of Leicester, University Road, Leicester LE1 7RH, UK

R.H.Silverman
Uniformed Services University of the Health Sciences, F.Edward Hebert School of Medicine, 4301 Jones Bridge Road, Bethesda, MD 20814-4799, USA

L.Vodinelich
Celltech Limited, 250 Bath Road, Slough, UK

Y.Watanabe
Institute for Virus Research, Kyoto University, Sakyo-ku, Kyoto 606, Japan

Contents

Abbreviations

2-5A	2′5′ oligoadenylate
AET	S-2-aminoethylisothiouronium bromide hydrobromide
BAP	bacterial alkaline phosphatase
BCDF	B cell differentiation factor
BCGF	B cell growth factor
b.f.u.	blast-forming unit
BHK	baby hamster kidney
BSA	bovine serum albumin
CB	cellulose buffer
c.f.u.	colony-forming unit
CHO	Chinese hamster ovary
CM	conditioned medium
ConA	concanavalin A
CPE	cytopathic effect
CPER	cytopathic effect reduction
CPG	controlled pore glass
CRP	C-reactive protein
CSF	colony-stimulating factor
%CV	% coefficient of variation
DEPC	diethyl pyrocarbonate
DHFR	dihydrofolate reductase
DI	defective interfering
DMEM	Dulbecco's modified Eagle's medium
DMSO	dimethyl sulphoxide
ds	double-stranded
DTH	delayed type hypersensitivity
DTT	dithiothreitol
EAT	Ehrlich ascites tumour
EBSS	Earle's balanced salt solution
EDF	eosinophil differentiation factor
EDTA	ethylenediamine tetraacetic acid
EIA	enzyme immunoassay
ELISA	enzyme-linked immunosorbent assay
EMCV	encephalomyocarditis virus
FACS	fluorescence-activated cell sorting
FCA	Freund's complete adjuvant
FCS	fetal calf serum
FITC	fluorescein isothiocyanate
GM	growth medium
GPC′	guinea pig complement
HAT	hypoxanthine–aminopterin-thymidine
HBS	Hepes-buffered saline
HBSS	Hank's balanced salt solution
Hepes	*N*-2-hydroxyethylpiperazine-*N*′-2-ethanesulphonic acid
HRP	horseradish peroxidase
HumE	human mammary epithelium
IFA	incomplete Freund's adjuvant
IFN	interferon

IL	interleukin
IRMA	immunoradiometric assay
IRP	international reference preparation
IVNAS	inhibition of viral nucleic acid synthesis
LAF	lymphocyte-activating factor
LAL	*Limulus* amoebocyte lysate
LF	lymphotoxin
LPS	lipopolysaccharide
LRP	laboratory reference preparation
LU	laboratory units
MAF	macrophage-activating factor
McAb	monoclonal antibody
2-ME	2-mercaptoethanol
MEM	minimal essential medium
MEM-G	glucose-enriched minimal essential medium
m.o.i.	multiplicity of infection
MTT	3-(4,5-dimethylthiazol-2-yl)-2,5-diphenyl tetrazolium bromide
NDV	Newcastle disease virus
NGS	normal goat serum
NP-40	Nonidet P-40
OPD	*o*-phenylene diamine
PAGE	polyacrylamide gel electrophoresis
PBMC	peripheral blood mononuclear cells
PBS	phosphate-buffered saline
PE	peritoneal exudate
PEG	polyethylene glycol
PEI	polyethyleneimine
PFC	plaque-forming cells
p.f.u.	plaque-forming unit
PHA	phytohaemagglutinin
Pipes	piperazine-*N*,*N'*-bis-2-ethanesulphonic acid
PMA	phorbol myristate acetate
PMN	polymorphonuclear cells
PMSF	phenylmethylsulphonyl fluoride
pOHPA	*p*-hydroxyphenylacetic acid
PPi	sodium pyrophosphate
PTH	phenylthiohydantoin
PWM	pokeweed mitogen
SAA	serum amyloid A
SAP	serum amyloid P
SDS	sodium dodecyl sulphate
SPF	specific pathogen free
SRBC	sheep red blood cells
SSC	standard saline citrate
TCA	trichloroacetic acid
$TCID_{50}$	50% tissue culture infectious dose
TdR	thymidine
TEAB	triethylammonium bicarbonate
TFA	trifluoroacetic acid

TMC	tonsillar mononuclear cells
TNF	tumour necrosis factor
TRF	T cell replacing factor
TSB	tryptic soy broth
VSV	vesicular stomatitis virus
WECM	WEHI conditioned medium

CHAPTER 1

Induction, production and purification of natural mouse IFN-α and -β

YOSHIHIKO WATANABE and YOSHIMI KAWADE

1. INTRODUCTION

Recombinant DNA technology has made it possible to produce large quantitites of various interferons (IFNs), using *Escherichia coli* or other cells harbouring the IFN genes. However, it is still necessary to have good systems of natural IFN production for several reasons. First, IFN molecules produced in artificial systems, especially in bacteria, may not be exactly identical to natural molecules. Thus, the molecules produced in *E. coli* will lack the sugar moiety, which is present in most species of natural IFN proteins. Also, differences could exist in the amino-terminal and carboxy-terminal structures, owing to technical reasons or to differences in post-translational processing in different cells (1–3). Further, disulphide bridges absent in natural IFN molecules could be formed in *E. coli*-derived molecules (2,4). Second, IFN-α in various mammals, as well as IFN-β in bovine and some other species, comprises multiple genes, and the natural IFN proteins represent a mixture of various subtypes (5); they will have different physiological activities from single subtype IFNs produced by recombinant DNA technology. Third, for the study of the molecular mechanisms of IFN gene expression, well-defined natural IFN-producing systems will be of interest.

In this chapter we deal with induction, production and purification of natural mouse IFN-α and -β. It must be noted that these two types of IFN are concomitantly induced in various mouse systems (6,7), in contrast to human systems in which either IFN-α or -β is produced as a sole or predominant IFN species. First we describe the procedure for induction and production of IFN-α, and -β in L cells; the line of L cell maintained in this laboratory (8) is distinct from the commonly used L929 line, and seems to rank among the best mouse IFN producers that can be found in the literature, such as C243 (9) and Ehrlich ascites tumour (EAT) cells (10). Then we describe our procedures for the purification and biochemical characterization of L cell IFN. Our purification method is very simple and efficient, using antibody affinity chromatography as the central feature. In addition, mention is made of how to separate IFN-α and -β from each other.

2. INDUCTION AND PRODUCTION OF NATURAL MOUSE IFN-α AND -β

2.1 **Cells and IFN inducers**

Mouse fibroblasts, such as L cells, are induced to produce IFN-α and -β when treated with virus or with double-stranded polyribonucleotide, such as polyinosinic acid:poly-

Table 1. Media and salt solutions.

1.	MEM-G[a]: Eagle's minimum essential medium (MEM) containing 60 μg/ml of kanamycin (Nissui Seiyaku Co., Tokyo, Japan), supplemented with 1 g/l of glucose (final 2 g/l).
2.	Growth medium (GM): MEM-G supplemented with 5−10% (v/v) calf serum.
3.	Maintenance medium (MM): MEM-G supplemented with 0.5% (v/v) calf serum.
4.	EBSS: Earle's balanced salt solution.
5.	PBSA: Dulbecco's phosphate-buffered saline, minus Ca^{2+} and Mg^{2+}.

[a]The glucose-enriched MEM can support our L cells in confluency better than regular MEM.

Table 2. Cultivation of mouse L cells.

1.	Grow the L cells in monolayer; their doubling time is about 20 h in GM, and their cell density becomes $1-2 \times 10^5$ cells/cm² at confluency.
2.	For subcultivation, drain the medium from the confluent cell sheet, cover the cells with a small volume of 0.2% trypsin and 0.02% EDTA in PBSA, and incubate for a few minutes at room temperature. Suspend the cells, now loosened from the vessel wall, by vigorous pipetting in GM, and dispense into new culture bottles with a split ratio of 1:6−1:10.
3.	The cells reach confluency after 3−4 days and stay healthy for several days longer.
4.	It is preferable that the subcultivation is done not long after cells reach confluency, but for IFN production the cultures are aged 2−3 days longer.

Table 3. Procedure for preparing Newcastle disease virus stock.

1.	Clean the shells of chicken eggs 10 days after fertilization with 70% ethanol and drill to make a small hole in the shell.
2.	Inoculate each egg with 0.1 ml of NDV solution (containing 10^5-10^6 p.f.u.[a] of NDV) diluted in PBSA, added if necessary with an antibiotic mixture (250 units/ml penicillin and 300 μg/ml streptomycin).
3.	Close the holes with paraffin, and incubate the eggs at 37°C for 2−3 days.
4.	Cut off the air sac part of the shell and harvest the allantoic fluid (usually ~10 ml per egg) by pipetting into appropriate centrifuge tubes.
5.	Clarify the fluid by centrifugation at 7700 *g* for 20 min (Sorvall Centrifuge RC2-B), and harvest the virus by ultracentrifugation at 70 000 *g* for 1 h (Hitachi Automatic Preparative Ultracentrifuge 55P-2; Hitachi, Ltd., Tokyo).
6.	Disperse the pellet in PBSA containing 0.1% BSA in one-tenth of the original volume, and gently homogenize it in a loose fitting Dounce homogenizer.
7.	Clarify the virus suspension by centrifugation at 7700 *g* for 10 min, dispense into small tubes and store at −70°C. About 1×10^{10} p.f.u. of NDV are obtained from one egg.

[a]The virus titre is assayed on primary chick embryo fibroblasts.

cytidylic acid (polyI:polyC). PolyI:polyC exhibits higher IFN-inducing activity in many cases when admixed with DEAE−dextran (11,12) or some other polycations. As for virus, Newcastle disease virus (NDV) and Sendai virus are commonly used. These inducers are effective also in various other cell types, such as lymphoid cells and macrophages. In this laboratory, NDV, Miyadera strain, is used for the IFN induction in L cells because of its higher potency than Sendai virus and polyI:polyC. Killed NDV can be used as an inducer in certain systems, and is convenient especially where live virus is to be avoided (note that the Miyadera strain is virulent in chickens), though our L cells require live NDV for optimal IFN induction. *Table 1* lists the media and salt solutions utilized. We use L cells grown in monolayer, as described in *Table 2*;

L cells that grow in suspension are also available and may be convenient in some cases. *Table 3* describes the procedure for preparing NDV stocks. Sendai virus can be prepared similarly to NDV.

2.2 **Determination of optimal conditions for IFN induction in L cells**

The optimal conditions for IFN induction must first be determined in individual systems, using small-scale cell cultures. Generally speaking, the cell density (or the culture 'age') and the dose of inducer appear to be the most important parameters that have to be examined at the outset. The small-scale IFN production system convenient for this type of experiment is as follows.

(i) Inoculate 1×10^5 L cells, suspended in 1 ml of growth medium (GM), into each well (2 cm^2) of a plastic 24-well tissue culture plate (Nunc), and cultivate for 2–3 days until they form a dense monolayer ($3-5 \times 10^5$ cells/ml).

(ii) Remove the culture fluid, and to each well add NDV in 0.5 ml of glucose-enriched minimal essential medium (MEM-G) containing 0.1% bovine serum albumin (BSA). Then incubate the cells for 1 h at room temperature or at 37°C. From experiments in which the input dose of virus was varied, the optimal dose for IFN production was found to be 10^7-10^8 plaque-forming units (p.f.u.) per well (input multiplicities of from 20 to 200 p.f.u./cell).

(iii) Remove the virus-containing medium and add 1 ml of MEM-G containing 0.1% BSA to each well.

(iv) After incubation at 37°C for 24 h, harvest the culture fluids, centrifuge for clarification, and acidify to below pH 2 by adding HCl to inactivate the NDV.

(v) Four days later, assay the fluids for IFN (see Section 3.3 and *Table 4*).

The culture fluid, upon optimal induction, will contain $5-10 \times 10^4$ international units (IU)/ml of IFN as a mixture of IFN-α and -β species.

The NDV multiplies only poorly in L cells (the titre in the harvested fluid being of the order of 10^7 p.f.u./ml), but it causes a strong cytopathic effect (CPE). If the CPE is not marked at the time of harvest, due to, for instance, inadvertantly low input of virus, the IFN yield will be low. At pH 2.0, it takes 4 days to inactivate NDV in the harvested fluid to below a detectable level.

2.3 **Kinetics of IFN induction in L cells after NDV infection**

To define the best time for harvesting L cell IFN, we examined the kinetics of IFN accumulation in the culture fluid after NDV challenge. As shown in *Figure 1*, the IFN activity appeared at 5–6 h in the medium, increased gradually thereafter and reached a plateau level at 16–20 h. We therefore harvest the culture fluid 20–24 h after NDV challenge.

The time course of IFN induction was examined also at the transcriptional level by Northern blot analysis using mouse IFN-β cDNA (13) as probe. The kinetics of IFN mRNA accumulation corresponded well to that of IFN accumulation in the culture fluid. Thus, the mRNA was detectable at 6 h post-infection, and its level increased until it reached a maximal level at 12–16 h; then the level decreased rapidly and became undetectable at 24 h (14).

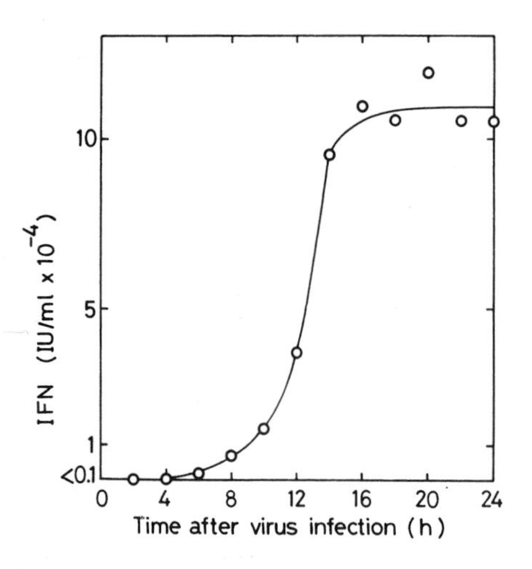

Figure 1. Time course of IFN production in L cells. Cumulative IFN activities in the culture fluid of L cells were examined at the indicated times after NDV challenge.

2.4 Production of mouse IFN-α and -β

The procedure for preparing a large amount of culture fluid (2.5–5 l) from NDV-infected L cells, containing a few milligrams of IFN protein, is as follows.

(i) Culture L cells in glass Roux bottles (cell monolayer area 150 cm^2 per bottle), each containing 50 ml of GM, until they form a densely packed monolayer. The cell number in a Roux bottle will then be $3-5 \times 10^7$ cells ($2-3 \times 10^5/cm^2$). In large-scale production, we use 50–100 bottles at one time, which are assembled in groups of six, by placing three bottles on an aluminium plate (18 × 34 cm) and securing one such set on top of another by rubber bands.

(ii) Discard the medium, add 10 ml of NDV diluted in MEM-G (10^8-10^9 p.f.u.) and incubate the cultures for 1 h at room temperature.

(iii) Aspirate the virus-containing medium into an acid trap, and wash the cell sheets twice with 15 ml of Earle's balanced salt solution (EBSS) to reduce extraneous proteins; this is important to obtain a crude IFN preparation with a high specific activity.

(iv) Add 50 ml of protein-free MEM-G containing 20 μM phenylmethylsulphonyl fluoride (PMSF), a protease inhibitor, to each bottle. The omission of protein does not reduce the level of IFN production as compared with medium containing protein such as 0.1% BSA and 5% calf serum.

(v) After incubation for 20–24 h at 37°C, inspect the cell sheets under a microscope for strong CPE, and harvest the culture fluid and clarify it by centrifugation (7700 *g* for 20 min; Sorvall Superspeed Automatic Refrigerated Centrifuge RC2-B) at 4°C.

(vi) Acidify the fluid to below pH 2 by adding 1 M HCl solution (final 0.03–0.06 M HCl) and store at 4°C for over 4 days to inactivate NDV. Add an appropriate detergent solution to the Roux bottles, and autoclave the bottles to solubilize cell debris and kill virus attached to the bottles.

(vii) Assay the harvested culture fluid for IFN activity and protein content. It will contain $5-10 \times 10^4$ IU/ml of IFN and approximately 50 μg/ml of protein.

Crude materials thus prepared can be stored at 4°C under the acidic condition for several months without any loss of IFN activity. Otherwise, the materials can be neutralized by adding NaOH solution, and stored frozen at −20°C also without inactivation.

3. PURIFICATION OF NATURAL IFN-α AND -β FROM L CELLS

3.1 **General comments**

Since IFN is often labile at protein concentrations below about 100 μg/ml, in common with various other proteins, it is necessary that the protein concentration of IFN preparations is not rendered too low during the entire process of IFN purification. Addition of protein such as BSA (1 mg/ml or more) greatly stabilizes IFN at low concntrations but this is of course not desirable in IFN purification. This means that one should start with a large amount of crude IFN for successful purification, so that the protein concentration of even the most highly purified samples will not become too low. We always utilize, as the starting material, at least 2 l of crude L cell IFN which contains a total of over 10^8 IU of IFN.

Repeated freezing and thawing of IFN samples should be avoided because it often results in extensive inactivation of the IFN activity. Buffer solutions are autoclaved before use to sterilize them. The whole purification procdure should be carried out at below 4°C.

Though various procedures for mouse IFN purification have been reported from this (15,16) and other laboratories (9,17), most of them consist of multiple steps and are time-consuming; the recovery of IFN activity has often been low. We describe here a greatly simplified method to purify mouse IFN-α and -β with high recovery of IFN activity. It consists essentially of two steps: (i) initial purification, mainly to concentrate IFN from the crude material, and (ii) immunosorbent affinity column chromatography.

3.2 **Determination of protein concentration**

Protein concentration is determined by either of the following two methods: (1) measuring the absorbance at 280 nm, and (2) measuring the binding of Coomassie Brilliant Blue G250 according to Sedmak and Grossberg (18). Method (1) is used under the assumption that $E_{280}^{1\%} = 10.0$. In method (2), we use BSA as standard protein.

3.3 **Determination of IFN activity**

Our method of IFN assay uses L cells as indicator cells and vesicular stomatitis virus (VSV) as challenge virus, and is based on the inhibition by IFN of the CPE caused by VSV infection. The procedure is indicatd in *Table 4*. The most essential points in this type of assay are as follows.

(i) The cell density of indicator cells should be chosen so as to give optimal virus sensitivity (strong CPE) and IFN sensitivity. Low cell density will give good CPE but low IFN sensitivity, whereas high cell density will give high IFN sen-

Table 4. Procedure for IFN assay.

1.	Inoculate 3×10^4 L cells in 0.1 ml of GM in each well of a 96-well microtitre plate (Nunc).
2.	After 10–20 h incubation at 37°C in a 5% CO_2-incubator, drain the GM by aspiration or by shaking the plate quickly.
3.	Immediately add IFN samples, serially diluted in MM, to individual wells (0.1 ml per well).
4.	Incubate the plates at 37°C for 8–20 h in a 5% CO_2-incubator.
5.	Withdraw the IFN-containing medium as in step 2.
6.	Without delay, add VSV in MM (containing ~100 $TCID_{50}$ in 0.1 ml) to each well.
7.	At 20–30 h post-infection, assess the CPE of VSV in each well either by observation under a microscope, or alternatively by staining the cells with crystal violet. The extent of the CPE is scored from zero to four: zero means no CPE or less than a tenth of the cell sheet destroyed, one means about a quarter destroyed, two means about a half destroyed, three means about three quarters destroyed, and four means over nine-tenths destroyed.
8.	The reciprocal of the dilution of the IFN sample that gives 50% protection of the cell sheet from the CPE (a CPE score of two) is taken to be the titre, and is expressed as experimental units per ml. This should be normalized to IU/ml to correct for variations in IFN sensitivity of the assay system from one assay to another; for this purpose, the international reference mouse IFN-α, β preparation (NIH-G002-904-511) or the equivalent must be titrated as the internal standard in each assay in parallel with the test samples.

sitivity but tend to make CPE hard to observe. In our system, 3×10^4 cells per well are found to be most appropriate.

(ii) The dose of challenge virus should be sufficient so that complete CPE is produced with certainty in control cultures not treated with IFN, although too high virus doses will cause low IFN sensitivities. We usually use about 100 $TCID_{50}$ (50% tissue culture infectious dose) of VSV as challenge dose; some investigators express the virus dose in p.f.u., but we believe $TCID_{50}$, as determined in the system employed, is better because it is more directly relevant to the assay. The IFN sensitivity of this assay system is such that one experimental unit/ml (protection of 50% of the cell sheet from viral CPE) equals 1–5 IU/ml. When in a hurry, the time of incubation of cells with the IFN samples before virus challenge may be shortened to 8 h without sacrificing the IFN sensitivity of the assay.

3.4 Concentration of IFN samples

Concentration of IFN samples often becomes necessary at various steps of purification. One of the most powerful methods is lyophilization, but it may occasionally bring about an extensive loss of activity of highly purified IFN preparations for unknown reasons. However when other proteins such as BSA are added as stabilizers to highly purified IFN samples, the IFN activity is well preserved. We find Aquacide II powder [Calbiochem-Behring; carboxymethyl (CM) cellulose with high molecular weights of ~500 000] convenient for concentration; the IFN sample solution is sealed in a dialysis tube, covered entirely with the dry powder and left in the cold until most of the aqueous content is sucked out by the powder. If necessary, the concentrated IFN sample in the same tube can then be dialysed against a new buffer. Care must be taken not to suck out the aqueous content completely, however, because the recovery of IFN activity from the tube may then become very low.

3.5 **Initial purification of L cell IFN**

For the initial purification of L cell IFN-α and -β we use either of two adsorbents, Blue–Sepharose and controlled pore glass (CPG), both endowed with a high capacity for protein binding.

3.5.1 *Blue–Sepharose*

Blue–Sepharose (Pharmacia; Sepharose conjugated with the dye Cibacron Blue F3G-A) can adsorb various proteins including mouse IFN-α and -β (19). Adsorb mouse IFN-α and -β from a solution of low ionic strength such as 10 mM Tris-HCl buffer (pH 7.6); using our crude L cell IFN with specific activities of $5-20 \times 10^5$ IU/mg protein, $2-10 \times 10^6$ IU of IFN activity are retained per 1 ml of gel volume. Then desorb IFN with a high ionic strength solution, such as 10 mM Tris-HCl (pH 7.6) containing 1 M NaCl and 50% ethyleneglycol, or the same Tris-HCl containing 3 M NaCl, at a flow-rate of 0.2–0.4 ml/min. The eluates have a specific activity not much above that of the crude material. Thus, this method is not very efficient for purification, but is occasionally useful for concentration of both crude and highly purified L cell IFN samples, owing to the high protein binding capacity of the column. Blue–Sepharose can be re-used after washing with a 1% Triton X-100 solution.

3.5.2 *Controlled pore glass*

Mouse IFN-α and -β can be adsorbed on CPG at neutral pH and eluted with an acidic buffer (20). We utilize CPG 350 or CPG 240 (Sigma) packed in a 9×1.4 cm column. (It may also be used as a suspension.) The binding capacity of the column is $1-3 \times 10^7$ IU of our crude L cell IFN per ml of packed CPG, corresponding to about 200–600 ml of the crude L cell culture fluid. After loading, wash the column extensively with Ca^{2+}- and Mg^{2+}-free Dulbecco's phosphate-buffered saline (PBSA), and elute the bound IFN with 0.02 M HCl solution containing 0.1 M NaCl, 50% ethyleneglycol and 20 μM PMSF at a flow-rate of 0.2–0.4 ml/min. The IFN eluted has a considerably elevated specific activity ranging from 1 to 6×10^7 IU/mg protein. A typical result is shown in *Table 5*.

The CPG is re-usable after washing in the following way.

(i) Transfer the CPG to a Pyrex glass tube which is stoppered at the bottom with a glass filter, and slowly pour about an equal volume of concentrated H_2SO_4 onto it, taking care to avoid excessive heating.

(ii) Allow the sulphuric acid to pass through the packed CPG slowly.

(iii) Wash the column with distilled water to regenerate clean colourless CPG.

Table 5. Purification of mouse L cell IFN.

Purification step	*Volume (ml)*	*Protein (mg)*	*Interferon (IU)*	*Recovery (%)*	*Specific activity (IU/mg protein)*
Culture fluid	1500	87	1.8×10^8	100	2.1×10^6
CPG	10	4	2.2×10^8	125	5.6×10^7
Antibody column	8	0.34	1.3×10^8	74	3.8×10^8

3.6 Immunosorbent affinity column chromatography

3.6.1 *Antibodies*

For IFN purification, monoclonal antibodies are ideal as affinity sorbents; the use of monoclonal antibodies for mouse IFN-β (21) and -γ (22) has been described. However, monoclonal antibodies to mouse IFN-α have only recently been obtained (Y.Watanabe, manuscript in preparation), and a procedure for purifying natural mouse IFN-α (which is a mixture of various subtypes) without loss has not yet been established.

Polyclonal antibodies (antisera) to mouse IFN-α and -β can readily be raised in xenogenic animals, such as rabbit (23) and sheep (24,25), if enough IFN protein is used as immunogen, and such antisera will be useful as reagents for purification, especially of natural IFN-α.

The immunogens used by many investigators contain both IFN-α and -β, and therefore the antisera obtained usually contain both anti-α and anti-β antibodies. The antibody titre is usually measured by neutralization of the anti-viral activity of IFN; it is highly desirable to do the neutralization tests against IFN-α and IFN-β separately, since the relative contents of anti-α and anti-β antibodies vary widely in different animals immunized even with the same IFN preparation (α,β mixture). Also, it must be noted that the relationship between the neutralization titre and the IFN-binding antibody content is not simple, as discussed elsewhere (26). Antibody titration in terms of IFN binding, rather than neutralization, is therefore desirable.

Since the IFN preparations used as immunogen are usually impure, the antisera obtained generally contain antibodies against the impurities and these should be removed as much as possible. This is done by extensive absorption of the antiserum with conceivable impurities, such as normal components of IFN-producing cells, inducer virus, and so on.

Here we describe an affinity sorbent made by using a high-titred sheep antiserum against L cell IFN, kindly provided by Drs Dalton and Paucker of Pennsylvania Medical College, Philadelphia, PA (25).

3.6.2 *Purification of antibody*

The sheep polyclonal antibody we use for the IFN purification is called MONA, and has high neutralization titres to both mouse IFN-α and -β (27); at a dilution of $1{:}10^5$, it neutralizes 10 units/ml of both mouse IFN-α and -β to 1 unit/ml or less. Purification of the antibody is carried out by absorption with L cells and with an 'impurity column' (16) in following way.

(i) Incubate the IgG fraction of 24 ml of the antiserum overnight at 4°C with L cells densely grown in 10 Roux bottles.

(ii) Load the supernatant onto the 'impurity column', and take the flow-through fraction; wash the column with 0.2 M acetic acid (pH 3) followed by PBSA.

(iii) Load the flow-through fraction again onto the washed impurity column. Repeat this procedure (13 times in all) until u.v.-absorbing materials at 280 nm become undetectable in the acid eluates from the column.

(iv) Subject the final flow-through fraction to DEAE–cellulose column chromatography, and concentrate the IgG in the flow-through fraction by precipitation with 50% ammonium sulphate solution.

In these processes, no significant loss of the IFN-neutralizing activity is observed.

3.6.3 *Conjugation of purified antibody to AffiGel 10*

The purified and concentrated antibody MONA can be conjugated to AffiGel 10 (BioRad) to make an affinity immunosorbent.

(i) After dialysis at 4°C against 1 l of 5 mM Hepes (pH 7.5)−0.1 M NaCl solution, mix the antibody (506 mg protein in 7.5 ml) with 20 ml wet volume of AffiGel 10, and shake gently at 4°C for 16 h in a reaction volume of 30 ml. After this reaction, 20% of the protein (102 mg) and about 5−7% of the IFN-neutralizing activity of the antibody preparation remain in the supernatant of the reaction mixture.
(ii) Block residual reactive sites of the gel by further incubation with 0.1 M monomethanolamine (pH 8) for 2 h at room temperature.
(iii) Pack the MONA-conjugated AffiGel 10 in an appropriate 10 × 2 cm cylindrical tube (which we call MONA-column) and sequentially wash with PBSA, 2 M NaCl in 0.2 M phosphate buffer (pH 7.6), 50% ethyleneglycol in PBSA, 0.1% Triton X-100 in PBSA, 0.2 M acetic acid−0.1 M NaCl solution and again PBSA.
(iv) Store the gel at 4°C in the presence of 0.02% NaN_3.

3.6.4 *Procedure for antibody column chromatography*

For efficient elution of adsorbed IFN from the MONA-column, an eluant with sufficiently low pH is required. We use 0.02 M HCl containing 0.1 M NaCl, 10% glycerol, 20 μM PMSF and, if necessary, 0.05% BSA.

The IFN-binding capacity is determined by using a small column (1 ml bed volume) and loading an excess of partially purified IFN. After washing extensively with PBSA until the protein in the effluent became almost undetectable by u.v. absorption, the adsorbed material is eluted with the acidic solution described above and assayed for IFN. From this experiment, the column was found to bind approximately 10^8 IU of IFN per ml of bed volume.

The procedure for large-scale purification of L cell IFN (10^8-10^9 IU or more at one time) using a MONA-column with a bed volume of 10−20 ml is similar to the above, but the column is more extensively washed before elution of IFN to remove proteins non-specifically bound to it, using the following in sequence:

(i) PBSA,
(ii) 1 M NaCl in 0.2 M phosphate buffer (pH 7.6),
(iii) 50% ethyleneglycol in PBSA,
(iv) 0.1% Triton X-100 in PBSA,
(v) PBSA again,

each with at least a 20-fold column volume and at a flow-rate of 0.2−0.4 ml/min. Also, BSA is omitted from the acid eluant. The elution of IFN from the column is monitored by measuring the u.v. absorbance (280 nm) of the eluate. This is a good measure of IFN protein, since now the eluted protein consists of pure or nearly pure IFN.

The MONA-column is re-usable many times without any significant loss of its purification power. No release of the conjugated antibody was detected in acid eluates as examined by SDS-polyacrylamide gel electrophoresis (SDS−PAGE) and silver staining.

3.7 **Purification of L cell IFN**

A typical result of purification of L cell IFN in two sequential steps using CPG and the MONA-column is given in *Table 5*. The entire procedure took 5 days. The recovery of IFN activity was high and the purified IFN was electrophoretically pure as described in the following section.

3.8 **Biochemical analysis of purified L cell IFN**

3.8.1 *SDS–polyacrylamide gel electrophoresis*

The purity and the molecular heterogeneity of purified L cell IFN can be investigated by using Laemmli's buffer system (28). We use 1 mm thick slab gels which consist of a 12% polyacrylamide-containing separation gel overlaid with a 4.5% stacking gel. Heat the samples for 5 min in 0.1% SDS-containing sample buffer in the presence or absence of 5% 2-mercaptoethanol and electrophorese at room temperature under constant current conditions (20 mA). After electrophoresis, detect protein bands in the gel by staining the gel with 0.125% Coomassie Brilliant Blue R250 dissolved in 10% acetic acid and 40% methanol solution, and destaining with 7.5% acetic acid and 5% methanol solution.

Protein contents of the bands are estimated by densitometry at 575 nm, as standardized by the bands of known amounts of BSA electrophoresed on the same gel. [For more highly sensitive detection of protein bands, we utilize a silver stain method (29) with the aid of Ag-Stain-'Daiichi' (Daiichi Pure Chemicals Co., Ltd., Tokyo).]

Distribution of IFN activity in the gel is determined as follows. Soon after the electrophoresis, cut out a gel lane and slice it into 1 mm wide pieces. Extract each piece with 1 ml of GM in a rubber-stoppered tube shaken at 4°C for 16 h, and assay the extracts for IFN. The recovery of IFN activity in such an extraction is generally quantitative. *Figures 2* and *3* show a typical SDS–PAGE analysis of L cell IFN purified as indicated in *Table 5*. Coincidence of the protein bands and the IFN activity is excellent, indicating high purity of the material.

3.9 **Molecular heterogeneity of L cell IFN**

The results of the SDS–PAGE analysis (*Figure 3*) revealed that NDV-induced L cell IFN comprised at least five molecular species with molecular weights of 35 000, 30 000, 24 000, 22 000 and 18 000 daltons. The first two species are the β-type and the other three the α-type, as indicated by neutralization tests using rabbit antisera specific to either mouse IFN-α or -β (23). Distinction of the two IFN types was also evident when their cross-species activity was examined on human fibroblast GM2405 cells. Thus, as indicated in *Table 6*, the anti-viral activities of the two species with higher molecular weights of 35 000 and 30 000 daltons were very low on the human cells, while those of the two with lower molecular weights of 24 000 and 22 000 daltons were relatively high (the species with the lowest molecular weight of 18 000 daltons was not investigated owing to its low content). These low and high cross-species activities are characteristic of β- and α-type mouse IFNs, respectively, as previously reported (30).

Previous work in this laboratory (31,16) showed two molecular species of L cell

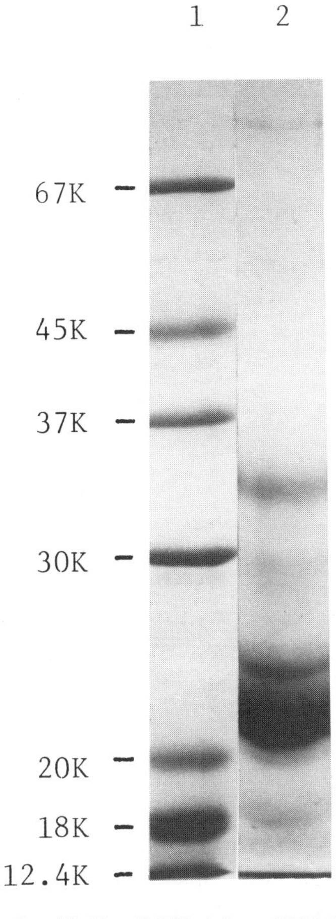

Figure 2. SDS−PAGE analysis of purified L cell IFN. 6.8 μg of IFN (3.3 × 10^6 IU), purified as in *Table 5*, were applied to **lane 2**. **Lane 1** shows a mixture of molecular weight markers (2.5 μg each): BSA (67 K), ovalbumin (45 K), glucose 6-phosphate dehydrogenase (37 K), carbonic anhydrase (30 K), soybean trypsin inhibitor (20 K), myoglobin (18 K) and cytochrome *c* (12.4 K). The gel was stained with Coomassie Brilliant Blue R250.

IFN which were distinguishable in electrophoretic migration and in their antigenicity; they were designated F (molecular weight 24 000 daltons, the α type) and S (36 000 daltons, the β type), but the presence of the minor β-type species (30 000 daltons) was not noted. This minor β-type species may correspond to the β-type subspecies detected in C243 cell- and EAT cell-derived IFN (32,33). This heterogeneity of IFN-β species is considered to be due to variations in the sugar moiety; note that there exist three potential N-glycosylation sites on the polypeptide (13). The presence in L cell IFN of at least three α-type species with molecular weights of 24 000, 22 000 (the major species) and 18 000 daltons is understandable in view of the existence of multiple IFN-α genes (5). Most of these mouse IFN-α subtypes are considered to be glycoproteins (30), in contrast to their human counterparts which are not glycosylated.

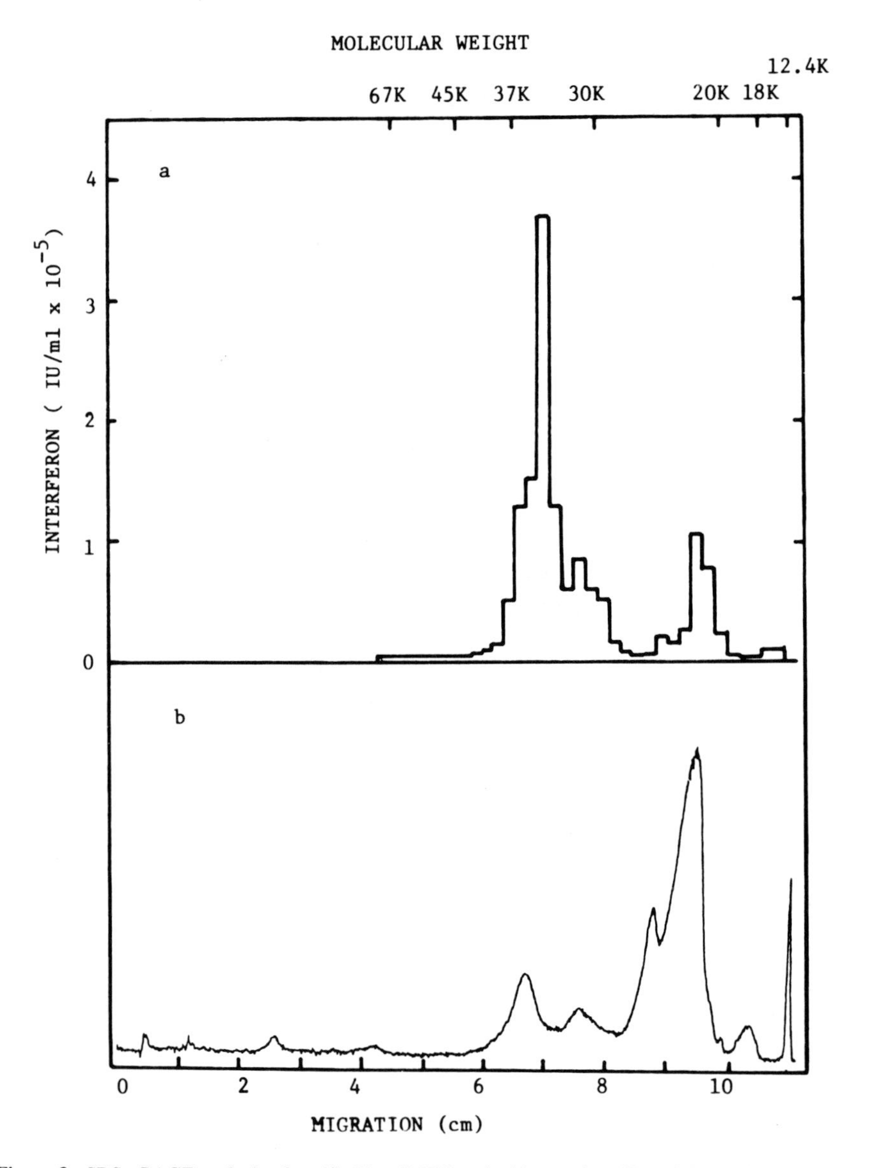

Figure 3. SDS–PAGE analysis of purified L cell IFN; coincidence of profiles of IFN activity and protein. **(a)** 3.4 μg of purified L cell IFN (1.63×10^6 IU) were co-electrophoresed in a lane neighbouring lane 2 in *Figure 2*, and the distribution of IFN activity was determined as described in Section 3.8.1. **(b)** The densitometer tracing of lane 2 in *Figure 2* at 575 nm, obtained with a double-beam recording microdensitometer (Joyce Loebl & Co. Ltd.).

3.10 Specific activities of L cell IFNs

Highly purified L cell IFN preparations, a mixture of IFN-α and -β species, are found to have a specific activity of $3-5 \times 10^8$ IU/mg protein. We can also estimate the values for each of the molecular species of L cell IFN separated by SDS–PAGE from the ratio of IFN activity to protein content of each band. As shown in *Table 6*, the two

Table 6. Characterization of four molecular species of mouse L cell IFN.

Molecular species	*Antigenicity*	*Species specificity (human/mouse, %)*	*Specific activity*[a] *(IU/mg protein)*
35 K	β	0.002	1.09×10^9
30 K	β	0.0035	5.1×10^8
24 K	α	1.7	4.6×10^7
22 K	α	2.7	8.3×10^7

[a]The IFN titres were determined in L cells.

IFN-β species possess relatively high values of 5.1×10^8 and 1.09×10^9 IU/mg protein, in agreement with the values previously reported for mouse IFN-β (9,16,17,33), whereas the two IFN-α species with molecular weights of 24 000 and 22 000 daltons exhibit specific activities of 0.5×10^8 and 0.8×10^8 IU/mg protein, respectively (the value for the α-type species with the smallest molecular weight was not determined because of its low IFN activity). These values for IFN-α are one order of magnitude lower than those estimated previously (9,16,17,33). The reason for this difference is not clear, and a number of possibilities may be considered. First, the present IFN-α might be contaminated by a large number of impurities that co-migrate with IFN-α in SDS−PAGE. This appears unlikely because the purified IFN preparations did not reveal any non-IFN proteins in non-SDS−PAGE at pH 4.3, and on monoQ and monoS column chromatography (data not shown). Second, our IFN-α might have undergone more extensive inactivation during purification and SDS−PAGE than our previous IFN-α. This also appears unlikely, in view of the good recovery of IFN activity during the whole procedure. Third, L cell IFN-α is a mixture of various subtypes, whose composition may well vary from one preparation to another. It seems likely that different IFN-α subtypes differ from each other in specific activity, and therefore the specific activity of natural IFN-α preparations may not necessarily be constant. Further studies seem necessary to settle this problem.

4. SEPARATION OF MOUSE IFN-α AND -β FROM EACH OTHER

As mentioned earlier, various mouse cells produce IFN-α and -β together; at present no convenient systems that produce either type of mouse IFN alone are known. It is therefore necessary to have a means of separating IFN-α and -β from each other.

As described above, they can be clearly separated by SDS−PAGE. This is a convenient and reliable method, but is limited to microgram quantities of IFN; loading larger amounts tends to sacrifice the resolution of the different species. Also, one may feel some doubt as to whether or not the IFN molecules exposed to SDS are really native (although the IFN activity is well preserved); further, the IFN preparations obtained inevitably contain some SDS, which could be troublesome for certain purposes.

To avoid the use of SDS, non-SDS−PAGE at pH 4.3 can be employed, but the resolution tends to be lower than that of SDS−PAGE, and the limitation about the quantity of IFN that can be applied is no less.

The most desirable means of IFN-α and -β separation involves the use of monoclonal antibodies. A monoclonal antibody to mouse IFN-β was reported to be able to purify mouse IFN-β in one step from polyI:polyC-induced L929 cell IFN containing both IFN-α and -β in comparable amounts (21). Monoclonal antibodies to mouse IFN-α have not been reported in the literature, but we recently obtained those to mouse IFN-α as well as those to mouse IFN-β, and are in the process of establishing their use as immunosorbents. In the case of IFN-α, it is unlikely that a single monoclonal antibody reacts with all the different naturally occurring subtypes. Therefore, methods of purifying natural IFN-α without loss of any particular subtype will probably require a combination of two or more monoclonal antibodies. One of our antibodies to IFN-α, however, can bind to most of our L cell IFN-α and seems to be a promising agent for its purification.

5. ACKNOWLEDGEMENTS

We thank Dr B.J.Dalton and the late Dr K.Paucker for their generous gift of sheep antibody MONA and A.Iwata for his excellent assistance in this work.

6. REFERENCES

1. Rinderknecht,E., O'Conner,B.H. and Rodriguez,H. (1984) *J. Biol. Chem.*, **259**, 6790.
2. Bodo,G. and Maurer-Fogy,I. (1986) In *The Biology of the Interferon System 1985.* Stewart,W.E.,II and Schellekens,H. (eds), Elsevier Science Publications B.V., Amsterdam, p. 59.
3. Nagata,K., Kikuchi,N., Ohara,O., Teraoka,H., Yoshida,N. and Kawade,Y. (1986) *FEBS Lett.*, **205**, 200.
4. Mark,D.F., Lu,S.D., Creasey,A.A., Yamamoto,R. and Lin,L.S. (1984) *Proc. Natl. Acad. Sci. USA*, **81**, 5662.
5. Weissmann,C. and Weber,H. (1986) *Prog. Nucleic Acid Res. Mol. Biol.*, in press.
6. Kawade,Y. (1983) In *Humoral Factors in Host Defense.* Yamamura,Y., Kishimoto,T., Hayashi,H., Honjo,T. and Osawa,T. (eds), Academic Press, New York, p. 175.
7. Yamamoto,Y. (1981) *Virology*, **111**, 312.
8. Kawade,Y. and Yamamoto,Y. (1981) In *Methods in Enzymology.* Pestka,S. (ed.), Academic Press, New York, Vol. 78, p. 139.
9. DeMaeyer-Guignard,J., Tovey,M.G., Gresser,I. and DeMaeyer,E. (1978) *Nature*, **271**, 622.
10. Taira,H., Broeze,R.J., Slattery,E. and Lengyel,P. (1980) *J. Gen. Virol.*, **49**, 231.
11. Stewart,W.E.,II (1979) *The Interferon System.* Springer-Verlag, Wien-New York.
12. Trapman,J. (1979) *FEBS Lett.*, **98**, 107.
13. Higashi,Y., Sokawa,Y., Watanabe,Y., Kawade,Y., Ohno,S., Takaoka,C. and Taniguchi,T. (1983) *J. Biol. Chem.*, **258**, 9522.
14. Higashi,Y. (1985) *Nucleic Acids Res.*, **13**, 5157.
15. Kawade,Y., Fujisawa,J., Yonehara,S., Iwakura,Y. and Yamamoto,Y. (1981) In *Methods in Enzymology.* Pestka,S. (ed.), Academic Press, New York, Vol. 78, p. 522.
16. Iwakura,Y., Yonehara,S. and Kawade,Y. (1978) *J. Biol. Chem.*, **253**, 5074.
17. Kawakita,M., Cabrer,B., Taira,H., Rebello,M., Slattery,E., Weideli,H. and Lengyel,P. (1978) *J. Biol. Chem.*, **253**, 598.
18. Sedmak,J.J. and Grossberg,S.E. (1977) *Anal. Biochem.*, **79**, 544.
19. DeMaeyer-Guignard,J. (1981) In *Methods in Enzymology.* Pestka,S. (ed.), Academic Press, New York, Vol. 78, p. 513.
20. Edy,V.G., Braude,I.A., DeClerq,E., Billiau,A. and DeSomer,P. (1976) *J. Gen. Virol.*, **33**, 517.
21. Vonk,W.P. and Trapman,J. (1983) *J. Interferon Res.*, **3**, 169.
22. Gribaudo,G., Cofano,F., Prat,M. and Landolfo,S. (1985) *J. Interferon Res.*, **5**, 199.
23. Yamamoto,Y. and Kawade,Y. (1980) *Virology*, **103**, 80.
24. Gresser,I., Tovey,M.G., Bandu,M.-T., Maury,C. and Brouty-Boyé,D. (1976) *J. Exp. Med.*, **144**, 1305.
25. Dalton,B.J. and Paucker,K. (1980) *Ann. N.Y. Acad. Sci.*, **350**, 332.
26. Kawade,Y. and Watanabe,Y. (1984) *J. Interferon Res.*, **4**, 571.
27. Kawade,Y., Watanabe,Y., Yamamoto,Y., Fujisawa,J., Dalton,B.J. and Paucker,K. (1981) *Antiviral Res.*, **1**, 167.
28. Laemmli,U.K. (1970) *Nature*, **227**, 680.
29. Oakley,B.R., Kirsch,D.R. and Morris,N.R. (1980) *Anal. Biochem.*, **105**, 361.
30. Fujisawa,J. and Kawade,Y. (1981) *Virology*, **112**, 480.
31. Yamamoto,Y. and Kawade,Y. (1976) *J. Gen. Virol.*, **33**, 225.
32. Kawade,Y., Aguet,M. and Tovey,M.G. (1982) *Antiviral Res.*, **2**, 155.
33. Taira,H., Broeze,R.J., Jayaram,B.M., Lengyel,P., Hunkapiller,M.W. and Hood,L.E. (1980) *Science*, **207**, 528.

CHAPTER 2

Induction, production and purification of murine gamma interferon

PETER A.KIENER and GEORGE L.SPITALNY

1. INTRODUCTION

Each of the species of interferons (IFN) can induce a state of anti-viral resistance in cells that possess specific receptors. In addition to possessing anti-viral activity, IFNs are known to mediate an array of pleiotropic effects on a variety of organ systems in the body. In recent years, one of the most intensively studied has been the effect of IFN on the immune system. IFN-γ has been shown to have the most profound effect of any of the IFN classes on the functioning of the immune system.

Molecular cloning and expression of human and mouse IFN-γ genes have provided a readily available source of protein for pre-clinical and clinical testing. However, because not all researchers have had access to either the clones or the recombinant proteins, it is necessary to have alternative sources of IFN. In this chapter we will describe methodologies for the production and purification of natural murine IFN-γ.

2. INDUCTION OF IFN-γ SYNTHESIS

In this section the conditions and reagents required for optimal induction of IFN-γ synthesis will be described. Direct measurement of induction of synthesis can be determined by quantitating intracellular levels of mRNA. This methodology will be discussed in Chapter 7.

2.1 **Cell source**

It is well established that T-lymphocytes are the major source of IFN-γ (1); however, limited evidence has also been published to suggest that B-lymphocytes can produce small quantities of IFN-γ (2). Even so, it is generally accepted by most investigators that T-cells are the primary source of IFN-γ.

There are two organ systems in the mouse where the number of T-cells present in the tissues make it practical to consider them for collection and induction of IFN synthesis. The first of these is the lymphoid system which is made up of a collection of interconnecting ducts that link a series of lymph nodes. The nodes are distributed throughout the animal's body; their exact position in the tissues can be found in a standard text book on laboratory animals (3). Approximately 40% of cells in the lymph nodes are T-lymphocytes; the bulk of the remainder are primarily B-cells along with a few macrophages and other non-lymphoid cells. Although a trained researcher can remove all of the major lymph nodes, the task remains labour intensive and time consuming, and one which should be seriously reviewed if either large numbers of mice are to be used or if this procedure will be performed on a regular basis.

The second source of cells, which is better suited for producing large quantities of IFN on a regular basis from large numbers of mice, is the spleen. The spleen is easily removed and contains large numbers of cells ($8 \times 10^7 - 10^8$), 30% of which are T-cells.

2.1.1 *Processing of tissues*

It is necessary to process either the spleens or lymph nodes to make the cells ready for stimulating IFN synthesis. Tissue preparation in this case refers to the process of dispersing the cells from organs to form a suspension of individual cells. To accomplish this it is necessary to disrupt the tissue capsule surrounding the cells. This can be achieved using a plunger from a 3 or 5 ml syringe and a piece of no. 7 stainless steel screen. The stainless steel screen is purchased in large sheets and can be cut to any size. Typically the screens are fashioned to fit over the opening of 50 ml tubes or 150 ml beakers. Both the screens and plungers should be sterilized before use.

(i) Over a 150 ml beaker, place one to three spleens, which have been asceptically removed from mice, on the screen and slice them into small pieces with scissors or a scalpel blade.
(ii) Wet the chunks of tissue with a small amount of cold tissue culture medium (typically we use RPMI 1640, which can be purchased from a variety of vendors).
(iii) Using a vertical motion, press the tissue chunks with the syringe plunger onto the stainless steel screen. The tissue should not be scraped across the screen since this tends to lower cell viability.
(iv) When this has been done 10–15 times, again wash the tissues with 5–10 ml of tissue culture medium. This will release the cells from the dispersed tissues, allowing the cells to be collected below in the beaker.
(v) Repeat this process until a whitish, gelatinous mass, which is the remains of the tissue capsule surrounding the spleen, is left on the screen.
(vi) Remove this residue with sterile forceps. Repeat the entire process with the remaining spleens.

A similar process is used with lymph nodes except that because the nodes are so much smaller than spleens it is not necessary to slice them into small pieces with scissors or scalpel.

The method describe above is useful for a small to moderate number of spleens (up to 20). However, when large numbers of spleens (20–100) are going to be used, the process is adapted to a more automated procedure by using a gentle blender to disperse the tissues to form a single cell suspension. We typically use spleens from 100–200 mice to make large quantities of natural IFN-γ.

(i) Treat spleens from 100 mice as above using a larger syringe plunger (50 ml) and only lightly disperse the cells.
(ii) Transfer the spleens to sterile plastic bags (Tekmar Co., Cincinnati, Ohio) and place in a Stomacher Lab Blender 80 (Tekmar Co.). This machine operates by gently pressing the tissues with paddles, which mechanically forces the cells out of the tissue capsule.

(iii) Subject the spleens to this gentle blending action for 3 min.

Following manual or automated dispersal of the spleens pass the cells through two or three layers of sterile gauze to remove large clumps of debris, and then wash them with tissue culture medium. No supplements (e.g. serum or anti-coagulant) are added to the tissue culture medium except for antibiotics. Divide the cells at the equivalent of approximately 10 spleens/50 ml tube and centrifuge (1500 *g*, 5 min), and then resuspend in 10–20 ml of medium and pipette up and down 10–20 times to break apart clumps of cells. Wash the cells again, resuspend in medium, count for total number and assess for viability by trypan blue dye exclusion. Typically, both of the procedures described yield cells with more than 95% viability.

The cells should be resuspended to $5 \times 10^6 - 2 \times 10^7$ cells/ml in medium. When large numbers of spleen cells are processed for the induction of IFN synthesis, up to 40 ml, at 2×10^7 cells/ml, can be distributed into 150 mm tissue culture dishes. If smaller numbers of spleen cells are being used, the following guidelines should be considered:

Diameter of tissue culture dish	Volume of fluid/well	Cells/ml
60 mm	6.0 ml	$1-2 \times 10^7$
35 mm	2.0 ml	$1-2 \times 10^7$
12 mm (24 well cluster)	1.0 ml	$5 \times 10^6 - 10^7$
96 well cluster	0.2 ml	5×10^6

2.1.2 *Use of enriched T-cell preparations*

There is a variety of methods available for enriching or purifying T-cells. The procedures used are numerous and are detailed in a variety of published reports.

It is important to note that we have found that enrichment of T-cells does not significantly improve the induction of IFN-γ synthesis. The greatest yield of IFN-γ has come from whole populations of spleen cells. In fact, when we have vigorously purified T-cells away from accessory cells (dendritic cells or macrophages) the yield of IFN-γ has fallen off dramatically. This topic will again be addressed below in the discussion of induction signals.

2.2 **Signals that induce IFN-γ synthesis — antigens**

One of the features that attracted immunologists to the study of IFN-γ is the association of its synthesis with the generation and expression of an immune response (1,4). Evidence has been presented showing that higher than normal levels of IFN-γ are induced by the development of a cell-mediated immune response (1,5), while little if any enhancement is detected during the development of an antibody response to an antigen (5). When administered to animals, an antigen is processed by accessory cells (macrophages and/or dendritic cells) and then presented by these accessory cells to antigen-binding lymphocytes. These lymphocytes respond by expansion of those cells whose antigen receptors on the surface match the configuration of the particular antigen. It should be noted that the absence of accessory cells from this mixture of cells virtually abolishes the immune response and the induction of IFN-γ synthesis.

For a variety of reasons, it would be labour intensive to study the induction of IFN-γ synthesis *in vivo*. Also, development of a *de novo* immune response *in vitro* is difficult to set up. The easiest and most reproducible system for studying induction of IFN-γ is to use spleen or lymph node cells from mice that are either undergoing an active immune response to antigen or have developed a state of memory immunity as a result of previous exposure to antigen.

Suspensions of single cells from immunized animals are placed in culture vessels. Preliminary studies need to be performed to determine both the optimal amount of antigen to add to the cells, and the timing of induction of synthesis. This is somewhat simplified since many events in this process are coincident with each other, so that if one knows when, for example, proliferation begins it is quite likely that induction of IFN-γ will occur at or about the same time.

Antigenic induction of IFN synthesis is a clonally restricted process; that is to say that only those cells that are genetically programmed to respond to that antigen will respond. This generally represents a very small percentage of the total population of lymphocytes in the spleen and so it would be rather difficult to detect induction of synthesis. This problem can be alleviated by employing compounds that are not so clonally restricted; these are described in the next section.

2.3 **Signals that induce IFN-γ synthesis — mitogens**

Unlike clonally restricted antigens, plant mitogens are polyclonal activators. Two mitogens are particularly effective for inducing IFN-γ synthesis; these are concanavalin (Con A) and phytohaemagglutinin (PHA). Both of these have been shown to be rather specific for activating T-cells and not B-cells or macrophages. While antigen induction will provide information about the selective number of cells responsive to a specific stimulus, exposure to Con A or PHA will reveal the IFN-γ synthesizing potential of the entire repertoire of T-cells.

It is important to note that not all sources of PHA and Con A are equivalent in their ability to induce IFN-γ synthesis. We have found that PHA-P from Burroughs-Wellcome (Research Triangle, NC) consistently induces the highest levels of IFN-γ; for Con A we have relied on purified Type IV (cat. number C-2010) from Sigma, St. Louis, MO. For our studies we have used PHA-P at 5 μg/ml or Con A at 2 μg/ml, regardless of cell number and volume of fluid. However, in each particular study it is best to perform a range-finding study to determine which concentration of mitogen induces maximum activity.

3. PRODUCTION OF IFN-γ

Following induction of IFN-γ synthesis, a variety of subcellular events take place ultimately leading to the secretion of the IFN-γ protein from the cell.

3.1 **Kinetics of production *in vitro***

Once the parameters for induction of IFN-γ synthesis have been determined, the kinetics of production can be assessed by measuring either the accumulation of IFN-γ over time or production per unit time. Obviously as IFN is produced it will accumulate in the culture medium, so the first measure is easily done by placing the cells in culture, stimulating with mitogen, collecting samples of the culture medium over time and testing for IFN activity. Previous studies have shown that near peak levels of IFN-γ accumulate in the surrounding medium in 18–24 h. Maintaining the cells for an addition 24–48 h does not increase the yield much more (1).

To measure production per unit time, it is necessary to collect all of the medium from the culture vessel, wash the cells once or twice with medium and then place the cells back in culture. This process is repeated for as many cycles as desired. This procedure is best carried out with parallel cultures so it is possible to keep a check on cell viability at each time point.

3.1.1 *Kinetics of production in vivo*

If desired, it is possible to measure the production of IFN-γ *in vivo*. Unlike *in vitro* production where antigen- or mitogen-induced synthesis can be employed, only antigen-induced IFN production can be used *in vivo*. This is because if PHA or Con A is administered to mice the mitogens will rapidly bind to irrelevant cell and tissue proteins, never reaching the appropriate site to stimulate T-cells. The best antigen to use in these studies is one that can induce a cell-mediated response since those that only induce an antibody response yield extremely low levels of IFN-γ.

It has been shown that peak production of IFN-γ *in vivo* coincides with peak development of immune response to the immunizing antigens. One of the first such systems studied was production of IFN-γ during a self-limiting infection with *Mycobacterium bovis* strain BCG (4). The development of immunity and maximum production of IFN-γ were shown to coincide with the peak delayed-type hypersensitivity (DTH) response to antigen injection subcutaneously. Thus, to gain a preliminary idea as to when IFN-γ might be optimally produced, an assessment of immunological parameters such as DTH should be measured for the antigen under study. Once this has been determined, it is also necessary to determine the amount of antigen that needs to be injected to induce IFN production. Generally, to induce maximal levels of IFN-γ *in vivo*, excessive quantities of antigen must be injected into the animals (4). It has been found that following antigen injection, peak levels of IFN-γ can be detected in the serum of animals in 3–5 h. The levels of IFN-γ in the circulation can reach up to 2000–4000 units/ml, but these levels are detectable for only a few hours. Unlike *in vitro* conditions where accumulation of IFN-γ occurs, the lymphokine can be detected in the serum of animals for only a few hours, due to protein catabolism and clearance by tissues.

3.2 **Procedures for enhancing production of IFN-γ**

As discussed above, mitogen induction of IFN-γ yields substantially higher levels of protein production than does antigenic stimulation. Again, this has to do with the fact that PHA and Con A can stimulate virtually all T-cells while antigens stimulate only those clones of T-cells that respond directly to antigenic stimulation. In this regard, mitogens induce 3- to 5-fold higher levels of IFN than do antigens.

It was discovered a few years ago that mitogen-induced production of IFN-γ could be substantially enhanced. It was determined that at the peak of a strong cell-mediated immune response mitogenic stimulation of spleen cells for IFN was enhanced 10- to 20-fold above normal. The particular system employed was a sublethal infection of

Listeria monocytogenes. This bacterial infection is known to be quite unique because a sublethal infection will induce an immune response that is exclusively cell-mediated with virtually no evidence of any antibody production. Thus, because cell-mediated immunity is so heavily involved in this disease, more T-cells are available for stimulation by PHA or Con A (1).

L. monocytogenes is a Gram-positive facultative intracellular bacterium. Most of the strains available from investigators and the American Type Culture Collection are capable of growing in mice. A sublethal (immunizing) dose of this bacterium is 10-fold below the LD_{50}. Thus, if an unknown strain of *Listeria* is obtained, it will be necessary first to determine the LD_{50}. We have found that the optimal time for harvesting spleen cells following injection of bacteria is at the peak of the immune response which is 6 days after inoculation of *Listeria*. The spleens are processed as described in Section 2.1.1 and the concentration of mitogen that is added to cell cultures is unchanged.

3.3 **Alternative sources of IFN-γ**

In addition to spleen cells as a source of IFN-γ, several studies have shown that interleukin-2 (IL-2) propagated clones of T-cells (6,7) and T-cell hybridomas can produce fairly substantial quantities of IFN-γ. The derivation of these clones or cell lines is a laborious process and is outside the scope of this chapter.

Some investigators have made their clones available to the general scientific public through direct contact or by making them available through the American Type Culture Collection.

4. PURIFICATION OF MURINE IFN-γ

The procedures that can be used to purify murine (Mu) IFN-γ are dependent on two main factors. The first is the source of the IFN-γ. Purification strategies may differ depending on whether the IFN is recombinant, in mouse serum or in supernatants derived from mouse spleen cells or T-cell lymphomas, since the level and type of contaminants will significantly vary from system to system.

The second factor that will govern the approach to the purification is the resources available to the investigator. Interferons have been purified by traditional biochemical procedures, by h.p.l.c. or f.p.l.c., and by immunoaffinity procedures. With the advent of bio-compatible h.p.l.c. and f.p.l.c., most low-pressure chromatography is readily adaptable to chromatography at medium or high pressure. The advantages of chromatography at higher pressure are 2-fold; higher resolution and shorter purification times.

We routinely use a single step immunoaffinity column that gives us high yields (30–40%) of high specific activity ($1-2 \times 10^8$ units/mg) MuIFN-γ. The column can be re-used many times provided care is taken to inhibit microbial growth. However, the procedure requires access to a source of the anti-MuIFN-γ monoclonal antibody (see below).

In these sections we will discuss procedures for immunoaffinity and biochemical purification of MuIFN-γ. Where appropriate we will mention any differences in strategies that may arise due to the source of IFN-γ. Since assays for IFN-γ are covered in a separate chapter, we will not discuss them here except to say that the purification is routinely followed by measuring the anti-viral activity in the fractions. A brief descrip-

Table 1. Purification of rat anti-MuIFN-γ monoclonal antibody.

Prepare the following buffers:

Loading buffer: 20 mM Tris-HCl, 25 mM NaCl, pH 7.2
Elution buffer: 20 mM Tris-HCl, 50 mM NaCl, pH 7.2

1. Clarify the ascites fluid by centrifugation at 16 000 *g* for 10 min.
2. Make the supernatant 50% saturated in ammonium sulphate by addition of solid (29.1 g/100 ml). Leave the suspension overnight at 4°C.
3. Recover the protein precipitate by centrifugation at 16 000 *g* for 20 min, and dissolve it in 25 ml of the loading buffer.
4. Dialyse the dissolved sample against two changes each of 2000 ml of the loading buffer.
4. Clarify the crude antibody by centrifugation at 16 000 *g* for 10 min, and load it onto a pre-equilibrated Affigel blue column (5 mg protein per 1 ml of gel).
6. Wash the column with 10 times the original sample volume of loading buffer.
7. Elute the antibody with the elution buffer. Collect 10 fractions, each the volume of the original sample.
8. Concentrate the active fractions 20-fold in an Amicon 8200 using a YM10 membrane.
9. Aliquot the fractions and freeze at −20°C, unless used immediately.

tion of the procedures that allow the characterization of the purity of the IFN-γ will also be given.

4.1 Immunoaffinity purification

We will not go into the details of the immunization and generation of the rat anti-MuIFN-γ monoclonal antibody, since they are beyond the scope of this chapter (8).

4.1.1 *Purification of monoclonal antibody*

The rat anti-MuIFN-γ hybridoma (available from the American Type Culture Collection) is grown as an ascites tumour in BALB/c nude mice. About 10^7 cells, grown in culture, are injected intraperitoneally; after 10−15 days the ascites fluid is collected and frozen until the antibody is purified. The purification scheme is given in *Table 1*. Briefly, make the ascites fluid to 50% saturation of ammonium sulphate and leave overnight at 4°C. Harvest the antibody by centrifugation and dissolve the pellet in loading buffer. Dialyse this solution against two changes of the same buffer and clarify by centrifugation. Measure the protein concentration of the dialysed antibody.

Set up a column of DEAE affigel blue (BioRad) and equilibrate with the loading buffer. The column is ready for use when the pH and conductivity of the buffer off the column is the same as that going onto the column. After the monoclonal antibody is applied, wash the column with 10 volumes (of the original sample volume) of the loading buffer and subsequently elute the antibody with the eluting buffer. Collect 10 fractions, each equal to the original sample volume. Throughout the purification, follow the protein concentration and activity (ability to neutralize IFN-γ in the anti-viral assay) of each fraction. Pool those samples possessing anti-MuIFN-γ activity and concentrate in an Amicon 8200 concentrator using a YM10 membrane. Freeze the antibody and store at −20°C until it is needed.

The purification of the antibody can be assessed by SDS−gel electrophoresis on a 7−15% gradient gel (*Table 6*). Usually these purification procedures yield 75−80% pure antibody, with a specific activity of about 3×10^6 neutralizing units/mg.

Table 2. Immobilization of rat anti-MuIFN-γ antibody.

Prepare the following buffer:

Coupling buffer: 0.1 M sodium borate, pH 8.5

1. Dialyse the purified antibody against a 100-fold excess of the coupling buffer, with two changes.
2. Wash the activated gel, Reactigel 6X (Pierce), once with the coupling buffer and use it immediately.
3. Couple the antibody to the gel at a ratio of 2 mg protein/ml of packed resin for 48 h at 4°C. Follow the coupling by measuring the protein removed from solution.
4. Incubate the mixture for an additional 48 h at 4°C to allow the hydrolysis of unreacted groups on the resin.
5. Wash the resin twice with two column volumes of the IFN-γ elution buffer (*Table 3*).
6. Check the resin to ensure it can remove IFN-γ activity from solution, and then store it at 4°C in PBS containing 0.01% azide.

Table 3. Immunoaffinity purification of MuIFN-γ.

Prepare the following buffers:

Wash 1: 10 mM sodium phosphate, 1 M NaCl, 0.1% Nonidet-P40, pH 7.4
Wash 2: PBS
Wash 3: PBS containing 0.5 M ammonium thiocyanate, 25% (v/v) glycerol, pH 7.4
Elution buffer: PBS containing 3.0 M ammonium thiocyanate, 25% (v/v) glycerol, pH 7.4
Dialysis buffer: PBS:glycerol, 1:1 (v/v)

1. Centrifuge the IFN-containing supernatants (1000–2000 ml) at 8000 *g* for 15 min.
2. Concentrate the cell-free supernatants 20- to 40-fold in an Amicon DC2, or equivalent, using a 10 000 molecular weight cut-off filter.
3. Clarify the concentrate by centrifugation at 12 000 *g* for 20 min.
4. Filter the supernatant through 0.22 μm sterile filters.
5. Apply the filtrate to the immunoaffinity column (12 × 1 cm) at a flow-rate of about 0.75 ml/min. Collect the percolate.
6. Re-apply the percolate to the column.
7. Wash the column sequentially with 100 ml of Wash 1, 100 ml of Wash 2 and 20 ml of Wash 3.
8. Elute the IFN-γ with 25 ml of elution buffer.
9. Dialyse the eluate against three changes, 500 ml each, of the dialysis buffer.
10. Store the MuIFN-γ at −20°C.

We should also point out that there are now several different columns available that allow the purification of monoclonal antibodies by h.p.l.c. We will not discuss them here.

4.1.2 *Immobilization of rat anti-MuIFN-γ*

The procedure for the immobilization of the anti-MuIFN-γ antibody is given in *Table 2*. Dialyse the purified antibody against 0.1 M sodium borate, pH 8.5, and then couple to Reactigel 6 (Pierce). Allow unreacted groups on the gel to hydrolyse by incubating in the buffer for an additional 48 h. Estimate the coupling efficiency by measuring the protein removed from solution. Wash the resin with the IFN-γ elution buffer (*Table 3*) and store in phosphate-buffered saline (PBS) containing 0.01% azide.

It is possible to use this immobilized antibody to deplete the IFN-γ specifically from samples which may have mixtures of lymphokines. This would then allow biological assays for other lymphokines in the samples to be carried out.

4.1.3 *Immunoaffinity chromatography*

The purification scheme for MuIFN-γ from spleen cells or T-cell lymphoma supernatants, in serum-free media, is given in *Table 3*. To summarize, concentrate the supernatants (10^6-10^7 units), centrifuge, filter and then apply to the affinity column. After washing the column to remove all non-bound protein, elute the IFN-γ with a PBS buffer containing 3 M ammonium thiocyanate and 25% glycerol (9).

It is possible to purify IFN-γ from serum on the same affinity column, but the serum should be diluted 4-fold with PBS prior to application to the column. The wash and elution steps are the same. It is likely that the life of a column used for purifying IFN from serum will be considerably shorter than that of a column used with spleen cell supernatants due to the action of proteases and the non-specific adsorption of proteins.

Two points should be noted. Firstly, the purified natural MuIFN-γ is very unstable and rapidly loses activity at 4°C or −20°C. However, it is possible to stabilize the IFN by dialysing it against PBS:glycerol (1:1, v/v), and storing it at −20°C, or by adding back 50% heat-inactivated serum and storing the mixture at 4°C. Under both conditions, the IFN-γ is stable for several weeks. Secondly, the protein concentration of the purified natural MuIFN-γ is very low and it requires the use of many units to detect the protein by conventional assays. Purified recombinant IFN-γ appears to be more stable and is available in quantities where protein concentration can be accurately measured.

4.2 **Biochemical chromatography of MuIFN-γ**

Traditional chromatographic procedures such as ion-exchange, isoelectric focusing, lectin affinity and hydrophobic interaction indicate that there are several species of IFN-γ present in mouse serum or from mixed populations of spleen cells (10,11). This probably reflects heterogeneity in the multimeric states of IFN and also different states of glycosylation. The IFN-γ from a cloned T-cell lymphoma appears to be more homogeneous, which will slightly simplify the purification procedure (12). While no purification procedure has been published for recombinant MuIFN-γ, it too should be more homogeneous.

We will discuss below a general purification scheme for IFN-γ; however one should bear in mind that this scheme selects for the predominant form of IFN-γ in the starting material. The specific activity of the IFN-γ obtained using the procedures outlined below will be about $2-4 \times 10^6$ U/mg, which is significantly lower than that obtained by immunoaffinity chromatography.

4.2.1 *Silicic acid chromatography*

This procedure allows both concentration and purification of the samples in one step. The procedure is outlined in *Table 4*. Briefly, dilute the IFN-γ-containing supernatants from a T-cell lymphoma or from spleen cells 1:1 with 10 mM sodium phosphate, pH 7.4. Adsorb the IFN-γ to sterile silicic acid, wash by centrifugation and resuspend in the wash buffer. To obtain the IFN-γ, treat the silicic acid pellet with the eluting buffer for 1 h. About a 10-fold purification can be achieved from this step.

IFN-γ in whole mouse serum should be diluted 10-fold with 10 mM sodium phosphate, prior to adsorption onto the silicic acid.

Table 4. Chromatography on silicic acid.

Prepare the following buffers:

Loading: 10 mM sodium phosphate, pH 7.4
Wash: 10 mM sodium phosphate, pH 7.4, containing 1.4 M NaCl
Elution: Wash buffer:ethylene glycol, 1:1 (v/v)

1. Centrifuge the IFN-γ containing supernatants at 8000 *g* for 15 min.
2. Dilute the supernatants 1:1 with the loading buffer.
3. Adsorb the IFN-γ to sterile silicic acid and stir this overnight at 4°C, using 20 mg of silicic acid/ml sample.
4. Centrifuge the suspension at 4000 *g* for 10 min and wash the pellet four times by resuspending it in four volumes of the wash buffer and centrifuging again.
5. Elute the IFN-γ by stirring the pellet for 1 h at 4°C in an equal volume of elution buffer and then sedimenting the silicic acid. Wash the pellet with one volume of elution buffer using the same procedure. Combine the last two samples.
6. Store the sample at −20°C until ready for the next column.

Table 5. Con A chromatography.

Prepare the following buffers:

Loading: 20 mM sodium phosphate, pH 7.4, containing 1 M NaCl
Elution: loading buffer containing 200 mM α-methylmannoside

1. Dialyse the eluate from the silicic acid against a 10-fold excess of the loading buffer, with two changes.
2. Concentrate the dialysate 5-fold in an Amicon 8200 concentrator using a YM5 membrane and apply it to a Con A Sepharose column (20 × 1.5 cm) which has been pre-equilibrated with the loading buffer, at 4°C.
3. Wash the column with loading buffer until no protein can be detected in the wash.
4. Elute the MuIFN-γ with the elution buffer. Collect 3 ml fractions and assay them.
5. Pool samples possessing anti-viral activity. For storage an equal volume of glycerol can be added and the sample kept at −20°C. This does not have to be removed before the sample is concentrated for the next column.

4.2.2 *Con A chromatography*

Most of the IFN-γ (~90%) in spleen cell and T-cell lymphoma supernatants is glycosylated and will bind to Con A (11). The glycosylation state of recombinant IFN will depend on the cell in which the lymphokine is produced. Recombinant interferon produced in *Escherichia coli* will not be glycosylated, but that produced in mammalian cells may be.

The purification protocol is given in *Table 5*. Dialyse the eluate from the silic acid against 20 mM sodium phosphate, pH 7.4, containing 1 M NaCl, concentrate in an Amicon 8200 with a YM5 membrane and apply to a Con A Sepharose (Pharmacia) column (20 × 1.5 cm). Wash the column with the loading buffer until no protein can be detected in the wash and then elute the IFN-γ with the loading buffer containing 200 mM α-methylmannoside.

4.2.3 *Gel filtration*

Concentrate the IFN-γ-containing fractions from the lectin column to a final volume of 1.5−2 ml using an Amicon 8010 with a YM5 membrane, and then apply them to

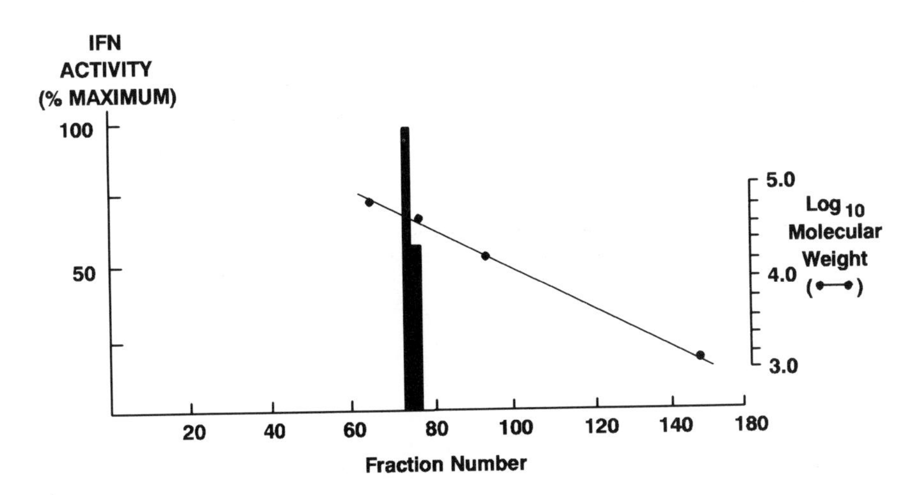

Figure 1. Gel filtration of MuIFN-γ from spleen cells on an Ultragel AcA 54 column. The anti-viral activity is expressed as a percentage of that found in the peak fraction. The molecular weight standards are bovine serum albumin, ovalbumin, myoglobin and vitamin B12.

Table 6. SDS–gel electrophoresis of MuIFN-γ.

Prepare the following solutions and buffers:

29.2% (w/v) acrylamide, 0.8% (w/v) bisacrylamide
Buffer 1: 1.5 M Tris-Cl pH 8.8, 0.4% (w/v) SDS
Buffer 2: 0.5 M Tris-Cl pH 6.8, 0.4% (w/v) SDS
Buffer 3: 0.125 M Tris-Cl pH 6.8, 4% (w/v) SDS, 20% (v/v) glycerol, 10% (v/v) 2-mercaptoethanol (optional)
Buffer 4: 0.025 M Tris, 0.192 M glycine, 0.1% SDS, pH 8.3
Extraction buffer: 90% (v/v) RPMI, 10% (v/v) fetal calf serum

1. Pour a 7–15% (linear) separating gel:

	7%	15%
Buffer 1	4.0 ml	4.0 ml
Acryl/Bis	3.75 ml	8.0 ml
Water	8.2 ml	3.95 ml
ammonium persulphate (10%, w/v)	50 μl	50 μl
TEMED	8 μl	8 μl

   Put 14.5 ml of the acrylamide solutions in the compartments of a Hoefer SG50 gradient maker; put the 15% acrylamide in the reservoir connected to the gel apparatus. After pouring the separating gel carefully overlay it with water.
2. Pour the stacking gel: 2.6 ml of Acryl/Bis; 4.0 ml of Buffer 2; 9.2 ml of water; 200 μl of ammonium persulphate, 20 μl of TEMED.
3. Heat the MuIFN-γ in an equal volume of Buffer 3 for 4 min at 90°C, and apply it to the wells of the gel in duplicate.
4. Run the gel in buffer 4 under constant current; 20 mA until the samples run into the separating gel, then 30 mA until the dye front reaches the bottom of the gel. Total run time is about 3 h.
5. Carefully dismantle the system and cut out the gel lanes. Stain one of each duplicate lane with Coomassie Blue, or silver stain; slice the other into 2 mm slices and incubate the slices overnight at 4°C with 0.5 ml of the extraction buffer.

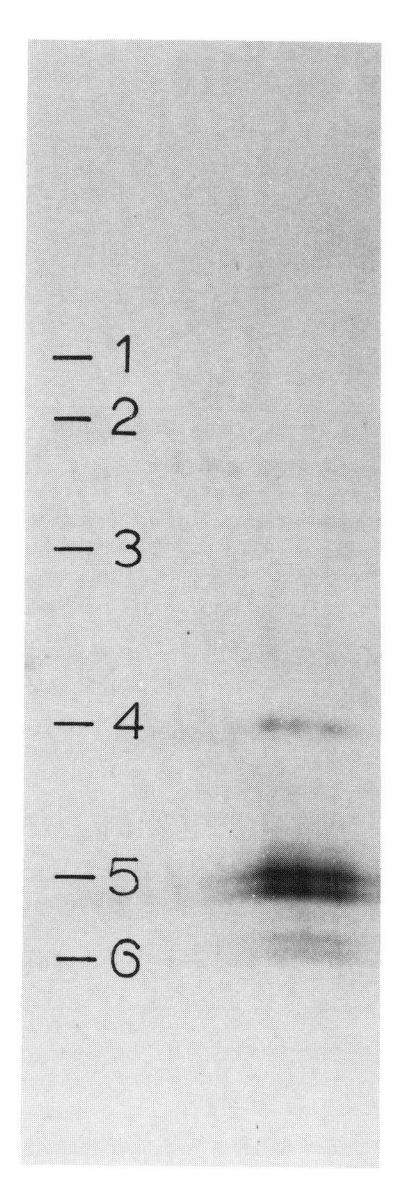

Figure 2. SDS−polyacrylamide gel electrophoresis of immunopurified MuIFN-γ on a 7−15% gradient gel. The gel is stained with Coomassie Blue. The standards are: **(1)** phosphorylase, 94 kd; **(2)** bovine serum albumin, 67 kd; **(3)** ovalbumin, 43 kd; **(4)** carbonic anhydrase, 30 kd; **(5)** lysozyme, 14.4 kd.

an Ultragel AcA 54 (LKB) gel filtration column (90 × 2.5 cm) that has been pre-equilibrated with PBS containing 15% glycerol and 1 M NaCl. Collect fractions (3 ml) and assay for protein concentration and anti-viral activity. The same column is calibrated with standard proteins. A typical column profile of MuIFN-γ from spleen cells is shown in *Figure 1*; the standard proteins are: bovine serum albumin, 67 kd; ovalbumin, 43 kd;

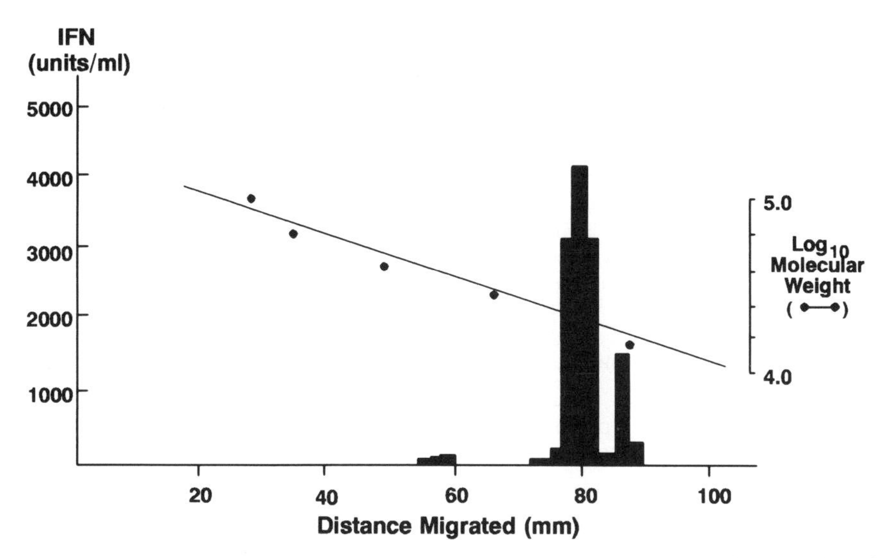

Figure 3. Anti-viral activity found in the eluates from gel slices of a duplicate lane of the gel shown in *Figure 2*.

myoglobin, 17 kd; and vitamin B12, 1350 daltons. These studies indicate that almost all the IFN-γ has a molecular weight of about 40 kd.

4.3 Characterization of the purity of MuIFN-γ

4.3.1 *SDS–gel electrophoresis*

It is possible to check the purity of the MuIFN-γ and identify the different forms present by analysing the samples by SDS–polyacrylamide gel electrophoresis (PAGE). Recently it has been shown that following SDS–PAGE, IFN-γ activity can be detected after elution of the protein from the gel. Thus protein bands can be eluted from the gel, assayed for biological activity and those containing IFN-γ activity identified.

The methods for the preparation and running of SDS–gels are well established so we will not go into significant detail here except where it is relevant to the MuIFN-γ. To get the best separation over a wide molecular weight range we use a discontinuous buffer system (13) with a 7–15% linear gradient gel and a 5% stacking gel in a Hoefer SE600 apparatus. The procedures given in *Table 6* are based on this.

Assays of the protein eluted from gel slices indicate that it is possible to extract 1–5% of the applied activity from the SDS–gel; thus, to detect any minor species of IFN-γ (1–10% of the total), it is necessary to load 50 000–100 000 units. *Figure 2* shows the protein staining profile of immunopurified MuIFN-γ; *Figure 3* shows the anti-viral activity eluted from the slices obtained from a duplicate gel. Note that it is possible to detect three species possessing anti-viral activity. This is an example of how useful this technique can be in identifying different forms of MuIFN-γ.

4.3.2 *Radiolabelling of MuIFN-γ*

To increase the sensitivity of detection of proteins in the IFN-γ preparations, it is also possible to radiolabel the sample. The labelling technique that preserves the most IFN-

Table 7. Iodination of MuIFN-γ.

1.	Dialyse IFN-γ (~10^6 units) against PBS:glycerol (1:1, v/v); use three changes of 500 ml each.
2.	Concentrate the IFN-γ in an Amicon 8010 or Centricon 10 to about 0.5 ml.
3.	Dry down 1 mCi (3.7 × 10^7 Bq) of Bolton and Hunter reagent (New England Nuclear) under nitrogen.
4.	Add the IFN-γ to the Bolton and Hunter reagent and leave for 6 h at 4°C.
5.	Stop the labelling by adding 100 μl of 30 mg/ml sodium bisulphite.
6.	Separate the iodinated protein from unbound label on a Sephadex PD 10 column equilibrated with PBS containing 2 mg/ml bovine serum albumin, 1 mM NaI.
7.	Dialyse the eluate against 500 ml of PBS:glycerol with four changes over 2 days. Assay the labelled protein and determine its radioactivity. Store the IFN-γ at −20°C until ready for use.

γ activity is the Bolton and Hunter procedure. The method, adapted for MuIFN-γ, is given in *Table 7.*

We have found that the pattern of iodinated protein bands in the gel following SDS electrophoresis is essentially the same as that seen in Coomassie blue- or silver-stained gels, confirming that there are no other major protein contaminants. The major advantage of radiolabelling the proteins is that, compared with conventional staining procedures, much less MuIFN-γ is needed for this analysis.

4.3.3 *Protein assays*

Protein assays are carried out according to established techniques. Routinely we used the micro dye binding assay (14) which will detect protein down to about 5 μg/ml. For increased sensitivity (down to 500 ng/ml) it is possible to label the protein with fluorescamine (15).

5. ACKNOWLEDGEMENTS

We would like to thank Bob Curry and George Rodgers for their expert technical assistance and Lee Canale for typing this chapter.

6. REFERENCES

1. Havell,E.A., Spitalny,G.L. and Patel,P.S. (1982) *J. Exp. Med.*, **156**, 112.
2. Hirt,H.M., Becker,H. and Krichner,H. (1978) *Cell. Immunol.*, **38**, 168.
3. Mishell,B.B., Shiigi,S.M., Henry,C., Chang,E., North,J., Gallily,R., Slomich,M., Miller,K., Marbrook,J., Parks,D. and Good,A.H. (1980) In *Selected Methods in Cellular Immunology.* Michell,B.B. and Shiigi,S.M. (eds), W.H.Freeman, San Francisco, p. 13.
4. Salvin,S.B., Younger,J.S. and Lederer,W.H. (1973) *Infect. Immun.*, **7**, 68.
5. Spitalny,G.L. (1986) In *Clinical Applications of Interferons and Their Inducers.* Stringfellow,D.A. (ed.), Marcel Dekker, Inc., New York, p. 83.
6. Marcucci,F., Walter,M., Krichner,H. and Krammer,P. (1981) *Nature*, **291**, 79.
7. Ratliff,T.L., Thomasson,D.L., McCool,R.E. and Catalond,W.J. (1982) *Cell. Immunol.*, **68**, 311.
8. Spitalny,G.L. and Havell,E.A. (1984) *J. Exp. Med.*, **159**, 1560.
9. Greenfield,R.S., Kiener,P.A., Chin,D.P., Curry,R.C. and Spitalny,G.L. (1986) *J. Leukocyte Biol.*, **38**, 75.
10. Stefanos,S., Weitzerbin,J., Huygen,K. and Falcoff,E. (1982) *J. Interferon Res.*, **2**, 447.
11. Havell,E.A. and Spitalny,G.L. (1984) *Arch. Virol.*, **80**, 195.
12. Gribaudo,G., Cofano,F., Negro-Ponzi,A. and Landolfo,S. (1984) *J. Interferon Res.*, **4**, 91.
13. Laemmli,U.K. (1970) *Nature*, **227**, 680.
14. Bradford,M.M. (1970) *Anal. Biochem.*, **72**, 248.
15. Bohlen,P., Stein,S., Dairman,W. and Undenfriend,S. (1973) *Arch. Biochem. Biophys.*, **155**, 213.

CHAPTER 3

Production and purification of recombinant mouse interferon-γ from *E. coli*

KIYOSHI NAGATA, OSAMU OHARA, HIROSHI TERAOKA, NOBUO YOSHIDA, YOSHIHIKO WATANABE and YOSHIMI KAWADE

1. INTRODUCTION

We describe our procedures for obtaining large amounts of purified recombinant mouse interferon-γ (rMuIFN-γ) from *Escherichia coli*. The production and purification of rMuIFN-γ includes the following steps:

(i) Isolation of cDNA encoding mature MuIFN-γ.
(ii) Construction of the expression plasmid for rMuIFN-γ.
(iii) Expression of rMuIFN-γ in *E. coli*.
(iv) Purification of rMuIFN-γ.
(v) Characterization of rMuIFN-γ.

Since many of the experimental procedures employed in these steps are now familiar to biochemists and have been described elsewhere (e.g. previous volumes in the '*Practical Approach*' series), only those procedures that are informative for the preparation of rMuIFN-γ produced in *E. coli* will be described here.

2. PRODUCTION OF rMuIFN-γ IN ESCHERICHIA COLI

2.1 Isolation of MuIFN-γ cDNA

2.1.1 *Induction of IFN-γ production in mouse B5 cells*

Natural MuIFN-γ is often prepared by inducing mouse spleen cells in culture with a T cell mitogen, such as concanavalin A (Con A). This system is, however, expected to be low in MuIFN-γ mRNA content, because the IFN titres in the fluids are usually low. We have utilized a T cell line designated B5, derived from C57BL/6 mouse spleen, for the mRNA preparation, which can produce high levels of MuIFN-γ (10^3-10^4 units/10^6 cells) when stimulated with Con A at concentrations of $5-10$ μg/ml (1).

B5 cells are grown in RPMI 1640 medium containing 10% (v/v) fetal calf serum and supplemented with a crude interleukin-2 (IL-2) preparation, such as culture supernatant of rat spleen cells stimulated with Con A (1); they reach a cell density of $2-3 \times 10^5$ cells/ml at saturation.

For MuIFN-γ induction, collect the B5 cells by centrifugation, re-seed in appropriate bottles at a cell density of 10^6 cells/ml in serum-free medium, and add 10 μg/ml Con A. To determine the timing of extraction of MuIFN-γ mRNA from the Con A-stimulated cells, the time course of MuIFN-γ production must be examined. As shown in *Figure 1*, the IFN activity appears in the culture fluid at 4 h after the addition of Con A and increases steeply after 10 h. We harvest Con A-stimulated B5 cells for preparing the

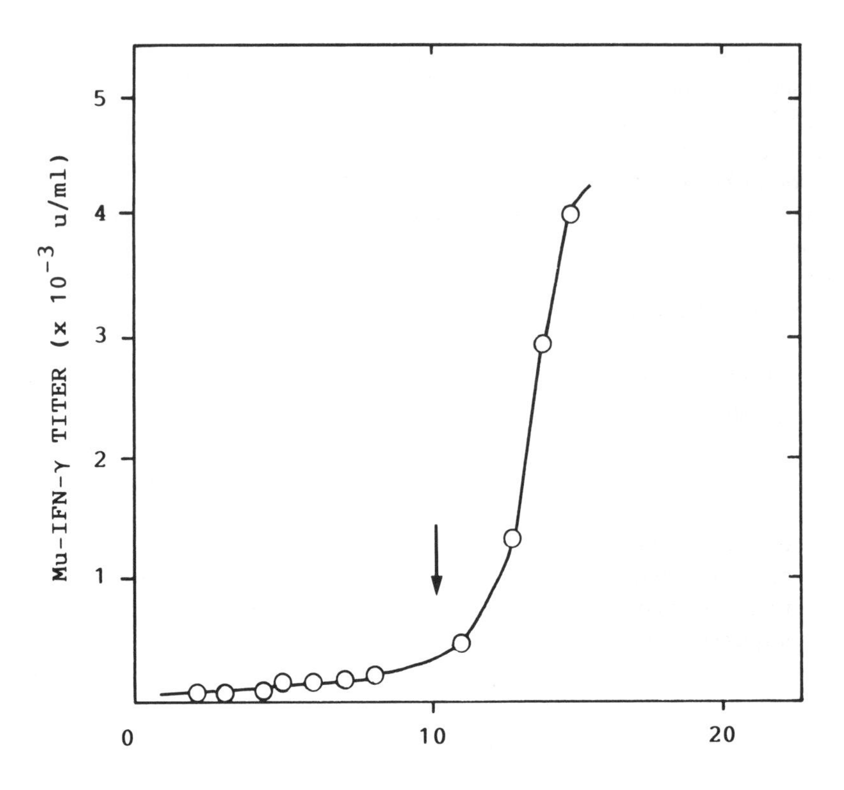

Figure 1. Time course of the production of MuIFN-γ in B5 cells after concanavalin A induction. For the preparation of MuIFN-γ mRNA, B5 cells are collected at the time indicated by the arrow.

whole cellular RNA at 10 h after the treatment (when RNA is extracted at 13 h, the content of MuIFN-γ mRNA is very low).

2.1.2 *Isolation of mRNA from induced B5 cells*

In general, RNA prepared from 10^8 cells is sufficient for the construction of a cDNA library containing about 10^5 clones by the cloning method of Okayama and Berg (2). In the present case, RNA is prepared from the induced B5 cells (4×10^9 cells) described above. RNA can be prepared either by the sodium dodecyl sulphate (SDS) – phenol method (3) or the guanidinium thiocyanate method (4). The RNA is chromatographed on a column of oligo(dT)-cellulose according to a standard method for the preparation of polyadenylated RNA [poly(A)$^+$ RNA]. The poly(A)$^+$ RNA obtained should be assayed in an *in vitro* protein synthesizing system (e.g. using wheat germ extract or rabbit reticulocyte lysate). If the poly(A)$^+$ RNA does not show any template activity *in vitro*, it should not be used to construct the cDNA library.

2.1.3 *Enrichment of MuIFN-γ mRNA*

The poly(A)$^+$ RNA is fractionated by sucrose density gradient centrifugation for enrichment of MuIFN-γ mRNA. Centrifuge the poly(A)$^+$ RNA sample through a 5–20% (w/v) sucrose gradient in 10 mM Tris-HCl (pH 7.6) – 1 mM ethylenediaminetetraacetic acid (EDTA) – 0.2 M NaCl for 16 h at 27 000 r.p.m. (89 000 *g*) with a Beckman SW 41 rotor (at 20°C). For the detection of fractions containing MuIFN-γ mRNA, the micro-injection technique of mRNA into *Xenopus laevis* oocytes (5) is convenient because biological assay of IFN-γ is very sensitive. If human IFN-γ cDNA is available, MuIFN-γ can be detected by Northern hybridization (6) or Southern hybridization after cDNA synthesis (7) using human IFN-γ cDNA as a probe. MuIFN-γ mRNA appears in the fractions with a sedimentation coefficient of about 17S.

2.1.4 *Construction and screening of the cDNA library*

(i) Construct the cDNA library from the fractionated poly(A)$^+$ RNA by the method of Okayama and Berg (2). To prepare the cDNA library, use *E. coli* K-12 strain HB101 as a recipient host of cDNA according to the method of Hanahan (8).

(ii) Since the complete nucleotide sequence of MuIFN-γ cDNA is available (9), screen the cDNA library by colony hybridization (10) using a synthetic oligonucleotide probe designed on the basis of its nucleotide sequence. Such an oligonucleotide probe should be designed to detect cDNA which carries the complete nucleotide sequence encoding the mature protein. Therefore, the oligonucleotide probe is made according to the nucleotide sequence encoding the amino-terminal portion of mature MuIFN-γ;

Mature MuIFN-γ NH_2- Cys - Tyr - Cys - His - Gly - Thr - - - - - -

TGT TAC TGC CAC GGC ACA

Probe (17-mer)

(iii) Label the oligonucleotide probe (17-mer) with ^{32}P with T4 polynucleotide kinase.

(iv) Examine about 40 000 clones in the cDNA library by high density colony hybridization (10) using this ^{32}P-labelled probe.

(v) Prepare the plasmids from several positive clones by a standard SDS – alkali method (11) and then analyse them using restriction enzymes.

(vi) Sequence the cDNA with the longest insert size by a chain termination method (12) using, for example, a DNA sequencing kit from Takara Shuzo Co., Kyoto, Japan. All positive clones obtained by this screening carry the full-length cDNA of MuIFN-γ. One of these clones (designated as pBMγ-E3) has been used for further experiments in our own laboratory.

2.2 Construction of the expression plasmid of MuIFN-γ cDNA

The two important requirements for producing a large amount of rMuIFN-γ in *E. coli* are:

(i) a strong promoter in *E. coli*;

(ii) the nucleotide sequence of the ribosome-binding site including the initiation codon.

The *trp* promoter and the nucleotide sequence of the ribosome-binding site shown below, which is the same as that reported by Simons *et al.* (13), are employed.

5′- - -ACGTAAAAGGGTATCGATACT**ATG**- - -3′

S.D. initiation codon

S.D. indicates the Shine-Dalgarno sequence (14), which is assumed to interact with the 3′ end of *E. coli* 16S rRNA. The nucleotide sequence encoding mature MuIFN-γ is therefore under the control of the *trp* promoter with the ribosome-binding site as shown above.

2.2.1 *ATG vector*

A plasmid carrying a strong promoter in *E. coli* and the ribosome-binding site including the ATG codon (the so-called ATG vector) is versatile for the construction of the expression plasmid of a cDNA, as reported by Nishi *et al.* (15). In our case, the ATG vector contains the *trp* promoter and the ribosome-binding site shown above. The use of the ATG vector is illustrated in *Figure 2*.

2.2.2 *Primer extension*

Figure 2 shows the construction scheme of the expression plasmid of rMuIFN-γ, which is essentially the same as that reported by Gray and Goeddel (9). The most important step in this construction scheme is primer extension which generates the DNA fragment containing the nucleotide sequence encoding mature MuIFN-γ as determined by Gray and Goeddel (9). To generate a DNA fragment without using a restriction cleavage site, primer extension is simple and expedient (16); the experimental protocol is given in *Table 1*. Although single-stranded DNA (e.g. M13 phage DNA) can be used as a template for primer extension, double-stranded DNA is used here because this allows the elimination of an additional cloning step. The oligonucleotide used as a primer is the same as that used in cDNA library screening. The DNA fragment is recovered from the gel as described elsewhere (17). According to the construction scheme, three DNA fragments are ligated with T4 DNA ligase.

2.2.3 *Screening of the expression plasmid of rMuIFN-γ*

Nuclease S1 treatment and primer extension sometimes produce unexpected DNA fragments due to incomplete removal of single-stranded tails or incorporation of the wrong bases during chain elongation. Since even a slight change in the nucleotide se-

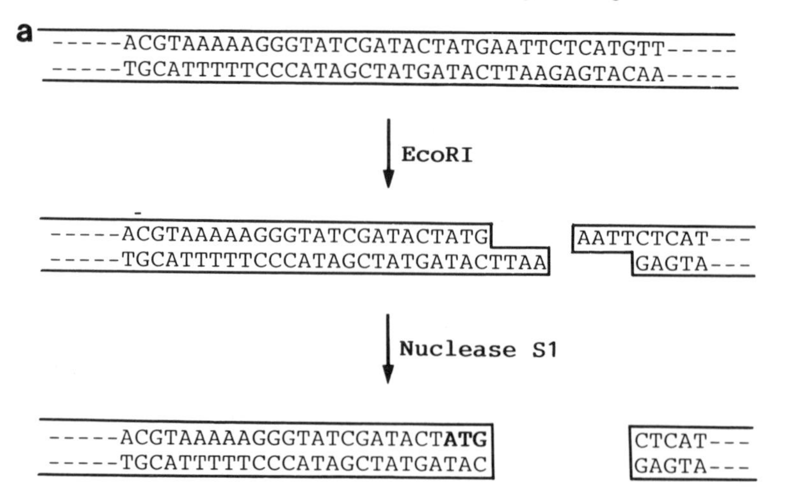

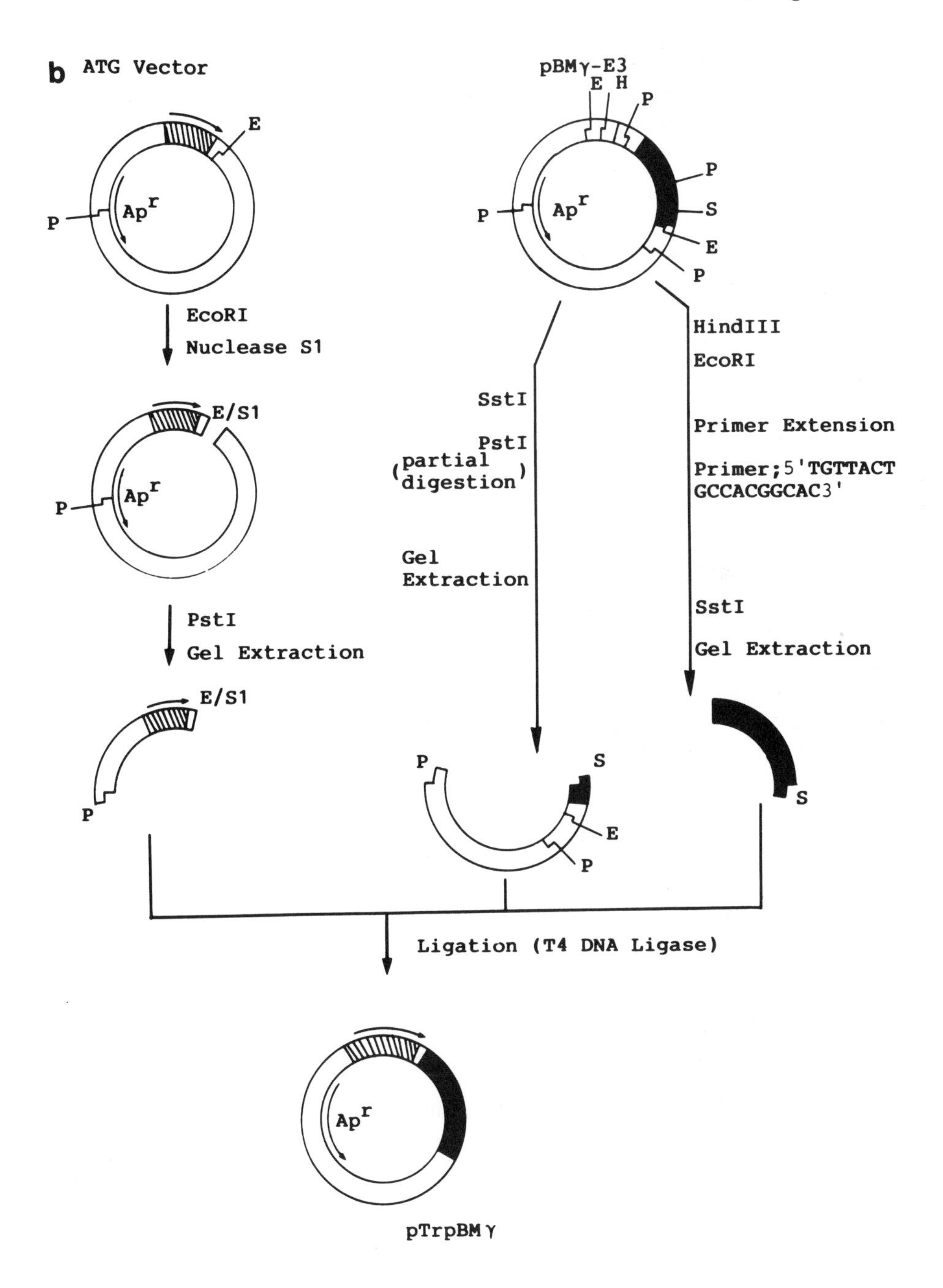

Figure 2. Construction scheme of the expression plasmid for mature MuIFN-γ. **(a)** Use of the ATG vector. The ATG vector is a pBR322 derivative, which contains the *trp* promoter and the ribosome-binding site as described in the text. **(b)** pBMγ-E3 is a plasmid carrying MuIFN-γ cDNA, which is isolated from the cDNA library constructed by the method of Okayama and Berg (2). The *trp* promoter and the nucleotide sequence encoding mature MuIFN-γ are depicted as hatched and solid areas, respectively. Curved arrows indicate the direction of transcription. The cleavage sites for the restriction enzymes are indicated as follows: E, *Eco*RI; H, *Hin*dIII; S, *Sst*I; P, *Pst*I. Ap^r: ampicillin resistance gene. ATG vector is tailored as shown in **a**.

Table 1. Primer extension procedure.

1. Digest pBMγ-E3 (~5 μg) with *Hin*dIII and *Eco*RI, then recover DNA fragments by phenol extraction followed by ethanol precipitation.
2. Dissolve DNA fragments in 10 μl of solution containing oligonucleotide primer[a] (17-mer, ~50 pmol).
3. Boil the resulting DNA fragment–primer solution for 3 min, and then cool it rapidly in a dry ice–acetone bath.
4. Add Klenow fragment (2 U), 1 mM 4dNTP[b], and 10× buffer[c] to this DNA fragment–primer solution to obtain a primer-extension mixtuer (20 μl) of the following final content:

Tris-HCl (pH 7.5)	7 mM
$MgCl_2$	7 mM
EDTA	0.1 mM
NaCl	20 mM
dATP, dCTP, dGTP, dTTP	0.25 mM each
DNA fragments of pBMγ-E3	~ 5 μg (~2 pmol)
primer (17-mer)	~50 pmol
Klenow fragment	2 U

5. Incubate the primer–extension mixture at 37°C for 3–4 h. Recover DNA fragments by phenol extraction followed by ethanol precipitation.
6. Digest the product of primer extension with *Sst*I.
7. Separate DNA fragments on 5% PAGE[d].
8. Recover the target DNA fragment from the gel.

[a]5′-TGTTACTGCCACGGCAC-3′.
[b]An equimolar mixture of dATP, dCTP, dGTP and dTTP.
[c]70 mM Tris-HCl–70 mM $MgCl_2$–1 mM EDTA–200 mM NaCl (pH 7.5).
[d]PAGE is carried out in 89 mM Tris–borate–2 mM EDTA (pH 8.0).

quence around the translation initiation site should not be tolerated, colony hybridization is used as a screening method to select only the clones harbouring the desired plasmid. In this screening, the probe contains the nucleotide sequence around the initiation codon in the desired plasmid which is underlined here:

5′- - - -ACGTAAAAAGGGTATCGATACTATG TGTTACTGCCACGGC- - - -3′

19-mer

Seven nucleotides at the 5′ end of the 19-mer correspond to the sequence derived from the ATG vector, and 12 nucleotides at the 3′ end encode the four amino acid residues at the amino-terminal end of mature MuIFN-γ. The positive clones obtained by this screening are likely to be those harbouring the desired expression plasmid with correct joining of the two DNA fragments; one is derived from the ATG vector and the other from primer extension. On the other hand, the clones giving only weak signals probably carry the plasmid with a subtle change in the nucleotide sequence around the junction between these DNA fragments. The plasmid obtained from the clones thus selected should be examined by DNA sequencing to exclude the possibility that the wrong bases have been added onto an elongating chain during the primer extension step. The results of DNA sequencing show the efficiency of this screening method. The expression plasmid we have obtained using these procedures is designated as pTrpBMγ.

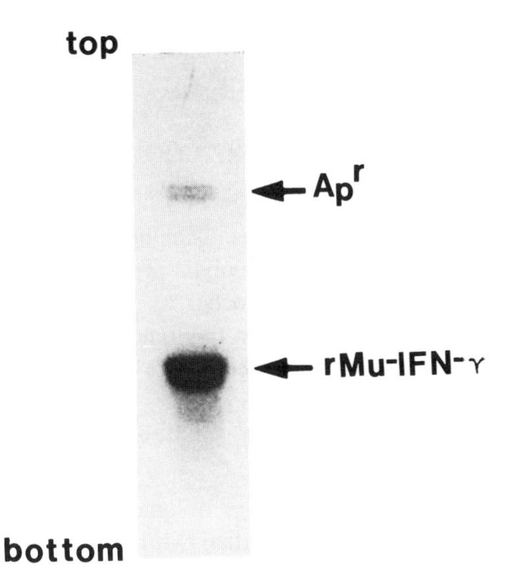

Figure 3. Analysis on a 15% SDS – polyacrylamide gel of proteins synthesized in a cell-free coupled transcription – translation system (see ref. 18). The template DNA was pTrpBMγ (2.5 μg/assay). Arrows indicate the bands of the product of the ampicillin resistance gene (Ap^r) and rMuIFN-γ. SDS – PAGE was carried out as described in Section 2.1.1, and the bands were detected by fluorography.

2.2.4 *Verification of the expression plasmid by a cell-free coupled transcription – translation system derived from E. coli*

The expression plasmid, pTrpBMγ, is subjected to a cell-free coupled transcription – translation system derived from *E. coli* (18), which is now available from Amersham International. In this system, proteins encoded by added DNA are synthesized and labelled with [^{35}S]methionine according to the supplier's instructions. The plasmid prepared by a simple SDS – alkali method without further purification (e.g. CsCl density gradient centrifugation or column chromatography) can give a tolerable result, although the purified plasmid always gives better results. *Figure 3* shows the pattern of SDS – polyacrylamide gel electrophoresis (SDS – PAGE) of the products in this system, which indicates that the constructed expression plasmid for rMuIFN-γ directs the synthesis of a polypeptide with the expected molecular weight of about 15 000. The products of the cell-free system can also be directly examined by biological assay of interferon activity (see Section 3.1.3) since biologically active IFN is synthesized in this system.

2.3 **Expression of rMuIFN-γ in *E. coli***

E. coli K-12 strain C600 is used for the expression of rMuIFN-γ. *E. coli* harbouring pTrpBMγ is grown overnight in 100 ml of LB medium (17) supplemented with 40 mg/l ampicillin, and then diluted into 10 l of M9 medium (17) containing 0.5% glucose, 0.5% casamino acids and 40 mg/l ampicillin. After culture at 37°C for 24 h under agitation and aeration, *E. coli* cells are harvested by centrifugation. The cell pellet should be stored frozen at $-70°C$ unless purification of rMuIFN-γ is performed immediately after the cultivation.

3. PURIFICATION AND CHARACTERIZATION OF rMuIFN-γ

In purifying rMuIFN-γ from *E. coli* cells, the following must be considered.

(i) The rMuIFN-γ does not form any inclusion bodies in *E. coli* cells at the expression level attained. Therefore, proteolytic processing in *E. coli* cells as well as at the early purification steps can occur.

(ii) As the anti-viral assay takes a long time and is not convenient for the actual purification process, other simple methods for detecting rMuIFN-γ are necessary throughout the purification process.

(iii) The purification process should be as simple and rapid as possible and must also be easy to scale up.

Our procedures for purifying rMuIFN-γ, designed according to these points, and some procedures for characterizing the purified rMuIFN-γ are described in this section.

3.1 Analytical procedures for the purification process

3.1.1 *SDS–polyacrylamide gel electrophoresis*

Assay for anti-viral activity is very time-consuming and not convenient for detecting rMuIFN-γ during the purification procedure. Therefore, SDS–PAGE should be used instead to detect rMuIFN-γ protein.

(i) Prepare 15% slab gels (1 mm thick, 12 × 14 cm) according to Laemmli (19).

(ii) Incubate the samples with 1% SDS and 5% 2-mercaptoethanol (2-ME) at 85–90°C for 5 min prior to electrophoresis.

(iii) Carry out electrophoresis for 3 h at 20 mA/gel after loading 1 μg of protein for each lane.

(iv) Stain the electrophoresed gels with a silver stain kit (for example, from Daiichi Chemicals), and then scan the gels immediately with a Soft Laser scanning densitometer with a Hewlett Packard 3390 A integrator (parameters: PK WD = 0.04, THRSH = 8, CHT SP = 10.0, AR REJ = 1000, AT↑2 = 7 or 8), or equivalent equipment.

The rMuIFN-γ can be identified as a protein band at the molecular weight of 15 kd. Molecular weight marker proteins which can be used are phosphorylase b (94 kd), bovine serum albumin (67 kd), ovalbumin (43 kd), carbonic anhydrase (31 kd), soybean trypsin inhibitor (21 kd) and α-lactalbumin (14.4 kd) (Pharmacia Fine Chemicals).

3.1.2 *Determination of protein concentrations*

Protein concentrations can be determined by the dye-binding method using a Protein Assay reagent (BioRad Laboratories).

(i) Add 5-fold diluted dye reagent (2.0 ml) to a sample solution (5–100 μl) placed in a 4-ml polystyrene cuvette.

(ii) Measure absorbance at 595 nm after 10 min. Use human serum albumin dissolved in 0.01% sodium azide at 1.00 mg/ml (determined by means of amino acid analysis) as a standard.

The standard plot (μg protein versus A_{595}) usually shows good linearity within the range from 2 to 20 μg of protein under these conditions.

3.1.3 *Anti-viral activity*

Only the final purified rMuIFN-γ is usually assayed for its anti-viral activity. The assay is done using microtitre plates by inhibiting the cytopathic effect of vesicular stomatitis virus on mouse L cells [a line designated LO (20)] using mouse α/β international reference IFN (NIH-G-002-904-511) as a tentative standard, because the international reference standard for MuIFN-γ has not been established yet.

3.2 **Purification procedure of rMuIFN-γ from *E. coli* cells**

3.2.1 *Buffers for purification*

A: 0.1 M Tris-HCl, 10% (w/v) sucrose, 5 mM EDTA, 0.2 M NaCl, pH 7.5.

B: 20 mM Tris-HCl, 5 mM benzamidine, 2 mM dithiothreitol (DTT), 10% (v/v) glycerol, pH 8.0.

C: 50 mM sodium phosphate, 10% (v/v) glycerol, 1 mM DTT, 5 mM benzamidine, 10% (w/v) ammonium sulphate, pH 7.2.

D: 50 mM sodium phosphate, 1 mM DTT, 5 mM benzamidine, 80% (v/v) ethylene-glycol, pH 7.2.

E: 50 mM sodium phosphate, 1 mM DTT, 50% (w/v) ammonium sulphate, pH 7.2.

F: Phosphate-buffered saline (PBS), 2 mM DTT, 5% (w/v) sucrose, pH 7.2.

3.2.2 *Cell disruption and ammonium sulphate fractionation*

All procedures are carried out at 4°C.

(i) Harvest the cells, suspend them immediately in four volumes of buffer A (400 ml for 100 g of wet cells) using a Polytron model PT 45/6 'OD' homogenizer, and then centrifuge the samples for 30 min at 10 000 r.p.m. in a Beckman J21 centrifuge with a JA10 rotor.

(ii) Resuspend the washed cells with four volumes of cold buffer A containing 50 mM benzamidine hydrochloride, 0.2 mM DTT and 0.2 mM phenylmethylsulphonyl fluoride (PMSF).

(iii) Pass the suspension through a Gaulin homogenizer (Model 15M-8TA) three times at 8000 p.s.i. to disrupt the cells. After each passage, cool the homogenate immediately to 6–8°C.

(iv) Centrifuge the homogenate for 30 min at 10 000 r.p.m. in a JA 10 rotor and discard the precipitate.

(v) Add solid ammonium sulphate to the supernatant up to 30% saturation (176 g of solid ammonium sulphate to every 1.0 litre of supernatant) and discard the precipitate again after centrifugation.

(vi) Add solid ammonium sulphate to the supernatant to make 85% saturation (400 g of solid ammonium sulphate to every 1.0 litre of supernatant).

(vii) Collect the resultant precipitate by centrifugation and store it at −80°C until further purification.

3.2.3 *Q-Sepharose column chromatography*

(i) Dissolve the ammonium sulphate precipitate in five volumes of buffer B.
(ii) Adjust the conductivity of the solution to that of the buffer B by means of diafiltration using an Amicon Model CH2 concentrator with two H1P10-20 hollow fibre cartridges (mol. wt cut-off at 10 000). (The solution is ultra-filtered with continuous addition of buffer B. The volume is kept constant throughout the filtration. About eight volumes of buffer B are necessary to adjust the conductivity by this procedure.)
(iii) Load the solution onto a column of Q-Sepharose fast flow (Pharmacia Fine Chemicals, about 4 ml of resin for every 100 mg of protein) pre-equilibrated with buffer B.
(iv) After washing the column with buffer B, carry out the elution with a 0−0.3 M NaCl linear gradient in buffer B (15 column volumes). rMuIFN-γ is eluted from the column at a concentration of NaCl of 0.05−0.15 M.
(v) Withdraw aliquots (20−100 μl) to determine the protein concentration and take protein samples (1 μg each) for SDS−PAGE from every two fractions.
(vi) After scanning the gels, pool the fractions containing rMuIFN-γ with a purity of higher than 50%.

3.2.4 *Phenyl−Sepharose column chromatography*

(i) Add solid ammonium sulphate to the pooled fractions from Q-Sepharose to make a 10% (w/v) concentration and load the sample directly onto a column of phenyl−Sepharose CL-4B (Pharmacia Fine Chemicals, about 30 ml of resin for every 100 mg of protein) pre-equilibrated with buffer C.
(ii) Carry out the elution with a linear gradient from buffer C to buffer D (eight column volumes). The rMuIFN-γ is eluted at around 60% ethylene glycol.
(iii) Withdraw aliquots (10−50 μl) to determine protein concentration and take protein samples (1 μg each) for SDS−PAGE from every two fractions.
(iv) After scanning the gels, pool the fractions with a purity of more than 95%.
(v) Salt out the pooled protein by dialysis against buffer E.

3.2.5 *Gel filtration on a Sephadex G-75 column*

(i) Collect the ammonium sulphate precipitate obtained as in Section 2.2.4 by centrifugation for 30 min at 10 000 r.p.m. in a JA10 rotor.
(ii) Dissolve the precipitate in buffer F (15−20 mg/ml) and load it onto a column of Sephadex G-75 pre-equilibrated with the same buffer to remove any traces of impurities.
(iii) Withdraw aliquots (5−20 μl) to determine the protein concentration and take protein samples (2 μg each) for SDS−PAGE from every two fractions.
(iv) In this step, pool the fractions without any visible impurities (which are often observed around 35 kd and 12 kd in gels).

Table 2. Results of purification of rMuIFN-γ from *E. coli*.

	Total protein (mg)	*Specific activity (units/mg protein)*	*Total activity (units)*	*Recovery (%)*
Cells	(100 g wet weight)			
Cell homogenate	10 170	2.0×10^5	2.03×10^9	100
Ammonium sulphate fractionation	7040	2.7×10^5	1.90×10^9	94
Diafiltration	6870	2.7×10^5	1.85×10^9	91
Q-Sepharose	460	1.6×10^6	0.75×10^9	37
Phenyl – Sepharose	149	2.4×10^6	0.36×10^9	18
Sephadex G-75	79	3.0×10^6	0.24×10^9	12

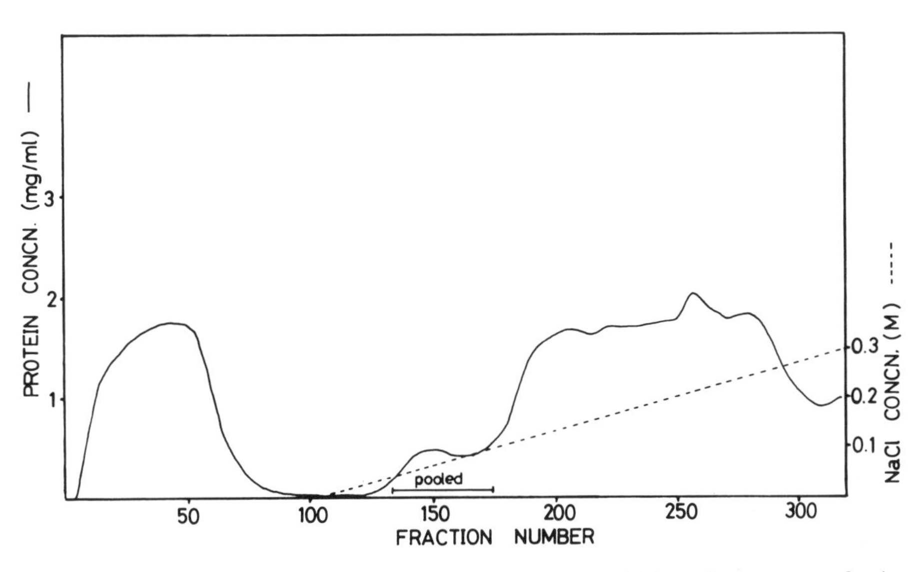

Figure 4. A typical elution profile from Q-Sepharose. Samples were withdrawn from every two fractions for protein concentration determination and SDS – PAGE. After scanning of silver-stained gels, fractions from 130 to 174 were pooled.

(v) Immediately subject the pooled G-75 fractions to sterile filtration with a Sterivex GS 0.22 μm filter unit (Millipore Corp.).

3.2.6 *An example of actual purification of MuIFN-γ*

Table 2 shows the results of purification of rMuIFN-γ. In general, about 80 mg of pure rMuIFN-γ protein is obtained from 100 g of wet cells with an overall recovery of 12%. The specific activities of some final preparation batches are $2-6 \times 10^6$ units/mg protein, increasing about 15 times from the starting cell homogenate and agreeing closely with that reported by Burton *et al.* (21). Typical elution profiles from Q-Sepharose, phenyl – Sepharose and Sephadex G-75 are shown in *Figures 4, 5* and *6*, respectively. *Figure 7* shows an example of the densitometry of SDS – PAGE patterns of chromatography fractions (fraction 136 of Q-Sepharose shown in *Figure 4*). SDS – PAGE patterns of rMuIFN-γ in each step are also shown in *Figure 8*.

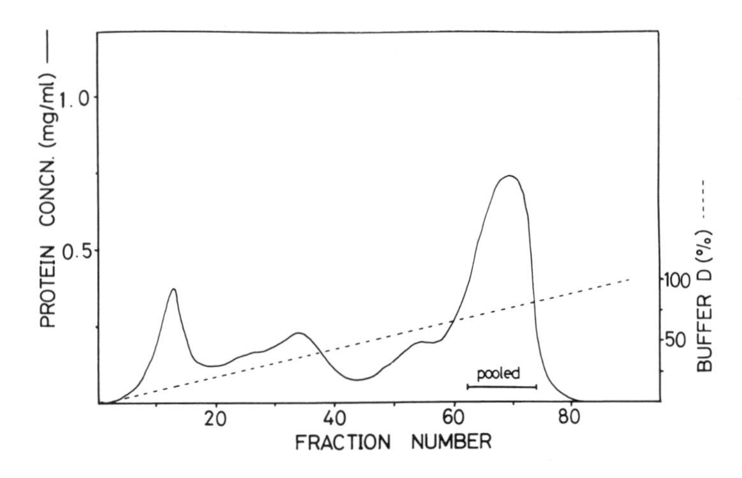

Figure 5. A typical elution profile from phenyl – Sepharose. Three peaks usually appeared and after scanning of the gels, fractions 62 – 74 were pooled.

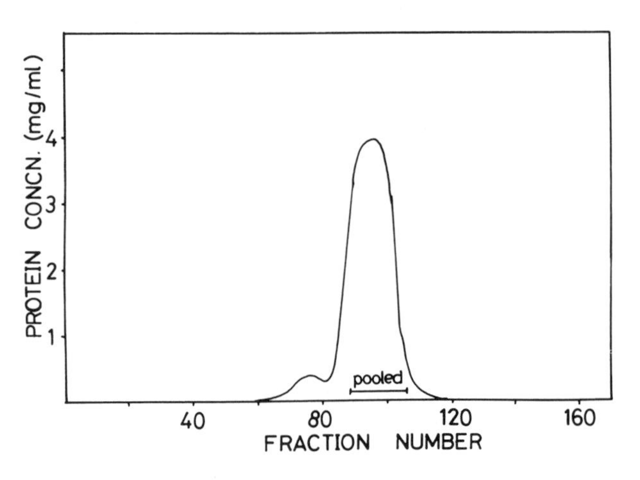

Figure 6. A typical elution profile from Sephadex G-75. After SDS – PAGE of the fractions, fractions 88 – 106 were pooled and immediately subjected to sterile filtration.

3.3 Procedures for characterization

After purification of a target recombinant protein from microorganisms, the protein must be examined to see whether it has the exact characteristics expected from the cDNA. A purity check of the final purified protein and some protein chemical characterization to confirm that the protein is derived from the same cDNA are essential for recombinant proteins. The complete protein sequence study, which requires much effort, does not necessarily seem to be essential.

3.3.1 *Procedures for SDS – PAGE*

(i) Using gradient slab gels (1 mm thick, 12 × 14 cm, separating gel from 27% acrylamide and 0.43% bis-acrylamide to 8% acrylamide and 0.25% bis-acryl-

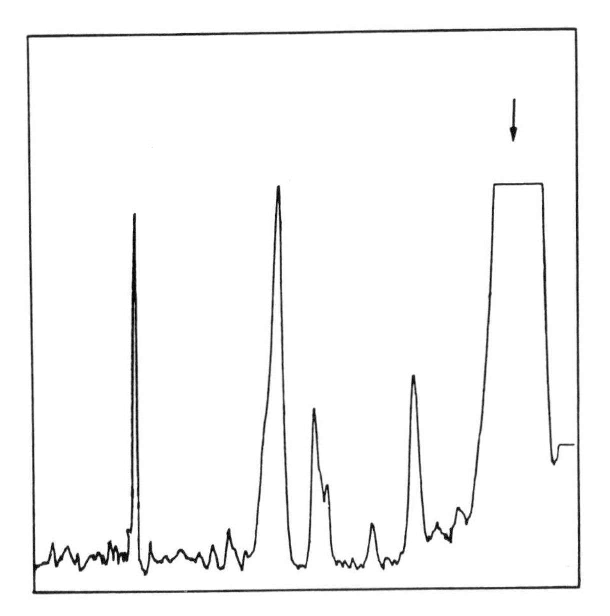

Figure 7. Densitometric analysis of an SDS – polyacrylamide gel of a Q-Sepharose chromatography fraction (fraction 136 shown in *Figure 4*). The arrow shows the position of rMuIFN-γ. The integrated data indicated 84.4% purity for this preparation.

amide) prepared according to the buffer system of Laemmli (19) check the purity of the purified rMuIFN-γ.

(ii) Apply samples (20 μg) previously incubated with 1% SDS and with or wihtout 5% 2-ME for 5 min at 85 – 90°C.

(iii) Carry out electrophoresis at 60 V for 16 h at room temperature.

(iv) After electrophoresis, stain the gels with 0.1% Coomassie Brilliant Blue R-250 dissolved in 15% acetic acid, 40% methanol for 1 h, and then de-stain them with 7.5% acetic acid, 20% methanol.

The conditions of densitometry are described in Section 3.1.1.

3.3.2 *Actual SDS – PAGE pattern*

Figure 9 shows an actual SDS – PAGE pattern of the purified rMuIFN-γ, and *Figure 10* shows its densitometry data. There is one major protein band at the molecular weight of 15 000 daltons and one faint band at the molecular weight of 30 000 daltons. The minor band increased when the samples were treated without 2-ME prior to electrophoresis. This finding and the molecular weight data suggest that the 30-kd band may be an undissociated dimeric form like that observed with natural human (22) and murine (23) IFN-γs. This undissociated dimeric form was less than 3% in the presence of 2-ME, and no other band of impurity was observed in the dye-stained gel.

3.3.3 *Procedures for analytical high performance liquid chromatography (h.p.l.c.)*

Analytical h.p.l.c. is performed with a column of TSK Gel TMS-250 (4 mm × 250 mm) for the purity check of the final preparations. Equilibrate the column with a 0.1% aqueous

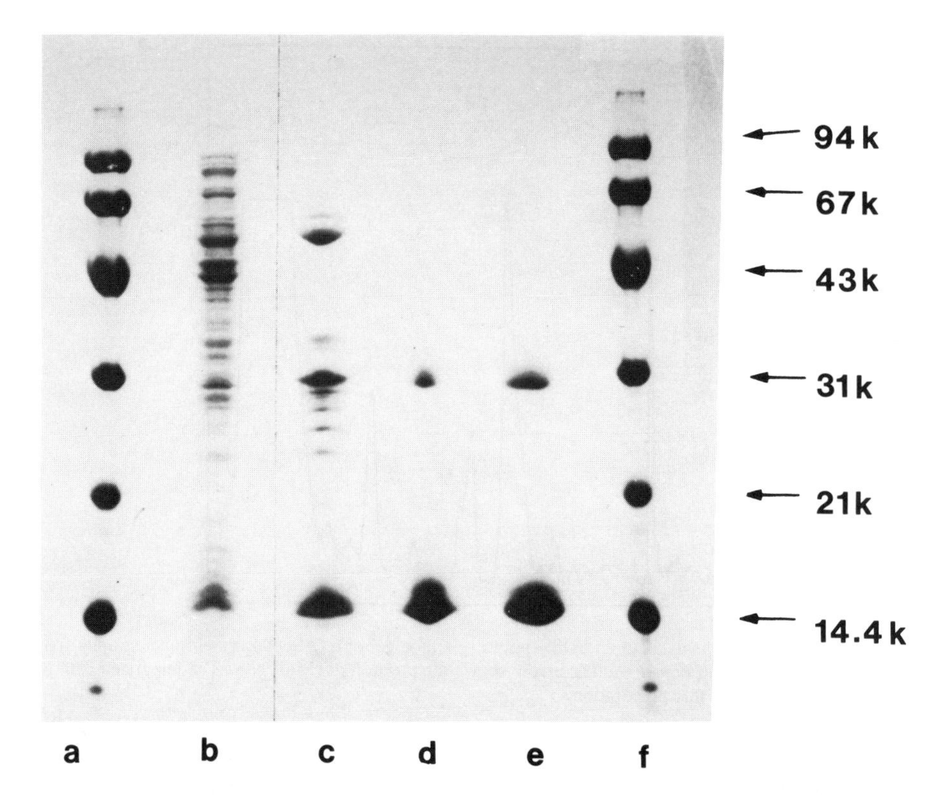

Figure 8. SDS – PAGE patterns of rMuIFN-γ preparations at each purification step. Samples (20 μg protein for each lane) were loaded onto a 15% 1 mm thick gel. The electrophoresed gel was stained with Coomassie Brilliant Blue R-250. **a** and **f**, molecular weight marker proteins; **b**, after ammonium sulphate fractionation; **c**, Q-Sepharose pooled fractions; **d**, phenyl – Sepharose pooled fractions; **e**, final Sephadex G-75 pooled fractions.

solution of trifluoroacetic acid (TFA) and elute it with a linear gradient of 0 – 100% isopropanol (2%/min) in 0.1% TFA at a flow-rate of 0.5 ml/min. Monitor the eluted materials at 220 nm.

3.3.4 *Actual h.p.l.c. profile*

Figure 11 shows an actual h.p.l.c. profile of the purified rMuIFN-γ. A single symmetrical peak appears, suggesting that the purity is almost 100%.

3.3.5 *Removal of salts, sugar and DTT from rMuIFN-γ preparations by reversed phase h.p.l.c.*

Salts, sugar and DTT should be removed from the rMuIFN-γ solution prior to the characterization procedures described in Sections 3.3.6 – 3.3.12. Reversed phase h.p.l.c. is useful for this purpose when there is not much protein sample.

(i) Load the rMuIFN-γ protein (up to 2 mg) dissolved in buffer F onto a reversed phase guard column (MPLC guard column RP-300, 4.6 mm internal diameter, 10 μm, Brownlee Labs Inc.) pre-equilibrated with 0.1% TFA at a flow-rate of 1.0 ml/min.

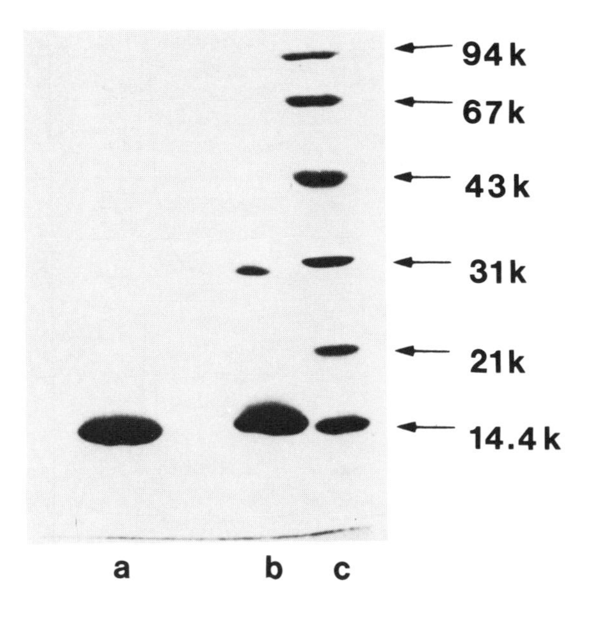

Figure 9. SDS – PAGE pattern of rMuIFN-γ. Purified rMuIFN-γ was treated with or without 2-ME and subjected to SDS – PAGE. The gel was stained with Coomassie Brilliant Blue R-250. Molecular weight marker proteins used were phosphorylase b (94 kd), bovine serum albumin (67 kd), ovalbumin (43 kd), carbonic anhydrase (31 kd), soybean trypsin inhibitor (21 kd) and α-lactalbumin (14.4 kd). **a** and **b**, 20 μg of rMuIFN-γ with **(a)** and without **(b)** reduction by 2-ME; **c**, marker proteins.

(ii) After a 5 min wash with 0.1% TFA, carry out the gradient elution with acetonitrile (0 – 100%, 10%/min) in 0.1% TFA.

(iii) Collect the rMuIFN-γ which appears in a single sharp peak and dry it *in vacuo*. The protein recovery is usually more than 90%.

3.3.6 *Procedures for amino acid analysis*

(i) Hydrolyse the de-salted samples (5 – 50 μg) in sealed, evacuated test tubes with 50 μl of 4 M methanesulphonic acid containing 0.2% 3-(2-aminoethyl)indole for 24 – 72 h at 110°C (24).

(ii) Adjust the hydrolysate to pH 2.2 by addition of 50 μl of 4 M NaOH and examine the samples with an amino acid analyser (e.g. Hitachi Model 835 amino acid analyser).

(iii) For the determination of the cysteine residues, oxidize the samples with performic acid prior to hydrolysis and recover the cysteine residue as cysteic acid.

3.3.7 *Results of an actual amino acid analysis of purified rMuIFN-γ*

Table 3 shows the amino acid composition of the final preparation. Most values agree well with the theoretical ones predicted from the cDNA sequence, while serine, cysteic

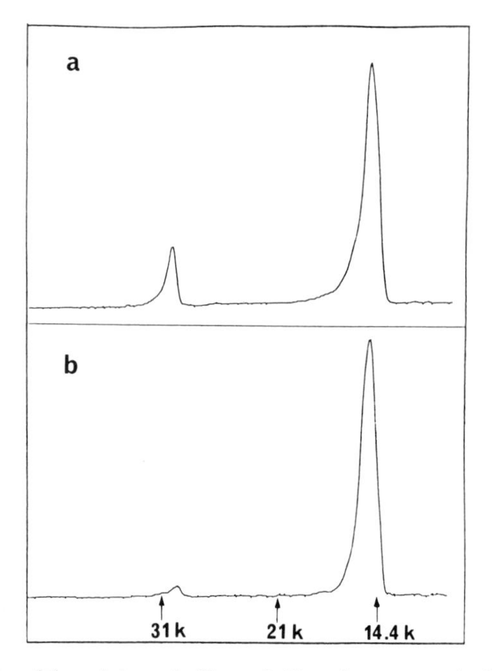

Figure 10. Densitometry of the gel shown in *Figure 9*. The gel was scanned with a Soft Laser scanning densitometer. **a**, lane b in *Figure 9*, 20 μg of rMuIFN-γ without 2-ME treatment; **b**, lane a in *Figure 9*, 20 μg of rMuIFN-γ with 2-ME treatment.

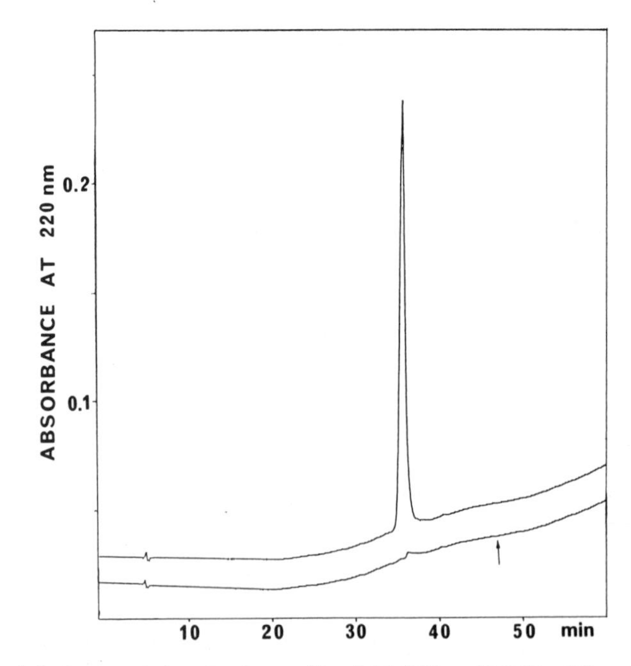

Figure 11. Analytical reversed phase h.p.l.c. profile of rMuIFN-γ. rMuIFN-γ (25 μg dissolved in 10 μl of PBS) was injected. Detailed conditions are described in Section 3.3.3. The arrow indicates the profile when 10 μl of PBS was injected.

Table 3. Amino acid composition of rMuIFN-γ.

	Observed[a]		*Based on cDNA*
Asp	17.0	(17)	17
Thr	2.9[b]	(3)	3
Ser	14.9[b]	(15)	16
Glu	17.0	(17)	17
Pro	2.1	(2)	2
Gly	3.3	(3)	3
Ala	5.4	(5)	5
Cys/2	1.7[c]	(2)	3
Val	7.8[d]	(8)	8
Met	2.0	(2)	2
Ile	11.5[d]	(12)	12
Leu	12.7	(13)	13
Tyr	2.7	(3)	3
Phe	8.9	(9)	9
Lys	10.0	(10)	10
His	2.9	(3)	3
Trp	1.9	(2)	2
Arg	5.9	(6)	8
Total		(132)	136

[a]Values in parentheses are the nearest integers.
[b]Values corrected to 0 h hydrolysis.
[c]Value of cysteic acid after performic acid oxidation.
[d]Values of 72 h hydrolysis.

acid and arginine are one or two residues less than the theoretical values, suggesting the presence of a truncated form.

3.3.8 *Reduction and S-carboxymethylation (25)*

(i) Dissolve de-salted and lyophilized rMuIFN-γ (1.5 mg) in 0.2 ml of 0.5 M Tris-HCl buffer (pH 8.0) containing 6 M guanidine hydrochloride and 2 mM EDTA, and then reduce it with 2 μl of 2-ME under a nitrogen atmosphere.
(ii) After incubation for 4 h at 37°C, add 6 mg of sodium iodoacetate dissolved in 50 μl of distilled water to the mixture.
(iii) After 10 min reaction at room temperature in the dark, subject the reaction mixture directly to h.p.l.c. as described in Section 3.3.5 to remove excess reagents. The reduced and S-carboxymethylated rMuIFN-γ (RCm-MuIFN-γ) is used for the further characterization described in Sections 3.3.9–3.3.13.

3.3.9 *Amino-terminal sequence determination*

Edman degradation of RCm-MuIFN-γ is performed as described by Iwanaga *et al.* (26) with slight modifications. Use the reagents specially prepared for protein sequence studies.

(i) Dissolve the de-salted and lyophilized RCm-MuIFN-γ (10 nmol) in 250 μl of DMAA buffer (mix 1.0 ml of N,N′-dimethylallylamine, 17 ml of pyridine and 12 ml of distilled water and adjust to pH 9.5 by addition of TFA).

(ii) Add 10 μl of phenylisothiocyanate, flush with nitrogen gas for 10 sec and mix vigorously.
(iii) After 30 min reaction at 50°C, extract the excess reagent three times with 0.3 ml of benzene.
(iv) Discard the organic phase after centrifugation for 3 min at 2000 r.p.m.
(v) Lyophilize the aqueous phase thoroughly, then add 50 μl of TFA, flush with nitrogen gas and mix vigorously.
(vi) After 10 min cleavage reaction at 50°C, remove the TFA by flushing with nitrogen gas at 50°C.
(vii) Suspend the residual material with 100 μl of distilled water and extract the anilino-thiazolinone derivatives three times with 0.25 ml of ethyl acetate.
(viii) Try the next cycle of Edman degradation with the aqueous phase after lyophilization.
(ix) Dry the organic phase by flushing with nitrogen gas at 50°C. Add 0.1 ml of 1.0 M HCl, flush with nitrogen and incubate for 5 min at 80°C.
(x) Dry the resultant phenylthiohydantoin (PTH) derivatives.
(xi) Identify the PTH derivatives by reversed phase h.p.l.c. with a Nucleosil C18 column (Nagel, 10 μm, 4 × 300 mm) pre-equilibrated with 2 mM sodium acetate buffer (pH 4.5) containing 41% acetonitrile.
(xii) Dissolve the dried PTH derivatives in 100 μl of the same buffer, and inject 5–20 μl for identification with monitoring at 254 nm.

3.3.10 *Carboxy-terminal sequence*

The carboxy-terminal sequence of purified rMuIFN-γ is characterized by digestion with carboxypeptidase P (27).

(i) Dissolve the de-salted RCm-MuIFN-γ (10 nmol) and DL-norleucine (20 nmol, as an internal standard) in 100 μl of 50 mM sodium acetate buffer (pH 3.7) containing 0.1% Triton X-100.
(ii) Add carboxypeptidase P (15 μl of 0.1 mg/ml solution) and incubate the mixture at 37°C.
(iii) Withdraw aliquots (20 μl) at appropriate times, dilute them with 380 μl of 0.2 M sodium citrate buffer (pH 2.2) and boil the mixture immediately for 2 min.
(iv) Examine it with an amino acid analyser and calculate the liberated amino acids as mol/mol protein using the values of DL-norleucine (2 nmol norleucine corresponds to 1 nmol RCm-MuIFN-γ protein).

Table 4. Liberation of amino acids from RCm-MuIFN-γ by carboxypeptidase P.

	Liberated amino acid (mol/mol protein) at		
	30 min	*60 min*	*180 min*
Lys	1.30	1.89	1.86
Arg	1.28	1.95	2.00
Leu	0.30	0.66	1.45
Ser	0.20	0.40	1.20
Glu	0	0	0.35
Pro	0	0	0.27

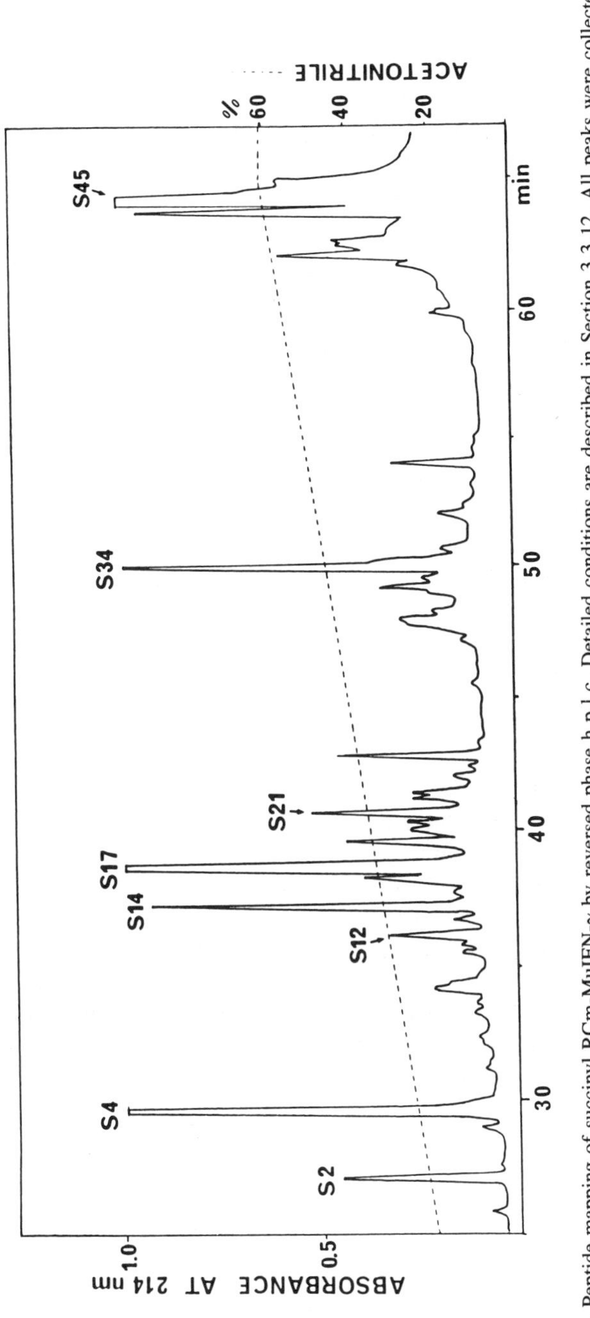

Figure 12. Peptide mapping of succinyl RCm-MuIFN-γ by reversed phase h.p.l.c. Detailed conditions are described in Section 3.3.12. All peaks were collected, re-chromatographed and then subjected to amino acid analysis.

Table 5. Amino acid compositions of peptide fragments from peaks in *Figure 12*[a].

	S2	*S4*	*S12*	*S14*	*S17*	*S21*	*S34*	*S45*	*Total*[c]
Asp		2.7 (3)		3.9 (4)		3.7 (4)	2.3 (2)	4.4 (4)	17
Thr			0.8 (1)				1.9 (2)		3
Ser	1.8 (2)			1.8 (2)		3.5 (4)	3.7 (4)	3.2 (3)	15
Glu		3.6 (4)	1.0 (1)	2.2 (2)	2.2 (2)	3.1 (3)	1.4 (1)	4.4 (4)	17
Pro		0.9 (1)			1.0 (1)				2
Gly			1.0 (1)			1.3 (1)		1.3 (1)	3
Ala		0.9 (1)		1.0 (1)			3.0 (3)		5
CmCys[b]			1.3 (2)						2
Val		1.8 (2)	0.7 (1)	1.6 (2)	1.7 (2)	1.2 (1)			8
Met							1.0 (1)	1.0 (1)	2
Ile			0.8 (1)	2.6 (3)	1.1 (1)	1.2 (1)	2.1 (2)	2.8 (4)	12
Leu	1.0 (1)			1.0 (1)	3.0 (3)	2.0 (2)	1.1 (1)	4.7 (5)	13
Tyr			0.8 (1)			1.0 (1)		1.0 (1)	3
Phe		0.9 (1)				1.0 (1)	4.0 (4)	3.0 (3)	9
Lys	1.8 (2)			0.9 (1)			3.5 (4)	2.8 (3)	10
His			0.8 (1)		1.0 (1)		1.1 (1)		3
Trp								1.5 (2)	2
Arg	2.0 (2)	1.0 (1)			1.0 (1)			2.0 (2)	6
Total residues	7	13	9	16	11	18	25	33	132
Recovery (%)	31	70	11	29	59	10	14	9	

[a]Samples were hydrolysed at 110°C for 24 h in 4 M methanesulphonic acid. Values in parentheses are estimated numbers of residues.
[b]Carboxymethylcysteine.
[c]Number of residues.

3.3.11 *Results of actual amino- and carboxy-terminal analyses*

When RCm-MuIFN-γ was subjected to Edman degradation, the single amino-terminal sequence of NH_2-carboxymethylcysteine (CmCys)-Tyr-CmCys-His-Gly- was observed to be consistent with that predicted from the cDNA sequence. *Table 4* shows the liberation of amino acids from RCm-MuIFN-γ by carboxypeptidase P, suggesting the sequence of -Ser-Leu-(Arg2, Lys2)-COOH as the carboxy terminus. Considering the theoretical protein sequence predicted from the cDNA and the results of amino acid composition and carboxy-terminal analysis mentioned above, the actual carboxy terminus of purified rMuIFN-γ is Lys132 with four residues from the native carboxy terminus (Arg133-Ser-Arg-Cys136) being completely absent.

3.3.12 *Procedures for peptide mapping of RCm-MuIFN-γ*

(i) Dissolve RCm-MuIFN-γ (0.56 mg) in 0.5 ml of 0.1 M sodium bicarbonate (pH 8.5) containing 6 M guanidine hydrochloride.
(ii) Add solid succinic anhydride (10 mg) five times every 10 min.
(iii) After dialysis against 50 mM ammonium bicarbonate (pH 8.0), add *Staphylococcus aureus* V8 protease (10 μg) to the solution.
(iv) After 4 h incubation at 37°C, subject the mixture directly to reversed phase h.p.l.c. with a Nucleosil C18 column (Nagel, 5 μm, 4.6 × 150 mm) at room temperature.

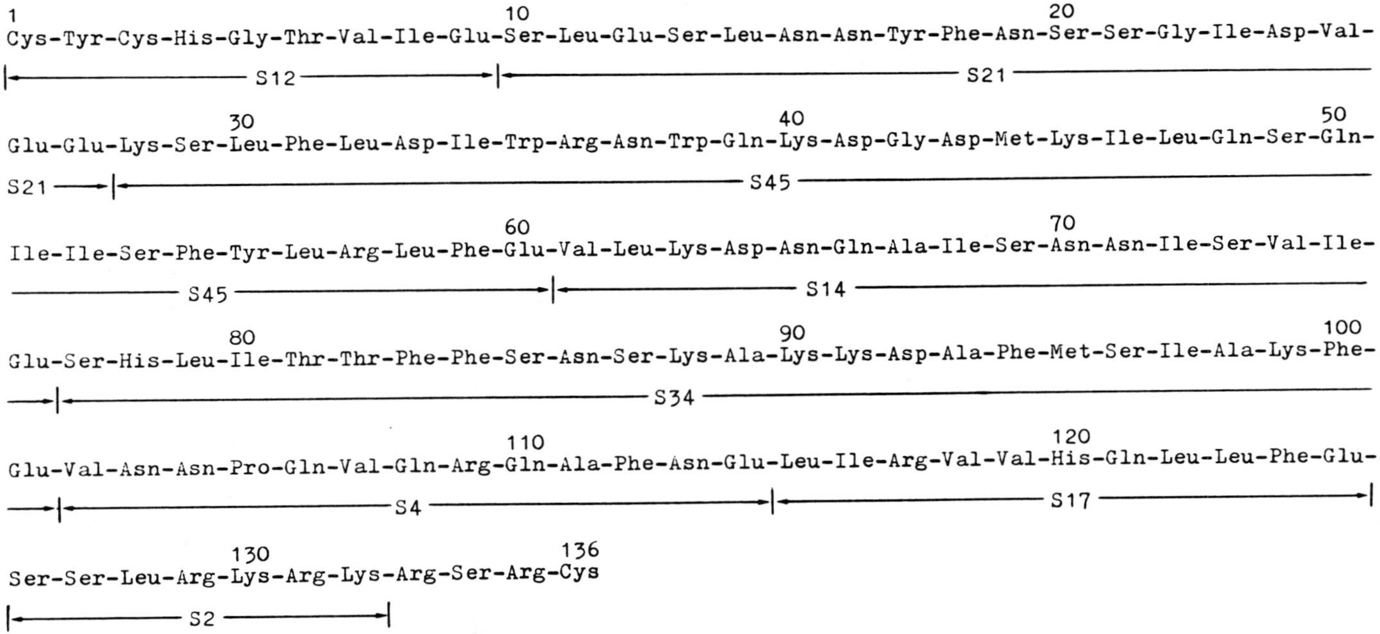

```
1                                       10                                      20
Cys-Tyr-Cys-His-Gly-Thr-Val-Ile-Glu-Ser-Leu-Glu-Ser-Leu-Asn-Asn-Tyr-Phe-Asn-Ser-Ser-Gly-Ile-Asp-Val-
|<--------------- S12 ------------->|<------------------------- S21 ----------------------------------

                30                                      40                                      50
Glu-Glu-Lys-Ser-Leu-Phe-Leu-Asp-Ile-Trp-Arg-Asn-Trp-Gln-Lys-Asp-Gly-Asp-Met-Lys-Ile-Leu-Gln-Ser-Gln-
S21 -->|<------------------------------------ S45 -------------------------------------------------

                                        60                                      70
Ile-Ile-Ser-Phe-Tyr-Leu-Arg-Leu-Phe-Glu-Val-Leu-Lys-Asp-Asn-Gln-Ala-Ile-Ser-Asn-Asn-Ile-Ser-Val-Ile-
----------- S45 ----------------------->|<------------- S14 ---------------------------------------

                80                                      90                                     100
Glu-Ser-His-Leu-Ile-Thr-Thr-Phe-Phe-Ser-Asn-Ser-Lys-Ala-Lys-Lys-Asp-Ala-Phe-Met-Ser-Ile-Ala-Lys-Phe-
-->|<------------------------------------------- S34 ----------------------------------------------

                                        110                                     120
Glu-Val-Asn-Asn-Pro-Gln-Val-Gln-Arg-Gln-Ala-Phe-Asn-Glu-Leu-Ile-Arg-Val-Val-His-Gln-Leu-Leu-Phe-Glu-
-->|<--------------------- S4 -------------------------->|<------------- S17 ---------------------->|

                130                     136
Ser-Ser-Leu-Arg-Lys-Arg-Lys-Arg-Ser-Arg-Cys
|<--------- S2 ------------>|
```

Figure 13. Assignment of peptide fragments obtained by the peptide mapping shown in *Figure 12* according to their amino acid compositions listed in *Table 5*.

(v) Carry out the elution by increasing the concentration of acetonitrile linearly from 0 to 60% (1%/min) in 0.1% TFA at a flow-rate of 1.0 ml/min, monitoring at 214 nm.

(vi) Collect all peaks, re-chromatograph the samples with the same column and then subject them to amino acid analysis to assign the peptide fragments.

3.3.13 *Results of peptide mapping and assignment of peptide fragments derived from RCm-MuIFN-γ*

RCm-MuIFN-γ was chemically modified with succinic anhydride to make it become more soluble and digested with Glu-specific *S. aureus* V8 protease. The peptide fragments were separated by reversed phase h.p.l.c. *Figure 12* shows the resultant peptide mapping pattern. All peaks were collected, re-chromatographed and subjected to amino acid analysis for assignment. Only eight peaks (S2, S4, S12, S14, S17, S21, S34 and S45) were assignable. The other small peaks could not be assigned because they were non-proteinaceous or too small to be analysed. The results of amino acid analyses (*Table 5*) suggest that peptides S12, S21, S45, S14, S34, S4, S17 and S2 correspond to Cys1-Glu9, Ser10-Glu27, Lys28-Glu60, Val61-Glu76, Ser77-Glu101, Val102-Glu114, Leu115-Glu125 and Ser126-Lys132, respectively (*Figure 13*). There was no detectable peptide containing the Arg133-Cys136 sequence. These results support the idea that purified rMuIFN-γ has the protein sequence predicted from the cDNA except for four residues being absent from the normal carboxy terminus.

3.3.14 *Some comments on the carboxy-terminal structure of rMuIFN-γ*

Purified rMuIFN-γ consisted of the protein sequence expected from the cDNA sequence with the exception of four amino acid residues from the native carboxy terminus. It is not clear whether the processing of the carboxy terminus occurred during fermentation or further purification. Considering that the cleavage site should be between Lys132 and Arg133, we have added some inhibitors of enzymes having trypsin-type specificity, such as benzamidine and PMSF, to the purification buffers to try to prevent proteolytic processing during the purification procedure. We have also tried to isolate rMuIFN-γ from the fresh *E. coli* cells harvested at 15 h instead of the usual 24 h, expecting that a small amount of full-sized rMuIFN-γ might be obtainable if the protein were processed gradually in the cells, but again, the carboxy terminus of the rMuIFN-γ obtained was completely truncated. Therefore, the processing seems to occur at an early stage of expression, although the possibility does remain that some full-sized protein exists but is removed during the purification procedure.

Rinderknecht *et al.* (28) reported that both 20-kd and 25-kd species of natural human IFN-γ derived from peripheral blood lymphocytes were heterogeneous at their carboxy termini, having forms which terminated at all residues from Gly130 to Met137 of the mature protein sequence predicted from the cDNA. It is not clear whether all forms are active nor which carboxy termini are predominant *in vivo*. The physiological meaning of the carboxy-terminal processing is also still unknown. Gray and Goeddel (9) reported that anti-viral activity can be maintained by genetically-engineered human IFN-γ lacking up to 10 amino acid residues from the normal carboxy terminus, which agrees with the above finding that natural human IFN-γ lacks at least nine carboxy-terminal amino acid residues but retains its activity. For natural murine IFN-γ, post-translational proteolytic processing was suggested by Gribaudo *et al.* (23). With respect to *E. coli*-derived IFN-γ, Rinderknecht and Burton tried limited proteolysis of human IFN-γ with trypsin and showed that about 12 carboxy-terminal amino acid residues were not important for anti-viral activity (29). That is also supported by the fact that protein homology between human and murine IFN-γs is around 40% and the characteristic difference is the omission of nine carboxy-terminal amino acid residues (9). In summary, carboxy-terminal regions of IFN-γs seem to be subject to attack by some proteolytic enzymes.

Some recombinant proteins might not necessarily be the same as those expected from the cDNA sequence, as shown here. However, the availability of recombinant proteins should be of help in understanding their biological and physiological roles and also offer opportunities for *in vitro* as well as *in vivo* studies.

4. ACKNOWLEDGEMENTS

We would like to thank Drs M.Shin for synthesizing the oligonucleotides, K.Matsumoto for performing the fermentation, and K.Sato and K.Sugita for helpful discussions. The technical assistance of M.Tamaki, E.Nakamura, Y.Fujii, N.Sakane, N.Kikuchi, H.Yonezawa and H.Tsuzuki is greatly appreciated.

5. REFERENCES

1. Watanabe,Y., Taguchi,M., Iwata,A., Namba,Y., Kawade,Y. and Hanaoka,M. (1983) In *The Biology of the Interferon System 1983.* De Maeyer,E. and Schellekens,H. (eds), Elsevier Science Publishers B.V., Amsterdam, p. 143.
2. Okayama,H. and Berg,P. (1982) *Mol. Cell Biol.*, **2**, 161.
3. Brawerman,G. (1974) In *Methods in Enzymology.* Moldave,K. and Grossman,L. (eds), Academic Press Inc., London and New York, Vol. 30, p. 605.
4. Chirgwin,J.M., Przybyla,A.E., MacDonald,R.J. and Rutter,W.J. (1979) *Biochemistry*, **18**, 5294.
5. Gurdon,J.B. and Wickens,M.P. (1983) In *Methods in Enzymology.* Wu,R., Grossman,L. and Moldave,K. (eds), Academic Press Inc., London and New York, Vol. 100, p. 370.
6. Thomas,P.S. (1983) In *Methods in Enzymol.* Wu,R., Grossman,L. and Moldave,K. (eds), Academic Press Inc., London and New York, Vol. **100**, p. 255.
7. Southern,E.M. (1975) *J. Mol. Biol.*, **98**, 503.
8. Hanahan,D. (1983) *J. Mol. Biol.*, **166**, 557.
9. Gray,P.W. and Goeddel,D.V. (1983) *Proc. Natl. Acad. Sci. USA*, **80**, 5842.
10. Hanahan,D. and Meselson,M. (1980) *Gene*, **10**, 63.
11. Birnboim,H.C. and Doly,J. (1979) *Nucleic Acids Res.*, **7**, 1513.
12. Sanger,F., Nicklen,S. and Coulson,R. (1977) *Proc. Natl. Acad. Sci. USA*, **74**, 5463.
13. Simons,G., Remaut,E., Allet,B., Devos,R. and Fiers,W. (1984) *Gene*, **28**, 55.
14. Shine,J. and Dalgarno,L. (1974) *Proc. Natl. Acad. Sci. USA*, **71**, 1342.
15. Nishi,T., Sato,M., Saito,A., Itoh,S., Takaoka,C. and Taniguchi,T. (1983) *DNA*, **2**, 265.
16. Goeddel,D.V., Shepard,H.M., Yelverton,E., Leung,D. and Crea,R. (1980) *Nucleic Acids Res.*, **8**, 4057.
17. Maniatis,T., Fritsch,E.F. and Sambrook,J. (1982) *Molecular Cloning. A Laboratory Manual.* Cold Spring Harbor Laboratory Press, New York.
18. Chen,H.-Z. and Zubay,G. (1983) In *Methods In Enzymology.* Wu,R., Grossman,L. and Moldave,K. (eds), Academic Press Inc., London and New York, Vol. 101, p. 674.
19. Laemmli,U.K. (1970) *Nature*, **227**, 680.
20. Kawade,Y. (1980) *J. Interferon Res.*, **1**, 61.
21. Burton,L.E., Gray,P.W., Goeddel,D.V. and Rinderknecht,E. (1985) In *The Biology of the Interferon System 1984.* Kirchner,H. and Schellekens,H. (eds), Elsevier Science Publishers B.V., Amsterdam, p. 403.
22. Yip,Y.K., Barrowclough,B.S., Urban,C. and Vilcek,J. (1982) *Proc. Natl. Acad. Sci. USA*, **79**, 1820.

23. Gribaudo,G., Cofano,F., Prat,M., Baiocchi,C., Cavallo,G. and Landolfo,S. (1985) *J. Biol. Chem.*, **260**, 9936.
24. Simpson,R.J., Neuberger,M.R. and Liu,T.-Y. (1976) *J. Biol. Chem.*, **251**, 1936.
25. Crestfield,A.M., Moore,S. and Stein,W.H. (1963) *J. Biol. Chem.*, **238**, 622.
26. Iwanaga,S., Wallen,P., Groendahl,N.J., Henschen,A. and Blombaeck,B. (1969) *Eur. J. Biochem.*, **8**, 189.
27. Yokoyama,S., Miyabe,T., Oobayashi,A., Tanabe,O. and Ichishima,E. (1977) *Agric. Biol. Chem.*, **41**, 1379.
28. Rinderknecht,E., O'Connor,B.H. and Rodriguez,H. (1984) *J. Biol. Chem.*, **259**, 6790.
29. Rinderknecht,E. and Burton,L.E. (1985) In *The Biology of the Interferon System 1984.* Kirchner,H. and Schellekens,H. (eds), Elsevier Science Publishers B.V., Amsterdam, p. 397.

CHAPTER 4

Production of human interferon-α

JOHN SHUTTLEWORTH

1. INTRODUCTION

Interferon (IFN) is expressed by virtually all cell types in response to a variety of inducing stimuli. For the purpose of producing human IFN-α for experimental or clinical use, considerations such as availability, convenience, yield and the nature of IFN expressed have over the years limited the choice of cell type and inducer. Human leukocytes obtained from 'buffy coats' were for a long time the most widely used source of human IFN-α. Following induction by paramyxoviruses, notably Sendai virus, yields of around 64 000 IU/ml can be obtained (1). The inconvenience of obtaining and handling useful quantities of leukocytes prompted the search for alternative sources of human IFN-α. After extensive screening the B-lymphoblastoid cell line Namalwa was found to produce the highest yields, of around 10^4-10^5 IU/ml when induced by Sendai virus (2). This system has been used extensively for IFN production (3) and is the subject of this section. Higher yielding cell lines such as NC-37, a sub-line of Raji lymphoblastoid cells, have been reported more recently (4). However the techniques described here are capable of producing equivalent yields of IFN ($>10^6$ IU/ml) from Namalwa cells and can readily be applied to alternative cell lines.

2. IFN PRODUCTION BY SENDAI VIRUS-INDUCED NAMALWA CELLS

Before describing the basic techniques and their applications, a brief consideration of the properties of this system would be useful.

2.1 Namalwa cells

Namalwa cells are an Epstein–Barr virus-transformed B-lymphoblastoid cell line derived from a Burkit lymphoma patient (2). The cells can be readily grown in suspension culture by a variety of methods depending on the demands of scale and equipment availability (stationary cultures, roller bottles, spinner flasks or small-scale fermenters). Reasons for the superior yield of IFN from Namalwa cells compared with other lymphoblastoid cell lines are not known.

2.2 Induction of IFN synthesis

The most effective and commonly used inducer of IFN synthesis in Namalwa cells is Sendai virus (2,5), a paramyxovirus containing a non-segmented, negative-stranded RNA genome. The related Newcastle disease virus (NDV) can also be used but results in lower yields of IFN. The effectiveness of Sendai virus as an IFN inducer has been

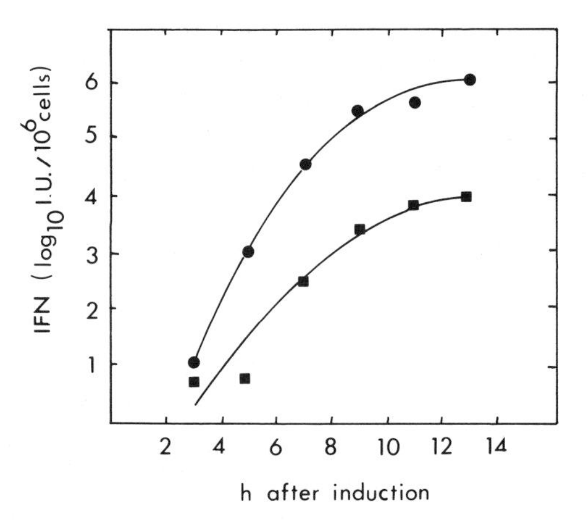

Figure 1. The time course of accumulation of IFN in butyrate-treated and untreated cells. Cells were incubated in maintenance medium containing butyrate (● – ●) or maintenance medium alone (■ – ■) and then induced. The cumulative IFN yield was assayed at various times after induction.

attributed to the presence of defective interfering (DI) particles in the stocks of virus used. The accumulation of DI particles in virus stocks passaged at high multiplicity in the allantoic cavity of fertile chicken eggs is thought to give rise to partially double-stranded RNA molecules in the infected cell. These molecules could act as efficient inducers of IFN gene expression (6).

2.3 IFN gene expression in Namalwa cells

Viral infection of Namalwa cells induces the coordinate expression of IFN-α and IFN-β genes (7). Approximately 10% of the IFN produced has been characterized as IFN-β (8). Several IFN-α genes are known to be expressed in Namalwa cells. From peptide sequencing and cDNA cloning data IFN-α1, IFN-α2, IFN-α3 and at least two other IFN-α components are present (9).

Following induction, IFN synthesis and secretion can first be detected after 2–3 h, peaks by around 10 h and continues for up to 20 h (see *Figures 1* and *2*). Greater than 90% of the IFN is secreted from the cell. Near maximal accumulation of IFN in the medium is achieved by around 10–12 h after which time the cell supernatant is usually harvested.

2.4 Factors affecting IFN production

Some discussion of the factors and treatments which are known to influence the IFN yield in Namalwa cells will be helpful. However it should be pointed out that even with the most rigorous standardization of techniques and reagents, variations in yield of 10-fold or more between experiments should be expected. The factors causing such fluctuation have not been identified.

2.4.1 *Culture conditions*

Most of the parameters relating to the growth and induction of Namalwa cells have been well characterized (5). Their effects on IFN production can be summarized as follows.

(i) *Cell density.* For most practical purposes achieving the highest concentration of IFN in the culture supernatant is more important than overall yield per cell. Although IFN yield per cell is greater at lower cell densities, concentrations of $2-5 \times 10^6$ cells/ml during IFN induction and production are used since they result in higher titres of IFN. At higher densities IFN yield drops dramatically, presumably because of nutrient depletion and pH destabilization. Similarly, cell density is not allowed to exceed 10^7 cells/ml during routine passage since induction of such saturation density cultures results in poor yields.

(ii) *Serum.* The serum requirements of Namalwa cells are not at all demanding. Cells will grow well in serum from a variety of sources, including human, bovine and porcine. Furthermore induction and production of IFN can be performed in serum-free medium with no consistent detectable effect on yields. The stability of IFN produced in serum-free culture may be reduced, hence if serum is omitted (for purification purposes) the addition of serum albumin is desirable.

(iii) *Scale of culture.* Yields of IFN are reduced as the scale of culture is increased, for reasons which are not entirely understood. It may be worthwhile to consider multiple small scale (<1 l) cultures if IFN production is contemplated on a modest scale. Spinner flask or roller bottle cultures of 100 ml to 1 l have been found reproducibly to give best results. Presumably the mixing and aeration of larger cultures imposes sub-optimal conditions for growth and induction, added to which mechanical damage is more likely to occur. Both situations manifest themselves in reduced growth rate and increased cell death.

2.4.2 *The inducer*

Different batches of egg-grown Sendai virus can vary greatly in their effectiveness as inducers of IFN production. One possible explanation for this is the ratio of infectious virus to DI particles (6). It is important to check new stocks of virus for their ability to induce IFN synthesis, in small-scale inductions, before embarking on costly and time-consuming large-scale experiments.

2.4.3 *Treatment of cells*

Various regimes of cell treatment, prior to and during induction of IFN synthesis, have been reported to increase IFN yields (for review see 10). Priming by the addition of small quantities of IFN before induction, and superinduction schedules based on sequential cycloheximide and actinomycin D treatment before and during induction, have little or no effect on IFN yields in Sendai-induced Namalwa cells (5,11). However, several agents are known to increase IFN yields when added to Namalwa cells before induction. These include butyrate and 5′-bromodeoxyuridine (12). Butyrate treatment is the most effective and reproducible, resulting in up to 100-fold enhancement of IFN yield. The kinetics of IFN production (see *Figure 1*) and the character of the IFN expressed are not altered by treatment. The effect is caused by increased transcription and accumulation of IFN mRNA in the treated cells (12).

A similar increase in IFN production (up to 30-fold) can be achieved by lowering the incubation temperature of cultures from 37 to 28°C at 7 h after induction (13). Low temperature treatment causes 'superproduction' of IFN by stabilizing IFN mRNA and prolonging the period of IFN production (see *Figure 2*).

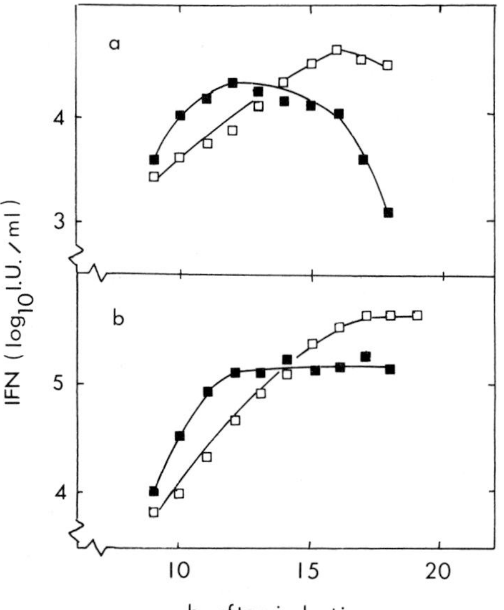

Figure 2. IFN production in cells incubated at 37 and 28°C. Cells were induced and incubated at 37°C for 7 h. **(a)** The rate of IFN synthesis (measured as the amount of IFN secreted during a 1 h period) and **(b)** the cumulative yield of IFN were measured in cells incubated at 37°C (■ – ■) or at 28°C (□ – □) from 7 h after induction.

Table 1. Preparation of growth medium.

	Final concentration
To 900 ml of medium RPMI 1640 add:	
1. 100 ml of adult or newborn bovine serum	10%[a]
2. 10 ml of sterile 200 mM glutamine	2 mM
3. 1 ml of 60 mg/ml penicillin	60 μg/ml
1 ml of 100 mg/ml streptomycin	100 μg/ml
4. Either10 ml of 5% $NaHCO_3$	0.05%
or 10 ml of 2 M Hepes pH 6.5	20 mM

[a]For maintenance medium add only 20 ml of serum to give a final concentration of 2%.

3. ROUTINE CULTURE OF CELLS

It is assumed that the reader has a basic knowledge of tissue culture techniques. It should be remembered that Namalwa cells are a transformed human cell line. Suitable precautions must be taken for handling.

3.1 Media

Namalwa cells are grown in medium RPMI 1640. This can be purchased either in solution or as a solid which is prepared according to manufacturers' instructions. Before use the medium is supplemented with serum, glutamine, antibiotics and buffer. The choice of buffer depends on the method of culture (see Section 3.2) therefore medium lacking sodium bicarbonate should be obtained. Hepes-buffered medium does not require gassing with 5% CO_2. Prepare growth medium, containing 10% serum, as shown in *Table 1*. Maintenance medium, used for IFN production, is prepared in the same way except that serum is added to a concentration of 2% instead of 10%.

Glutamine is used to supplement the medium because this component is labile and may become growth limiting if any loss has occurred during storage. Namalwa cells prefer the slightly lower pH of 6.5. The medium indicator should be an orange/red colour at this pH, if not adjust with a few drops of HCl (for Hepes buffer) or by bubbling CO_2 (for bicarbonate).

3.2 **Culture methods**

3.2.1 *Stationary cultures*

A variety of methods can be used to grow Namalwa cells. Although the cells are normally grown in suspension, it is often impractical to culture small volumes (<100 ml) in anything other than stationary cultures (for example, when growing up cells from liquid nitrogen storage). The cells will grow well in stationary culture if:

(i) a large surface area is provided;
(ii) the volume of culture is limited to approximately 1/10 the capacity of the vessel.

Cultures of up to 100 ml can therefore be grown in glass (medical flats or flow bottles) or sterile tissue culture plastic flasks of up to 1 l capacity. Either Hepes- or bicarbonate-buffered medium can be used.

3.2.2 *Spinner cultures*

Cultures of virtually any volume can be grown in flasks or bottles provided with mechanical agitation. On a laboratory scale indirectly driven magnetic stirrer bars are the most convenient method of mixing. The speed of mixing is limited to that which is sufficient to maintain a uniform suspension of cells since there is a danger of mechanical damage to the cells. Temperature regulation is conveniently provided by setting up cultures in a hot room.

(i) Cultures grown in Hepes-buffered RPMI 1640 grow well in stirred flasks provided that the culture volume is limited to approximately 1/5 the volume of the vessel in order to ensure adequate aeration.
(ii) Larger cultures (up to 40 l) can be grown in bicarbonate-buffered RPMI 1640 contained in glass bottles (up to 50 l capacity). The vessel is supplied with 5% CO_2:95% air which is bubbled through the culture via glass or stainless steel tubes.

3.2.3 *Roller cultures*

Namalwa cells can be grown satisfactorily in roller bottles using Hepes-buffered RPMI 1640, provided that:

(i) the volume of the culture does not exceed 1/5 the capacity of the vessel;
(ii) the speed of the roller is sufficient to maintain the cells in suspension. Some machines do not rotate fast enough to prevent cells from settling out of suspension.

3.3 **Cell culture**

3.3.1 *Cell density*

Namalwa cells should be maintained as actively dividing, exponentially growing cultures. Since the cells are transformed and grow in suspension no contact inhibition occurs. Instead, as cells reach a density approaching 10^7 cells/ml growth ceases and cell death occurs as a consequence of media depletion and low pH (the cells glycolyse very actively and rapidly acidify the medium). Conversely cells at densities less than 10^5 cells/ml divide very poorly unless provided with conditioned medium. The ideal cell density for routine culture is around 10^6 cells/ml.

3.3.2 *Monitoring*

The most convenient way of monitoring cultures is to count cells under the microscope using a Neubauer chamber in the presence of the vital dye trypan blue. This provides not only an estimation of cell density but also an indication of the state of the cells in culture. Healthy, live cells should appear regular and spherical in shape and exclude dye. Cultures containing irregular cells or greater than 10% dead cells indicate problems with media or culture conditions. The doubling time for Namalwa cells is around 48 h at 37°C and this provides another means of assessing the state of the culture.

3.3.3 *Passaging*

Set up cultures at a density of 10^6 cells/ml. Once the culture has reached a density of 2×10^6 cells/ml (after ~48 h) add an equal volume of fresh pre-warmed growth medium and if necessary:

(i) divide the culture between two vessels; or
(ii) transfer to a larger vessel; or
(iii) discard surplus.

4. PRODUCTION OF IFN

The previous sections have discussed most of the background information required to set up IFN production using Namalwa cells. Once healthy cultures in sufficient quantity are available the following protocols can be used to produce IFN.

4.1 **Butyrate treatment**

The option of using butyrate treatment to increase IFN yield has been introduced in Section 2.4.3. Details for the preparation of sodium butyrate and the treatment of Namalwa cells are shown in *Tables 2* and *3*. Sodium butyrate can be added directly to cells at the appropriate density in growth medium, omitting *Table 3*, steps 1 and 2. The resulting increase in IFN production after induction is then usually lower than that obtained using the full protocol.

Table 2. Preparation of sodium butyrate stock solution.

1.	Add 18 ml of butyric acid to 982 ml of medium RPMI 1640 to give a final concentration of 200 mM.
2.	pH the solution to pH 6.5 using NaOH pellets.
3.	Store as aliquots at −20°C.

Table 3. Butyrate treatment of cells.

1.	Remove the cells from the growth medium by centrifugation at 750 *g* for 10 min.
2.	Resuspend the cells in maintenance medium at a density of 10^6 cells/ml (maintenance medium is prepared as in *Table 1*).
3.	Add 1/200 volume of stock 200 mM sodium butyrate to give a final concentration of 1 mM.
4.	Incubate the cells at 37°C for 48 h.
5.	Induce the cells with Sendai virus as in *Table 4*.

Table 4. Induction of IFN synthesis.

1.	Remove the cells from the growth medium, or maintenance medium containing sodium butyrate, by centrifugation at 750 *g* for 10 min.
2.	Resuspend the cells in pre-warmed maintenance medium at a density of 2×10^6 cells/ml.
3.	Add Sendai virus at a rate of 100 haemagglutinating units/10^6 cells.
4.	Incubate immediately at 37°C for 12–20 h.
5.	Remove the cells from the medium by centrifugation at 750 *g* for 10 min.
6.	Discard the cells, adjust the supernatant to pH 2 by adding concentrated HCl.
7.	Incubate the supernatant at 4°C for 24 h to inactivate the virus.
8.	Clarify the supernatant by centrifugation at 8000 *g* for 10 min.
9.	Store at −20°C or below.

Table 5. Low temperature treatment of cells.

1.	Proceed as in *Table 4*, steps 1–3.
2.	Incubate the cells immediately at 37°C for 7 h.
3.	Reduce the temperature of culture rapidly to 28°C and incubate for a further 14 h.
4.	Recover the supernatant and proceed as in *Table 4*, steps 5–9.

Butyrate treatment inhibits DNA synthesis and cell division hence cell numbers should remain constant during treatment. If cultures are semi-synchronous replication sometimes occurs if butyrate is added just before or during cell division.

4.2 Induction of IFN synthesis

The induction of IFN production is detailed in *Table 4*. The exact ratio of virus:cells giving maximum induction may vary between different batches of virus. The figure of 100 haemagglutinating units per 10^6 cells is only a guide. Virus titre depends on the method of assay and may vary between laboratories. The optimum ratio should be determined in small-scale inductions with each batch of virus.

The time at which the supernatant is harvested is reasonably flexible. Synthesis and secretion of most of the IFN is complete by around 12 h after induction, but cultures can be left for up to 20 h. No degradation is detected during this period, but prolonged incubation is not desirable since some cell death occurs resulting in increased protease activity in the medium as induced cultures age.

IFN proteins are stable at pH 2 and full activity is recovered once the pH returns to neutral. This provides a useful method of inactivating any Sendai virus persisting in the supernatant which could interfere with subsequent biological assays. The IFN in the cleared supernatant is stable when stored at −20°C or lower, but repeated freeze–thaw cycles should be avoided.

4.3 Low temperature treatment

Low temperature treatment, described in *Table 5*, provides another option for increasing IFN yield. Its success depends on reducing the temperature of the culture uniformly to 28°C at precisely 7 h after induction. This time point is critical since it immediately precedes the time at which maximum levels of IFN mRNA accumulate in the cells (15). Unless some method can be devised to rapidly reduce the temperature of large culture volumes this procedure will only be successful with small to medium scale production.

5. REFERENCES

1. Cantell,K., Hirvonen,S. and Koistinen,V. (1981) In *Methods in Enzymology.* Pestka,S. (ed.), Academic press, New York, Vol. 78, pp. 499.
2. Strander,H., Mogensen,K.E. and Cantell,K. (1975) *J. Clin. Microbiol.*, **1**, 116.
3. Finter,N.B. (1982) *Texas Rep. Biol. Med.*, **41**, 175.
4. Adolph,G.R., Bodo,G. and Swetly,P. (1981) *Antiviral Res.*, **1**, 275.
5. Johnston,M.D., Fantes,K.H., Finter,N.B. and Chir,B. (1978) In *Human Interferon.* Stinebring,W.R. and Chapple,P.J. (eds), Plenum Press, New York, p. 61.
6. Johnston,M.D. (1981) *J. Gen. Virol.*, **56**, 175.
7. Shuttleworth,J., Morser,J. and Burke,D.C. (1983) *Eur. J. Biochem.*, **133**, 399.
8. Havell,E.A., Yip,Y.K. and Vilcek,J. (1977) *J. Gen. Virol.*, **38**, 51.
9. Allen,G. and Fantes,K.H. (1980) *Nature*, **287**, 408.
10. Stewart,W.E.,III, (1979) *The Interferon System.* Springer Verlag, Vienna and New York.
11. Morser,J., Flint,J., Graves,H.E., Baker,P.N., Colman,A. and Burke,D.C. (1979) *J. Gen. Virol.*, **44**, 231.
12. Shuttleworth,J., Morser,J. and Burke,D.C. (1982) *J. Gen. Virol.*, **58**, 25.
13. Morser,J. and Shuttleworth,J. (1981) *J. Gen. Virol.*, **56**, 163.

CHAPTER 5

Production of recombinant interferons by expression in heterologous mammalian cells

A.G.MORRIS and G.WARD

1. INTRODUCTION

Enormous amounts of a protein such as interferon (IFN) may be produced by the use of a bacterial (or other microorganism) expression system (see Chapter 1). However, for many purposes, such products are not ideal because post-translational modification of the protein which occurs in mammalian cells may not occur (or may occur in a different way) in microorganisms. Furthermore, purification of the product from microorganisms may well not be straightforward, and the presence of bacterial components (even in trace amounts) such as lipopolysaccharides must be rigorously excluded especially if the material is to be used for immunological experiments. Hence the present great interest in the use of mammalian cell-based expression systems. Yields from these (up to 1 mg/l) are usually much less than from bacterial systems, but this is not a serious disadvantage in the case of proteins such as IFNs with their enormous specific activities. Furthermore, with at least some of these systems, the IFN is continuously secreted from the cells and so may be continuously harvested. This chapter describes two systems we have used to express IFN genes, one using the expression vector pSVL in COS cells, the other using $pKSV_{10}$ in conjunction with pSV_2-$DHFR^+$ in Chinese hamster ovary (CHO) $DHFR^-$ cells. The former gives transient expression; yields are low, but can be quickly obtained. The latter gives continuous expression which may be amplified to very high levels by selection techniques; but takes time.

We make the assumption that a cloned cDNA for the gene of interest is available, with secretion signal sequence intact and with suitable enzyme linker sequences present at the ends. Description of the methods for obtaining such a clone is obviously beyond this article as are details of the tissue culture techniques employed. The examples we shall use are MuIFN-γ expressed in COS cells and HuIFN-γ expressed in CHO cells. Obviously the methods are generally applicable and we are currently using them for expression additionally, of MuIFN-α_2 and MuIFN-β; in principle, any lymphokine gene (provided it is not toxic) can be expressed in the same way.

All the cell types (mammalian and bacterial) described below are available from the authors.

2. COS CELL EXPRESSION

The principle of this technique is to transfect COS cells (1) with a plasmid containing the SV40 origin of replication and the IFN or lymphokine gene under the control of

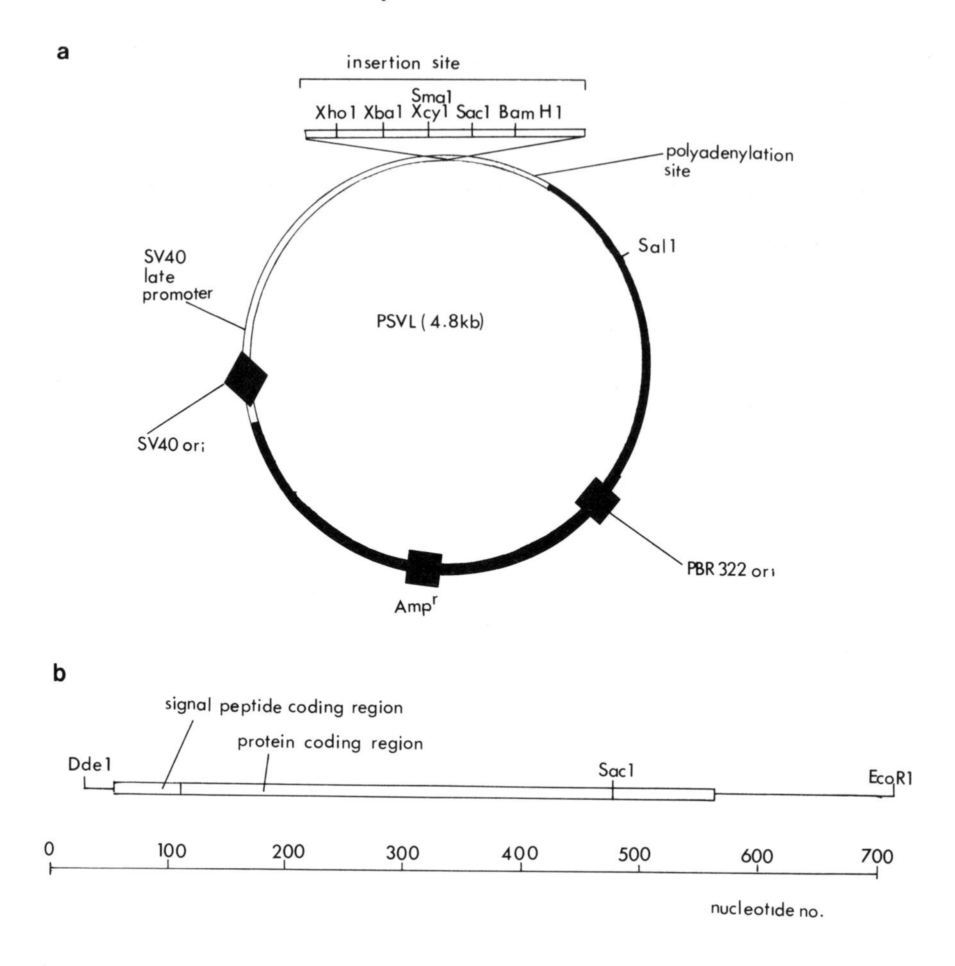

Figure 1. Maps of (**a**) pSVL, (**b**) MuIFN-γ. In (**a**) pBR322 sequences are indicated by solid, SV40 by open blocks. Nucleotide numbering of (**b**) based on (3).

a suitable promoter. COS cells are transformed with a defective SV40 and contain the SV40 T-antigen: the plasmid will therefore replicate to a high copy number and the gene is therefore expressed at a high level (2). COS cells are available from the American Type Culture Collection [ATCC: CRL 1650 (COS 1) or 1657 (COS 7)]. A suitable plasmid vector is pSVL available from Pharmacia (catalogue no 27-4509-01). *Figure 1* shows a simplified restriction endonuclease map of pSVL and also of the MuIFN-γ cDNA we have used (isolated by G.Ward and K.Siggens).

(i) Ligate a blunt-ended subclone of the MuIFN-γ cDNA molecule (*Figure 1*) containing the entire protein coding region into the *Sma*I site of the plasmid pUC 8 (available from Pharmacia; catalogue no 27-4916-01), transform this into *Escherichia coli*, grow up and prepare bulk plasmid DNA as described in Chapter 7.

(ii) Isolate the insert after double digestion at the *Eco*RI and *Bam*HI sites on either side of the insert by agarose gel electrophoresis as described in Chapter 7.

Table 1. Use of the Klenow fragment of *E. coli* DNA polymerase I to convert DNA fragments with protruding 5′ termini to blunt ends.

1. Following the digest with a restriction enzyme that generates protruding 5′ termini, either agarose gel purify a particular DNA fragment[a] or, if only one DNA fragment is produced, extract once with phenol/chloroform[b] and ethanol precipitate by adjusting the solution to 0.3 M NaCl using a stock solution of 3 M NaCl and adding 2 vols of ethanol at −20°C.
2. Spin down the ethanol precipitate in an Eppendorf centrifuge for 10 min. Wash the pellet with 80% ethanol and re-centrifuge. Decant off the ethanol and dry briefly *in vacuo*. Dissolve the DNA in 19 μl of water (autoclaved, double-distilled).
3. Prepare the following solutions:
   (a) 10 × nick translation (NT) buffer
   0.5 M Tris-HCl, pH 7.2
   0.1 M $MgSO_4$
   500 μg/ml bovine serum albumin (BSA Pentax Fraction V)
   (b) 10 × dNTP mix[c]
4. Set up the following reaction:

DNA restriction fragment	19 μl (1−2 μg of DNA)
10 × NT buffer	2.5 μl
10 × dNTP mix	2.5 μl

   Klenow fragment of *E. coli* DNA polymerase I[d] 1 μl (Amersham Cat No. T.2140Y; 2−8 units/μl)
5. Mix and incubate at room temperature for 1 h.
6. Add 1 μl of 0.5 M EDTA to stop the reaction.
7. Extract once with phenol/chloroform and ethanol precipitate as before.

[a]See Chapter 7, Table 4.
[b]The phenol in this 1:1 mixture with chloroform is re-distilled at 160°C to remove contaminants. This re-distilled phenol is stored frozen in aliquots at −20°C. In preparing fresh batches of phenol/chloroform, an aliquot of phenol is melted in a water bath at 65°C and extracted several times with an equal volume of 10 mM Tris-HCl (pH 7.5), 300 mM NaCl, 1 mM EDTA until the pH of the aqueous phase reaches pH 7. An equal volume of chloroform is then added along with 8-hydroxyquinoline to a final concentration of 0.1%. This is stored at 4°C, wrapped in foil, for up to 1 month.
When extracting a DNA-containing solution with phenol/chloroform the volume of the solution is usually increased to 300−500 μl and then an equal volume of phenol/chloroform is added. After vigorous shaking, the two phases are separated by centrifuging in an Eppendorf centrifuge for 2 min.
[c]See Chapter 7, Table 5.
[d]Enzymes used here and in subsequent tables are all commercial preparations obtainable, for example from Amersham International.

(iii) The resulting 640-bp fragment cannot be directly ligated into pSVL, hence blunt-end ligation is necessary. Since both ends of the 640-bp fragment have protruding 5′ termini, these need to be blunt-ended using the Klenow fragment of *E. coli* DNA polymerase I as in *Table 1*. Digest pSVL at the *Sma*I site in the multi-cloning site of pSVL vector DNA, which generates blunt ends, and dephosphorylate these ends (*Table 2*).

(iv) Ligate together the two as described in *Table 3*, mixing 50−100 ng of the 640-bp fragment with 200−300 ng of the vector DNA.

(v) Transform a suitable strain of *E. coli* such as *E. coli* HB101, with aliquots of the ligated DNA following the procedure described in *Table 4*.

The pSVL plasmid carries the ampicillin resistance gene for selection of transformants. Once single colonies of transformants have grown up they can be screened by preparing minipreparations as described in *Table 5*. A single restriction digest with *Sac*I identifies a recombinant plasmid and also gives the orientation in which the insert has ligated

Table 2. Dephosphorylation of vector DNA using calf intestine alkaline phosphatase.

1. Linearize 2–5 μg of plasmid DNA[a] by digestion to completion with the appropriate restriction enzyme. Extract once with phenol/chloroform and ethanol precipitate by adjusting the solution to 0.3 M NaCl using stock 3 M NaCl and adding 2 vols of ethanol at −20°C. (If necessary, the vector DNA is blunt-ended at this stage.)
2. Spin down the ethanol precipitate in an Eppendorf centrifuge for 10 min. Wash the pellet with 80% ethanol and re-centrifuge. Decant off the ethanol and dry briefly *in vacuo*. Dissolve the DNA in 10 μl of 10 mM Tris-HCl, pH 8.0.
3. Prepare the following buffers:

10 × CIP buffer	10 × TNE
0.5 M Tris-HCl, pH 9.0	100 mM Tris-HCl, pH 8.0
10 mM $MgCl_2$	1 M NaCl
1 mM $ZnCl_2$	10 mM EDTA
10 mM spermidine	

4. Add 5 μl of 10 × CIP buffer and 31 μl of water (autoclaved, double-distilled) to the DNA. Add 1 unit of calf intestine alkaline phosphatase (Boehringer, Mannheim, cat. no 713-023) by adding 4 μl of a freshly diluted batch of enzyme at a concentration of 0.25 units/μl[b].
5. Incubate at 37°C for 30 min. Add a second aliquot of enzyme and incubate for a further 30 min. If the vector DNA to be dephosphorylated has blunt ends or recessed 5′ termini then it is incubated for 15 min at 37°C followed by 15 min at 56°C. A second aliquot of enzyme is then added and the incubations at both temperatures repeated.
6. Add 40 μl of water, 10 μl of 10 × TNE and 5 μl of 10% SDS. Heat to 68°C for 15 min. This inactivates the enzyme.
7. Extract twice with phenol/chloroform and twice with chloroform/iso-amyl alcohol (24:1 v/v).
8. Precipitate the DNA with ethanol as before. Spin down, wash and dry the DNA pellet as before. Dissolve the DNA in 10 mM Tris-HCl (pH 8.0), 1 mM EDTA.
9. Run an aliquot on a 0.8% agarose – TBE minigel[c] along with known amounts of marker DNA (such as phage λ DNA) to get an estimate of the concentration of dephosphorylated vector DNA used in subsequent ligation reactions.

[a]The DNA is isolated as described in Chapter 7, Table 3.
[b]This represents a vast excess required to dephosphorylate this amount of DNA since 1 unit of calf intestine alkaline phosphatase will remove the terminal phosphates from 0.1 nmol of 5′ ends of DNA (0.1 nmol of 5′ ends of a 4-kb linear DNA molecule in 160 μg of DNA). The enzyme is diluted down with 10 mM Tris-HCl (pH 8.0), 1 mM EDTA and discarded after use, since its activity is reduced when stored in this buffer.
[c]See Chapter 7, Table 4.

into the vector. There is a unique *Sac*I site in the multi-cloning site of pSVL, on the 3′ side of the *Sma*I insertion site. There is also a unique *Sac*I site in the published sequence of the MuIFN-γ cDNA molecule at position 466 (3). The MuIFN-γ subclone inserted into pSVL spans the published sequence from position 62 to position 706. Thus a *Sac*I digest of a recombinant plasmid will produce a small fragment of size, approximately equal to 240 bp if the insert is in the 5′→3′ orientation. Therefore, set up digests of 10 μl of several minipreparations of plasmid DNA and analyse fragments produced on a 1.5% agarose–TBE minigel.

(i) Pick out a clone containing the insert in the 5′→3′ orientation and make a glycerol stock from the remaining overnight culture from which the plasmid minipreparation had been made. This is achieved by simply adding an equal volume of autoclaved 80% glycerol, mixing and storing the glycerol stock at −20°C.
(ii) Use this glycerol stock to grow up a bulk culture of the recombinant bacteria.

Table 3. Ligation reactions.

1. Prepare 10 × ligation buffer
   0.66 M Tris-HCl, pH 7.5
   50 mM $MgCl_2$
   50 mM dithiothreitol
   10 mM ATP
   1 mg/ml bovine serum albumin
   This is used for both blunt-end and sticky-end ligations.
2. Mix the vector DNA[a], insert DNA[b], 10 × ligation buffer and T4 DNA ligase (Amersham cat no T.2010Y) in a final volume of 50 μl and incubate at 16°C overnight. (If the DNA fragment has cohesive ends this time may be reduced to 4−6 h.)
3. Stop the ligation reaction by adding 1 μl of 0.5 M EDTA, and store it at −20°C. It is not necessary to extract or ethanol precipitate the DNA before transformation into *E. coli*.

The efficiency of ligating a blunt-ended fragment to blunt-ended, linear plasmid DNA is poor for three reasons. First, the K_m for the activity of T4 ligase on blunt-ended DNA is nearly 100 times greater than its K_m on DNA with cohesive ends. Thus, ligation of blunt-ended DNA requires high concentration of enzyme and a high concentration of DNA ends (>1 μM). Large amounts of the fragment to be cloned are therefore needed. Second, during the blunt-end ligation, a fraction of the plasmid will recircularize — hence the need for efficient dephosphorylation of the vector. Third, because of the high concentration of the fragment to be cloned, some recombinant plasmids will contain more than one insert.

For cohesive end ligations 1 μl of Amersham T4 DNA ligase is added; this is increased to 2−3 μl for blunt-end ligations. This represents an excess of enzyme added to the reaction since the enzyme is supplied at a concentrated activity: 1−5 pyrophosphate exchange units/μl = 125−625 ligation assay units/μl.

[a]Usually 200−300 ng of vector DNA is added to each ligation reaction giving a concentration of vector DNA of 4−6 μg/ml.

[b]The amount of insert DNA to add depends on its size and the size of the vector. A ratio of insert to vector that favours intermolecular ligation at the expense of intramolecular ligation is achieved with a molar ratio of 3:1 insert to vector.

(iii) Isolate a stock of purified recombinant plasmid by the method described in Table 3, Chapter 7.
(iv) Check the purity of this preparation by removing a small aliquot to run on an agarose minigel.
(v) Digest 2−3 μg with *Sac*I to ensure the correct clone has been grown up, that is one containing the MuIFN-γ insert in the 5′→3′ orientation. Run this digest on the same minigel.
(vi) Make a sterile stock of the plasmid for use in subsequent COS cell transfections by re-ethanol precipitating a fraction of the preparation and dissolving the resulting pellet at a concentration of 1 μg/μl in autoclaved double-distilled water, under sterile conditions.

The COS cells, which are cultured in any suitable animal cell medium (e.g. Eagle's minimal essential medium, MEM) supplemented with fetal calf serum (FCS) are transfected with the recombinant plasmid as follows.

(i) Add the DNA to the COS 7 cell line in tissue culture in a solution of DEAE−dextran which facilitates entry of DNA into cells.
(ii) Prepare an autoclaved solution of 10 mg/ml DEAE−dextran (Pharmacia) in water.
(iii) Grow up COS cells in 75 cm^2 plastic tissue flasks to near confluence.

Table 4. Transformation of *E. coli* HB101.

1. Prepare LB medium[a].
2. Prepare agar plates containing the appropriate antibiotic as follows. Add 20 g of Difco agar per litre of LB medium. Sterilize by autoclaving. Allow to cool to 55°C before adding the appropriate antibiotic from a sterile, concentrated stock solution[b]. Pour the agar directly from the bottle, allowing 30–35 ml per 85 mm Petri dish, and allow to harden. Remove condensation by placing the plates in a drying cabinet for 10–15 min.
3. Grow up an overnight culture of *E. coli* HB101 by inoculating 10 ml of LB medium with a small quantity of a glycerol stock of the bacteria, kept at −20°C. This is incubated at 37°C overnight with vigorous shaking.
4. The next day inoculate 50 ml of L broth in a 250 ml conical flask with 0.5 ml of the overnight culture. Incubate at 37°C until the optical density of the culture at 600 nm has reached 0.4. This usually takes about 2–3 h.
5. Chill the culture on ice for 10 min.
6. Centrifuge 20 ml of the cell suspension in a plastic Universal tube at 3000 r.p.m. for 10 min at 4°C (chill spin centrifuge).
7. Discard the supernatant. Resuspend the cell pellet in 10 ml of ice-cold 50 mM $CaCl_2$ (prepared from a sterile stock of 1 M $CaCl_2$) and place on ice for 60 min.
8. Re-centrifuge as before. Decant off the supernatant and resuspend the competent cells in 1 ml of 50 mM $CaCl_2$ (ice-cold). This preparation of competent cells is enough for 10 transformations as 100 μl is added to each ligation sample.
9. Add the DNA to the 100 μl of competent cells in a final volume of 50 μl of 10 mM Tris-HCl (pH 8.0), 1 mM EDTA. Mix and store on ice for 30 min[c].
10. Transfer to a water bath at 42°C for 2 min (heat shock step).
11. Add 850 μl of L broth to each tube and incubate at 37°C for 60 min.
12. Plate out 50 μl of the transformation mixture per agar plate (containing the appropriate antibiotic) using a sterile loop.
13. Spin down the remaining 950 μl in an Eppendorf centrifuge for 1 min. Decant off most of the supernatant, leaving a small volume in the bottom of the tube in which to resuspend the bacterial pellet. Plate this out on a separate plate as a back-up to the first.
14. Invert the plates and incubate at 37°C overnight. Colonies should appear in 12–16 h.
15. Store the plates at 4°C, while individual colonies are screened for recombinant plasmid.

[a]See Chapter 7, Table 3.
[b]Most vectors contain either the ampicillin or the tetracycline antibiotic resistance gene. Fresh stocks of these antibiotics are prepared each time they are used. Ampicillin is dissolved in water at a concentration of 100 mg/ml; tetracycline at a concentration of 15 mg/ml. These are then filter sterilized through a 0.45 μm membrane filter and added to L broth at a 10^{-3} dilution giving final concentrations of 100 μg/ml of ampicillin or 15 μg/ml of tetracycline.
[c]Up to 40 ng of DNA dissolved in 10 mM Tris-HCl (pH 8.0), 1 mM EDTA can be added in each transformation reaction. This usually represents approximately 1/10 of the volume of the ligation reaction that is set up. The remainder of the ligation mixture is stored at −20°C in case the transformation has to be repeated with a different amount of DNA.

(iv) Aspirate off the medium and wash the cell sheet twice with 10 ml of serum-free medium.
(v) Add to the cells 5 ml of transfection mix containing DEAE–dextran at a concentration of 0.2 mg/ml in serum-free medium, plus the recombinant plasmid at a concentration of 2 μg/ml (i.e. 10 μg of DNA is added per flask).
(vi) Incubate for 3 h.
(vii) After this time, aspirate off the transfection mix and wash the cell sheet carefully with 10 ml of serum-free medium.

Table 5. Minipreparations of plasmid DNA by alkaline lysis.

1.	Prepare LB medium solutions, A, B and C[a].
2.	Inoculate 5 ml of LB medium containing the appropriate antibiotic with a single bacterial colony. Incubate at 37°C overnight with vigorous shaking.
3.	The next day pipette 1.5 ml of the culture into an Eppendorf tube. Centrifuge for 1 min in an Eppendorf centrifuge. Store the remainder of the overnight culture at 4°C.
4.	Aspirate off the medium, leaving the bacterial pellet as dry as possible.
5.	Resuspend the pellet by vortexing in 100 μl of ice-cold solution A containing 5 mg/ml lysozyme (the lysozyme should be dissolved in solution A just prior to use).
6.	Leave at room temperature for 5 min.
7.	Add 200 μl of ice-cold, freshly prepared solution B. Do not vortex. Mix the contents of the tube by inverting rapidly two or three times. Leave to stand on ice for 5 min.
8.	Add 150 μl of ice-cold solution C. Vortex gently in the inverted position for 10 sec. Leave to stand on ice for 5 min.
9.	Centrifuge for 5 min in an Eppendorf centrifuge at 4°C.
10.	Transfer the supernatant to a fresh tube.
11.	Extract once with phenol/chloroform and once with chloroform/iso-amyl alcohol[b].
12.	Add two volumes of absolute ethanol. Mix by vortexing and leave at room temperature for 2 min.
13.	Spin down the ethanol precipitate for 5 min in an Eppendorf centrifuge.
14.	Wash the pellet by adding 1 ml of 70% ethanol, vortexing briefly and re-centrifuging for 5 min.
15.	Discard the ethanol and dry the pellet briefly *in vacuo*.
16.	Re-dissolve the DNA pellet in 50 μl of 10 mM Tris-HCl (pH 8.0), 1 mM EDTA containing DNase-free pancreatic RNase (20 μg/ml). Vortex briefly[c].
17.	Store these minipreparations of plasmid DNA at −20°C.
18.	Usually 10 μl of these preparations is removed for a restriction digest. The appropriate buffer and a few units of the restriction enzyme are added and it is incubated at the appropriate temperature for 1− h.
19.	Analyse the digests by agarose gel electrophoresis.

[a]See Chapter 7, Table 3.
[b]See *Table 1*, footnote b.
[c]RNase that is free of DNase is prepared by dissolving pancreatic RNase (RNase A) at a concentration of 10 mg/ml in 10 mM Tris-HCl (pH 7.5), 15 mM NaCl. This is then heated to 100°C for 15 min and allowed to cool slowly at room temperature before dispensing into aliquots and storing at −20°C.

(viii) Add 20 ml of medium supplemented with serum to the cells and incubate for about 2 days.

(ix) Harvest the cell supernatant by pouring it off into a plastic Universal tube, spinning off floating cells in a bench centrifuge and decanting off the supernatant into a fresh Universal tube. Store this at 4°C before assaying for IFN.

(x) Add back fresh medium (+ serum) and repeat harvesting the supernatant daily for 3−4 days.

(xi) At the end of this time, discard the flask since there is only transient expression of the MuIFN-γ.

3. CHO CELL EXPRESSION

Again expression is driven by a SV40 promoter, but this time the early promoter which does not need T-antigen for its activity. The aim is to generate a cell line containing the integrated cDNA sequences under the control of the SV40 promoter and to obtain high yields by selectively amplifying these sequences. This is done by co-transfecting

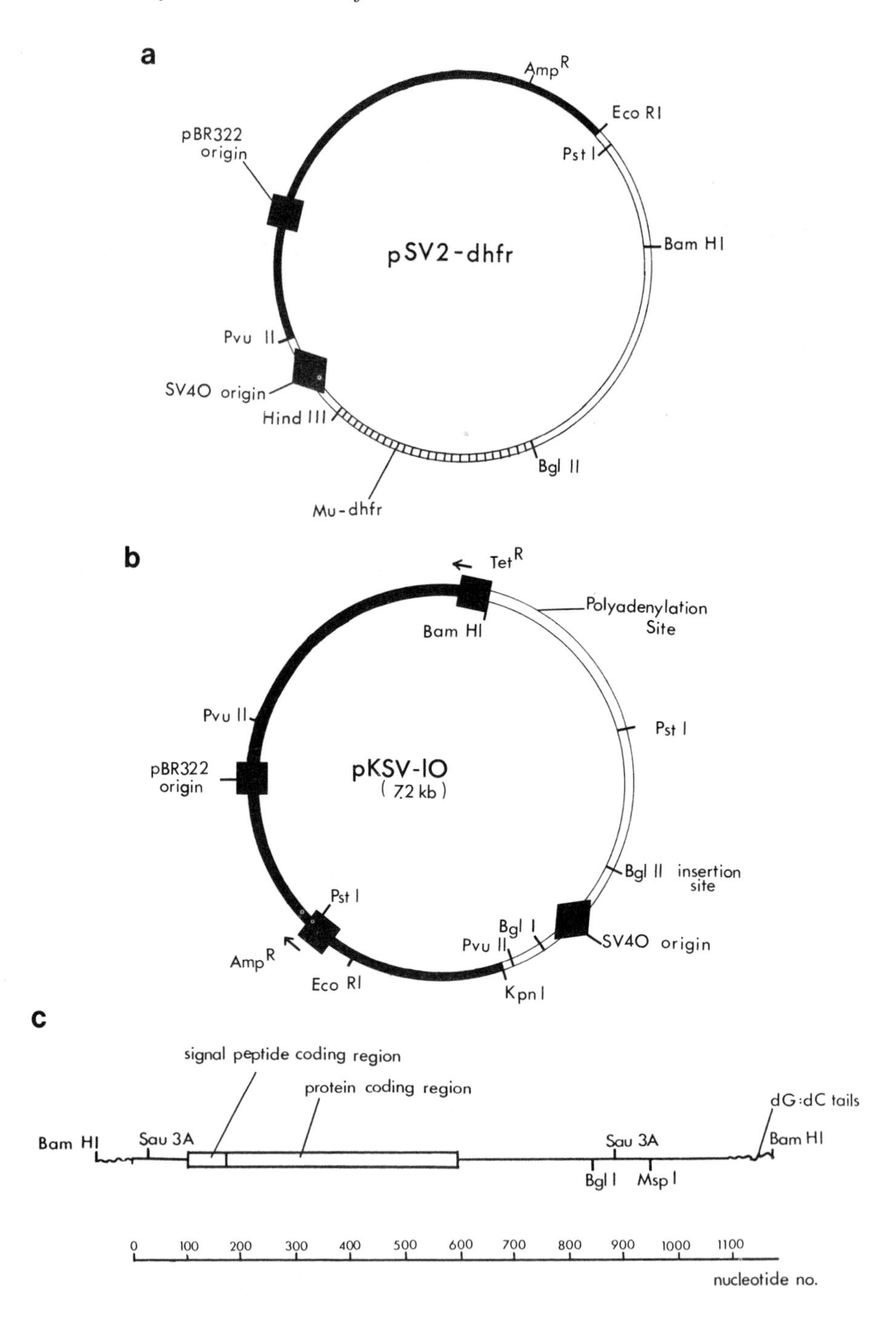

Figure 2. Maps of (**a**) pSV_2-DHFR, (**b**) $pKSV_{10}$, (**c**) HuIFN-γ. In (**a**) and (**b**) pBR322 sequences are indicated by solid, SV40 by open blocks. Nucleotide numbering of (**c**) based on (9).

the $pKSV_{10}$ (Pharmacia catalogue no 27-4926-01) vector containing the ligated IFN cDNA (4) (not in itself selectable) together with pSV_2-DHFR (5) into CHO cells lacking dihydroxyfolate reductase (DHFR) (6); pSV_2-DHFR contains the cDNA for the murine DHFR gene (again, under SV40 promoter control) and so stable co-transformants (containing both vectors integrated) can be obtained by selecting for the $DHFR^+$ phenotype. Amplification may be achieved by treating the transformants with an inhibitor of DHFR, methotrexate (7). Maps of $pKSV_{10}$, pSV_2-DHFR and the HuIFN-γ cDNA clone used are shown in *Figure 2*.

3.1 Construction of expression plasmid

As with COS cell expression the first step is ligation of the expression vector (pKSV10) with the cDNA to be expressed.

(i) Isolate a 846-bp *Sau*3A fragment of the HuIFN-γ cDNA (8) (see *Figure 2*), containing the entire protein coding region, by digesting the full length cDNA clone (cloned into the *Bam*HI site of pBR322) first with *Msp*I and then with *Sau*3A, ensuring that the required fragment is readily separated on an agarose gel.

(ii) Since *Sau*3A and *Bgl*II generate complementary cohesive ends, clone the HuIFN-γ *Sau*3A fragment directly into the *Bgl*II site insertion site in $pKSV_{10}$ (see *Figure 2*).

(iii) Digest the $pKSV_{10}$ vector DNA with *Bgl*II and dephosphorylate with alkaline phosphatase as before. Being a 'sticky-ends' ligation, the efficiency of ligation should be high. $pKSV_{10}$ carries both the ampicillin and tetracycline resistance gene for selection of *E. coli* after transformation.

(iv) Isolate several minipreparations of plasmid DNA and set up a *Bgl*I digest of $10-20$ μl of each to screen for the correct orientation of the insert in the vector.

There is a unique *Bgl*I site at position 854 in the HuIFN-γ cDNA sequence. The *Sau*3A fragment spans the sequence from position 32 to 878. There is also a *Bgl*I site (though not unique) at position 5107 in the vector (70 bp from the 5′ end of the insertion site). Thus, if the insert is in the $5' \rightarrow 3'$ orientation in the vector, a *Bgl*I digest will generate a fragment of about 890 bp in size. This should be compared against a control digest of vector DNA not containing an insert.

As before, prepare a sterile stock from a clone with the HuIFN-γ cDNA insert in the $5' \rightarrow 3'$ orientation in $pKSV_{10}$ for subsequent transfection into CHO cells.

All the other steps are essentially analagous to those described in Section 2.

Table 6. Hepes[a]-buffered saline (HBS).

0.28 M NaCl
0.01 M KCl
0.003 M Na_2HPO_4
0.012 M glucose
0.05 M Hepes
Adjust to pH 7.1 with NaOH and filter sterilize. Use tissue culture grade Hepes.

[a]Hepes = N-2-hydroxyethylpiperazine-N′-2-ethane sulphonic acid.

3.2 Transfection of CHO DHFR$^-$ cells

CHO DHFR$^-$ cells, which are unable to synthesize nucleotides, need to be cultured in a medium containing added nucleotides, such as Ham's F12. This is done by the calcium phosphate precipitation method as follows.

(i) Dissolve about 20 μg of pSV$_2$-DHFR DNA and 5 μg of pKSV$_{10}$ containing the IFN cDNA in 0.5 ml of Hepes-buffered saline (*Table 6*) in a clear plastic 5 ml bottle.

(ii) Bubble N_2 gently through the solution and add dropwise 0.5 ml of 0.28 M $CaCl_2$ solution. An opalescent precipitate should form more or less immediately.

(iii) Leave to stand for 30 min, then add to 5×10^5 CHO DHFR$^-$ cells in a 75 cm^2 plastic tissue culture flask (seeded the previous day in 10 ml of Ham's F12 medium supplemented with 10% FCS). Swirl to mix. Leave for 1–2 days.

(iv) Wash the cells carefully and add 10 ml of fresh Ham's F12 medium with 10% FCS.

(v) After 2–3 days, trypsinize the cells and resuspend in 3 ml of medium.

(vi) Re-seed 2 ml of this suspension into a 150 cm^2 plastic flask in about 50 ml of selection medium. The selection medium is αMEM supplemented with 10% dialysed FCS (Gibco) which contains no nucleotides; DHFR$^-$ cells will die, but successfully transformed DHFR$^+$ cells will survive and proliferate and will, at a fairly high frequency, also contain and express the IFN-γ gene and IFN-γ will be secreted into the medium.

(vii) Serially dilute the remaining 2 ml 10-fold in selection medium to give 20 ml aliquots at 10, 10^2, 10^3 dilution and seed these in 200 μl aliquots into a 96-well microtitre tray. The aim of this is to clone the transformed cells. Since the transformation frequency is not known in advance, a range of dilutions needs to be employed.

(viii) After 1–3 weeks, clones of cells should be visible in the bulk culture, and on the microtitre trays. Grow up the bulk culture, test for IFN production and freeze down at the earliest possible opportunity.

(ix) Scan the microtitre trays under a low-power microscope and mark those wells containing single colonies.

(x) As soon as they are large enough, trypsinize, transfer to 25 cm^2 flasks, grow up, test for IFN production and freeze down.

3.2.1 *Amplification of expression*

Clones producing IFN-γ should then be subjected to methotrexate (synthetic aminopterin) selection to amplify DHFR expression which will often result in co-amplification of the non-selected gene. Methotrexate (Sigma Chemicals) should be made up at 400 μM by dissolving initially in 0.1 M NaOH and diluting down to 400 μM with distilled water; keep frozen in aliquots. Add to the selection medium at 50 nM; after several passages, the cells will grow rapidly at that methotrexate concentration and the IFN production should be assayed. Repeat the selection at 250 nM, and 1250 nM (i.e. 5-fold increases). Often, the production of IFN-γ rises as the cells become resistant to methotrexate; for example, a clone we have isolated which before selection produced 10–100 units IFN per ml, after selection produced $5-10 \times 10^4$ units/ml ($\sim$1 mg/l).

4. REFERENCES

1. Gluzman,Y. (1981) *Cell*, **23**, 175.
2. Gray,P.W., Leung,D.W., Pennica,D., Yelverton,E., Najarian,R., Simonsen,C., Derynck,R., Sherwood,P., Wallace,D., Berger,S., Levinson,A. and Goeddel,D. (1982) *Nature*, **295**, 503.
3. Gray,P. and Goedell,D.V. (1983) *Proc. Natl. Acad. Sci. USA*, **80**, 5842.
4. Siggens,K., Wilkinson,M., Boseley,P., Slocombe,P., Cowling,G. and Morris,A. (1984) *Biochem. Biophys. Res. Commun.*, **119**, 157.
5. Subramani,S., Mulligan,R. and Berg,P. (1981) *Mol. Cell. Biol.*, **1**, 854.
6. Chasin,L. and Urlaub,G. (1980) *Proc. Natl. Acad. Sci. USA*, **77**, 4216.
7. Dijkema,R., Van der Meide,D.H., Pouwels,P.H., Caspars,M. and Schellekens,H. (1985) *EMBO J.*, **4**, 761.
8. Devos,R., Cheroutre,H., Taya,Y., Degrave,W., Von Heuverswyn,H. and Fiers,W. (1982) *Nucleic Acids Res.*, **10**, 2487.

CHAPTER 6

Biological assays for interferons

JOHN A.LEWIS

1. INTRODUCTION

For many years the study of interferons (IFNs) was the private domain of virologists, but with the advent of extensive biochemical and molecular biological studies many non-virologists are faced with the need to assay for the presence of these molecules. Sometimes it is necessary to standardize IFNs for use in experiments (titres supplied with IFNs are not always correct and some IFN preparations lose their potency), to characterize and titrate materials suspected of being IFNs or to test cells for their sensitivity to IFNs. This article describes simple assays that can be readily adapted to any particular system with minor modifications. These procedures have been worked out by many investigators over the years and references to some of the original publications are provided though no attempt to do justice to the many workers in the field has been made. For an excellent discussion of factors governing the choice of a procedure see Finter (1).

Until recently, all assays for IFNs have been based on their anti-viral properties and the procedures described here reflect this fact. The most accurate but tedious procedure is a virus yield reduction assay which involves plaque assays of virus progeny from cells treated with dilutions of the IFN preparation. A simple biochemical version of this is provided by measuring the inhibition of viral RNA synthesis. A convenient and simple procedure is to measure the ability of dilutions of the IFN to protect cells against the cytotoxic effects of viruses. Using microtitre plates this assay can cope with many samples or cell lines and provides results in 2–4 days. A qualitative assay which provides a direct measure of the inhibition of viral protein synthesis is to radiolabel proteins being synthesized in virus-infected, IFN-treated cells and analyse them by polyacrylamide gel electrophoresis. The availability of monoclonal antibodies has permitted the use of radioimmunoassay techniques for human IFN-α and -γ. Kits for performing such assays are available commercially [e.g. from Cell-Tech, Culture Products Division, CellTech Ltd, 244 Bath Road, Slough SL1 4DY, UK (the Cell-Tech kit for human IFN-γ is marketed in the USA by Bethesda Research Labs., Gaithersburg, MD, 20877) or, for Hu-IFN-γ, from Centocor, 244 Great Valley Parkway, Malvern, PA 19355]. As these kits are supplied with instructions their use will not be described here. IFNs and antibodies to them can be obtained commercially from several sources (Lee Biomolecular Research Laboratories Inc., 11211 Sorrento Valley Road, San Diego, CA 92121; Interferon Sciences, Inc., 783 Jersey Avenue, New Brunswick, NJ 08901; Sigma Chemical Co. Inc., P.O. Box 14508, St Louis, MO 63178; Amgen Biologicals, 1900 Oak Terrace Lane, Thousand Oaks, CA 91320; Schwarz/Mann, 56 Rogers Street, Cambridge, MA 02142).

2. DEFINITIONS AND UNITS

IFNs are defined as proteins or glycoproteins which interact with cells in a relatively species-specific fashion to induce an anti-viral state which protects the cell from infection by a wide range of virus classes. The development of this anti-viral state depends upon macromolecular synthesis (RNA and protein) by the treated cell and decays on removal of IFNs over a period of 24–72 h. The IFNs have been classified according to the nature of the producing cells and by their chemical properties, especially stability or lability at pH 2. More precise identification of suspected activity as a *bona fide* IFN can now be achieved by specific neutralization with antisera raised against different classes of IFN. In deciding that an unknown material can indeed be considered as an IFN these factors should be tested.

As with all bio-assays, the end-point obtained in a given experiment is subject to variations due to differences in the reagents involved, in particular the cell strains and virus stocks. In order to ensure a degree of uniformity, all titrations of IFNs should be compared with standards provided free of charge by The World Health Organization through the National Institute for Biological Standards and Control, London and the National Institutes of Health, Bethesda, Maryland. It is strongly recommended that the relationship between laboratory units and the reference standard be established and re-tested at frequent intervals. A crude preparation of IFN can be purchased or prepared (see ref. 2), aliquoted and compared with the reference standard. In future experiments this 'laboratory standard' can be used for standardization purposes and the reference material used only to re-calibrate the assay at periodic intervals. Crude preparations are usually very stable.

In general, one unit of IFN is defined as the reciprocal of the dilution at which 50% protection against virus infection is obtained. This can be viewed in several ways: in a virus yield reduction assay it is the point at which the virus yield is reduced to 50% that obtained with cells not exposed to IFN; in a cytotoxicity assay it is the point at which 50% of the cells are protected from the viral cytopathic effects. The accuracy of such determinations depends on the dilution increments and the number of replicates but normally a 2-fold difference in titre is of marginal significance.

In the following sections I describe procedures which can be used routinely with mouse fibroblastoid L-929 cells. The same protocols work with several other mouse cell strains and with the human HeLa and Wish lines although we have experienced problems with some cultures of the latter which seem to be refractory to IFN-γ. This problem seems to be related to the passage number of the cells. Since these experiments involve animal viruses it is important that the investigator should ensure that he/she understands the hazards posed by these organisms and that he/she complies with local regulations for their use. In the United Kingdom, vesicular stomatitis virus (VSV) can only be used in specified containment laboratories and Semliki Forest virus (1) or Sindbis virus (3) have been used in its place by several investigators. All contaminated materials should be autoclaved. Work surfaces should be swabbed with Chlorox and then with water after experiments. When working with virus suspensions avoid forming an aerosol (e.g. by blowing out from a pipette) and never mouth pipette. Wash hands frequently. Ascertain that the type of culture hood used is appropriate (e.g. do not use a horizontal laminar flow hood which blows air from the working zone towards the investigator).

3. DETERMINATION OF INTERFERON TITRES

If it is necessary to determine an accurate titre for a preparation of IFN the effect of multiple serial dilutions of the sample on viral replication is determined. This can be achieved using several parameters to measure viral replication including the yield of infectious virus, viral RNA synthesis or cytopathic effects (cell killing) induced by lytic viruses.

3.1 Virus yield reduction assay

The classical, and most accurate, procedure is to measure the yield of virus from cultures treated with IFN. This assay is time-consuming and relatively slow especially if several samples are to be titred. With experience it yields very accurate and reproducible results. The same procedure can be used to compare the effect of IFN preparations on different cell lines.

3.1.1 *Virus production by interferon-treated cells*

(i) Seed L-929 cells in 6-well cluster dishes in normal growth medium (3 ml of 3×10^5 cells/ml) and incubate at 37°C until confluent, normally overnight. Alternatively, use 6 cm culture dishes containing 5 ml of growth medium.
(ii) Dilute IFN preparations in growth medium according to the expected titre of the preparation. It is best to have several 2-fold dilutions on either side of the expected 1 unit/ml point.
(iii) Remove the medium from the cell cultures and replace it with 2 ml of the IFN dilutions (3 ml for 6 cm dishes). Incubate at 37°C for 18 h.
(iv) Remove the IFN dilutions with gentle suction (avoid touching the cell monolayer). Rinse the cells once with 2 ml of phosphate-buffered saline (PBS, see *Table 1*) pre-warmed to 37°C and remove the wash thoroughly. Add gently to the centre of each well 0.25 ml of virus diluted in PBS to a concentration of 8×10^7 plaque-forming units (p.f.u.)/ml (0.4 ml of 1×10^8 p.f.u./ml for 6 cm dishes). Avoid touching the cell layer with the pipette. Incubate the dishes at 37°C for 60 min and rock them periodically throughout this period to ensure even distribution of the innoculum.
(v) Remove the virus innoculum with a Pasteur pipette and discard into Chlorox or an autoclavable container. Gently rinse the monolayers with 2 ml of PBS (run the PBS down the side of the well rather than dropping it directly on the cells to avoid dislodging any cells). Remove the wash and replace it with 2 ml of growth medium by again running it down the side of the well.
(vi) Incubate the infected cultures for 24 h at 37°C and inspect them in an inverted microscope. Note the degree of cytopathic effect in each well. If the cells which were not exposed to IFN show a complete cytopathic effect proceed to harvest the virus; if cytopathic effects are incomplete continue the incubation.
(vii) Freeze the dishes by placing them on the bottom of a −70°C freezer or a block of dry ice. Then thaw rapidly at 37°C and repeat the freeze−thaw cycle.
(viii) Collect the fluid from the wells and transfer it aseptically to sterile, capped centrifuge tubes. Centrifuge at 1000 *g* for 5 min at 4°C (to eliminate cell debris). Collect the supernatant and store at −70°C.

Table 1. Formulae of solutions used in the text.

1. Phosphate-buffered saline (PBS)
80.0 g of NaCl
2.0 g of KCl
11.5 g of Na_2HPO_4 (anhydrous)
2.0 g of KH_2PO_4
0.1 g of phenol red
1000 ml of water.
Adjust to pH 7.3
2. SNU lysis mix
0.5% (w/v) SDS
1.0% (v/v) Nonidet P-40
6 M urea
3. Methyl violet dye mix
0.5% (w/v) methyl violet
0.9% (w/v) NaCl
2.0% (w/v) formaldehyde
50.0% (v/v) ethanol
4. Gel lysis buffer
10 mM Tris-HCl pH 7.5
15 mM NaCl
1.5 mM $MgCl_2$
1% (v/v) Nonidet P-40 or Triton X-100
0.25% (w/v) sodium deoxycholate
1 mM PMSF
15 units/ml trasylol (aprotinin)
5. SDS lysis buffer
1% (w/v) SDS
50 mM Tris-HCl pH 6.7
1 mM PMSF
15 units/ml trasylol
6. Gel sample cocktail
3% (w/v) SDS
150 mM Tris-HCl pH 6.7
30% (v/v) glycerol
300 μg/ml bromophenol blue

In this type of assay, several rounds of infection take place. For some viruses this can lead to an increase in sensitivity of the overall assay since IFN produced in response to infection can further protect against subsequent rounds of infection. This will be most marked in those cultures which are partially protected against infection (by the IFN pre-treatment) since the virus normally shuts off synthesis of host proteins. To avoid such complications it may be worth carrying out a single cycle assay in which the infection is allowed to proceed only for the duration of one round of replication (usually 8–10 h) before harvesting. Obviously in this case lower titres of virus are achieved.

3.1.2 *Plaque assay of virus yield*

In preparation for plaque assaying the virus, grow a stock of the cells to be used. For VSV one can routinely use L-929 cells but it is also possible to use Vero (African green monkey) cells. L-929 cells can also be used for Semliki Forest virus (1). Mengo virus [or encephalomyocarditis virus (EMC)] is also titred on L-929 cells. Until experienced, it is best to limit the number of samples to be handled in one day to six (54 dishes!). It is wise to plan the day for virus to be added to the cells in advance as development of medium size plaques takes 2 days with VSV, although Mengo plaques are usually ready to be stained after 1 day. Reovirus plaques take 3–5 days to develop and it is best to initiate this assay at the beginning of a week with dishes seeded the previous week.

(i) For each sample to be titred prepare nine confluent cultures of the indicator cells in 6 cm dishes.

(ii) Rapidly thaw the samples to be tested at 37°C and then keep them at 0°C until used. They can be re-frozen at −70°C in case the assay has to be repeated. Prepare a series of five sterile tubes for each sample and label them −2, −4, −6, −7, −8. To the first three tubes add 1.0 ml of PBS; to the others add 0.9 ml. Mix the virus sample thoroughly (but avoid vigorous vortexing) and transfer 10 μl to the tube labelled −2 (giving a 100-fold dilution). Discard the pipette tip and mix the dilution by vortexing gently. With a new pipette tip transfer 10 μl of this dilution to the tube labelled −4, mix as before and, with a new tip, pipette 10 μl to the tube labelled −6. The next dilutions are 10-fold steps and so 100 μl of the −6 is transferred to the −7 tube and so on. Be sure to mix thoroughly between each step and use clean pipettes at each stage. The dilutions actually used will be dictated by the virus titres obtained and, for some virus/cell systems, higher dilutions may be needed (as with Mengo virus in L-929 cells for example).

(iii) Void the medium from the dishes to be used and add 0.25 ml of the highest dilution (−8) to three dishes. Next add 0.25 ml of the preceding dilution (−7) to three more dishes and finally 0.25 ml of the −6 dilution to the remaining dishes. Add 0.25 ml of PBS to a further set of dishes as a control. Incubate the dishes at 37°C for 1 h.

(iv) In the meantime prepare the overlay. Equilibrate 50 ml of molten 1.0% agarose (Seakem ME from FMC Corp., Marine Colloids Division, Rockland, ME 04841) at 45°C. The agarose can be prepared either at the time of use by autoclaving 1 g of agarose in 50 ml of water or by melting pre-sterilized, solidified solution in a microwave oven (3 min at medium setting). In the latter case use a large glass bottle (e.g. 500 ml) to prevent bumping. Also equilibrate 2 × MEM at 37°C. This medium concentrate can either be purchased as a 10 × concentrate (Gibco or Flow Laboratories) or prepared by dissolving the contents of a 1 litre sachet of MEM powdered formula in 500 ml of water. If the concentrate is to be stored, add only half the normal amount of sodium bicarbonate to prevent precipitation of salts in the cold.

(v) Immediately before adding to the plates, combine equal volumes of the 2 × MEM and 1% agarose. Add serum to 5% final concentration. Remove the virus innocula and add 7 ml of agarose overlay. It is necessary to work fast to avoid the stock of agarose gelling but avoid air bubbles forming in the plates. It is best to handle only 100 ml of overlay at a time (i.e. 12 dishes) keeping the concentrated stocks at the appropriate temperatures until needed.

(vi) Leave the plates at room temperature for 10 min to solidify and then place them *inverted* in a CO_2 incubator. Plaques of VSV take 2 days to develop while Mengo virus may need only 1 day. For reovirus, which takes 3–4 days to develop, a further 5 ml of agarose overlay is added after 48 h.

(vii) Plates can be stained by adding 2 ml of 0.03% Neutral red (available as a 10 × stock from Gibco) on top of the agarose and incubating at 37°C for 2 h. Plaques are then visible as clear areas amid the red-stained viable cells. In this case the plaques must be counted immediately and the plates cannot be stored. Alternatively, cool the plates to room temperature for 30 min and remove the agarose with a broad scalpel blade or flat spatula. First rim around the agarose to separate from the edges of the dish and then insert the tip of the instrument under the agarose and quickly flip upwards collecting the disc in an autoclave bag. To the dish add 2 ml of 1% crystal violet in 5% ethanol. Stain for 5 min and then remove and rinse with running tap water. Plaques are clear areas amid the purple-stained cell monolayer.

(viii) Count the number of plaques on dishes where the individual plaques can be clearly seen. Hold the dish inverted over a light box and spot each plaque with a marker pen as it is counted. Avoid dishes (dilutions) with many merging plaques. Calculate the average number of plaques (N) for each useable dilution. The virus titre is given by:

$$\text{Titre} = N \times \text{dilution factor} \times 4$$

(ix) The virus yield can then be plotted against the interferon concentration and the 50% end-point estimated.

3.2 Incorporation of [^{3}H]uridine

A fairly rapid biochemical assay has been developed (4) using the capacity of IFN to inhibit viral RNA synthesis. The latter can be distinguished from any host RNA synthesis by its resistance to inhibition by actinomycin D.

(i) Seed cells in a 24-well cluster dish (3×10^5 cells in 1.0 ml). The number of wells depends on the dilutions required. A set of cells should also be seeded for the standard preparation. Each point should be run in duplicate or triplicate. Grow to confluency (usually overnight).

(ii) Remove the medium and replace with serial dilutions of IFN. As this assay is rather cumbersome it is best to use 2-fold steps starting near the expected end-point value (e.g. 20 reference units/ml). Incubate for 18–24 h.

(iii) Remove the medium from each well by shaking into a pan containing sterile gauze or paper tissues.

(iv) To cell control wells add 0.2 ml of PBS. To all other wells add 0.2 ml of virus diluted in PBS to 2×10^7 p.f.u./ml. Incubate at 37°C for 1 h.

(v) Remove the innocula and rinse the cells gently with warm PBS. Add 0.25 ml of medium containing 5 μg/ml actinomycin D (purchased from Calbiochem as sterile vials containing 200 μg of solid and reconstituted just before use in 0.8 ml

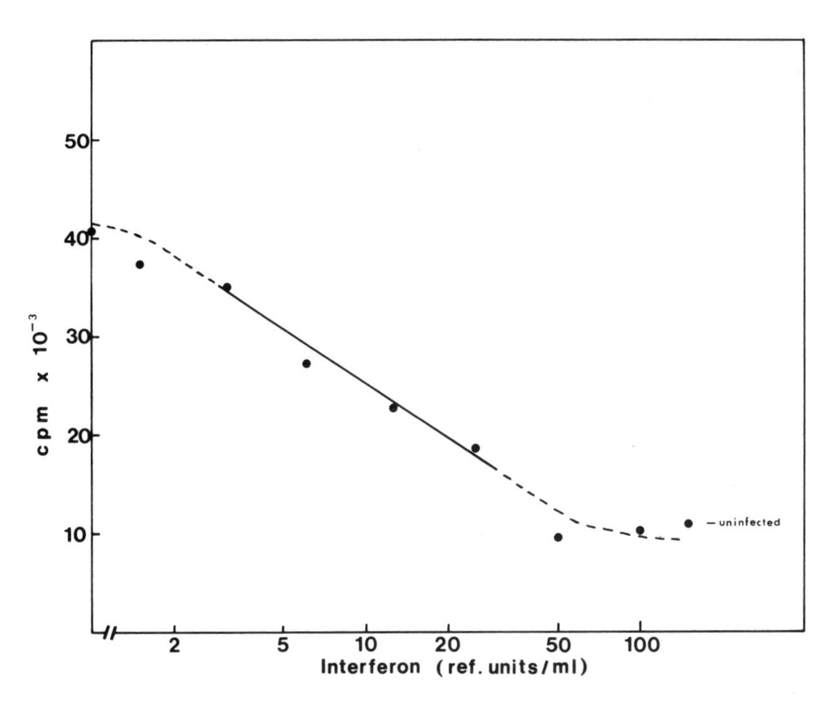

Figure 1. Titration of mouse interferon by measuring the decrease in viral RNA synthesis. The assay was performed as described in Section 3.2 using mouse L-929 cells treated for 18 h with the doses of interferon indicated and then infected with VSV (10 p.f.u./cell).

of PBS to provide a working stock of 250 μg/ml). Incubate for 2 h at 37°C.

(vi) Remove the medium from all wells and replace it with 0.2 ml of medium containing 5 μg/ml actinomycin D and 5 μCi/ml (1.85×10^5 Bq/ml) of [^{3}H]uridine (20 Ci/mmol; 7.4×10^{11} Bq/mmol). Incubate for 3 h.

(vii) Remove the radioactive medium and place the dishes on ice. Wash each well twice with 1 ml of ice-cold PBS. Remove as much as possible of the last wash.

(viii) Add 0.5 ml of SNU mix (*Table 1*) followed by 0.5 ml of 20% trichloroacetic acid (TCA). The SNU buffer is an extremely effective lysis buffer.

(ix) With a pipette tip mix and transfer the lysates to centrifuge tubes kept on ice [plastic microcentrifuge tubes or disposable glass culture tubes (12 × 75 mm) can be used]. Leave at 0°C for 30 min and centrifuge at 1000 *g* for 5 min at 4°C. Remove and discard the supernatant. Wash the pellet once with 1 ml of 10% TCA and centrifuge again.

(x) Dissolve the pellet in 0.5 ml of 0.1 M NaOH and transfer quantitatively to a scintillation vial. Add 5 ml of a suitable scintillation cocktail (e.g. Aquasol 11, ACS or Liquiscint) and count the radioactivity.

(xi) Alternatively, the precipitate formed in step (viii) can be collected on glass-fibre discs (Whatman GF/C) in a filtration manifold. Wash the discs three times with cold 10% TCA, once with 95% ethanol and dry under an infra-red lamp. Radioactivity can then be determined using a suitable scintillation cocktail (Econofluor, PPO/POPOP or Betafluor).

An example of a titration using VSV is shown in *Figure 1*. This procedure is considerably shorter than the virus yield reduction assay and may appeal to biochemists. However, it does require the use of radioisotopes which are relatively expensive and require licensing. These disadvantages may be overcome by the reduction in labour time. The procedure is slightly less sensitive than virus yield reduction and cytopathic effect assays.

3.3 **Micro-titre assays of cytopathic effects**

This type of assay is probably the simplest and most convenient for obtaining an accurate titre of IFN. It has been described by Armstrong (5) and by Havell and Vilcek (6). A rapid procedure has also been described by Familietti *et al.* (7). Two variations are described which can be used according to whether many cell lines are being tested (I) or whether many IFN preparations are being titred (II).

3.3.1 *Type I assay*

(i) Seed duplicate rows of a 96-well microtitre dish with 0.1 ml of cell suspension (3×10^4 cells). Incubate until the monolayers are confluent (usually overnight).

(ii) In a new 96-well dish prepare serial dilutions of the IFN preparation(s) in the appropriate growth medium. This step can be carried out under non-sterile conditions and the open dish sterilized by exposure to u.v. light for 20 min in a culture hood. In this case the sterilized dilutions should be placed in a CO_2 incubator for 30 min to readjust the pH. It is convenient to use a multi-channel pipetting device (e.g. Titertek, obtainable from Flow Laboratories) to simplify the dilution procedure. For 2-fold dilutions pipette 125 μl of growth medium to each well except row B which receives 250 μl of the starting dilution of IFN (e.g. 200 units/ml). Serially transfer 125 μl from row B to row C and mix by pipetting up and down several times. Continue down the dish to row G. Rows A and H will serve as cell and virus controls, respectively, and receive only medium. For 3-fold step dilutions proceed as above but pipette 160 μl to rows C–G. Transfer 80 μl from row B to row C, etc.

(iii) Shake the medium from the dish of test cells into a pan containing sterile gauze or paper tissues. Using the multi-channel pipettor transfer 100 μl from the wells of the IFN dilution plate to the corresponding wells of the cell plate. Incubate for 18–24 h.

(iv) Add to all wells, except those in row A, 25 μl of virus at a concentration of 4×10^5 p.f.u./ml and continue the incubation until a strong cytopathic effect is observed in the virus control wells (row H). This usually takes 48 h but depends on the virus.

(v) Shake out the medium into a tray containing Chlorox and add 100 μl of methyl violet dye solution (*Table 1*) to each well. After 5–10 min shake the dye out in a sink and wash the dish by repeatedly plunging into tap water.

(vi) The titre can be estimated by eye from the dilution at which the staining intensity is midway between the cell and virus control levels. For quantitative analysis, elute the dye into 100 μl of ethyleneglycol monomethyl ether for 60 min on a shaking device and combine the contents of duplicate wells in a tube containing

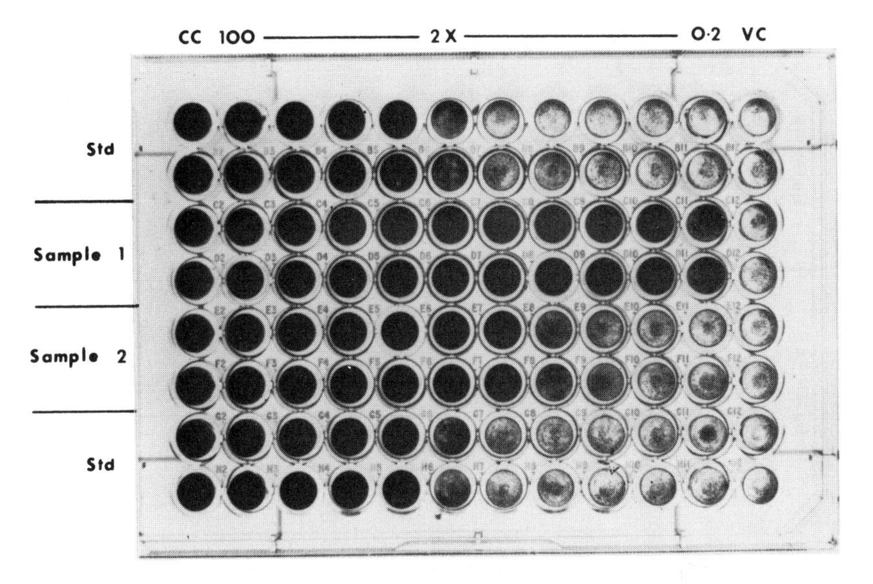

Figure 2. Titration of mouse interferons by the cytopathic effect assay. The assay was performed as described in Section 3.3.1 using mouse L-929 cells and two preparations of fibroblast interferon (induced in L-929 cells by Newcastle disease virus). A standard preparation (Std) is assayed on the same plate for comparison. The N.I.H. standard was used at 100 reference units/ml (well 2) and subsequent 2-fold dilutions horizontally across the plate. The end-point occurs in well 6 (6.25 reference units/ml) in this particular experiment. Samples 1 and 2 were diluted 20-fold initially (well 2) and serially 2-fold across the plate with end-points in wells 11 and 8, respectively.

0.8 ml of solvent. Read the optical density at 550 nm in a spectrophotometer using a 1.0 ml cuvette.

3.3.2 *Type II assay*

(i) Prepare IFN dilutions as above except at concentrations twice the desired final values.
(ii) Transfer 100 μl of each dilution to a new dish in the same format as described above.
(iii) Add to each well 100 μl of cell suspension (5×10^5 cells/ml) and incubate for 18–24 h.
(iv) Proceed as described in step (iv) above.

An example of an assay of two IFN preparations together with a reference standard is shown in *Figure 2*. The actual end-point for the standard occurs at a dose of 6.25 reference units/ml indicating that the assay underestimates the potency of the preparations by a factor of 6.25. The end-point for sample 1 occurs at a dilution of 10 240-fold. The titre of this preparation is therefore 64 000 reference units/ml. Similarly, sample 2 reaches an end-point between the 640- and 1280-fold dilutions and thus has a titre of 6000 reference units/ml. Several factors determine the end-point. Usually the end-point in a type I assay extends to a greater dilution than a type II assay, possibly reflecting the tendency of aged, confluent cultures to show higher sensitivity to IFN. The multiplicity of virus infection also affects the end-point as seen in *Figure 3*; high m.o.i.

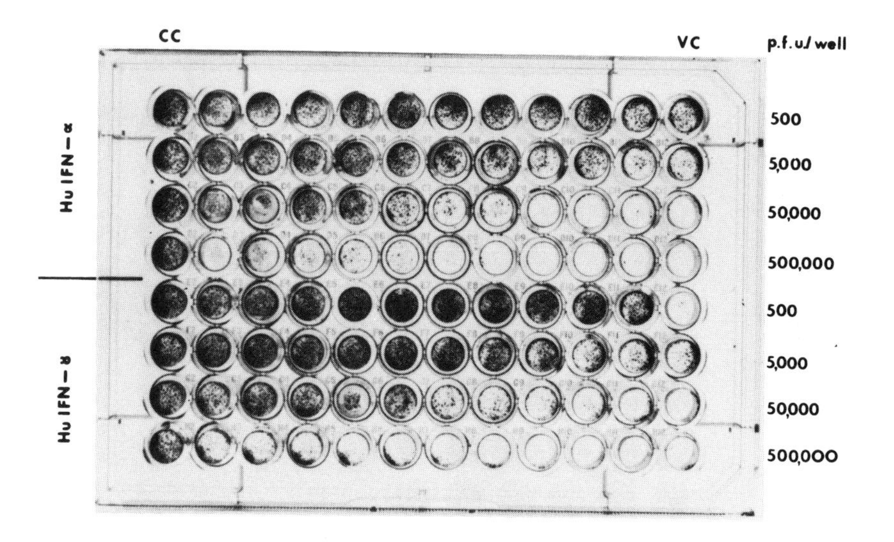

Figure 3. Effect of the multiplicity of infection on the end-point in a cytopathic effect assay. The assay was performed as described in the text with human IFN-α and -γ on human Wish cells. After treatment with serial 3-fold dilutions of the interferons (starting from 1 $\times$ 10^4 units/ml in well 2), virus was added as indicated. The wells were stained 24 h after infection.

values give less sensitive assays but the time of development of cytopathic effect is reduced by high input multiplicity so a balance can be drawn between the sensitivity and rapidity required. In general, the end-point scored by visual observation in a microscope or of the stained plate corresponds well to the value obtained by the more cumbersome spectrophotometric procedure and this may be reserved for experiments requiring exactitude. An example of a titration is shown in *Figure 4* for an IFN preparation and a standard.

4. ASSAY FOR INTERFERON SENSITIVITY

4.1 Cytopathic effect assay

If it is necessary to determine whether a cell line is sensitive to IFN one can perform a rapid assay using the cytopathic effect of lytic viruses such as VSV, Mengo, EMCV or Semliki Forest virus.

(i) Seed the cell lines of interest in three wells of a 24-well cluster dish (3 $\times$ 10^5 cells in 1.0 ml). Also seed a positive control (e.g. L-929 or HeLa cells). Grow to confluency (usually overnight).
(ii) Remove the medium and replace it with 0.5 ml of normal growth medium in two of the wells (a and b) and medium containing IFN (e.g. 100 units/ml) in the third well (c). Incubate for 18–24 h.
(iii) Remove the medium from all three wells. To well (a) add 0.2 ml of PBS. To wells (b) and (c) add 0.2 ml of virus diluted in PBS to 2 $\times$ 10^7 p.f.u./ml. Incubate at 37°C for 1 h. Remove the innocula and add 1 ml of growth medium.
(iv) Incubate for 24–48 h and observe the cytopathic effect. When over 95% of the cells in well (b) are killed, remove the medium, rinse the wells once with PBS and stain with methyl violet dye solution (see *Table 1*).

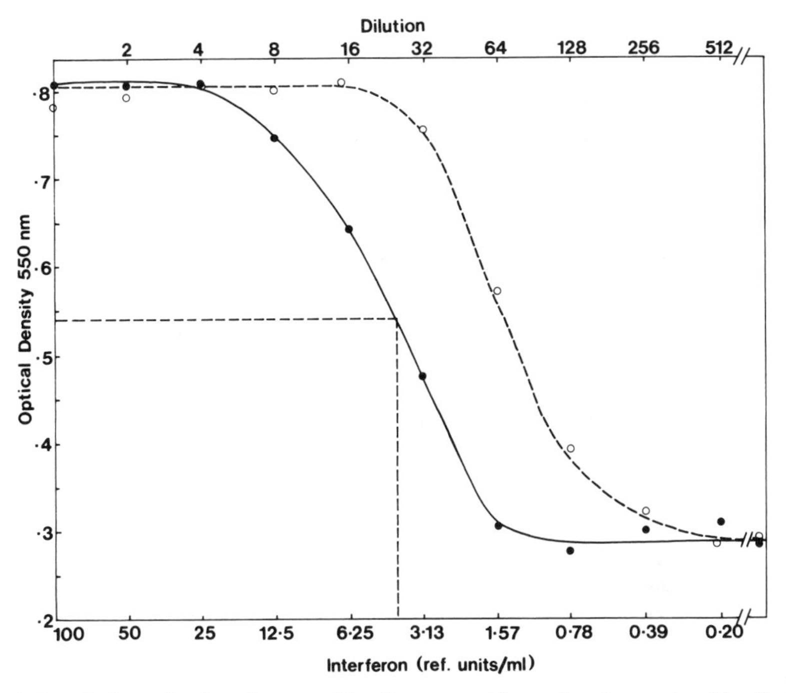

Figure 4. Quantitative estimation of a cytopathic effect assay. After performing a cytopathic effect assay as described in the legend to *Figure 2*, dye retained by the monolayers was eluted and the optical density determined. The bottom scale shows the reference units/ml for the standard preparation. The top scale indicates the dilution steps. The end-point is taken as the optical density mid-way between the virus and cell controls. Standard preparation ●——●; interferon sample ○ – – – ○.

4.2 Assay of viral protein synthesis

This assay provides some indication of the step at which IFN prevents viral replication. It is not a rapid procedure since it necessitates analysis by gel electrophoresis.

(i) Seed four wells of a 24-well cluster dish with each cell line to be tested and allow to grow to confluency.

(ii) Remove the growth medium and rinse each well with 1.0 ml of pre-warmed PBS. Remove and replace with either 0.1 ml of PBS (cell control) or 0.1 ml of virus diluted to 5×10^7 p.f.u./ml to give an m.o.i. of 10. Some virus preparations may require greater input multiplicities which can be determined empirically.

(iii) Incubate at 37°C for 1 h. Rock the dishes occasionally.

(iv) Remove the innocula and add to each well 0.25 ml of growth medium containing 5 μg/ml of actinomycin D (see Section 3.2).

(v) Incubate for 4 h at 37°C. For most viruses this is sufficient to permit development of maximal viral protein synthesis. In some cases, for example reovirus, a longer period of infection is required. (In the case of reovirus allow the infection to proceed overnight before labelling.)

(vi) Remove the medium and rinse the monolayers with 0.5 ml of pre-warmed medium lacking methionine. Replace with 0.25 ml of medium lacking methionine and supplemented with dialysed (to remove amino acids) serum and [^{35}S]methionine (10 μCi/ml; 3.7×10^5 Bq/ml: 1000 Ci/mmol; 3.7×10^{13} Bq/mmol). The

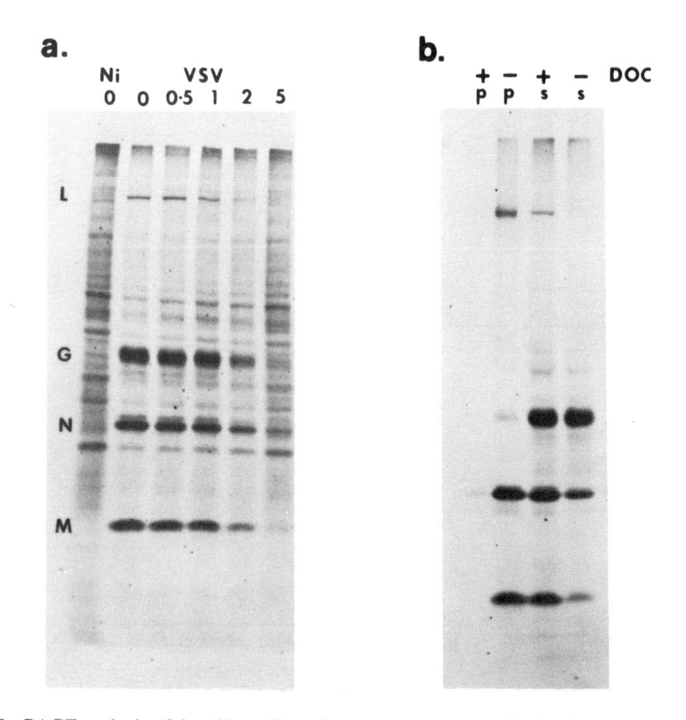

Figure 5. SDS−PAGE analysis of the effect of interferon treatment on synthesis of viral proteins in infected cells. (**a**) Mouse L-929 cells were treated for 20 h with low doses of interferon, infected with VSV (10 p.f.u./cell) and labelled with [^{35}S]methionine as described in Section 4.2. An autoradiograph of the SDS−PAGE analysis is shown. Virus infection inhibits the synthesis of host proteins; compare non-infected cells (Ni) and untreated cells (0). Inhibition of virus protein synthesis is detectable with 1−2 units/ml of interferon and essentially complete at 5 units/ml. Host cell synthesis is also restored at this dose. (**b**) Some viral proteins are not extracted from the cytoskeleton unless ionic detergents are employed in the lysis procedure. VSV-infected L-929 cells were lysed in the gel lysis buffer (*Table 1*) with (+) or without (−) sodium deoxycholate. Lysates were centrifuged and the pellets (p) and supernatants (s) analysed by SDS−PAGE and autoradiography.

low level of methionine present during the labelling period does not affect the rate of protein synthesis, at least with L-929 and HeLa cells. Unlabelled methionine can be included to about 10 μM without greatly changing the level of incorporation.

(vii) Incubate for 1 h at 37°C.

(viii) Remove the radioactive medium from the wells and place the dish on ice. Immediately wash the cells twice with 1 ml of ice-cold PBS and remove as much as possible of the second wash.

(ix) Add to each well 0.25 ml of gel lysis buffer (*Table 1*). Scrape the cells into the buffer using a rubber policeman or pipette tip. Transfer the lysate to a pre-chilled microcentrifuge tube and centrifuge at 5000 *g* for 5 min at 4°C. Transfer the supernatant to a new tube.

For complete extraction of viral proteins it is important to include an ionic detergent such as deoxycholate to disrupt the cytoskeleton (see *Figure 5b*). Note the presence of VSV L, N and M proteins in the pellet of lysates prepared without deoxycholate and their corresponding under-representation in the supernatant fraction. The G pro-

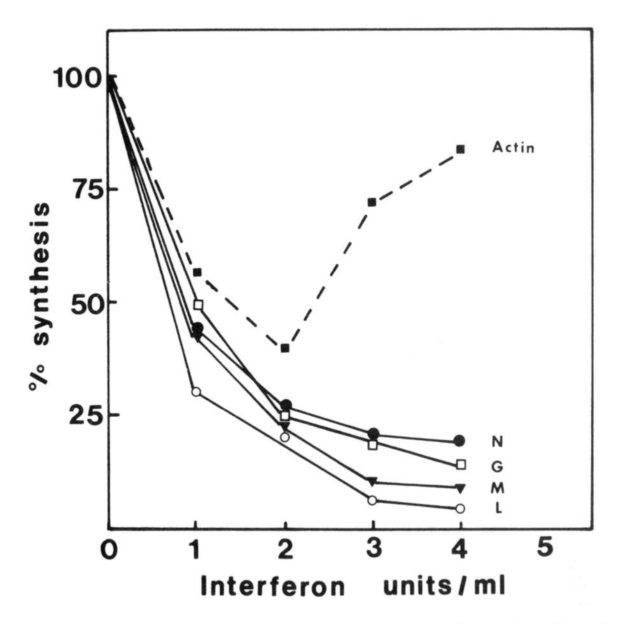

Figure 6. Quantitation of virus and host protein synthesis. An autoradiograph such as that shown in *Figure 5* was scanned with a densitometer and the areas under peaks corresponding to VSV viral proteins L, M. G and N and the host protein actin were integrated.

tein behaves differently, probably owing to its being synthesized by ribosomes associated with the rough endoplasmic reticulum. Alternatively, a total lysate can be prepared using sodium dodecylsulphate (SDS). This procedure yields a lysate containing the cellular DNA freed of its associated proteins forming a viscous clot which can make further handling a little difficult.

For preparation of total cell lysate:

(i) Add to each well 0.25 ml of SDS lysis buffer. This lysate is used directly without centrifugation.

(ii) Add to the lysate (prepared as in steps ix or x above) 125 μl of gel sample cocktail (*Table 1*) and store at -20°C for analysis by SDS−polyacrylamide gel electrophoresis (PAGE). Under these conditions, if 25−50 μl of sample is loaded in one lane of a slab gel, viral proteins can be clearly visualized after autoradiography at room temperature for 3 days. Enhancement of signal by fluorography is not usually necessary. If required, quantitation can be achieved by densitometry in which case it is important to ensure that exposures are within the linear response range of the film. Pre-flashing the film (8) improves both the sensitivity and linearity of response.

An example of the results obtained with such an assay is shown in *Figure 5a*. Quite low doses of IFN (1−2 reference units/ml) are sufficient to provide detectable reduction in viral protein synthesis, indicating that either transcription or translation have been affected. *Figure 6* shows the quantitation of a similar experiment and it can be seen that viral protein synthesis is reduced by 50% at 1−2 reference units/ml of IFN. However, although virus yield is reduced by 1000- to 10 000-fold by larger doses (e.g. 100 units/ml) the reduction of protein synthesis tends to plateau off at between 1 and

10% of the control value. In line with this discrepancy some authors have reported that virus released from IFN-treated cells is abnormal in its virion protein composition (9, 10) and consequently is of low infectivity. Host protein synthesis is progressively restored, although at very low doses of IFN the synthesis of actin is actually inhibited more than in infected, untreated cells.

5. PREPARATION OF VIRUS STOCKS

5.1 **Preparation of vesicular stomatitis virus**

VSV belongs to the rhabdovirus class and is a lipid-enveloped, negative strand virus (11). It infects a wide range of animal cells and is easily grown. Humans are not generally susceptible to infection although some individuals have developed cold-like symptoms after exposure. As mentioned earlier, the use of VSV is strongly controlled in the United Kingdom and Semliki Forest virus can be used as a substitute.

Virus stocks are prepared in L-929 or BHK (baby hamster kidney) cells as follows using minimal essential medium (with Earle's salts), penicillin (100 units/ml), streptomycin (50 μg/ml) and 10% heat-inactivated newborn (or fetal) calf serum.

(i) Seed cells in 10 cm culture dishes and grow to confluency.
(ii) Prepare a dilution of VSV in PBS at 3×10^6 p.f.u./ml (m.o.i. = 0.1 p.f.u./cell). Remove the medium from the cell monolayers, rinse once with pre-warmed PBS and then add 1.0 ml per dish of the diluted virus. It is important to use a low input multiplicity to avoid the generation of defective-interfering virus particles.
(iii) Adsorb the virus at 37°C for 1 h. Remove the innoculum and gently rinse the monolayers once with warm PBS. Add 5 ml of growth medium containing 2% serum.
(iv) Incubate the cultures for 24 h at which time a strong cytopathic effect should be observed.
(v) Freeze the dishes rapidly at −70°C and thaw at 37°C. Repeat the freeze−thaw cycle. Collect the culture medium aseptically in a sterile centrifuge tube and centrifuge at 2000 *g* for 10 min at 4°C.
(vi) Collect the supernatant, aliquot in suitable amounts (e.g. 1.0 ml) and store at −70°C.
(vii) Thaw one vial of the virus stock and titre by plaque assay as described in Section 3.1.2.

VSV is relatively heat labile and storage at −20°C is not recommendable. Once thawed it is best to discard any unused virus as the titre of re-frozen virus drops significantly (~2-fold after one freeze−thaw cycle). Also avoid mechanical stress such as vigorous vortexing or ultrasonication.

5.2 **Preparation of Mengo or EMC virus**

Mengo and EMC viruses are closely related members of the picornavirus group. They do not possess a lipid envelope and are positive strand RNA viruses. They pose relatively low risk to humans although there is some evidence in animals for a relationship between infection with EMCV and development of diabetes. Exercise caution as with all viruses. Virus stocks can be grown in L-929 cells.

(i) Grow cells to confluency in 10 cm culture dishes.
(ii) Dilute virus stock to 3×10^7 p.f.u./ml in PBS. Remove the culture medium from the monolayers and add 1.0 ml of the diluted virus. Adsorb at 37°C for 1 h.
(iii) Remove the innoculum and wash the monolayer with pre-warmed PBS. Replace with 5 ml of growth medium containing 2% serum and incubate for 24 h.
(iv) The cells should show a strong cytopathic effect. Freeze and thaw the dishes twice and centrifuge the culture fluid as described in Section 5.1 (v).
(v) Store aliquots frozen at −70°C (although Mengo virus is also stable at −20°C). Titre the stock preparation by plaque assay as described in Section 3.1.2.

6. ACKNOWLEDGEMENTS

I thank Ms Afroza Huq for her expert technical assistance. The data shown in this chapter stem from research supported by grants to the author from the National Institutes of Health and the National Science Foundation.

7. REFERENCES

1. Finter,N.B. (1973) In *Interferons and Interferon Inducers*. Finter,N.B. (ed.), Elsevier-North Holland Publishing Co., New York and Amsterdam, p. 135.
2. Kawade,Y. and Yamamoto,Y. (1981) In *Methods in Enzymology*. Pestka,S. (ed.), Academic Press Inc., London and New York, Vol. 78, p. 139.
3. Ho,M., Tan,Y.H. and Armstrong,J.A. (1972) *Proc. Soc. Exp. Biol. Med.*, **139**, 259.
4. Allen,P.T. and Giron,D.J. (1970) *Appl. Microbiol.*, **20**, 317.
5. Armstrong,J.A. (1971) *Appl. Microbiol.*, **21**, 723.
6. Havell,E.A. and Vilcek,J. (1972) *Antimicrob. Agents Chemother.*, **2**, 476.
7. Familietti,P.C., Rubinstein,S. and Pestka,S. (1981) In *Methods in Enzymology*. Pestka,S. (ed.), Academic Press Inc., London and New York, Vol. 78, p.387.
8. Laskey,R.A. and Mills,A.D. (1975) *Eur. J. Biochem.*, **56**, 335.
9. Maheswari,R.K., Demsey,A., Mohanty,S.B. and Friedman,R.M. (1980) *Proc. Natl. Acad. Sci. USA*, **77**, 2284.
10. Esteban,M. (1984) *Virology*, **133**, 220.
11. Wagner,R.R., Thomas,J.R. and McGowan,J.J. (1984) In *Comprehensive Virology*. Fraenkel-Conrat,H. and Wagner,R.R. (eds), Plenum Press, New York, Vol. 19, p. 223.

CHAPTER 7

Quantitation of IFN mRNA

KENNETH W.SIGGENS

1. INTRODUCTION

Studying the expression of interferons (IFNs) in the many different systems sometimes requires the use of a quantitative assay, not for a particular IFN protein, but rather for the corresponding mRNA of that IFN. This situation may arise, for example, in quantitating an IFN type for which there are no neutralizing antibodies available, but for which a suitable DNA probe exists. Also it may be of interest to study the expression of IFN in cells such as teratocarcinoma cells. These produce IFN in response to viral stimulation only after differentiation of the cells has been induced; it is possible, therefore, using nucleic acid hybridization to analyse the expression of IFN mRNA in these cells to determine whether IFN production in such cells is regulated at transcriptional or translational levels. Finally in animal model systems it is possible to study the expression of the IFN genes in specific cells or tissues of the animal with a view to identifying the cell sites induced, for example, following a viral infection.

An earlier volume in the Practical Approach Series (1) covered methods relevant to the isolation of eukaryotic mRNA and translation using *in vitro* and *in vivo* systems. Although these techniques have been used in the past for quantitation of IFN mRNA, the advances in DNA cloning techniques have allowed progress to be made in the quantitation of specific mRNAs. The methods described in this chapter relate to the quantitation of IFN mRNA by dot blot hybridization using ^{32}P-labelled DNA probes. There are now cDNA or genomic DNA clones available for all known types of human IFN and the majority of murine IFNs; a number of cloned genes have also been isolated for bovine and rat IFNs. *Table 1* presents a partial list, including references, of the IFN sequences which have been cloned to date. Access to the majority of these IFN

Table 1. Cloned IFN genes: references.

	IFN type	Cloned cDNA sequences	Cloned genomic sequences
(a) Human	IFN-α	2–5	6–9
	IFN-β	10–13	14–16
	IFN-γ	17,18	19,20
(b) Murine	IFN-α	21,22	21,23,24
	IFN-β	25,26	
	IFN-γ	27 and K.W.Siggens (unpublished)	27
(c) Rat	IFN-α	28	28,29

Table 2. Suppliers of chemicals and specialized equipment.

Guanidinium isothiocyanate	Fluka
Ficoll mol. wt 400 000	Pharmacia
Polyvinylpyrrolidone	Sigma
Bovine serum albumin (fraction V)	Sigma
2′-deoxynucleoside 5′-triphosphate (dATP, dCTP, dGTP, dTTP)	Pharmacia P-L Biochemicals Inc.
Deoxynucleoside 5′-[α-^{32}P]triphosphate 3000 Ci/mmol	Amersham
DNA polymerase I 'Klenow fragment'	Amersham
Nitrocellulose membrane (S&S BA85)	Schleicher & Schuell
S&S Minifold (SRC-96)	Schleicher & Schuell
BRL HYBRI.DOT system	Gibco-BRL
Agarose gel electrophoresis systems	Gibco-BRL
	Pharmacia
	Cambridge Biotech. Labs
Nick translation kit	Amersham
	Gibco-BRL

Commercial sources of chemicals and equipment are provided only as a guide. Other commercial sources may be superior.

DNA clones may generally be obtained by consultation of the publications. For details of the supplies of chemicals and equipment used in this chapter, see *Table 2*.

2. PREPARATION OF RADIOLABELLED DNA PROBES

The basis for the quantitation of a mRNA species is the ability to hybridize specifically a radiolabelled DNA sequence to its complementary mRNA sequence. In general, the levels of IFN mRNA present in a mRNA population of a cell represent sequences present in low abundance, therefore this puts constraints on the methodology used to quantitate these mRNAs. Firstly, since it is necessary to hybridize probes to sequences present at zero to several hundred copies per cell, it is essential to label DNA probes to a high specific activity, capable of allowing sufficient sensitivity for quantitation of the hybridization. The method presented is capable of generating DNA probes labelled to very high specific activity ($>10^9$ c.p.m./μg) allowing detection of at least 1 pg of mRNA in the sample. Non-specific binding of the labelled probe to either sequences present in the RNA samples, or to the nitrocellulose filter itself, must be minimized since this lowers the signal-to-noise ratio which affects the quantitation of the signals. Non-specific binding of the DNA probes can be kept to a minimum by using 'pure' complementary probe sequences representing only sequences complementary to the mRNA of interest and containing no flanking plasmid sequences. This necessitates cutting the plasmid containing the IFN sequence with enzymes which release the IFN sequence, or parts of it, and then purifying the DNA fragment of interest.

Most laboratories when asked to provide cloned IFN sequences for hybridization work supply either (i) small quantities of the plasmid DNA containing the sequence of interest, or (ii) a bacterial stock culture of the clone containing the plasmid. In either case the initial work is to generate a larger supply of the plasmid DNA in order to allow purification of the IFN-specific sequence. If purified plasmid DNA is supplied, it is first necessary to introduce the DNA into a suitable *Escherichia coli* host. Methods for transformation of DNA into *E. coli* and selection of cloned colonies harbouring the plasmid of interest are beyond the scope of this article and the reader is referred to other detailed protocols (30).

2.1 Plasmid purification

The provision of a bacterial stock allows the preparation of plasmid DNA by the method described in *Table 3*. The plasmid-containing bacterial culture is grown in medium containing an antibiotic which permits growth as a result of the presence of the appropriate antibiotic resistance gene on the plasmid vector. Most commonly used plasmid vectors (e.g. pBR322, pAT153) can be amplified to a higher copy number per cell by incubating the cultures in the presence of chloramphenicol which allows plasmid replication to continue whilst stopping cell multiplication thus generating a higher copy number per cell. The method in *Table 3* typically yields greater than 1 mg of plasmid per litre of culture.

The concentration and yield of plasmid DNA obtained after step 15, *Table 3*, should be measured spectrophotometrically. Determine the absorbance at 260 nm of the plasmid DNA solution or a dilution of this stock; calculate the concentration on the basis that a solution of DNA at a concentration of 50 μg/ml gives an $O.D._{260}$ equal to one.

2.2 Isolation of DNA fragments for use as DNA probes

Determine which restriction enzyme digest will yield a DNA fragment corresponding to the IFN-specific sequence then process an aliquot of the plasmid DNA for digestion and subsequent isolation of the fragment by agarose gel electrophoresis. Typically about 50 μg of plasmid DNA is digested with the appropriate restriction enzyme(s) in a final volume of about 100 μl using the buffer conditions supplied with the enzyme; prepare a 10-fold concentrated buffer solution to add to the reaction mixture at one tenth volume. For 50 μg of plasmid DNA use 20–40 units of restriction enzyme and incubate the digest for 4–5 h to ensure complete digestion of the DNA. Determine the state of the digest by analysing a small aliquot of the reaction (corresponding to ~0.5 μg of DNA) on an agarose minigel electrophoresis system, which allows rapid analysis of the digest. If the restriction digestion is incomplete, a further aliquot of the enzymes should be added, the incubation continued and then the products analysed by electrophoresis.

The DNA probe fragment is then isolated by preparative agarose gel electrophoresis, as described in *Table 4*. Prepare a standard agarose gel using a slot former generating wells approximately 30 mm wide; if wide sample well combs are not available then smaller wells may be used and the sample loaded into a larger number of wells. Load the restriction digest into the outer wells of the gel; for example if 50 μg of plasmid DNA was digested this can be loaded into the outermost wells at the sides of the gels, corresponding to a total track width of approximately 60 mm. The exact details of this stage can be varied widely, but the main objectives are to enable easy insertion of the DE81 paper strips and to keep the size of the DE81 strips required as small as possible by loading the DNA into as narrow a gel width as possible while maintaining reasonable resolution by the agarose.

The DNA fragment required is recovered from the gel by electrophoresis onto a strip

Table 3. Isolation of plasmid DNA[a].

1. Prepare the following solutions:

   LB (Luria–Bertani) medium

   10 g of Bacto-tryptone
   5 g of Bacto yeast extract
   10 g of NaCl

   Add distilled water to 950 ml. Adjust the pH to 7.2 with 1 M NaOH and add water to 1 l final volume. Sterilize the medium by autoclaving (15 p.s.i. for 20 min)

   Solution A

   50 mM glucose
   25 mM Tris-HCl pH 8.0
   10 mM EDTA

   Autoclave at 10 p.s.i. for 15 min and store at 4°C.

   Solution B

   0.2 M NaOH
   1% SDS

   This solution should be freshly made just prior to use from stock solutions of 1 M NaOH and 10% SDS

   Solution C

   5 M potassium acetate, pH 4.8

   This is prepared by mixing 60 ml of 5 M potassium acetate, 11.5 ml of glacial acetic acid and 28.5 ml of water.
2. Inoculate 5 ml of LB medium containing the appropriate antibiotic for the plasmid being used. Incubate at 37°C overnight with vigorous shaking.
3. The following morning inoculate 50 ml of LB medium (containing antibiotic) in a 500 ml flask with 0.2 ml of the overnight culture. Incubate at 37°C with vigorous shaking until the culture reaches $O.D._{600} = 0.6$.
4. Inoculate 25 ml of this culture into 500 ml of LB medium (containing antibiotic) pre-warmed to 37°C in a 2 l flash. Incubate for exactly 2.5 h at 37°C with vigorous shaking. The $O.D._{600}$ of the culture will reach approximately 0.4. Add 0.85 ml of a solution of chloramphenicol (100 mg/ml in ethanol). The final concentration of chloramphenicol is 170 μg/ml[b]. Incubate at 37°C with vigorous shaking for 12–16 h.
5. Harvest the bacterial cells by centrifugation at 4000 *g* in an angle-rotor for 10 min at 4°C. Discard the supernatant and allow the pellets to drain by standing the centrifuge pots at an angle in ice. Remove the drained supernatant.
6. Resuspend the pellet from a 500 ml culture in 10 ml of solution A containing 5 mg/ml lysozyme (lysozyme should be dissolved in solution A just prior to use). Transfer this to a 100 ml autoclaved glass bottle. Stand the suspended cells at room temperature for 5 min.
7. Add 20 ml of freshly prepared solution B and mix by gently inverting the bottle several times. Let the mixture stand on ice for 10 min.
8. Add 15 ml of ice-cold solution C and mix the contents by inverting the bottle sharply several times. Let the mixture stand on ice for 10 min.
9. Transfer the contents of the bottle to a Beckman SW27 polyallomer tube (or equivalent) and centrifuge on a Beckman SW27 rotor (or equivalent) at 20 000 r.p.m. for 20 min at 4°C[c].
10. Remove the supernatant into two 30 ml Corex tubes (~18 ml per tube). Add 0.6 volumes of isopropanol to each tube. Mix well and let the tubes stand at room temperature for 15 min to precipitate the DNA. Recover the DNA by centrifugation in a suitable swing-out rotor at 12 000 *g* for 30 min at room temperature.
11. Discard the supernatants and wash the pellets with 70% ethanol at room temperature — add the ethanol to each pellet, pour off and allow the tubes to drain briefly. Dry the pellets in a vacuum desiccator for the minimum time required to remove the liquid.

12. Dissolve the pellets in 20 ml total of 10 mM Tris-HCl pH 8.0, 1 mM EDTA. For every 1.0 ml of solution, add 1.0 g of solid CsCl and mix gently to dissolve. Next add 0.8 ml of 10 mg/ml ethidium bromide for every 10 ml of CsCl–DNA solution. Mix well, and then transfer to a tube suitable for centrifugation in a Beckman Type-50 vertical rotor (or equivalent). Fill the remainder of the tube with light paraffin oil. Centrifuge at 45 000 r.p.m. for 24 h at 20°C.
13. After centrifugation, view the DNA bands using u.v. illumination (305 nm). Collect the DNA band nearest to the bottom of the tube into a glass universal using a syringe needle inserted into the side of the tube. This is closed circular plasmid DNA.
14. Remove the ethidium bromide by extracting the DNA sample 4–5 times with iso-amylalcohol. Each time add an equal volume of iso-amylalcohol, shake for 1 min, then allow the phases to separate. Transfer the lower aqueous phase each time to a clean glass universal. Dialyse the aqueous phase against three lots of 5 l of 10 mM Tris-HCl pH 8.0, 1 mM EDTA (1 h at room temperature each time), to remove CsCl.
15. Precipitate the plasmid DNA by adjusting the solution to 0.3 M NaCl using a 3 M NaCl stock and adding 2 vol of ethanol at −20°C. Store DNA at −20°C overnight and collect DNA by centrifugation (step 10). Wash the pellet in 70% ethanol (−20°C) by adding ethanol and centrifuge a second time at 12 000 *g*. Discard the ethanol and dry the pellet *in vacuo*. Resuspend the DNA in a small volume of autoclaved water.

---

[a]Adapted from reference 31.
[b]For some bacteria containing plasmids, this concentration of chloramphenicol seems to cause lysis of the cells. If this occurs as evidenced by a loss of turbidity by the following morning, the problem can usually be overcome by decreasing the final concentration of chloramphenicol in the culture to 20 μg/ml.
[c]Using the quantities given there will be a slight excess of solution to fill the tube, therefore this is usually discarded.

of DE81 ion-exchange paper from which it is then eluted and processed. This method of recovery of DNA from agarose gels gives a yield of 40–70% and the DNA is of a high quality in terms of its ability to be labelled; other methods of recovery of DNA from agarose gels are available (30).

Since the amounts of DNA recovered from gels are of the order of a few micrograms it is not generally possible to measure the absorbance at 260 nm of these samples, and the DNA concentration must be determined by another means. The simplest method is to electrophorese a small amount of the DNA (estimated to be 50–200 ng) on an agarose minigel system, in parallel with known amounts of any DNA sample available for which the worker has an accurate concentration derived by spectrophotometric determination. Standard DNA should cover the range 50–500 ng. After visualization by ethidium bromide staining, examination of the fluorescent intensity allows an estimation of the amount of DNA in the sample of interest. This method gives a reasonably accurate estimation of the concentration of the gel-eluted DNA fragment.

2.3 Labelling of DNA probes

To achieve a high level of sensitivity from hybridization analysis of IFN mRNA it is necessary to generate high specific activity probes. Probes with a specific activity of approximately 5×10^9 d.p.m./μg can be generated by random primed synthesis of DNA from a DNA probe template.

Briefly, the double-stranded DNA to be used to generate a radioactive probe is denatured and mixed with random oligodeoxynucleotide hexamers which on a probability basis will contain some sequences complementary to hexanucleotide sequences in the 'template' DNA. The hexamers can base pair at these sites and then act as primers

Table 4. Agarose gel purification of DNA fragments[a].

1. Prepare the following solutions:

   10 × TBE

   108 g of Tris
   55 g of boric acid
   9.5 g of EDTA

   Add water to 1 l
   This 10-fold concentrated gel buffer is diluted as required for electrophoresis.

   Loading buffer

   This is prepared by mixing:

   5 ml of glycerol
   5 ml of 10 × TBE
   0.02% (w/v) bromophenol blue
2. Prepare a standard horizontal agarose gel using 1.0% (w/v) agarose in 1 × TBE; use a slot former generating loading wells approximately 30 mm wide.
3. Add 1 vol of loading buffer to 4 vol of a restriction enzyme digest of the plasmid of interest. Load the samples into the two outer wells and electrophorese at an appropriate voltage either overnight or for a shorter period of time.
4. When the bromophenol blue has electrophoresed 50–75% of the length of the gel remove the gel from the tank and stain the DNA bands by immersing the gel in 10 mM Tris-HCl pH 7.5, 1 mM EDTA containing 0.5 μg/ml ethidium bromide. **CARE**, wear gloves when handling ethidium bromide solutions or gels which have been in contact with ethidium bromide.
5. Visualize the bands by placing the gel on a u.v. transilluminator (long wavelength u.v. to minimize damage to DNA). Using a sterile scalpel cut a slit completely through the gel just in front of the DNA band required. It is easiest if the DNA is electrophoresed on the outer gel track enabling the slit to be cut from the edge of the agarose to a few millimetres beyond the inner edge of the DNA band.
6. Cut a piece of Whatman DE81, DEAE–cellulose paper to a width equivalent to the gel thickness and length several millimetres longer than the DNA band to be eluted. Insert the DE81 paper into the slit, and then squeeze the gel together again. Replace the gel into the electrophoresis tank and electrophorese at 100 V for approximately 15 min. Re-examine the gel on the transilluminator and observe the extent of transfer of DNA onto the DE81 paper. Continue electrophoresis until all of the DNA fragment of interest has transferred from the agarose onto the paper.
7. Remove the strip of DE81, blot it dry on 3 MM filter paper and using u.v. light to visualize the DNA, trim away all excess DE81 paper.
8. Elute the DNA from the paper by placing it in a small bottle and add 20 μl/mm^2 of elution buffer (10 mM Tris-HCl pH 7.5, 1.5 M NaCl 1.0 mM EDTA). Add three small glass beads and vortex until the paper is finely dispersed. Incubate the mixture at 37°C for 2 h.
9. Pierce the bottom of a 2.0 ml microcentrifuge tube with a fine needle and pack the lower one third of the tube with siliconized glass wool. Separate the eluted DNA from the DE81 paper by centrifuging the mixture through the glass wool — support the microcentrifuge tube in the top of a 15 ml Corex tube using the upper end cut from a plastic 5 ml syringe barrel. Add the paper/buffer mix and centrifuge at 4000 *g* for 2 min. Repeat until all of the DNA solution has been centrifuged. Wash the microcentrifuge tube and contents by adding 1.0 ml of elution buffer and centrifuging.
10. Pass the centrifuged DNA solution through a 0.45 μm cellulose acetate membrane filter (Schleicher and Schuell, FPO30/20). Wash the filter by passing 1.0 ml of elution buffer through. Pool the filtrates.
11. Add three volumes of butan-1-ol[b], and shake for 1 min. Centrifuge at 4000 *g* for 5 min. Remove and discard the excess butanol in the upper layer then remove the aqueous phase from below avoiding contamination with any remaining butanol. Transfer the DNA to a siliconized Corex tube and precipitate by adding 2 vols of ethanol. Store at −20°C overnight.
12. Recover the DNA by centrifugation (*Table 3*, step 10).
13. Wash the DNA precipitate by adding 70% ethanol to the tube and centrifuge again as in step 12.
14. Pour off the wash ethanol carefully, drain the tube onto tissue and then dry the sample *in vacuo* for the minimum time required.
15. Resuspend the DNA in 100 μl of water.
16. Determine the concentration of DNA by running a small aliquot on a gel and comparing the intensity of the visualized band with known standard amounts of DNA. Store the DNA at −20°C.

[a]Adapted from reference 32.
[b]Saturate butan-1-ol with water by shaking a mixture of butan-1-ol and water; leave the phases to separate. The upper layer is water-saturated butanol.

Table 5. Random primed labelling of DNA probes[a].

1. Prepare the following solutions:

   10 × RP buffer

   Mix the following:

   500 μl of 1 M Tris-HCl pH 7.5[b]
   167 μl of 3 M NaCl[b]
   100 μl of 1 M $MgCl_2$[b]
   70 μl of 1 M 2-mercaptoethanol
   50 μl of 10 mg/ml nuclease-free BSA
   113 μl of autoclaved double-distilled water

   Store in aliquots at −20°C.

   10 × dNTP mix

   2 mM of each unlabelled dNTP
   Prepare the dNTP mix from stock 100 mM sterile solutions of each dNTP (use autoclaved double-distilled water).

   Primer DNA

   Prepare a stock solution of oligodeoxyribonucleotide hexamer DNA (Pharmacia, product number 27-2166-01) at 1 mg/ml in sterile water[c].

2. Pipette 50 ng of DNA in 7 μl of water into a 0.5 ml microcentrifuge tube. Denature the DNA by placing the tube in a boiling water bath for 2.5 min and then chilling the tube rapidly in an ice-water slurry.

3. Add the following components on ice:

10 × RP buffer	2.0 μl
10 × dNTP mix	2.0 μl
Primer DNA	2.0 μl
Sterile water	2.0 μl
[^{32}P]dCTP, 10 μCi/μl	4.0 μl
Klenow fragment, DNA Pol I (7 units/μl)	1.0 μl

   Mix gently, centrifuge the tube briefly in a microfuge and incubate the reaction at 19°C overnight.

4. Stop the reaction by adding 1 μl of 10% SDS and 20 μg of unlabelled denatured salmon sperm DNA (see *Table 8*).

5. Set up a Sephadex G-75 column equilibrated with 10 mM Tris-HCl pH 7.5, 150 mM NaCl, 1 mM EDTA in a plastic disposable pipette. Apply the reaction mix from step 4 to the column and elute the DNA with equilibration buffer. The DNA elutes ahead of free dNTP, therefore its progress can be followed using a hand-held radiation monitor. When the DNA is near to eluting from the column, start collecting 200−300 μl fractions into 1.5 ml microfuge tubes[d]. Count the Cerenkov radiation in these fractions using a scintillation counter and pool the fractions corresponding to the DNA peak (labelled DNA is present in the first peak eluted, unincorporated [^{32}P]dNTP is in the second peak if elution is carried that far).

6. Determine the yield of radiolabelled probe by Cerenkov counting a small aliquot of the pooled fraction. Store the radiolabelled probe inside a lead container at −20°C.

[a]Adapted from reference 33.
[b]Stock solutions are autoclaved at 15 p.s.i. for 20 min.
[c]One absorbance unit at 260 nm is equivalent to 20 μg/ml.
[d]An estimation of the elution status of the unincorporated dNTP can be obtained by adding a very small amount of bromophenol blue to the sample prior to application to the column. The blue colour of the dye co-elutes with the dNTP thereby giving a visual clue to the progress of elution of the column.

for the synthesis of new DNA strands complementary to the template sequences; the polymerization of addition of deoxynucleotides is catalysed by DNA polymerase I ('Klenow' fragment). Inclusion of a radiolabelled nucleotide together with the other three unlabelled nucleotides results in incorporation of the isotope into the newly synthesized DNA sequences. The detailed protocol based on a published procedure (33) is described in *Table 5* and the following points should be noted.

(i) Using a single [α-^{32}P]deoxynucleotide triphosphate with a specific activity of 3000 Ci/mmol allows the generation of DNA probes with a theoretical specific activity of about 5×10^9 d.p.m./μg.

(ii) The reaction is efficient and the incorporation of isotope is generally 70–90% of the input. Uisng 50 ng of DNA and 40 μCi of isotope will result in the synthesis of approximately 13 ng of labelled DNA probe at an incorporation of 75% of the input isotope. This is sufficient probe for hybridization to approximately 260 cm^2 of nitrocellulose filter carrying the RNA samples to be analysed.

(iii) The reaction can be scaled up for situations where larger amounts of probe are required. For example, 100 ng of DNA can be labelled using the same reaction volumes, with the exception that the [α-^{32}P]dNTP should be increased to 80 μCi and the sterile water should be omitted from step 3 to compensate for the volume change.

(iv) The percentage incorporation of isotope is conveniently assayed by binding a small sample of labelled DNA to Whatman DE81 paper (free dNTP does not bind under the wash conditions used).

  (a) Add 1 μl of the reaction mixture to 9 μl of water.
  (b) Spot 2 μl aliquots onto four small squares of DE81 paper.
  (c) Dry two filters without further processing to determine the total radioactivity present.
  (d) The other two filters should be treated as follows: 4×5 min washes in 0.5% Na_2HPO_4; 2×1 min washes in water; 2×30 sec washes in 100% ethanol; dry the filters.

  Count the filters in scintillation fluid and calculate the percentage incorporation into the DNA.

(v) To obtain the highest specific activity, the radiolabelled isotope should be used as near as possible to the supplier's reference date. In any case the author does not use isotope older than 2 weeks from the reference date.

(vi) Due to the high specific activity of the probe, the length of the labelled sequences will be reduced rapidly as a result of strand breakages and will reduce to very small fragments within several days. Therefore, prepare the radiolabelled probe either immediately before it is required or about 24 h in advance. Older probes may, however, be employed successfully.

2.3.1 *Alternative labelling strategies*

If a commercial kit system is preferred for the generation of probes then nick translation (34) kits are available from several suppliers. Using nick translation the specific activity of the probes will generally be lower ($<10^9$ d.p.m./μg DNA) than using the method described above. However, despite the lower specific activity nick-translated probes can be used successfully in the quantitation of IFN mRNA.

3. ISOLATION OF RNA FROM CELLS

The major criteria for isolation of RNA from cells are that the method is rapid, easy and capable of giving quantitative recovery of non-degraded RNA. Many methods exist for the isolation of RNA from cells (1), but it is the experience of the author that the most suitable for processing multiple samples is that based upon lysis of the cells in the presence of the chaotropic agent guanidinium isothiocyanate which denatures

Table 6. Isolation of RNA[a].

1. Prepare the following solutions:

   Lytic buffer

   4 M guanidinium isothiocyanate
   1 M 2-mercaptoethanol
   20 mM sodium acetate, pH 5.0

   Use stock 1 M sodium acetate, pH 5.0 to prepare the buffer.
   Dissolve the guanidinium isothiocyanate and then add 2-mercaptoethanol. Filter the solution through a 0.45 μm filter unit. Store at room temperature.

   Caesium chloride

   5.7 M CsCl

   Prepare the solution and if necessary adjust it until the refractive index is 1.405. Filter the solution through a 0.45 μm filter unit just before use.

   5 mM Tris-HCl, pH 7.5

   Prepare a stock solution of 1 M Tris-HCl, pH 7.5 and autoclave. Dilute from this stock solution using DEPC-treated water[b] to prepare 5 mM Tris. Reserve these solutions entirely for RNA work.
2. Harvest the cells and centrifuge in a plastic Universal tube.
3. Discard the supernatant. Add 3.5 ml of lytic buffer to each sample and vortex for 10–20 sec to lyse the cells. Shake the tube hard. Vortex for a further 10–20 sec.
4. Pipette 1.5 ml of CsCl into the required number of centrifuge tubes. The tubes used should be polyallomer (not sterilized) suitable for use with the Beckman SW 50.1 rotor or MSE 6 × 5.5 rotor (or equivalent).
5. Layer the cell lysate onto the CsCl cushion. Centrifuge at 137 000 *g* for 22 h at 21°C.
6. After centrifugation aspirate off the liquid taking care to avoid the small RNA pellet at the bottom of the tube. Leave the tube inverted and then carefully wipe the inside with a tissue avoiding the RNA pellet.
7. Resuspend the pellet in 150 μl of 5 mM Tris-HCl pH 7.5 by pipetting vigorously and disturbing the pellet with the tip of the micropipette. Transfer the resuspended RNA to a 1.5 ml microcentrifuge tube. Add a second 150 μl of 5 mM Tris-HCl to the ultracentrifuge tube and resuspend any residual RNA. Transfer this RNA to the same 1.5 ml tube.
8. Extract the RNA solution with an equal volume of butanol:chloroform[c]; shake the mixture for 1 min then separate the phases by centrifugation for 2 min.
9. Remove the upper aqueous phase to a clean tube and extract it twice more with butanol:chloroform.
10. Adjust the final aqueous phase to 0.3 M NaCl and precipitate the RNA with 2.5 volumes of ethanol at −20°C overnight.
11. Recover the RNA by centrifugation in a microcentrifuge for 10 min. Wash the pellet by adding 0.6 ml of 70% ethanol (−20°C) and centrifuge for 5 min. Drain the pellet and dry briefly *in vacuo*.
12. Resuspend the RNA in 100 μl of DEPC-treated water. Store at −20°C or −70°C.

[a]Reference 35.
[b]DEPC treat water by adding 0.1% (v/v) diethylpyrocarbonate to double-distilled water and shake well. Allow to stand overnight and then autoclave to destroy the DEPC.
[c]Butan-1-ol and chloroform in the ratio 1:4.

proteins and thereby inactivates ribonucleases. In order to recover the RNA from the cell lysate, the method relies upon ultracentrifugation of the RNA through a CsCl cushion, which efficiently separates RNA from the DNA and protein (35). The RNA, under the conditions used, sediments to the bottom of the centrifuge tube and forms a small pellet, while the DNA and protein components remain in solution in the supernatant. The protocol is described in *Table 6*.

Using the ultracentrifuge rotors suggested allows isolation of RNA from a convenient number of cells. An important consideration in determining how many cells to process is that the extraction procedure should yield sufficient total RNA for subsequent analysis by hybridization with a ^{32}P-labelled probe. Generally, 10–20 μg of total RNA is required for each RNA dot and since duplicate samples are the minimum requirement, the cell number used should be capable of generating at least 50 μg of RNA. A further consideration, with regard to the numbers of cells which should be processed, is that the CsCl cushions used for isolation of the total RNA from the cell lysates can be overloaded in terms of the amount of DNA present in the cell lysate. Too much DNA as the result of lysis of a larger number of cells results in the DNA sedimenting nearer to the bottom of the tube, thereby contaminating the RNA pellet.

As a general guide lysis of $(1-1.5) \times 10^8$ lymphocytes in 3.5 ml of lytic buffer and centrifugation of this lysate on a single 1.5 ml cushion of CsCl yields approximately 100 μg of RNA (higher yields of RNA result if lymphocytes are stimulated with a mitogen). For adherent cells such as fibroblasts there may be approximately 2×10^5 cells/cm^2 when confluent, and 75 cm^2 of confluent cells yields approximately 100–200 μg of RNA. These figures are given only as guidelines and the optimum number of cells per centrifuge tube should be determined empirically.

As in the handling of any RNA samples, following centrifugation it is important to avoid RNase contamination of the solutions or plasticware used. The following precautions should be followed.

(i) All plasticware should be autoclaved at 15 p.s.i. for 20 min.
(ii) Glassware should be heated at 200°C for 3 h.
(iii) Tris-containing solutions should be autoclaved as in (i) and used solely for the purpose of RNA work.
(iv) Other aqueous solutions should be treated with diethylpyrocarbonate (DEPC) as in *Table 6*.
(v) Plastic disposable gloves should be worn to prevent RNase contamination from the skin.
(vi) In step 6, *Table 6*, the inside of the centrifuge tube is wiped with tissue after aspiration of the supernatant to remove any contaminating solution containing RNase activity which may reactivate in the aqueous solution used for resuspension. Avoid allowing the resuspension solution coming into contact with the tube side walls as far as possible.

After recovery of the RNA from the centrifuge tube, steps 6–12, resuspend the final material in 100 μl of DEPC-treated water and determine the concentration and yield of RNA by measuring the absorbance at 260 nm. For RNA, one absorbance unit at 260 nm is equivalent to a concentration of 40 μg/ml. For storage of the RNA, if it is to be used within a few weeks for analysis by dot blot hybridization, it can be frozen at −20°C or preferably at −70°C. For longer term storage the material should be left in a precipitated form in ethanol at −20°C; the RNA may be either left precipitated at step 10 until required for further analysis or may be re-precipitated after step 12 by adjusting the salt concentration to 0.3 M NaCl, adding 2.5 volumes of ethanol and placing the sample at −20°C.

4. QUANTITATION OF IFN mRNA BY HYBRIDIZATION

4.1. Application of RNA samples to nitrocellulose filters

Several manufacturers market manifolds suitable for applying samples to nitrocellulose filters and the author has successfully used the 96-sample Hybri Dot manifold (Gibco-BRL). A procedure for applying RNA samples using a dotting manifold is detailed in *Table 7*. The nitrocellulose membrane used is suplied by Schleicher & Schuell (S&S BA85); nitrocellulose filters produced by other manufacturers may have different characteristics in terms of background levels of binding of probes and RNA retention on the filters, and therefore an empirical approach is the only means to determine whether other types are more or less suitable. An important point in using nitrocellulose filter type BA85 is that the filter must be equilibrated in 20 × SSC for at least 2 h to ensure that non-specific binding of the probe is avoided and RNA retention is maximized. Furthermore, if the filter does not wet rapidly and evenly when lowered onto water (step 1, *Table 7*) the filter should be discarded and replaced with a fresh piece since this usually leads to background problems after hybridization.

To enable an assessment of the sensitivity of the hybridization and to allow trouble-shooting of the system in the absence of any detectable signal following hybridization, it is necessary to include standards on each filter. The easiest method is to dot known amounts of unlabelled DNA, the same as that which is being used to generate the radio-labelled probes. A suitable range to apply is 0.1, 0.01 and 0.001 ng of DNA which should be denatured at 100°C before application to the filter. If the protocols are working successfully then 0.001 ng of standard DNA should be detected following autoradiography for 24 h and the background level of detection should be virtually zero.

Table 7. Dot blot analysis of RNA.

1. Wet a piece of nitrocellulose filter, cut to size, in water then soak the filter in 20 × SSC[a] for 2 h. Assemble the filter in a dot blot manifold (e.g. Hybridot, Gibco-BRL). Wear gloves when handling nitrocellulose to avoid contact with the skin.
2. Denature 20 μg or less of RNA in 100 μl volume by heating to 65°C for 10 min and rapidly chilling the tube in an ice-water bath. This is most conveniently done in 1.5 ml microcentrifuge tubes. Add 50 μl of ice-cold 20 × SSC and mix. Keep the samples on ice.
3. Denature the DNA 'standards' in 100 μl volume by boiling for 5 min and rapidly chilling. Add 50 μl of ice-cold 20 × SSC and mix. Keep the samples on ice.
4. Apply the samples to the nitrocellulose filter, assembled in the apparatus, using a gentle vacuum to draw the solution through the filter. Do not include any washing step.
5. After applying all samples, remove the filter and blot it dry between 3 MM paper. Allow the filter to air dry then bake it at 80°C for 2 h in a vacuum oven. Store the filter at room temperature between 3 MM paper.

[a]1 × SSC is 0.15 M NaCl, 15 mM trisodium citrate.

4.2 Hybridization of radiolabelled probes

Many different procedures exist for hybridizing DNA probes to samples on nitrocellulose filters, although most are only slight variations of a few basic methods. The protocol described in *Table 8* has been used to analyse IFN mRNAs from human and murine sources (36,37). One important aspect of the conditions used is that the hybridization should proceed as near to completion as possible; since the kinetics of the hybridization reaction are difficult to predict from theoretical considerations, the suggestions for volumes of solutions used should be adhered to closely until the worker has a feel for the procedure and can better judge the effects of altering some of the reaction parameters.

Labelling of the DNA by random priming generates approximately $(1-2) \times 10^7$ Cerenkov c.p.m. of ^{32}P-labelled probe; this is sufficient for hybridization to RNA samples on a total area of filter of approximately 260 cm^2. The amount of hybridization buffer used for a filter should be calculated as in *Table 8*, and the amount of radiolabelled probe should be altered such that 10^6 Cerenkov c.p.m./ml is used. These quantities of buffer and probe per cm^2 of filter allow hybridization to proceed to completion during an overnight incubation at 42°C, thereby generating maximum sensitivity for the IFN mRNA.

The post-hybridization washing procedure for the filter is that which has been used for analysis of IFN-γ mRNA in samples derived from human and murine lymphocytes. It should be noted that the final wash in $0.5 \times$ SSC at 65°C (step 9) is not the most stringent conditions at which the wash can be performed while maintaining the stability of the IFN-γ DNA$-$RNA hybrid on the filter. However, the author has found this level of stringency to be ideal for elimination of non-specific binding of IFN-γ probes to non-complementary sequences. If, however, a DNA probe complementary to one of the many IFN-α sequences were being used, it would be recommended to try increasing the stringency of the final wash to eliminate hybrids formed between the probe sequence and other very closely related IFN-α sequences. This may be achieved by either reducing the salt concentration to $0.1 \times$ SSC and washing at 65°C, or by increasing the washing temperature to 68°C for example. Unfortunately the degree of stringency required is best determined empirically for the particular probe sequence being used, but these suggestions can be used as the basis for determination of optimal washing conditions. This problem of stringency of washing will only arise with closely related nucleotide sequences as typified by some members of the IFN-α multigene families. For work on 'unique sequences' the protocol described yields satisfactory results.

Radioactive dots should be detected by autoradiography and a typical example of an autoradiograph is shown in *Figure 1*. This shows samples of RNA isolated from mitogen-stimulated lymphocytes and probed using an IFN-γ cDNA sequence. It was possible to detect 0.01 ng of standard DNA after a 20 h exposure; 0.001 ng of the standard DNA was detectable on the original autoradiograph but the photographic reproduction fails to show this. An autoradiograph such as this may be quantitated by densitometric scanning of the dots, and this must be carried out on autoradiographs of varying exposure times to eliminate the possibility of saturation of the film by overexposure to high levels of radioactivity. Alternatively, individual dots may be cut from

Table 8. Quantitation of IFN mRNA by hybridization of radiolabelled probe.

1. Pre-hybridization buffer

   50% de-ionized formamide[a]
   5 × SSPE[b]
   5 × Denhardt's solution[c]
   100 μg/ml denatured salmon sperm DNA[d]
   1 μg/ml poly(A)

   Hybridization buffer

   50% de-ionized formamide
   5 × SSPE
   5 × Denhardt's solution
   100 μg/ml denatured salmon sperm DNA
   1 μg/ml poly(A)
   1 × 10^7 Cerenkov c.p.m. of denatured labelled probe[e]
2. Float the nitrocellulose filter on the surface of 6 × SSC; when wet immerse the filter and soak for 5 min.
3. Transfer the filter to a heat-sealable plastic bag and add 200 μl/cm^2 filter of pre-hybridization buffer. Squeeze out as much air as possible from the bag and heat seal the open end of the bag with a commercial bag sealer. Incubate the bag, clipped to a glass plate, submerged in a shaking water bath at 42°C for 4 h.
4. Remove the bag from the water bath and cut off a corner. Squeeze out as much pre-hybridization buffer as possible.
5. Prepare the hybridization buffer just before it is required and denature the ^{32}P-labelled probe before addition to the buffer. Using a Pasteur pipette add 75 μl of hybridization solution per cm^2 of filter, containing 10^6 c.p.m./ml of probe. Squeeze out as much air as possible and re-seal the cut edge. Incubate the bag submerged (clipped to a glass plate) in a non-shaking water bath at 42°C for 24 h.
6. Remove the bag from the water bath and open it by cutting along the edges. Wearing plastic gloves, transfer the filter immediately to 500 ml of 3 × SSC in a plastic box at room temperature.
7. Shake the box gently at room temperature for 10 min.
8. Pour away the liquid and replace it immediately with a further 500 ml of 3 × SSC. Do not allow the filter to dry out at any stage. Shake the box gently for a further 10 min.
9. Transfer the filter to a plastic box contianing 500 ml of 0.5 × SSC at 65°C in a shaking water bath. Wash the filter for 30 min.
10. Repeat step 9.
11. Remove the filter from the wash solution and blot dry on Whatman 3 MM paper.
12. Allow the filter to air dry, then stick it to a piece of 3 MM paper using adhesive labels. Mark several spots around the border of the filter using radioactive ink[f] to allow alignment of the autoradiograph. Wrap the whole paper plus filter in 'Saran' wrap or other suitable thin plastic film and expose it to X-ray film at −70°C in the presence of a calcium tungstate intensifying screen (e.g. Du Pont Cronex Lightning Plus or Fuji Mach 2).
13. To quantitate the amount of DNA bound to each dot, either (a) scan the autoradiographic image of the dot using a densitometer, or (b) cut the filter into squares using a grid pattern identical to the layout of the manifold. It is important that each area of filter corresponding to a dot plus its surrounding area is of the same size to minimize errors resulting from background radiation bound to the filter. Each dot can be quantitated by liquid scintillation counting.

[a]De-ionize formamide by stirring 5 g of Amberlite MB-3 mixed bed ion-exchange resin with 100 ml of formamide for 30 min at room temperature. Filter the formamide and store it in aliquots at −20°C until use.
[b]20 × SSPE is prepared by dissolving 174 g of NaCl, 27.6 g of $NaH_2PO_4.H_2O$ and 7.4 g of EDTA in 800 ml of water. Adjust to pH 7.4 with NaOH then increase the volume to 1l. Sterilize by autoclaving.
[c]Stock 50 × Denhardt's solution is prepared by dissolving 1 g of Ficoll (mol. wt 400 000), 1 g of polyvinylpyrrolidone, 1 g of BSA (Fraction V) in water to 100 ml total volume. Store in aliquots at −20°C.
[d]Prepare denatured salmon sperm DNA by dissolving the DNA (Sigma Type III, sodium salt) in water at 10 mg/ml. Shear the DNA by passing it 10 times through a 19-gauge needle. Boil the DNA for 10 min, cool rapidly and store at −20°C. Before adding to pre-hybridization or hybridization buffers, denature again by heating the aliquot to 100°C then chill rapidly on ice.
[e]Denature the radiolabelled probe by heating to 100°C for 10 min then chill rapidly on ice.
[f]Prepare radioactive ink by mixing a small amount of a ^{32}P-labelled compound with permanent black ink. Use sufficient c.p.m. to register on the autoradiograph as a small dot after the appropriate exposure.

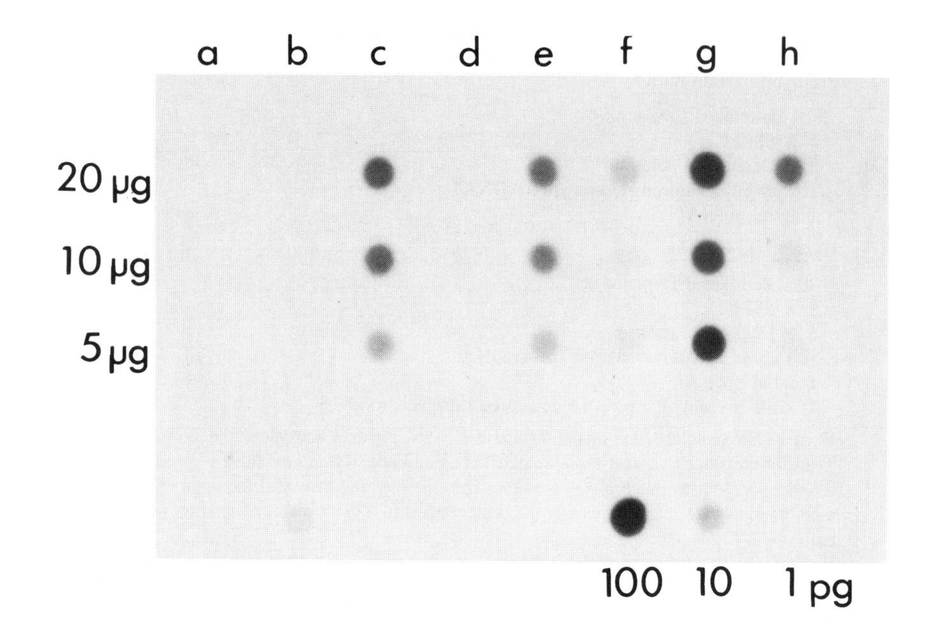

Figure 1. Dot blot analysis of mitogen-induced lymphocyte RNA samples. Human peripheral blood lymphocytes were induced with either Staphylococcal enterotoxin A (SEA) or with SEA plus the tumour promoter mezerein (37). Cells were incubated and harvested for RNA isolation as indicated below. Some samples as indicated were 'superinduced' for IFN-γ mRNA by treating the cells with an inhibitor of protein synthesis and harvesting the cells 4 h post-inhibition. Total RNA (20, 10 and 5 μg) was analysed by dot blot hybridization. Standard DNA was dotted as indicated (100, 10, 1 pg). **(a)** non-induced, **(b)** SEA-induced for 16 h, **(c)** SEA + mezerein-induced for 16 h, **(d)** SEA-induced for 20 h, **(e)** SEA + mezerein-induced for 20 h, **(f)** SEA-induced for 16 h then superinduced with puromycin for 4 h, **(g)** SEA + mezerein-induced for 16 h then superinduced for 4 h, **(h)** independent experiment, lymphocytes induced for 4 h with SEA. Autoradiograph exposed for 20 h at -70°C using a Du Pont Cronex intensifying screen (K.W.Siggens, unpublished data).

the filter using a square grid pattern corresponding to the exact layout of the hybridization manifold. Ensure that each area cut out is exactly the same size in order to keep the background counts the same for each sample.

Figure 1 illustrates one aspect of the hybridization method which should be noted. For each RNA sample, 20, 10 and 5 μg were dotted to titrate the IFN-γ mRNA signal. As can be seen there is a non-linear relationship between the level of the signal observed in a particular RNA sample and the amount of RNA applied to the dot. This may be due to saturation of the binding capacity of the nitrocellulose with high levels of RNA. However, the relative level of signal between different RNA samples is independent provided the same amount of RNA is dotted. This observation demonstrates the need to assay several times and to use a titration of the amount of RNA applied to ensure quantitative data. Clearly these requirements are not necessary for situations in which the assay is being used simply to determine the absence or presence of a particular mRNA species.

5. CONCLUDING REMARKS

The techniques described in this chapter have been used in studies on the expression of the IFN-γ gene in lymphocytes (36,37). Analysis of the type I IFNs has also been carried out in the same laboratory using these techniques.

Although in detailing the methods for the isolation of RNA, the emphasis was upon isolation from culture cells, the guanidinium isothiocyanate–CsCl method is directly applicable to isolation of RNA from animal tissues. The method may be adapted to isolation from tissues by homogenizing the tissue in the lytic buffer to ensure total disruption of the cells. The only other consideration is determining a suitable ratio of weight of tissue to volume of lytic buffer to enable quantitative recovery of RNA. Other publications refer to the use of a similar technique for the isolation of RNA from tissues (38,39).

A problem using the described methods arises when it is necessary to analyse a complex population of IFN mRNA types, as for example in the expression of multiple IFN-α mRNAs following viral induction of gene expression. Although it should be feasible to assay individual mRNA species by using stringency conditions which allow only the stability of true hybrids, this may involve a lot of extra analysis to be absolutely certain of the specificity of the hybridization signal. In such circumstances, it may be more practical to use an alternative strategy for quantitation of specific mRNAs using S1 analysis. An excellent example of the application of this technique is contained in an article by Hiscott *et al.* (40) in which multiple species of IFN mRNAs were analysed using S1 analysis. Since the methods employed for S1 analysis are beyond the scope of this current chapter, the reader is referred to the above paper for detailed methodology.

6. REFERENCES

1. Hames,B.D. and Higgins,S.J., eds (1984) *Transcription and Translation – A Practical Approach.* IRL Press, Oxford and Washington, D.C.
2. Goeddel,D.V., Leung,D.W., Dull,T.J., Gross,M., Lawn,R.M., McCandliss,R., Seeburg,P.H., Ullrich,A., Yelverton,E. and Gray,P.W. (1981) *Nature*, **290**, 20.
3. Mantei,N., Schwarzstein,M., Streuli,M., Panem,S., Nagata,S. and Weissmann,C. (1980) *Gene*, **10**, 1.
4. Streuli,M., Nagata,S. and Weissmann,C. (1980) *Science*, **209**, 1343.
5. Slocombe,P., Easton,A., Boseley,P. and Burke,D.C. (1982) *Proc. Natl. Acad. Sci. USA*, **79**, 5455.
6. Lawn,R.M., Adelman,J., Dull,T.J., Gross,M., Goeddel,D. and Ullrich,A. (1981) *Science*, **212**, 1159.
7. Lawn,R.M., Gross,M., Houck,C.M., Franke,A.E., Gray,P.W. and Goeddel,D.V. (1981) *Proc. Natl. Acad. Sci. USA*, **78**, 5435.
8. Ullrich,A., Gray,A., Goeddel,D.V. and Dull,T.J. (1982) *J. Mol. Biol.*, **156**, 467.
9. Brack,C., Nagata,S., Mantei,N. and Weissmann,C. (1981) *Gene*, **15**, 379.
10. Taniguchi,T., Ohno,S., Fujii-Kuriyama,Y. and Muramatsu,M. (1980) *Gene*, **10**, 11.
11. Derynck,R., Content,J., De Clercq,E., Volkaert,G., Tavernier,J., Devos,R. and Fiers,W. (1980) *Nature*, **285**, 542.
12. Goeddel,D.V., Shepard,H.M., Yelverton,E., Leung,D. and Crea,R. (1980) *Nucleic Acids Res.*, **8**, 4057.
13. Siggens,K.W., Slocombe,P., Easton,A., Boseley,P., Meager,A., Tinsley,J. and Burke,D. (1983) *Biochim. Biophys. Acta*, **741**, 65.
14. Tavernier,J., Derynck,R. and Fiers,W. (1981) *Nucleic Acids Res.*, **9**, 461.
15. Ohno,S. and Taniguchi,T. (1981) *Proc. Natl. Acad. Sci. USA*, **78**, 5305.
16. Gross,G., Mayr,U., Bruns,W., Grosveld,F., Dahl,H.M. and Collins,J. (1981) *Nucleic Acids Res.*, **9**, 2495.
17. Gray,P.W. Leung,D.W., Pennica,D., Yelverton,E., Najarian,R., Simonsen,C.C., Derynck.R., Sherwood,P.J., Wallace,D.M., Berger,S.L., Levinson,A.D. and Goeddel,D.V. (1982) *Nature*, **295**, 503.
18. Devos,R., Cheroutre,H., Taya,Y., Degrave,W., Van Heuverswyn,H. and Fiers,W. (1982) *Nucleic Acids Res.*, **10**, 2487.
19. Gray,P.W. and Goeddel,D.V. (1982) *Nature*, **298**, 859.
20. Taya,Y., Devos,R., Tavernier,J., Cheroutre,H., Engler,G. and Fiers,W. (1982) *EMBO J.*, **1**, 953.
21. Shaw,G.D., Boll,W., Taira,H., Mantei,N., Lengyel,P. and Weissmann,C. (1983) *Nucleic Acids Res.*, **11**, 555.
22. Kelley,K.A., Kozak,C.A., Dandoy,F., Sor,F., Skup,D., Windass,J.D., De Maeyer-Guignard,J., Pitha,P.M. and De Maeyer,E. (1983) *Gene*, **26**, 181.
23. Zwarthoff,E.C., Mooren,A.T.A. and Trapman,J. (1985) *Nucleic Acids Res.*, **13**, 791.

24. Kelley,K.A. and Pitha,P.M. (1985) *Nucleic Acids Res.*, **13**, 805.
25. Higashi,Y., Sokawa,Y., Watanabe,Y., Kawade,Y., Ohno,S., Takaoka,C. and Taniguchi,T. (1983) *J. Biol. Chem.*, **258**, 9522.
26. Skup,D., Windass,J.D., Sor,F., George,H., Williams,B.R.G., Fukuhara,H., De Maeyer-Guignard,J. and De Maeyer,E. (1982) *Nucleic Acids Res.*, **10**, 3069.
27. Gray,P.W. and Goeddel,D.V. (1983) *Proc. Natl. Acad. Sci. USA*, **80**, 5842.
28. Dijkema,R., Pouwels,P., Reus,A. and Schellekens,H. (1984) *Nucleic Acids Res.*, **12**, 1227.
29. Dijkema,R., van der Meide,P.H., Pouwels,P.W., Caspers,M., Dubbeld,M. and Schellekens,H. (1985) *EMBO J.*, **4**, 761.
30. Maniatis,T., Fritsch,E.F. and Sambrook,J. (1982) *Molecular Cloning. A Laboratory Manual.* Cold Spring Harbor Laboratory Press, New York.
31. Birnboim,H.C. and Doly,J. (1979) *Nucleic Acids Res.*, **7**, 1513.
32. Dretzen,G., Bellard,M., Sassone-Corsi,P. and Chambon,P. (1981) *Anal. Biochem.*, **112**, 295.
33. Feinberg,A.P. and Vogelstein,B. (1983) *Anal. Biochem.*, **132**, 6.
34. Rigby,P.W.J., Dieckmann,M., Rhodes,C. and Berg,P. (1977) *J. Mol. Biol.*, **113**, 237.
35. Morser,J., Flint,J., Meager,A., Graves,H., Baker,P.N., Colman,A. and Burke,D.C. (1979) *J. Gen. Virol.*, **44**, 231.
36. Siggens,K.W., Wilkinson,M.F., Boseley,P.G., Slocombe,P.M., Cowling,G. and Morris,A.G. (1984) *Biochem. Biophys. Res. Commun.*, **119**, 157.
37. Siggens,K.W., Tinsley,J.M. and Morris,A. (1985) *Eur. J. Immunol.*, **15**, 1079.
38. Chirgwin,J.M., Przybyla,A.E., MacDonald,R.J. and Rutter,W.J. (1979) *Biochemistry*, **18**, 5294.
39. Kaplan,B.B., Bernstein,S.L. and Gioio,A.E. (1979) *Biochem. J.*, **183**, 181.
40. Hiscott,J., Cantell,K. and Weissmann,C. (1984) *Nucleic Acids Res.*, **12**, 3727.

CHAPTER 8

Antibodies against interferons: characterization of interferons and immunoassays

A.MEAGER

1. INTRODUCTION

All interferons (IFNs) are proteins and, since the structures of individual types (α, β, γ) are different and show marked dissimilarities in amino acid sequence among species, antibodies may be readily raised against them in heterologous species. Thus, antibodies against human IFN may be raised in animals ranging from cynomolgus monkeys to mice; antibodies against mouse IFN may be raised in rats, rabbits, sheep, etc. Specific antibodies are extremely useful research tools for the characterization, purification and quantification of IFN. This chapter sets out to describe how to produce successfully high titred type-specific, including monoclonal, antibodies to IFN, and then how to apply them to the characterization of different IFN types (IFN neutralization assay) and for setting up immunoassays for IFN quantification.

2. PREPARATIONS OF IFN

The production of purified preparations of IFN-α, -β and -γ from natural sources is described elsewhere in this volume. IFN-β and -γ types and many IFN-α subtypes are also available from recombinant DNA technological processes and can be highly purified.

In general, the amino acid sequences of recombinant DNA IFN molecules are identical to, or very similar to, those of IFN molecules produced naturally. However, recombinant DNA technology allows ready changes in amino acid sequence to be made which may improve expression/stability in foreign host producer cells (e.g. substitution of Cys-17 by Ser-17 in human IFN-β prevents mismatch of disulphide bonds) and modify properties of the resultant IFN proteins beneficially for clinical use. Such IFN analogues (muteins) may show antigenic differences to the 'parent' molecules and may also show increased immunogenicity in the homologous, human or animal species.

Thus, nowadays, there is potentially a very large selection of highly purified IFNs for antibody production, not counting sub-molecular synthetic polypeptides which may also be used. It is always recommended that the purest IFN preparations available be used to immunize animals to avoid raising antibodies to protein contaminants. This is true even for the Kohler and Milstein (1) technique for monoclonal antibody (McAb) production in which hybridomas producing McAbs to protein contaminants are screened out and discarded, since the frequency of McAbs against IFNs is in general very low.

3. IMMUNIZATION OF ANIMALS

3.1 **Quantity of IFN**

Sufficient quantities of a particular IFN preparation must be available for immunization to continue over several months (or years). Larger animals, for example sheep and goats, will in general require larger amounts of antigen than small animals, such as rats, mice and guinea pigs, but this correlation is far from watertight. In the author's experience, sheep and cattle efficiently produce antibodies to heterologous IFN, and the amounts of antiserum are very much larger than can be obtained from rabbits and smaller animals. As a rough guide, mice will require between 50 and 100 μg of IFN per dose, rabbits 100–200 μg and sheep 200–500 μg; it is, of course, legitimate to give bigger doses but this may be wasting precious material (e.g. tolerization) and in any case only relatively small volumes may be given to small animals such as mice.

3.2 **Preparation and injection of antigen**

The first inoculation of antigen must be done in such a way as to prime the animal's immune system for subsequent antibody production. To do this effectively with IFNs, which are relatively weak immunogens, the IFN must be combined with adjuvant. Freund's complete adjuvant (FCA) is most suitable for this purpose, but care must be taken to ensure that the IFN is completely emulsified with FCA just prior to inoculation. This emulsification may be achieved by repeatedly squirting the IFN–FCA mixture (1:1 or 1:2) out of a syringe needle or by blowing air through the mixture, for example by forcing the end of a plastic syringe hard against the bottom of the vial or bottle containing the IFN–FCA mixture and rapidly pumping up and down. Animals are then injected with the emulsified mixture either intramuscularly (i.m.), intradermally (i.d.) or subcutaneously (s.c.). The latter two routes of injection often result in unhealing lesions and should be avoided; the i.m. route usually leads to adequate priming. In the case of small animals, such as mice, injections can be made into the four footpads (0.05 ml/footpad). Swelling of the footpads some days later is generally indicative of a good immune response.

3.3 **Booster injections**

When to give the second booster injection to obtain maximum antibody production depends on many factors, but to obtain *high affinity* antibodies this booster should be delayed for several weeks following the primary injection. Adjuvant is no longer necessary and IFN preparations are normally injected i.m. mixed with incomplete Freund's adjuvant (IFA), the oil base of FCA. Following boosting in this manner, specific antibody production should be induced in the animal and reach a maximum approximately 14 days later.

3.4 **Antiserum preparation**

(i) Bleed the animal from an appropriate vein and collect the blood into a clean container. Do not use anti-coagulant.

(ii) Allow the blood to clot at room temperature (or 30°C), break the clot to release serum, and chill at 4°C for several hours. The cold temperature shrinks the clot and allows the antiserum to be transferred by pipette to a new container, preferably a centrifuge tube.

(iii) Clarify the antisera by centrifugation, for example 5000–10 000 *g* for 10–20 min at 4°C (or room temperature) and then further purify or store frozen until required (see ref. 2).

3.5 Maintenance of antibody production

Animals may be boosted by regular injection of IFN plus IFA (i.m.) at 4–6 week intervals to increase antibody production. A plateau of antibody production will eventually be reached, and then further boosting will only maintain production at this level. Boosting injections in mice for McAb production are either intravenous (i.v.) or intraperitoneal (i.p.) without IFA 3 days before fusion with meyloma cells is carried out. The reader is referred to Section 7 for more details and other aspects of hybridoma technology.

4. FOLLOWING ANTIBODY PRODUCTION

4.1 Antibody assays

Specific antibody concentrations in antisera are normally determined in assays designed to measure antibody binding to antigen. Such assays can be performed by a variety of means and methods for anti-IFN immunoglobulin not dependent upon the biological activity of IFN. Nevertheless, one of the easiest, and routine, methods for assessing anti-IFN concentration is to assess the capacity of antisera or purified immunoglobulin (Ig) preparations containing anti-IFN to neutralize the biological activity, for example the anti-viral activity, of the IFN type used as immunogen. In effect, antibody concentration is turned into its potency to neutralize anti-viral activity, and the neutralizing titre is assessed in anti-viral assays in a similar fashion to those of IFN titres (see Chapter 9). The neutralizing titre will be the dilution of antiserum or Ig preparation which reduces IFN anti-viral activity by an arbitrary, but fixed, amount to produce an observable end-point. For convenience, a 10-fold (1 $\log_{10}$) reduction in IFN anti-viral activity is often chosen because in this way a titre of 10 observed or laboratory units (LU)/ml of an IFN preparation reduced to a titre of 1 LU/ml by the antiserum produces the same end-point as that chosen for IFN assays, that is 50% cytopathic effect (CPE), or 0.5 $\log_{10}$ reduction in virus yield (3). Units of IFN activity are not converted into IU in this case because of attendant theoretical difficulties with calculations [for explanation see Kawade (3)]. IFN antibody neutralization assays may be performed by the methods set out in Chapter 9, Section 4.1, with the exceptions that it is the antiserum or Ig preparation which is serially diluted and that the assays are preceded by an incubation stage where antibody serial dilutions are mixed with a fixed amount of IFN. This is illustrated by the following procedure based on the CPE reduction (CPER) method.

4.1.1 *IFN antibody neutralization assay*

(i) Add cell growth medium to wells of a microtitre plate (0.055 ml/well) and then to the top row (row A) add the antisera or Ig preparation (0.025 ml/well) to be diluted.

(ii) Using a multichannel micropipette mix the contents of the first well and transfer

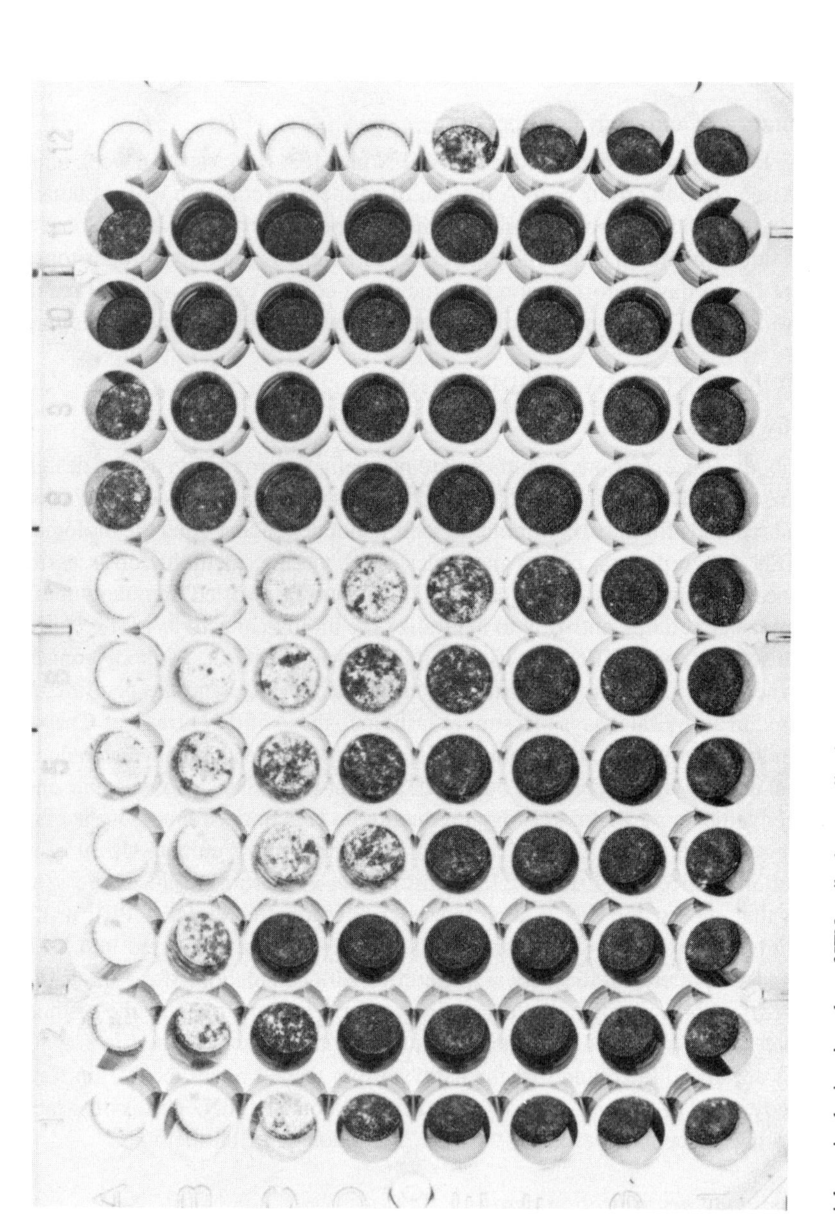

Figure 1. Gentian violet-stained microtitre plate of IFN antibody neutralization assays carried out by the CPE reduction method. Twelve different anti-IFN-γ Ig preparations, including McAbs, are depicted showing their relative neutralizing capacities against IFN-γ, fixed at 10 LU/ml. Neutralizing titres are equivalent to the reciprocals of the dilutions in the wells where CPE is 50%, i.e. at 1 LU/ml.

0.025 ml to the next row (row B). Change the tips and repeat the dilution procedure through to the last row (row H): this generates a 0.5 $\log_{10}$ series.

(iii) Add the fixed amount of IFN (0.025 ml/well). This IFN must be pre-diluted to an extent such that it will give the desired result following antibody neutralization, and this requires some experience of both the sensitivity and precision of the assay by the operator. As a rough guide, the aim in this assay should be to add 0.025 ml containing 80 LU/ml to each well to give approximately 25 LU/ml in the antibody dilution–IFN mixture.

(iv) Incubate the microtitre plate at 37°C for 1 h. Seed the wells with cells, for example Hep2/c, A549, appropriate to the anti-viral assay, 0.1 ml/well, 1×10^5 cells/well. The IFN is diluted in this step to roughly 10 LU/ml and the initial antibody dilution is 1:10. This IFN level should give virtually complete protection to cells against virus challenge.

(v) Proceed as outlined for the CPER method (Chapter 9, Section 4.1). The stained assay plates (see *Figure 1*) may be read by eye or data plotted on a semi-log plot following dye extraction and measurement of optical densities. The antibody neutralizing titre is the dilution of antiserum or Ig preparation at which the CPE is 50% (equivalent to 1 LU/ml) or at the point of median absorbance on the scale of dye uptake measurements that range between those of the cell controls and the virus controls.

The above method represents a rather rough and ready approach for estimating neutralizing capacities of antisera and Ig preparations. Assay sensitivity is an important determining factor in estimating neutralizing titres since no correction of IFN LU to IU is made and no internationally recognized anti-IFN antisera standards have been developed. The USA NIH does however hold a number of anti-IFN antisera of assigned neutralizing titre which can be used to calibrate assays in some cases (4).

An alternative formula for calculating neutralizing capacity of an antiserum in terms of the units of IFN neutralized per ml of antiserum is as follows:

$$\text{Units IFN neutralized/ml} = \frac{\text{IFN concentration neutralized} \times \text{Final IFN volume (ml)} \times \text{Antiserum dilution (reciprocal required)}}{\text{Volume of antiserum used}}$$

Example: 0.1 ml of antiserum + 0.1 ml of IFN at 10 LU/ml after mixing
1:2000 dilution of the 0.1 ml antiserum reduced the 10 LU/ml to 1 LU/ml
Final volume of IFN = 0.1 ml + 0.1 ml = 0.2 ml

$$\text{Units IFN neutralized/ml antiserum} = \frac{(10-1) \times 0.2 \times 2000}{0.1} = 36\,000$$

Note that the same value is obtained by finding the product of LU IFN neutralized/ml and the final dilution of antiserum used. In the example, 9 LU/ml were neutralized by a final antiserum dilution of 1:4000, thus $9 \times 4000 = 36\,000$. Note also that values of neutralizing capacity calculated this way are 9-fold higher than as estimated in (v) above.

4.2 Antibody levels following immunization

Neutralizing IFN-type specific antibodies should start increasing following the second booster injection. They reach a peak level some 2 weeks following this boost and then

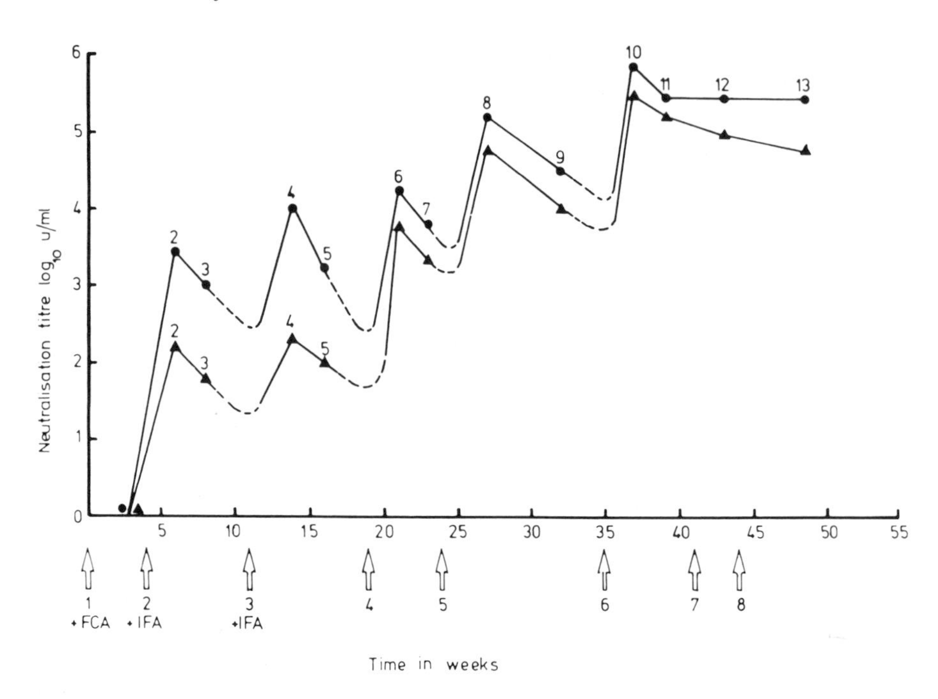

Figure 2. The production of anti-recombinant DNA HuIFN-γ (*E. coli*) in a sheep (H53) followed by IFN antibody neutralization tests. ●- - -● versus recombinant DNA HuIFN (*E. coli*); ▲- - -▲ versus natural HuIFN-γ (human leukocyte).

decline slowly. A further (3rd) injection several weeks later then starts off another round of antibody production with peak levels, often greater than the first, following around 14 days after injection of IFN. The animal may not be boosted again until antibody production levels have once more declined. A classic sawtooth profile of rises and falls of antibody production should eventually be generated (*Figure 2*) where neutralizing antibody titres may approach 10^6 units/ml of antiserum. Maximum neutralizing antibody concentrations rarely go above 10^6 units/ml of antiserum, and in many cases much lower maximum levels, 10^3-10^5 units/ml antiserum, are only ever achieved. For obvious reasons, the most potent anti-IFN antisera are the most desirable, but any antiserum of neutralizing capacity greater than 10^3 units/ml antiserum will have its uses.

5. SPECIFICITY OF ANTIBODIES TO IFN

5.1 Introduction

This section will deal only with the specificity of polyclonal anti-IFN antisera produced by animals. The specificity of McAbs against IFN will be examined in Section 7. Generally speaking, the three types of IFN, -α, -β and -γ are antigenically distinct so that antibodies raised against one of them will not cross-react with the other two. For example, rabbit anti-IFN-β will not cross-react with or neutralize IFN-α or IFN-γ. However, in some cases, for example where leukocyte of lymphoblastoid IFN have been used as immunogens, antisera may contain antibodies cross-reacting with and neutralizing the minor IFN components, IFN-β, in both leukocyte and lymphoblastoid IFN (*Table 1*, sheep 51). Antisera against recombinant DNA IFN-α subtypes may be

Table 1. Neutralizing titres of sheep polyclonal HuIFN type-specific antisera.

Sheep no.	*Immunogen*	*HuIFNs (log_{10} neutralizing titre/ml antiserum)*							
		αD	$\alpha 2$	αLe	αLy	β	$r[Ser^{17}]\beta$	γ	$r\gamma$(*E. coli*)
49	$\alpha 2$	4.5	6.0	3.5	4.0	<1	<1	<1	<1
51	αLy	5.0	6.0	5.0	5.0	2.7	nd	<1	nd
52	αD	4.7	3.3	3.0	2.7	<1	nd	<1	nd
53	rγ(*E. coli*)	<1	<1	<1	<1	<1	<1	4.9	6.0
60	$r[Ser^{17}]\beta$	<1	<1	<1	<1	3.2	3.7	<1	<1

readily raised in animals; these contain antibodies which do not cross-react with IFN-β or -γ, but which do cross-react to varying degrees with other, different rDNA IFN-α subtypes. For example, sheep anti-recombinant DNA IFN-α_2 is highly neutralizing for recombinant DNA IFN-α_2 but significantly less so for recombinant DNA IFN-α_D, which is 83% homologous to IFN-α_2 (*Table 1*). The converse is also true (*Table 1*). Nevertheless, no polyclonal anti-IFN-α subtype antiserum will recognize only that subtype used as immunogen, because the diversity and multitude of antibodies produced against all available epitopes on the IFN-α molecule in these circumstances is sufficient to effect some recognition of closely related (structurally) IFN-α subtype molecules and possibly also those of other animal species. A similar situation will often be found for the *Escherichia coli*-derived, non-glycosylated recombinant DNA IFN counterparts of natural IFN molecules, for example IFN-γ, where the former may be considered to be highly structurally related, but antigenically distinct, from the latter. For example, sheep anti-recombinant DNA IFN-γ antiserum neutralizes the non-glycosylated (homologous) DNA IFN-γ to a greater extent than the natural IFN-γ derived from human lymphocytes (*Figure 2* and *Table 1*, sheep 53).

5.2 IFN type identification

It follows from the above section that IFN type-specific polyclonal antisera may be used to identify the type(s) of IFN in samples of unknown IFN constitution by using IFN antibody neutralization assays. It is only necessary to test the IFN sample, suitably diluted, against potent antisera specific for each of the three IFN types in order to ascertain which is the major IFN type present. Such 'identity' assays are however more complicated if two types of IFN are present in the sample [e.g. supernatants from human PBL of immune donors stimulated with influenza A virus contain both IFN-α and IFN-γ (5)]. In these cases, neutralization will appear to be partial if the two IFN types are present in roughly equal proportions. If two IFN types are suspected, judicious combination of two different IFN type-specific antisera will often produce complete neutralization and confirmation of the identity of the two IFN types present. Alternatively, a relatively high concentration of one IFN type-specific antiserum may be used in order to neutralize all the anti-viral activity of the IFN type recognized by this antiserum and the residual IFN activity due to the other type present titrated in an IFN anti-viral assay. By applying the relevant antiserum for the second IFN type, the relative proportions of each IFN type present may be arrived at. For example:

	No antiserum	Anti-IFN-α	Anti-IFN-β
IFN titre (U/ml)	3200	630	2600

It should be noted that such an interpretation of the relative proportions of two IFN types will only be valid in anti-viral assays where the anti-viral activities of the IFN types is additive, for example IFN-α + IFN-β. Expect complications to arise if one IFN type potentiates the activity of another type (synergistic action) in the anti-viral assays employed for assessing either neutralization or estimates of residual IFN activity following antiserum treatment. For instance IFN-α and IFN-γ in combination frequently show synergistic action (6).

6. ANTIBODY–ANTIGEN BINDING TESTS

The major way of detecting antibodies is by binding to antigen. Most methods for doing this rely on having ample supplies of purified antigen, which in the case of IFN is not always possible although the supply situation has improved, largely due to the availability of recombinant DNA IFN, in recent years. Virtually all antibody–antigen binding tests are adaptations of the sandwich immunoassay method. In this case, the IFN type-specific antibody is sandwiched between antigen (IFN), usually attached to a solid phase, and either a second antibody or a protein such as protein A from *Staphylococcus aureus* which recognizes certain IgG isotypes. The second antibody or protein A will either be radiolabelled with [^{125}I]iodine or cross-linked to an enzyme able to effect a colour change reaction with appropriate substrates in order that the IFN type-specific antibody may be quantified. Such immunoassays will detect both neutralizing and non-neutralizing antibodies against IFN (non-neutralizing antibodies are usually found in anti-IFN antisera in the initial few weeks following immunization; synthetic polypeptides homologous to various regions of IFN molecules may also induce only non-neutralizing antibodies).

6.1 Enzyme linked immunosorbent assay

One of the most commonly used methods for detecting and quantifying antibody–antigen binding is the enzyme-linked immunosorbent assay (ELISA). An example of an ELISA is outlined below.

(i) Dilute highly purified IFN (>95% IFN protein) with phosphate-buffered saline (PBS) to 10 μg/ml and coat overnight on U-bottomed polyvinyl chloride microtitre plates (any plastic micro-ELISA plate will do), 0.05 ml per well, after prior treatment of the wells with 0.2% glutaraldehyde for 2–3 h.

(ii) Remove excess IFN by flicking out, wash the wells three times with 0.5% Tween 20–PBS, and add anti-IFN antisera dilutions (0.05 ml/well).

(iii) Following incubation at 37°C for 1 h in a humidified box, remove excess (unbound) antibodies. Wash the wells extensively with 0.05% Tween 20–PBS, and add a 1:500 dilution (or as appropriate) of anti-rabbit (or whatever animal species the anti-IFN antiserum was raised in) immunoglobulin linked to horseradish peroxidase (HRP) (0.05 ml in 2% heat-inactivated calf serum–PBS per well).

(iv) After a further 1 h incubation at 37°C, wash the wells firstly with 0.05% Tween 20–PBS and secondly with 0.1 M sodium citrate–phosphate buffer, pH 5.0 and allow the colour to develop (yellow) in the wells in the dark by the addition of freshly prepared substrate, *o*-phenylenediamine (1 mg/ml) in 0.1 M citrate-phosphate buffer containing 0.006% H_2O_2 (0.05 ml/well).

(v) Stop the reaction after 15–20 min by the addition of 1 M H_2SO_4, 0.05 ml/well, and measure the optical density at 450 nm in a Titertek Multiscan or other micro-ELISA plate reader. The titre of anti-IFN antiserum against IFN is taken from the semi-log plot of antiserum dilutions against OD_{450} readings as the reciprocal of the dilution at 50% maximum absorbance.

The *o*-phenylenediamine in step (iv) may be substituted by 5-amino salicylic acid (0.8 mg/ml in 0.05 M potassium phosphate buffer, pH 6.0 containing 0.001 M EDTA and 0.006% H_2O_2). The colour in this case will be brown, the reaction is stopped by addition of 0.3 M NaOH, and optical densities are measured at 492 nm. Note that both these substrates may be carcinogenic and should be handled with care.

The sensitivity of this ELISA may be increased by replacing step (iii) above by the following. After removal of anti-IFN antibodies and washing, add a 1:500 dilution of biotinylated anti-rabbit (e.g. Amersham International) (or animal) IgG at 0.05 ml/well. After 1 h incubation at 37°C, remove the biotinylated antibodies, wash the wells extensively with 0.05% Tween 20–PBS, and add a 1:500 dilution of streptavidin–biotinylated HRP complex (e.g. Amersham International) (0.05 ml in 2% heat-inactivated calf serum–PBS per well). Carry out the incubation for 30 min at 37°C before proceeding as in step (iv).

The final stage enzyme-linked reagents may also be replaced by ^{125}I-labelled reagents, but it will then be necessary to cut out the individual wells of the microtitre plate for counting in a γ radiation counter. Polyvinyl chloride plates are recommended for this purpose as they can be sliced with a hot wire.

The major drawback of the solid-phase ELISA is the very real possibility of non-specific reactions. With relatively low dilutions (e.g. 1:100) of antiserum, it is difficult to wash away all of the non-specifically bound antibodies, resulting in the occurrence of a positive colour reaction the well(s) at that antiserum dilution. It is clear therefore that control serum dilutions should be included in these ELISA; best is the pre-bleed serum of the animal subsequently immunized to produce antiserum, but control serum of the same animal species, and preferably the same breeds, may be suitable. Optical density measurements for control serum dilutions should be deducted from the optical density measurements for the equivalent antiserum dilutions when estimating antibody titres. Other controls, such as omission of the first and second antiserum should also be included in ELISA.

7. PRODUCTION AND SCREENING FOR MONOCLONAL ANTIBODIES AGAINST IFN

IFN-type specific polyclonal antisera are variable in both quantity and quality. Specificity has improved with the availability of highly purified IFN preparations for use as immunogens, but antibodies may still be raised against protein contaminants. The latter problem may be overcome by using sub-molecular synthetic polypeptides, homologous in amino acid sequences to particular parts or regions of IFN molecules, as immunogens. Synthetic polypeptides also confer the advantage of producing antibodies of predetermined specificity. For example, if an IFN N-terminal polypeptide is used, the antibodies developed against it will naturally react only with the N-terminal region of the native IFN molecule. This is all very well in theory, but in practice it is, in the author's experience, quite difficult to obtain a good anti-peptide antiserum which will also react well with the native IFN molecules (7).

A more profitable route of obtaining highly specific antibodies, without the attendant presence of antibodies against protein contaminants, is through hybridoma technology and the production of McAbs. It takes time and effort to establish hybridoma methodology, but the resultant anti-IFN McAbs, which are of invariant quality and which may be mass-produced, are generally worth it. Each McAb will bind to a single antigenic determinant or epitope on the IFN molecule.

To illustrate the method for the production of anti-IFN McAb producing hybridomas, only one version is presented. The reader is invited to look elsewhere for other versions and extra information (see ref. 8).

7.1 Cell fusion and establishing hybridomas

(i) Immunize BALB/c mice with IFN as outlined in Section 3 and follow the development of anti-IFN antibodies either by ELISA or by neutralization assays. The antibody binding or neutralizing titre should be relatively high (10^3-10^4 units/ml antiserum) before a fusion is contemplated.

(ii) Following the time at which high (maximum) antibody titres are found, allow sufficient time (3 weeks or more) for antibody titres to be falling off again before injecting IFN (in buffered saline) i.v. to re-boost antibody production.

(iii) Three days after the intravenous boost, remove the spleen from the selected mouse under sterile conditions. Wash the spleen in PBS and make a few cross-cuts with sharp scissors. Transfer the spleen to a glass homogenizer with a loose-fitting Teflon pestle, add 3–4 ml of serum-free RPMI medium, and release the spleen cells by pushing and pulling the pestle in a twisting manner. Remove the pestle and let lumps and debris settle for 1 min before transferring the cell suspension to a 50 ml conical-bottomed centrifuge tube. Add a further 15 ml of RPMI medium and spin down the cells at 1000 r.p.m. for 10 min at room temperature.

(iv) Transfer $4-5 \times 10^7$ exponentially growing NSO mouse myeloma cells to a 50 ml conical-bottomed centrifuge tube and spin down at 1000 r.p.m. for 10 min at room temprature.

(v) Decant the supernatant from the spleen cells and resuspend them in 20 ml of serum-free RPMI medium. Make a cell count; a mouse spleen should contain $50-200 \times 10^6$ cells. Also decant the growth medium from NSO cells and resuspend them in 20 ml of RPMI medium.

(vi) Mix NSO and spleen cells together at a ratio of 1:5 (the ratio can be varied). Spin down and decant the supernatant leaving as little as possible covering the cells.

(vii) Disperse the cells around the sides of the conical bottom of the centrifuge tube and add 2 ml of 50% polyethylene glycol (PEG; mol. wt 1540) over 1 min with gentle shaking. Leave the fusion mixture a further 1 min at room temperature and then dilute out the PEG with 18 ml of serum-free RPMI medium, adding this slowly at first, and mixing by gentle shaking or swirling.

(viii) Spin down the cells at 1000 r.p.m. for 10 min at room temperature. Decant the supernatant and gently resuspend the cells in growth medium, RPMI plus 20% fetal calf serum or 15% CLEX (Dextran Products) and hypoxanthine–aminopterin–thymidine (HAT), to a volume appropriate for the number of wells to be filled. [For the preparation of HAT make a 100 × stock solutions of hypoxanthine (136 mg/100 ml) and thymidine (75 mg/100 ml) using 1 M NaOH to dissolve the hypoxanthine. Make a 100 × stock solution of aminopterin (1.8 mg/100 ml) and use 1 M NaOH to dissolve (ref. 8).] For 24-well cluster plates, add 1 ml of cell suspension (up to 2×10^6 cells/ml)/well. Incubate the cluster plate at 37°C in a humidified, CO_2 incubator.

(ix) After 6–7 days, add a further 1 ml of HAT containing growth medium, then feed every 2–3 days by removing 1 ml of medium and replacing with 1 ml of HAT medium.

(x) After 10–14 days, when all parent unfused NSO cells have died, change to HT medium (no aminopterin) and examine microscopically daily to monitor hybridoma colony growth.

(xi) Commence screening individual wells containing colonies when the medium becomes persistently yellow.

7.2 Screening procedures for hybridomas producing anti-IFN McAbs

There are several screening methods available for the detection of anti-IFN McAbs present in hybridoma supernatants. For example, the solid-phase ELISA described in Section 6 may be readily adapted to screen hybridoma supernatants. Here antiserum dilutions are replaced by hybridoma supernatants from individual wells and bound anti-IFN McAbs (positive wells) detected by anti-mouse immunoglobulin conjugated to HRP (or β-galactosidase, or alkaline phosphatase) following reaction with substrate. However, in the author's experience, this screening method is plagued by non-specific reactions leading to numerous false-positives being found. The ELISA method does not miss true positive wells, but more reliable screening methods, one of which is described below, are available.

7.2.1 *Immunoprecipitation of radiolabelled antigen*

(i) Mix 0.025 ml of each supernatant to be tested with 0.025 ml of ^{125}I-labelled purified IFN ($4-10 \times 10^3$ c.p.m.) in Eppendorf microcentrifuge tubes and incubate for 2 h at 37°C. [IFN should be labelled with ^{125}I using either the Bolton–Hunter reagent (Amersham) or by Enzymo bead Reagent (Biorad) since chloramine T is strongly oxidative and denatures IFN, especially IFN-γ. Most McAbs against IFN appear to recognize only undenatured, biologically active IFN molecules, and so denaturation is to be avoided.]

(ii) Add 0.02 ml of normal mouse serum diluted 1:10 in PBS and 0.1 ml of sheep anti-mouse Ig or $F(ab')_2$ diluted 1:5 in PBS and incubate for 30 min at 37°C and 1–2 h at 4°C.

(iii) Spin down the precipitate in a microcentifuge, remove the supernatant (it is normally unnecessary to wash the pellet) and count the radioactivity per tube in a γ radiation counter.

Positive, anti-IFN McAb-containing wells, should have a signal-to-noise (c.p.m. bound by supernatant: c.p.m. bound by medium or irrelevant supernatant) ratio of 2 or greater. Signal-to-noise ratios from positive wells are frequently greater than 10, and this solution phase immunoprecipitation technique is virtually foolproof. It will detect both non-neutralizing and neutralizing McAbs.

7.2.2 *Immunoprecipitation of unlabelled antigen*

(i) Mix 0.025 ml of each supernatant to be tested with 0.025 ml of an IFN preparation (IFN need not be pure) containing 10^3-10^4 IU/ml in Eppendorf microcentrifuge tubes and incubate for 2 h at 37°C.

(ii) Add 0.02 ml of normal mouse serum diluted 1:10 in PBS and 0.1 ml of sheep anti-mouse Ig or $F(ab')_2$ diluted 1:5 in PBS and incubate for 30 min at 37°C and 1–2 h at 4°C.

(iii) Spin down the precipitate in a microcentrifuge, remove the supernatant and wash the pellet twice with PBS (0.2 ml/tube) followed by re-centrifugation.

(iv) Dissolve the pellet in 0.05 ml of 0.1 M glycine-HCl buffer, pH 2.2 for IFN-α and IFN-β, or 0.05 ml of 0.1 M ammonium hydroxide solution, pH 10.7–10.8 for IFN-γ.

(v) Titrate the IFN released in conventional IFN anti-viral assays. Wells containing IFN-specific McAbs should give IFN titres of greater than 10% of the initial IFN titre.

This method has the decided disadvantage of being long-winded, that is average 3 days. It will detect both non-neutralizing and neutralizing McAbs, but in the case of the latter some neutralization of released IFN may occur as the pH of the diluted IFN–McAb mixture readjusts to neutral in the anti-viral assay. This leads to diminished, lower than expected, IFN titres.

7.2.3 *Neutralization assays*

These are not recommended for initial screening because of the length of time required for IFN antibody neutralization assays to be completed, that is 3 days. Only strongly neutralizing McAbs against IFN will be reliably detected by this method since McAb concentrations are low in hybridoma supernatants. It should also be pointed out that McAbs against IFN-α may only neutralize the anti-viral activity of certian IFN-α subtypes (*Table 2*), and may show no effect against IFN-α subtype mixtures such as leukocyte IFN. McAbs against IFN-α which bind to the whole spectrum of IFN-α sub-

Table 2. Neutralizing titres of ascitic fluids containing monoclonal antibodies against (a) human recombinant DNA IFN-α_2, (b) human IFN-γ and (c) human recombinant DNA IFN-γ (*E. coli*).

Monoclonal antibody	*Maximum neutralization titres* (log_{10})*/ml against*						
	αD	$\alpha 2$	αLe	αLy	β	γ	$r\gamma$(*E. coli*)
(a) MT4/E4	2.0	5.5	<1	<1	<1	<1	<1
MT3/B4	<1.0	5.5	<1	<1	<1	<1	<1
(b) 4DC8F10	<1	<1	<1	<1	<1	3.5	2.0
4SB3	<1	<1	<1	<1	<1	2.0	2.0
(c) NIB42	<1	<1	<1	<1	<1	3.7	5.5
NIB45	<1	<1	<1	<1	<1	<1.0	2.0

Many McAbs show much greater specificity to the IFN used as immunogen than to related subtypes or even to the same type where post-translational modification is different. See refs 9 and 10.

types have rarely been found. Thus, the neutralization assay is probably best used as a back-up method during the initial screening process, allowing the investigator to decide which McAbs are potentially neutralizing against IFN (see Section 4.2 for details).

7.3 Cloning of positive hybridomas

Once positive wells have been identified, it is essential to clone the cells present in these wells in order to select only those hybridomas secreting anti-IFN McAbs. Cloning is best achieved by a soft-agar technique, as outlined below.

(i) Make up a 0.5% agar–growth medium mixture, for example 90 ml of RPMI growth medium plus 10 ml of molten 5% Difco agar, 40–50°C and add 12.5 ml to each 9 cm plastic Petri dish to be used. Allow the agar–growth medium to cool and set at room temperature.
(ii) Make three serial 10-fold dilutions of the cells from positive wells in RPMI growth medium.
(iii) Mix 1 ml of each serial dilution with 1 ml of agar–growth medium and transfer to a Petri dish containing an agar base.
(iv) Leave to set and then incubate at 37°C for 7–10 days in a humidified CO_2 incubator, or until the colonies become large enough to pick (30–100 cells/colony).
(v) Using a Pasteur pipette, pick the colonies out of the agar (view the colonies with an inverted microscope) and transfer each colony to a separate well of a plastic cluster plate containing RPMI growth medium (0.5–1 ml).
(vi) Grow clones and feed the wells with fresh growth medium when appropriate, for example every 2–4 days as the medium turns yellow.
(vii) When cells are 75–100% confluent, test for McAb secretion as previously described and expand the positive wells (clones).
(viii) Choose two or three positive clones from each of the originally positive wells, grow up to mass culture levels, and then freeze sufficient numbers of ampoules for re-growing cells at later times. Storage should be in liquid nitrogen.
(ix) The 'best' clones may also be injected into pristane-treated BALB/c mice (8), when a sufficient number of hybridoma cells are available ($5-10 \times 10^6$ cells/mouse), to produce ascites. The ascitic fluid is tapped from the abdomen at 5–10 days (or when the abdomen shows signs of enlarging) following injection.

Ascitic fluid produced by mice injected with anti-IFN McAb-producing hybridoma cells will contain on average 1000-fold greater amounts of McAb than the hybridoma cell medium supernatant. Therefore, it should be highly diluted for quantification and neutralization tests. A good neutralizing McAb should neutralize IFN at a $1{:}10^5$ dilution of ascitic fluid. However, less neutralizing McAbs will often be found, for which the degree of neutralization of IFN is independent of antibody concentration, unlike the situation with polyclonal neutralizing antibodies, above certain antibody levels (11). This in effect means that with some anti-IFN McAbs it will be impossible to neutralize completely the anti-viral effect of IFN, since the IFN–McAb complexes themselves retain some activity (*Table* 2).

8. IMMUNOASSAYS

Anti-viral and other biological assays for IFN are relatively lengthy in terms of the time necessary to produce results. They also require that a dose–response curve for each IFN sample be produced which inevitably involves lots of dilutions. Immunoassays, not based on the biological activity of IFN, offer a rapid and highly reproducible way of quantifying IFN, but there are some reservations concerning the validity of the results they produce, which will be discussed later. Immunoassays can be performed in a variety of ways, but generally require either pure IFN or antibody against pure IFN or preferably McAb against IFN. In the following section, only immunoassays involving anti-IFN antibodies and McAbs will be discussed and these will be illustrated by paricular examples of the immunoradiometric assay (IRMA) and the ELISA.

8.1 **The immunoradiometric assay**

The method outlined below is essentially that described by Secher (12). IFN is estimated by firstly attaching it to an immobilized antibody, which can be polyclonal or monoclonal, and secondly detecting it through binding of an [^{125}I]McAb to another antigenic site on the IFN molecules (*Figure 3*). The amount of [^{125}I]McAb bound is proportionate to the amount of IFN if the [^{125}I]McAb is in excess. Some purification of anti-IFN antibodies is required to produce the necessary reagents and this may be readily achieved.

(i) Add to high-titred polyclonal IFN type-specific antiserum or ascitic fluid containing IFN type-specific McAb solid ammonium sulphate to 35% saturation, roughly 2 g per 10 ml, and precipitate Ig in the cold (4°C).
(ii) Spin down the Ig precipitate at 5000 r.p.m. for 10 min, discard the supernatant, and re-dissolve the precipitate in 1/5th original volume of PBS.
(iii) Dialyse overnight against PBS and then fractionate by chromatography on Ultrogel ACA54 (or Sephacryl 200) using 0.1 M phosphate buffer, pH 7.2, containing 0.5 M NaCl as the eluting buffer. Analyse fractions for anti-IFN Ig and pool the peak fractions.

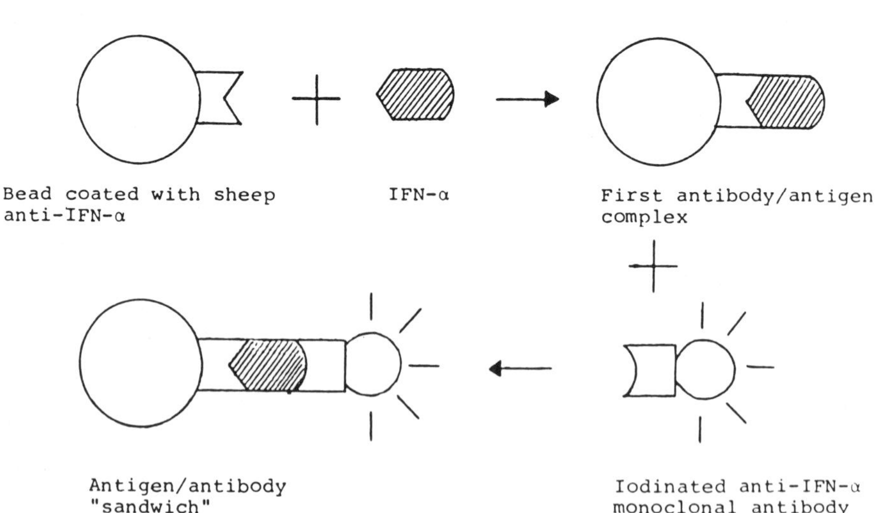

Figure 3. The basics of the immunoradiometric assay (IRMA).

(iv) Further purify McAbs by either protein A – Sepharose (or agarose) chromatography (13), by DEAE – Affigel Blue chromatography (14) or by h.p.l.c. (15).

(v) Dialyse pooled polyclonal anti-IFN-Ig and McAb-containing peak fractions against PBS, estimate protein by the Lowry method (16), aliquot and store frozen at −20°C.

(vi) Radioiodinate the anti-IFN McAb by the chloramine T method (17) as follows.

  (a) To 10 μg of McAb Ig preparation add 10 μl of 0.1 M sodium phosphate buffer, pH 7.4, and 10 μl (100 μCi) of [^{125}I]Na in the same buffer.

  (b) Add 10 μl of freshly prepared chloramine T (5 mg/ml distilled water), mix and agitate the tube contents for 45 sec.

  (c) Stop the reaction by the addition of 50 μl of L-tyrosine (0.4 mg/ml in sodium phosphate buffer).

  (d) Pass the iodination mixture through a disposable Dowex or Sephadex chromatography column, and elute the [^{125}I]McAb with 2 ml of PBS containing bovine serum albumin (BSA) (2 mg/ml).

  (e) Store the [^{125}I]McAb at 4°C with a drop of 10% sodium azide as preservative.

Note that there are many variations of the chloramine T radioiodination method (18) and investigators may have to play around with conditions to obtain optimal labelling. Over-exposure of some McAbs to chloramine T, which is highly oxidizing, can destroy most of their antigen-binding activity and render them useless for IRMA. It pays therefore to keep exposure times as short as possible.

(vii) Prepare immobilized antibody as follows: dilute polyclonal anti-IFN Ig preparation or anti-IFN McAb to 200 μg protein/ml in PBS. [If an anti-IFN McAb is used, it must have been previously demonstrated to bind to a different antigenic determinant to that of the [^{125}I]McAb (19), although this may not be a problem where the IFN to be estimated is a natural dimer, e.g. IFN-γ (20).] Add 100 or so etched polystyrene balls (6.5 mm diameter) to 20 ml of diluted antibody and leave at 4°C overnight. Remove anti-IFN and wash the beads five times with 0.1% BSA – PBS and store in this buffer at 4°C.

(viii) Block a few hundred Luckham LP4 (or equivalent) plastic tubes with 0.5% BSA – PBS by completely filling the tubes and standing them overnight at 4°C.

(ix) Remove by aspiration all of the 0.5% BSA – PBS from the LP4 tubes required for the assay (the others may be stored full at 4°C until required).

(x) Make serial dilutions of an IFN reference preparation (either an international reference preparation or a laboratory reference preparation) in culture medium or 0.5% BSA – PBS sufficient to generate a good calibration curve. Ideally, there should be 7 – 10 points on the curve spanning the entire sensitivity range of the assay. Calibration can be in either anti-viral or international units or, if the IFN concentration in weight is known in nanograms or picograms.

(xi) Pipette 200 μl of reference preparation dilutions or unknown IFN samples in duplicate into LP4 tubes, and add one bead, first blotted to dryness with filter or tissue paper, to each tube. The IFN dilution or sample should just cover the top of the bead. Remember to include negative controls, for example 200 μl of 0.5% BSA – PBS.

(xii) Incubate for at least 4 h at 4°C and then remove the liquid from each tube by

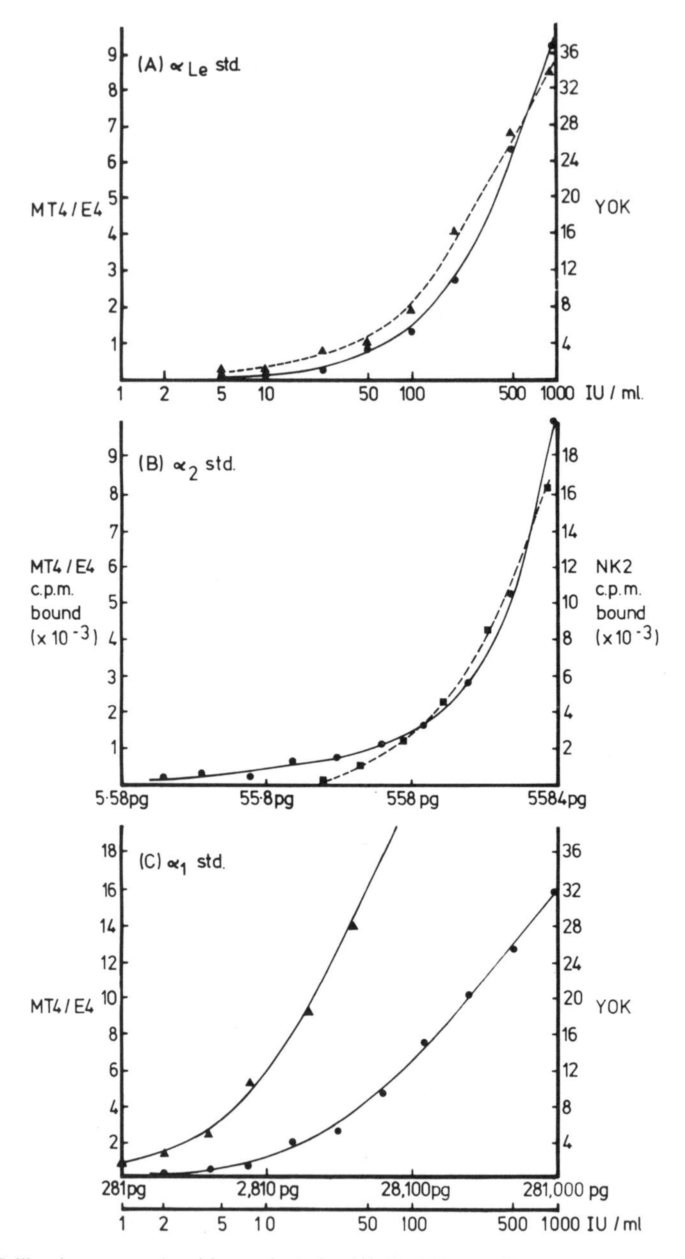

Figure 4. Calibration curves (semi-$\log_{10}$ plots) for **(A)** HuIFN-α_{Le}, **(B)** HuIFN-α_2 and **(C)** HuIFN-α_1 reference preparations as obtained by IRMA. ●- - -● [^{125}I]MT4/E4 McAb in **A**, **B**, **C**, ▲- - -▲ [^{125}I]YOK McAb in **A** and **C**, ■---■ [^{125}I]NK2 McAb in **B**. Note that in **C**, because MT4/E4 binds to HuIFN-α_1 more weakly than YOK, there is a marked difference in sensitivity ranges of the two IRMA.

aspiration and wash the beads in the tubes three times with 0.5% BSA–PBS containing 0.5% Tween 20 (e.g. 3 × 1 ml). Ensure all excess washing medium is completely removed by aspiration.

(xiii) Pipette 200 μl of [^{125}I]McAb, previously diluted in 0.5% BSA–PBS to contain

$1-2 \times 10^5$ c.p.m./0.2 ml, into each tube and incubate overnight at 4°C.

(xiv) The next day, remove [^{125}I]McAb solution by aspiration and repeat the washing steps.

(xv) Count the beads in a γ radiation counter.

(xvi) The data can be plotted as c.p.m. bound (minus negative control; non-specific binding) versus IFN dilution (titre) in $\log_{10}-\log_{10}$ or semi-$\log_{10}$ plots to give the calibration curve. Titres of unknown IFN samples may then be simply interpolated from the calibration curve (*Figure 4*).

Commercial versions of this IRMA are available as kits from Celltech, Boots-Celltech (HuIFN-α and -γ) and Centocor (HuIFN-γ only). These and other laboratory versions have sensitivities equivalent to or better than the conventional anti-viral assays, that is they will measure down to 1 IU/ml at least. Top of the range is around 1000 IU/ml, except for the very sensitive Centocor kit for IFN-γ determinations, which is based on two McAbs (20), where the effective range is from 0.1 to 100 IU/ml. All IRMA assays suffer from 'endedness'; in other words they are relatively less accurate at the bottom and top ends of their sensitivity ranges where the intra-assay values for percentage coefficient of variation (%CV) approach, or are greater than, 10%. In the middle-range, where the standard curve is linear, %CV is normally much below 10%. Inter-assay precision is also very good.

8.3 The ELISA method

ELISA do not depend on ^{125}I-labelled antibodies, which decay rapidly in terms of radioactivity, and thus their increased shelf-life offers some advantage over IRMA. However, making them work as reproducibly as IRMA is not always easy. Their detection limits often approach, but never better, those obtained in IRMA. Obviously, there is no single ELISA method; there are literally dozens of combinations of reagents and conditions which may be used. Most approaches rest upon the antibody capture technique used in IRMA where, in this case, the first antibody is attached to the wells of micro-ELISA plates. Antigen is incubated in these antibody-coated wells and then, following washing steps, detected by a second anti-antigen Ig linked to an enzyme producing a colour reaction with an appropriate substrate (*Figure 5*). Further layers on the sandwich may be applied to increase assay sensitivity, for example antigen is detected by unlabelled mouse anti-antigen McAb which is itself detected by anti-mouse Ig linked to enzyme. The method outlined below is an example of the antibody layering technique which, although it has several washing steps, does not require highly purified antibodies and which, in the author's experience, works adequately and reproducibly for IFN-α and IFN-γ estimations.

(i) Purify relevant polyclonal anti-IFN antiserum by ammonium sulphate precipitation and gel filtration.

(ii) Dilute purified anti-IFN Ig 1:100 in PBS and coat onto the wells of microtitre ELISA plates, 0.05 ml per well, for 2 h at 37°C in a humidified box.

(iii) Block the remaining sites in the wells with 0.5% BSA−PBS, 0.15 ml per well, for 16 h (overnight) at 4°C.

(iv) Remove excess Ig and blocking buffer by flicking out and wash the wells a few times with 0.05% Tween 20−PBS (0.2 ml/well). Following the last wash, add

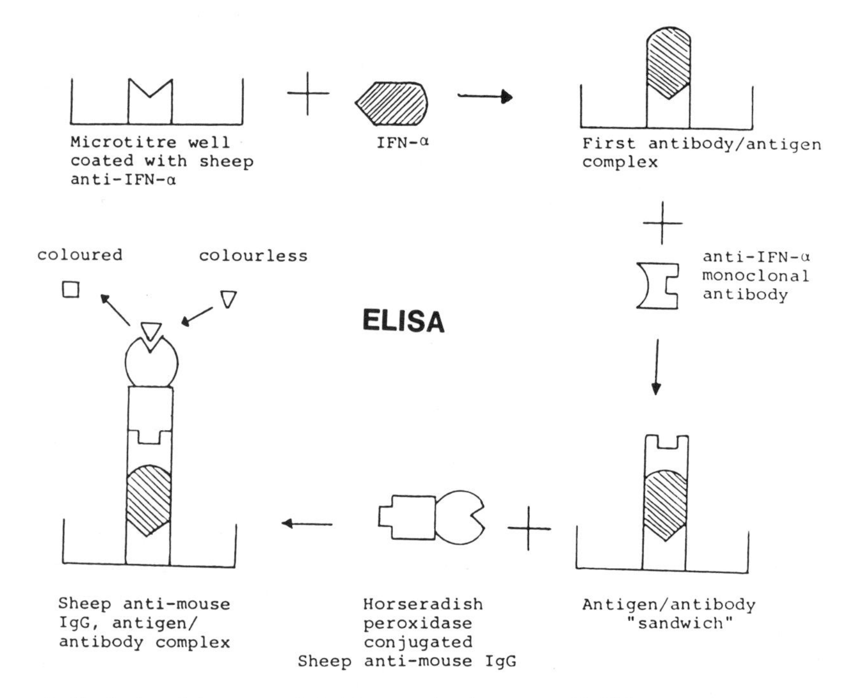

Figure 5. The basics of the enzyme-linked immunoabsorbent assay (ELISA).

unknown IFN samples and serial dilutions of an IFN reference preparation (0.05 ml per well) and incubate for 1 h at 37°C.

(v) Wash the wells again four times with 0.05% Tween 20–PBS and add anti-IFN McAb (0.05 ml/well) in 0.5% BSA–PBS. (Anti-IFN McAb need not be purified; crude, undiluted hybridoma supernatant is suitable or 1:100–1:1000 dilutions of unpurified ascitic fluid can be used. McAb Ig will be present at 10–50 μg/ml.)

(vi) Incubate for 1 h at 37°C, repeat the washing and add biotinylated sheep anti-mouse Ig (Amersham International) 1:500 in 0.5% BSA–PBS, 0.05 ml per well.

(vii) Incubate for a further 1 h at 37°C, repeat the washing and add streptavidin–biotinylated HRP complex (Amersham International), 1:500, in 0.5% BSA–PBS, 0.05% per well.

(viii) Incubate for 30 min at 37°C, wash three times with 0.05% Tween 20–PBS and twice more with 0.1 M citrate–phosphate buffer, pH 5.0.

(ix) Remove all traces of the final wash and add substrate [see Section 6.1, step (vi)].

(x) Develop the colour in the dark and read optical densities in a Titertek Multiscan or other micro-ELISA plate reader.

(xi) Plot absorbances (OD_{450}) against IFN dilutions, IU/ml or ng/ml (log_{10} scale) and read off values for unknown IFN samples from the calibration curve (*Figure 6*).

The ELISA method described above is reproducible, if conditions are rigidly adhered to, is reasonably sensitive (detection limit of 20 IU/ml for IFN-α, 2 IU/ml for IFN-γ) and very economical — the commercial reagents supplied as 2 ml batches are suffi-

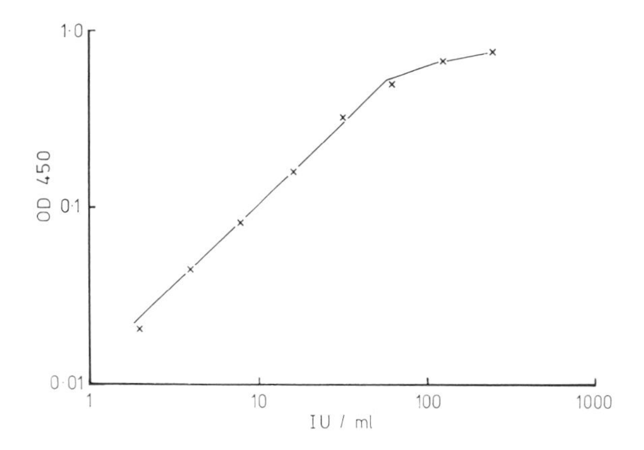

Figure 6. Calibration curves ($\log_{10} - \log_{10}$ plot) for a HuIFN-γ reference preparation as obtained by ELISA.

cient for 200 micro-ELISA plates. Non-specific binding is low, and often negligible.

Detection limits and sensitivity ranges do depend on the concentration of the second anti-IFN McAb. Also rabbit polyclonal anti-IFN Ig, as first antibody, generally gives more sensitive assays than if sheep or calf anti-IFN Ig is used. Substitution of the polyclonal antibody by an anti-IFN McAb, binding to a different epitope than the second anti-IFN McAb, may also improve sensitivity. Note however that use of *mouse* anti-IFN McAb as first antibody precludes use of the ELISA method just described. In this situation, the second anti-IFN McAb should itself be conjugated with HRP, see ref. 21.

8.4 Validation

Biological activity *per se* is not measured in immunoassays; they measure immunoreactive mass. However, many antibodies against IFN bind only to structurally intact or native IFN molecules, that is they bind to discontinuous epitopes which are destroyed by denaturation. The basic assumption is therefore that native IFN molecules as recognized by anti-IFN are also all biologically active so that a relationship between immunoreactive mass and biological activity holds. This assumption is difficult to prove, although in practice it has been found that loss of biological activity is often paralleled by a loss of immunoreactivity to certain anti-IFN McAbs (20,22,23). In any case, the relationship will only be strictly valid if the IFN contained in the reference preparation and the IFN contained in the unknown samples are of identical molecular structure, that is type and/or subtype. Therefore, ideally, every immunoassay should be calibrated only with the IFN type or subtype to be estimated in the unknown sample. In practice this is relatively simple to do when isolated recombinant DNA IFN types or subtypes are being measured (22), but very much more difficult when culture supernatants or sera containing natural IFN mixtures are being estimated. For example, human leukocyte IFN contains several IFN-α subtypes in unknown proportions; these subtypes will be structurally distinct and have varied specific activities with respect to anti-viral activity. A single anti-IFN-α McAb will not usually bind every IFN-α subtype, and if it

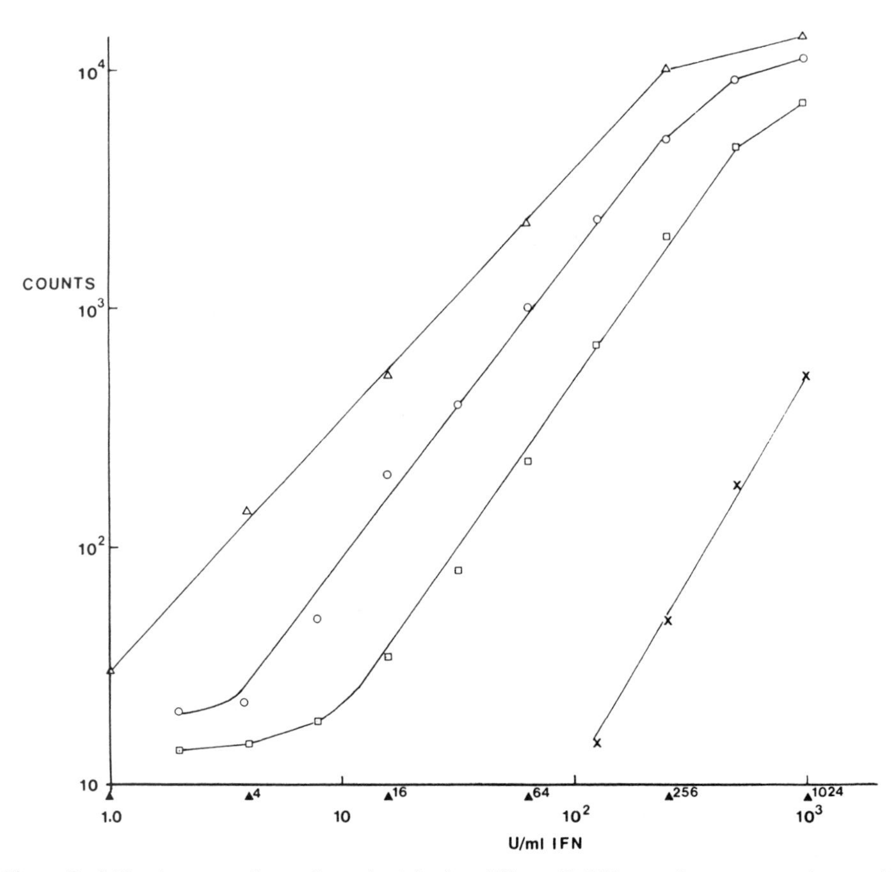

Figure 7. Calibration curves ($\log_{10}-\log_{10}$ plots) for four different HuIFN-α_{Le} reference preparations as obtained by YOK IRMA. Note the extreme variation in sensitivity due to the different IFN-α subtype compositions of the HuIFN-α_{Le} preparations; X- - -X, this preparation was purified by affinity chromatography by NK2-Sepharose, and thus was enriched in those IFN-α subtypes recognized by NK2 McAb. Such subtypes are poorly recognized by YOK McAb. [The author wishes to thank Celltech (UK) Ltd, and Boots-Celltech Diagnostics Ltd, both of Slough, Berkshire, UK for providing reagents and IRMA kits.]

does the differences in binding levels among subtypes are quite marked. This produces an intractable situation in which a single IFN-α reference preparation is not sufficient to calibrate an immunoassay in which unknown IFN-α samples, containing IFN-α subtypes in differing proportions, are tested (see *Figures 7* and *8*). The results from such assays would at best be approximate, and at worst totally inaccurate.

A further cause that may affect the results of immunoassays, especially for IFN-β and IFN-γ, is the variable degree of post-translational modification that IFN molecules receive in the producing cells. In the extreme case, IFN-β and -γ produced by *Escherichia coli* are non-glycosylated and therefore have structures different from those of glycosylated IFN-β and -γ produced by eukaryotic cells. For IFN-γ, the proportions of the three or more components, non-glycosylated, partially glycosylated and fully glycosylated molecules, produced by eukaryotic (human and mammalian) cells may also vary. Again it is recommended that immunoassays are calibrated with the

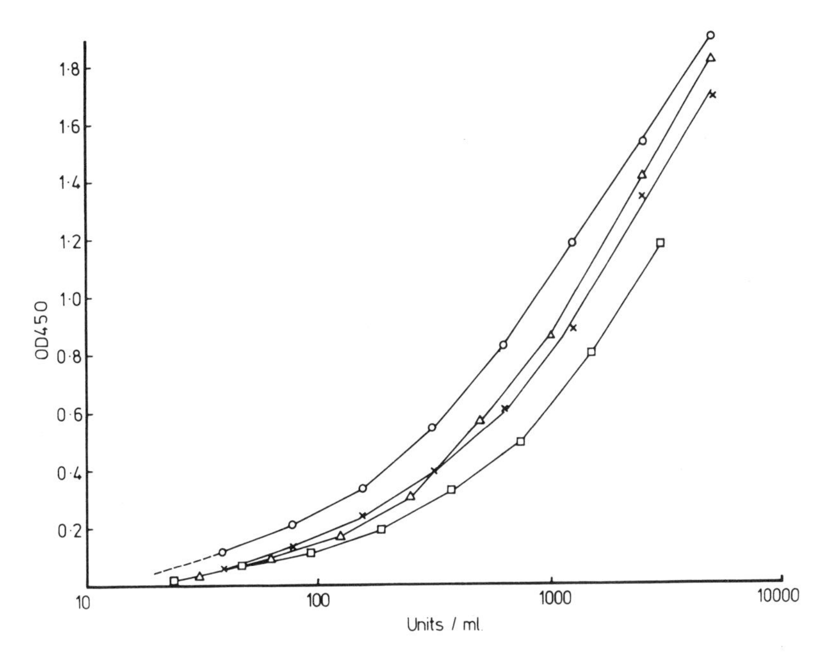

Figure 8. Calibration curves (semi-$\log_{10}$ plots) for the same four HuIFN-α_{Le} reference preparations of *Figure 7* as obtained by ELISA using a McAb (LO-22) showing broad cross-reactivity among HuIFN-α subtypes. The author thanks Dr K.Berg, University of Copenhagen, Denmark for providing the LO-22 McAb.

IFN reference preparations most like the IFN in the test specimens (*Figure 9*).

Assuming that antibody binding to IFN is not grossly affected by other non-IFN substances present in samples being tested (it sometimes is), it should be possible to produce results in good to fair agreement with results derived from anti-viral assays. There is however one further complication that should be considered. Biological activity *per se* is infrequently due only to the intrinsic functionality of the IFN molecules, but is often largely influenced by other substances present in IFN samples, which modify anti-viral effects. Such substances may potentiate or inhibit the anti-viral action of IFN. Even mixtures of two or more IFNs may lead to synergistic action (6). Immunoassays cannot monitor these 'external' influences on the biological activity of IFNs; they will only 'see' biological activity in terms of degree of structural integrity or 'nativeness' of IFN molecules. Thus, the results obtained from immunoassays suggest what the anti-viral titres ought to be if they were solely dependent on the presence of IFN, although the closer in composition of all substances contained in the IFN reference preparation, used to calibrate the immunoassay, to that of all substances contained in the unknown IFN samples, the less is the discrepancy likely to be. This emphasizes again that in the performance of immunoassays the importance of assaying like with like is paramount; unknown IFN samples whose compositions diverge significantly from that of the IFN reference preparation may produce discordant results, that is the immunoassay titre differs markedly from the bioassay titre.

In summary, the following criteria should be implemented to obtain valid and ac-

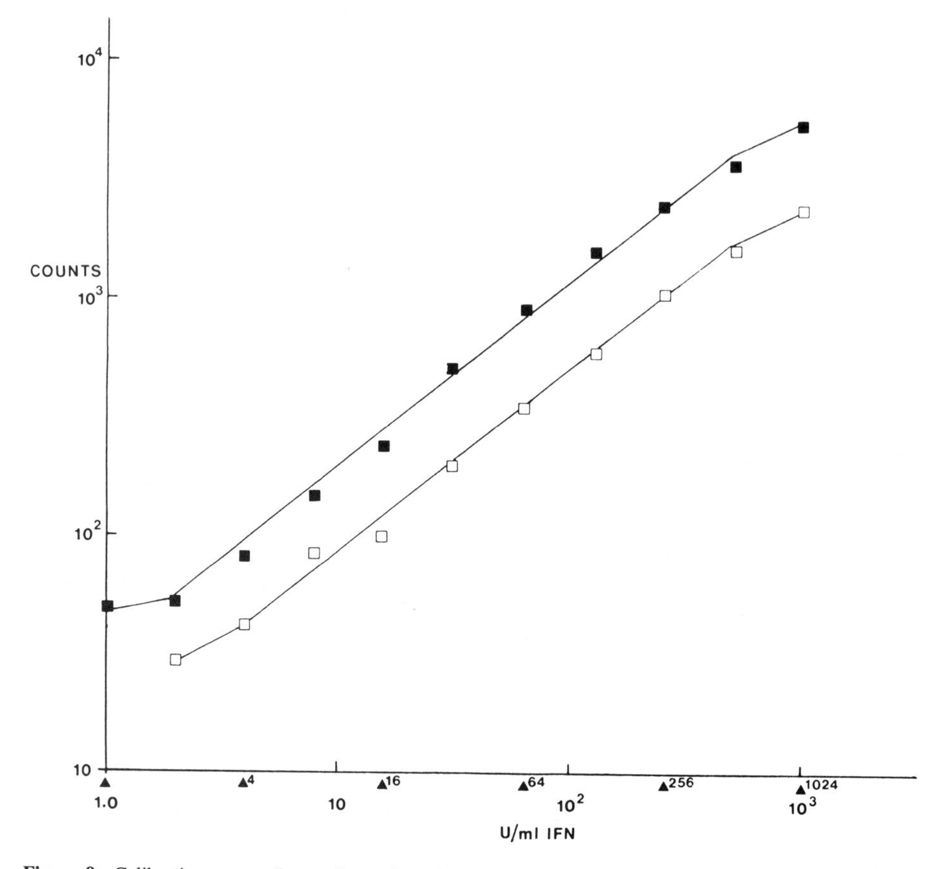

Figure 9. Calibration curves ($\log_{10} - \log_{10}$ plots) for a natural IFN-γ reference preparation (■- - -■) and a recombinant HuIFNγ (*E. coli*) reference preparation (□- - -□) as obtained by IRMA (Boots-Celltech Diagnostics Ltd). Note that the non-glycosylated recombinant DNA IFN-γ titrates at a lower sensitivity, based on anti-viral units, than the glycosylated IFN-γ suggesting a difference in affinity of the McAb for the two forms.

curate (equal to titres obtained from bioassays) results from immunoassays calibrated in anti-viral units (IU/ml).

(i) Identity of IFN type or subtype in reference preparation and samples to be estimated.
(ii) Identity of composition of molecular forms of IFN (e.g. non-glycosylated molecules) in reference preparation and samples to be estimated.
(iii) Identity of composition of all non-IFN substances in reference preparation and samples to be estimated, for example common diluent.
(iv) No interference of non-IFN substances in the immunoassay.

The fulfillment of all these criteria will often be difficult to achieve, and sometimes difficult to verify, and thus the results of immunoassays should be interpreted accordingly, that is, with caution. It should be remembered that immunoassays are alternative assays, but they must not be considered as substitutes for biological assays unless they can be properly validated.

9. REFERENCES

1. Köhler,G. and Milstein,C. (1975) *Nature*, **265**, 495.
2. Bailey,G.S. (1984) In *Methods in Molecular Biology. Vol. 1. Proteins.* Walker,J.M. (ed.), Humana Press, Clifton, New Jersey, p. 295.
3. Kawade,Y. (1980) *J. Interferon Res.*, **1**, 61.
4. Standardization of Interferons (Report of a WHO informal consultation (1983) *World Health Organization Technical Report Series*, No. 687, p. 35.
5. Ennis,F.A. and Meager,A. (1981) *J. Exp. Med.*, **154**, 1279.
6. Fleischmann,W.R.,Jr., Georgiades,J.A., Osborne,L.C and Johnson,H.M. (1979) *Infect. Immun.*, **26**, 248.
7. Leist,T., Titmas,R., Parti.S. and Meager,A. (1985) *Mol. Immunol.*, **22**, 929.
8. Kennett,R.H., McKearn,T.J. and Bechtol,K.B. (eds) (1980) *Monoclonal Antibodies.* Plenum Press, New York and London.
9. Exley,T., Parti,S., Barwick,S. and Meager,A. (1984) *J. Gen. Virol.*, **65**, 2277.
10. Meager,A., Parti,S., Barwick,S., Spragg,J. and O'Hagan,K. (1984) *J. Interferon Res.*, **4**, 619.
11. Kawade,Y. and Watanabe,Y. (1985) *Immunology*, **56**, 489.
12. Secher,D.S. (1981) *Nature*, **290**, 501.
13. Ey,P.L., Prowse,S.J. and Jenkin,C.R. (1978) *Immunochemistry*, **15**, 429.
14. Bruck,C., Portetelle,D., Glineur,C. and Bolton,A. (1982) *J. Immunol. Methods*, **53**, 313.
15. Burchiel,S.W., Billman,J.R. and Alber,T.R. (1984) *J.Immunol. Methods*, **69**, 33.
16. Lowry,O.H., Rosebrough,N.J., Farr,A.L. and Randall,R.J. (1951) *J. Biol. Chem.*, **193**, 265.
17. Hunter,W.M. and Greenwood,F.C. (1962) *Nature*, **194**, 495.
18. Bailey,G.S. (1984) In *Methods in Molecular Biology. Vol. 1. Proteins.* Walker,J.M. (ed.), Humana Press, Clifton, New Jersey, p. 325.
19. Staehelin,T., Stahli,C., Hobbs,D.S. and Pestka,S. (1981) In *Methods in Enzmology.* Pestka,S. (ed.). Academic Press, London and New York, Vol. 78, p. 589.
20. Chang,T.W., McKinney,S., Liu,V., Kung,P.C., Vilcek,J. and Le,J. (1964) *Proc. Natl. Acad.Sci. USA*, **81**, 5219.
21. Gaastra,W. (1984) In *Methods in Molecular Biology. Vol. 1. Proteins.* Walker,J.M. (ed.), Humana Press, New Jersey, p. 349.
22. Protzman,W.P., Minnicozzi,M., Jacobs,S.L., Surprenant,D.I., Achwartz,J. and Oden,E.M. (1985) *J. Clin. Microbiol.*, **22**, 596.
23. Sedmak,J.J., Siebenlist,R. and Grossberg,S.E. (1985) *J. Interferon Res.*, **5**, 397.

CHAPTER 9

Quantification of interferons by anti-viral assays and their standardization

A.MEAGER

1. INTRODUCTION

'Chemists and physicists know what they are measuring, biochemists think they know what they are measuring, but biologists certainly do not know what they are measuring'. The provenance of this quotation is unknown to the author, but when it comes to measuring the biological effects of interferon (IFN) there is a large element of truth in the latter part of the quotation! The hallmark of IFNs is their ability to induce anti-viral activity; IFNs themselves are not anti-viral. An anti-viral state in a cell is only established following contact with IFN through specific cell surface receptors. The underlying mechanism(s) whereby the IFN–IFN receptor complex signals to the cell nucleus to develop the anti-viral state remains enigmatic. Nevertheless, it is this cellular response to IFN and the subsequent inhibition of the replication of an infecting virus, which can be measured, which form the basis of IFN assays. The degree of inhibition of virus replication is related to the potency of an IFN preparation or sample; that is a measure of the degree to which an individual IFN preparation is able to protect cells of a particular type against productive infection by a virus of a certain type. Potency measurements (estimates) are relative, not absolute, and depend on:

(i) the proportion of IFN molecules of an IFN preparation which are biologically active in the anti-viral assay;
(ii) the sensitivity of the cell line to IFN (this is also dependent on the type of IFN being assayed);
(iii) The sensitivity of the virus to the anti-viral state established in the assay cells.

Cell culture media and conditions add a further dimension to the complexity and variability of anti-viral assays. The potency of IFN preparations has therefore been quantified by the arbitrary assignment of anti-viral units defined in particular virus–cell systems, usually where some kind of end-point can be observed, and expressed as the reciprocal of the dilution of an IFN preparation required to produce the end-point. It is thus necessary that individual laboratories develop and establish IFN reference preparations of assigned potencies to monitor the sensitivity of their assays and to allow comparison of potency estimates obtained from assays performed on different occasions. The World Health Organization (WHO) has established a number of IFN international reference preparations (or standards), the use of which allows individual laboratories to calibrate their assays in relation to the assigned potencies of these international reference preparations (IRPs) and thus to compare the results of their assays with other laboratories.

By definition all anti-viral assays to measure IFN are highly selective and operator-selective in the phenomenon (parameter) investigated and quantified as proportionate to IFN activity. Indeed they are 'horses for courses', but do not bet on them living up to their reputation all the time; like the best 'animals' they can fail! Thus, the purpose of this chapter is to explain to the researcher how to perform anti-viral assays which are reliable and produce good quantitative data. It is not intended that instruction for every assay method be given, nor that theory is extensively reviewed, as these are adequately covered elsewhere (1–3). Instead, a rather selective approach is made based on a common strategem of the use of the microtitre plates for setting and working up assays.

2. DEFINITIONS

The definitions outlined in this section are common to most biological assays.

(i) *Precision.* This is the demonstration that similar results can be obtained sequentially on different occasions in measurements (estimations) of the same IFN preparation; that is the degree of reproducibility of the assay. The observed precision will be reflected by the stability to which the sensitivity of the assay system can be maintained and will also depend on the accuracy with which individual manipulations, for example dilutions, and the determinations of the end-point can be achieved.

(ii) *End-point.* This is the arbitrarily selected point at which a significant reduction in virus growth or effect is apparent. In most IFN assays the end-point is chosen to be 50% of the maximum inhibitory effect attained (observed).

(iii) *Titre.* This expresses the potency of an IFN preparation in terms of the highest dilution of that preparation which produces an end-point.

(iv) *Unit.* This is the amount (concentration per ml) of IFN necessary to produce an end-point. It is usually related in proportionate terms to a reference preparation having an assigned potency.

3. POINTS TO CONSIDER WHEN SETTING UP ANTI-VIRAL ASSAYS

3.1 Handling of samples

Samples of IFN containing biological fluids or culture media are readily produced or obtainable (see Chapter 2). Many of these will have been produced at 37°C, a temperature at which most IFN types are relatively unstable. At temperatures at or above 37°C, virtually all IFNs lose biological activity in a progressive manner. It is therefore inadvisable to maintain IFN preparations or samples at 37°C for any longer than necessary, and this applies particularly to samples previously frozen and then thawed in 37°C water baths.

Frequently IFN samples will have to be stored, and this will necessitate a decision as to whether the samples should be chilled and kept at 4°C (2–8°C) or frozen at −20°C before assay. As a general rule, IFN samples containing serum or other proteins in high concentrations are perfectly stable at 4°C, whilst highly purified IFN preparations containing low amounts of protein (<1 mg/ml) are often unstable at this temperature. IFN-γ, and especially the non-glycosylated variety produced from recombinant *Escherichia coli*, is probably the least stable of the IFNs in this respect. Storage at 4°C does not prevent growth of mould contaminants if present and therefore prolonged storage at this temperature is not advised if microbiological sterility is uncertain. The majority of IFN samples may be safely frozen at −20°C or −70°C for long periods without any perceptible deterioration in biological activity. It is best however to ensure, where possible, that the IFN solution is protein rich before freezing, as protein has protective effects, for example against enzymatic degradation and against adsorption of IFN to the walls of the container. Frozen samples should not be thawed too rapidly, or held at 37°C, before being assayed. Repetitive freezing and thawing of the same sample is also to be avoided; serum-containing samples will readily form precipitates and the pH of bicarbonate-buffered samples will become increasingly alkaline under these circumstances. Care should be taken to maintain IFN-γ preparations at a pH close to, or slightly above, neutrality as activity may be lost rapidly at pH values less than 6.5. Such pH may readily occur if for example the IFN preparation is stored in high CO_2 concentrations (e.g. on dry ice), or even under tissue culture conditions.

3.2 Handling IFN reference preparations and preparation of laboratory IFN reference preparations

When an ampoule of an international IFN reference preparation (or standard) arrives, usually by post, it should be handled with respect, but not regarded with awe! The ampoule will contain a white or yellowish lyophilized powder and its IFN content is fixed and stable at room temperature. However, it is advisable to store the ampoule at −20°C or 4°C until such time as it is required for calibration of IFN assays.

Virtually all IFN reference preparations come as sealed glass ampoules and these should be carefully opened in order that no material inside the ampoule is lost. One millilitre of chilled sterile water, measured with the greatest accuracy possible, is then added and the powder dissolved. Normally the process of dissolution is rapid, but it is recommended that the ampoule contents are gently shaken, or mixed using a pipette, before transferring the contents to other containers. At this stage it is best to transfer the solution to 10 sterile vials, adding 0.1 ml to each, and to freeze nine vials at −20°C or −70°C. The IFN solution remaining unfrozen should be diluted 1:10 (or as desired) in the serum-containing medium in which the anti-viral assays are to be performed. This diluted IFN should be stable to freeze−thawing, but it is not recommended that this is done on a continual basis until it is all used up. It is far better to go on to a vial of undiluted IFN reference preparation after the first diluted preparation has been frozen and thawed once or twice or has been stored at 4°C for longer than a couple of months.

A list of the currently, or soon to be, available IFN IRPs is given in *Table 1*. These may be obtained from the Research Resources Branch, National Institutes of Health, Bethesda, MD 20205, USA where indicated NIH, or from the National Institute for Biological Standards and Control, Holly Hill, Hampstead, London NW3 6RB, UK where indicated NIBSC. Note that the assigned potencies of the majority of IRPs are

Table 1. International IFN reference preparations.

Code number	*Interferon*	*Assigned potency (IU/ampoule)*	
67/18	Chick	80	NIBSC
69/19	Human leucocyte, IFN-α	5000	NIBSC
Ga23-902-530	Human leucocyte, IFN-α	12 000	NIH
Ga23-901-532	Human lymphoblastoid, IFN-α	25 000	NIH
Gx01-901-535	Human recombinant, IFN-αA	9000	NIH
82/576	Human recombinant, IFN-α2	t.b.a.[a]	NIBSC
83/514	Human recombinant, IFN-αD	t.b.a.	NIBSC
G-023-902-527	Human fibroblast, IFN-β	10 000	NIH
Gb23-902-531	Human fibroblast, IFN-β	t.b.a.	NIH
Gxb01-901-535	Human recombinant, IFN-β	t.b.a.	NIH
Gxb02-901-535	Human recombinant, [Ser^{17}]IFN-β	t:b.a.	NIH
Gg23-901-530	Human leucocyte, IFN-γ	4000	NIH
82/587	Human leucocyte, IFN-γ	t.b.a.	NIBSC
Gxg01-901-535	Human recombinant, IFN-γ	t.b.a.	NIH
G-002-904-511	Mouse L-cell IFN-α/β	12 000	NIH
Gu02-901-511	Mouse IFN-α/β	t.b.a.	NIH
Ga02-901-511	Mouse IFN-α	t.b.a.	NIH
Gb02-902-511	Mouse IFN-β	t.b.a.	NIH
Gg02-901-533	Mouse IFN-γ	t.b.a.	NIH
G-019-902-528	Rabbit IFN-β	10 000	NIH

[a]To be assigned: these preparations were included in a 1985 international collaborative study, the results of which are expected shortly. See also refs 5 and 6.

sufficiently high enough to allow dilution even for relatively insensitive anti-viral assays; the exception is the 67/18 chick interferon IRP with an assigned potency of 80 international units (IU) per ampoule.

Before calibrating anti-viral assays, individual laboratories should prepare a sufficiently large stock of a laboratory IFN preparation having similar characteristics to the IFN contained in the IRP. In some instances this is easily achieved; for example, human IFN-β prepared from diploid fibroblasts will be virtually identical to the human IFN-β contained in the IRP. However, in other cases it may not be possible to prepare an IFN which is identical to that contained in the IRP. For example, the human leukocyte IFN-α IRP (69/19; *Table 1*) contains a mixture of IFN-α subtypes in unknown proportions and it is uncertain and probably unlikely that human leukocyte IFN prepared from different sources of buffy coats has exactly the same proportions of IFN-α subtypes to 69/19. For similar reasons, human lymphoblastoid IFN is non-identical to human leukocyte IFN and a separate human lymphoblastoid IFN IRP has been developed (*Table 1*). There may also be special problems relating to calibration of assays in which IFNs derived from *E. coli*, especially β and γ types which are not glycosylated by *E. coli*, are determined. It is advisable in these cases, and indeed in general, to choose the IFN IRP with IFN characteristics closest to that being studied until new rDNA IFN IRPs are available (see *Table 1* and refs. 5,6).

The laboratory IFN reference preparation should be made up at a suitable concentration (e.g. $10^3 - 10^4$ IU/ml) in a similar way to the IFN IRP and its long-term stability checked. Once a stable preparation has been prepared (or is available from outside

Table 2. Cell lines for interferon assays.

Cells	*Origin*	*Morphology*	*Source*
HEp2	Human epidermoid carcinoma, larynx	Epithelial	ATCC, cat. no. CCL23
WISH	Human amnion	Epithelial	ATCC, cat. no. CCL25
A549	Human lung carcinoma	Fibroblastic	NIH
GM-2504	Human skin (trisomic chr21)	Fibroblast	HGMCR, cat. no. GM-2504E
GM-2767	Human skin (trisomic chr21)	Fibroblast	HGMCR, cat. no. GM-2767B
MDBK	Bovine kidney	Epithelial	ATCC, cat. no. CCL22
EBTr	Embryonic bovine trachea	Fibroblastic	ATCC, cat. no. CCL44
L-929	Mouse connective tissue	Fibroblast[a]	ATCC, cat. no. CCL1

Abbreviations: ATCC, American Type Culture Collection, Rockville, MD, USA; HGMCT, Human Genetic Mutant Cell Repository, Camden, NJ, USA; NIH, National Institutes of Health, Bethesda, MD, USA.
[a]Different sublines show variable morphology.

sources) several hundred vials containing small volumes (< 1.0 ml) should be frozen away at −20°C and −70°C for future use. The calibration of the laboratory IFN reference preparation using the appropriate IFN IRP will be described in Section 4.1.1.

3.3 Choosing the cells and virus for anti-viral assays

Virtually all cells and cell lines show some sensitivity to IFNs, but experience has shown that relatively few lines are suitable for use in anti-viral assays. For obvious reasons continuous cell lines with stable phenotypes have advantages over primary cell cultures and diploid fibroblasts of limited lifespan. However, most continuous cell lines are heteroploid in chromosome number and show phenotype drift in serial passage; the latter may result in a significant change in sensitivity to IFN. Despite this drawback, which can to a large degree be overcome by maintaining a large number of ampoules of cells in liquid nitrogen and reconstituting these from time to time, continuous cell lines are the most useful for most types of anti-viral assays. Generally speaking it is best to choose cell lines of human origin for the assay of human IFN, cells of mouse origin for the assay of mouse IFN, and so on. However, certain IFNs, for example human IFN-α, may be assayed in heterologous cell lines. The most commonly used cell lines for the assay of human and mouse IFN are listed in *Table 2*. It must be pointed out here that there will often co-exist a number of sublines or strains of a particular cell line, frequently a result of different individual treatments in different laboratories (i.e. cell culture techniques and conditions), and that some sublines are to be preferred to others.

The choice of viruses which are sensitive to the anti-viral action of IFN is wide, but for anti-viral assays that choice is restricted to but a handful of viruses. The case for having only a few so-called 'challenge' viruses rests mostly on the experience of dedicated IFN researchers which has been gleaned over the years since the discovery of IFN in 1957. Moreover, from the point of view of standardizing anti-viral assays and being able to compare in a meaningful way results obtained in different laboratories, it is also a good idea to cut down the number of challenge viruses.

In the author's experience the following viruses have characteristics appropriate for use as challenge viruses in anti-viral assays:

(i) Encephalomycarditis virus (EMCV).
(ii) Vesicular stomatitis virus (VSV).
(iii) Semliki Forest virus (SFV) (alternative — Sindbis virus).

EMCV replicates well in virtually all cell lines and is safe to use, being non-pathogenic for man. VSV is also relatively safe to use although, due to its pathogenicity for cattle, its use may be restricted. SFV has a slight question mark about its safety, but it has a wide host range and therefore is an extremely useful challenge virus. The reader is referred to (4) for details of the growth, storage and assay of these viruses.

3.4 Choosing the assay method

Precision or reproducibility should be the prime factor in deciding which assay method to use, but in the real world economics may be of prime importance. Certainly the cheapest and, by now, the most common method in use is the cytopathic effect reduction (CPER) assay. In this type of assay cells in microtitre plates are protected with serial IFN dilutions, then challenged with a cytopathic virus, and cultured (incubated) until the cytopathic effect (cell killing) observed in untreated cells (virus control) is maximum. At this point in time, the end-point for each IFN sample, usually arbitrarily assigned to be the well in which 50% of the cells are living, may either be determined microscopically or determined following staining of the cells with a vital dye. After staining, the assays may be read visually or the dye re-extracted from the cells with solvents and the optical densities of each well determined spectrophotometrically. Such assays produce sharply increased CPE as the IFN is diluted out and thus it is not always easy to construct satisfactory dose–response curves. Usually only two or three points are on the rectilinear portion of the dose–response curve and parallelism of the slopes of dose–response curves generated by different IFN samples is not easily demonstrable (N.B. parallelism of slopes is a requirement for the validation of a biological assay). Nevertheless, the CPER assay provides an economic, relatively quick and easy to perform (see Section 4.1), sensitive and fairly reliable assay for measuring the potencies of IFN samples, particularly for screening large numbers of IFN samples.

Modification of the CPER assay to a plaque-reduction assay may readily be achieved by reducing the multiplicity of infection (m.o.i.) of the challenge virus and by using suitable overlay materials. The end-point chosen for this type of assay is usually the plate or well where the number of plaques is 50% of the virus control. Plaque-reduction assays may be carried out at the microtitre plate scale, but the number of plaques per monolayer will be reduced accordingly. This method offers few, if any, advantages over the CPER assay and counting plaques is very labour intensive.

Other assays to measure the anti-viral effect of IFN rely on the measurement either of infectious virus or particular viral components, for example haemagglutinin, viral enzymes, viral RNA, following IFN treatment. They are generally more expensive to perform, both in terms of resources and reagents, and often involve two steps. The infectious virus yield-reduction assay is an example of the two stage assay. Here the initial stage is much the same as that of the CPER method; however the number of progeny viruses generated following virus challenge is titrated in a further assay of either the plaque or CPE assay type. The progeny virus may also be quantified by alternative means such as haemagglutination assays (2). The end-point for yield reduction assays is normally chosen as the 0.5 $\log_{10}$ reduction in virus yield.

Some viruses, for example SFV, lend themselves to ready quantification of intracellular viral RNA production. This can be achieved by measuring incorporation of [^{3}H]uridine into viral RNA in actinomycin-treated cells. IFN will decrease the incorporation of [^{3}H]uridine into viral RNA and this forms the basis of the inhibition of viral nucleic acid (IVNAS) assay. The end-point in this instance corresponds to the IFN dilution inhibiting incorporation of [^{3}H]uridine into viral RNA by 50%. Both this assay and those of the yield-reduction type generate superior dose–response curves to the CPER assay. However, the former are relatively more cumbersome and labour intensive, as well as being more expensive, than CPER assays. Since in the author's opinion they are also no more reliable or reproducible than CPER assays, justification for their widespread use and/or use on a routine basis requires careful consideration.

3.5 Assay design

The design of assays is important and should be carefully considered in relation to the types of IFN, the numbers of samples and their volumes and the degree of statistical confidence limits required. The aim in every case should be to provide satisfactory quantitative data that may be referred back to and which is comparable on an international basis with results published by other laboratories. All assays should be carefully calibrated, first by using the IRP and secondly by using the laboratory reference preparation (LRP), the latter being included in every assay subsequently performed for a particular IFN type. Therefore, space must be alloted in every assay for the LRP; where microtitre trays are used the LRP should be included on every tray. Space must also be found for both cell controls (no IFN, no virus) and virus controls (no IFN, plus virus). Obviously, IFN titres derived from a dose–response curve where each point of the curve is derived from a single reading will be less statistically meaningful and more error prone than where each point is the averaged result of duplicate, triplicate or quadruplicate measurements for each dilution. Samples should therefore be run in duplicate (or greater) where possible; sample size may however sometimes preclude this option. Part of the art of setting up IFN assays is to guess the titres of the samples before proceeding. Thus where sample size is limiting, but 'guestimates' of sample titre indicate an initial dilution is possible, say 1:5 or 1:10, without losing the end-point of the assay, then it is recommended that the diluted sample be titrated in duplicate (or greater) rather than exhausting the sample in a single titration. Such a policy may also allow re-titration of the diluted sample on a second (or further) occasion. In fact, repeating assays is good scientific practice. It confirms not only the reproducibility of the assay method, but also provides further quantitative data from which, when analysed together with that of the preceding assay(s), geometric mean titres can be calculated, leading to increased statistical validity of the results.

Probably it goes without saying that samples of different IFN types should not be run in parallel in the same assay, for instance, together on the same microtitre tray. Assay results will only be valid when 'like' is compared with 'like' and the dose–response curves are parallel. Again educated guesses play a role in deciding which samples are included on a microtitre tray when the type(s) of IFN present in the samples is unknown or obscure.

4. ASSAY PROTOCOLS

4.1 Cytopathic effect reduction (CPER) assays

4.1.1 *The basic schedule*

See *Table 3* for useful cell–virus combinations for CPER assays.

(i) Add culture medium (normally cell growth medium containing 5 or 10% calf serum) to the microtitre plate appropriate to the intended serial dilution of the IFN sample, for example 0.1 ml/well for 1:2 dilution, 0.055 ml for 0.5 $\log_{10}$ dilution.

(ii) Assign wells for cell controls and virus controls.

(iii) Add IFN sample, or known dilution thereof, to the first well(s) of row A (or B, see later) of the microtitre plate, for example 0.1 ml sample/well if $\log_2$ (1:2) dilutions are intended, 0.025 ml/well if 0.5 $\log_{10}$ (1:3.2) steps are intended. It is recommended that all such wells have their IFN samples added before further

Table 3. Useful cell–virus combinations for interferon assays.

Assay type	*Cell–virus combination*	*IFN types*	*Comments*
CPER and plaque reduction assays	HEp2 + EMCV	Human IFN-α, -β, -γ	More sensitive to HuIFN-α and -β than to HuIFN-γ
	A549 + EMCV	Human IFN-α, -β, -γ	Much more sensitive to HuIFN-α and -β than to HuIFN-γ
	WISH + EMCV or + VSV	Human IFN-α, -β,-γ	
	MDBK + VSV	Human IFN-α and bovine IFNs	Not suitable for HuIFN-β and -γ
	EBTr + SFV	Human IFN-α	Not suitable for HuIFN-β and -γ
	L929 + EMCV	Mouse IFN-α, -β, -γ	Only some sublines of L929 are suitable
Infectious virus yield-reduction assays	Stage 1 combinations same as for CPER assays		
IVNAS assays	Human diploid fibroblasts, e.g. FS7 or GM2504, GM2767 or WISH + SFV or VSV	Human IFN-α, -β, -γ	Only moderate to low incorporation in virus controls on some occasions
	MDBK or EBTr + SFV or VSV	Human IFN-α and bovine IFNs	High incorporation in virus controls
	L929 + SFV	Mouse IFN-α, -β, -γ	

dilutions are made from row A to H (or B to G) rather than individually diluting samples.

(iv) Using a multichannel micropipette (up to 12 channels, if available) mix the contents of the first row of the dilution series by repeatedly taking up and expelling liquid from the tips and transfer 0.1 ml from row A to row B for $\log_2$ (1:2) dilutions, or 0.025 ml for 0.5 $\log_{10}$ (1:3.2) dilutions.

(v) At this stage it is best, because of a certain amount of carryover of liquid on the outside surface of the tips, to discard the original tips and continue the dilution with new tips. However, economics may prevent the profligate use of tips that this procedure requires. Mix well contents of row B thoroughly by repeatedly taking up and expelling liquid from the tips and then take up the dilution volume and transfer it to the wells of row C.

(vi) Repeat the procedure indicated in (v) until the last well, row H, of the dilution series is reached. After mixing, take up the dilution volume from the final row of wells and discard.

(vii) Trypsinize the cells and resuspend them in an adequate volume of growth medium sufficient to fill all the wells of all the microtitre plates used in the assay. To produce monolayers quickly, relatively high cell densities, for example $5-10 \times 10^5$/ml, will be required. The cells should however be present as a single cell suspension; aggregates are to be avoided. Trypsin must be removed as far as it is possible. As a rough guide, a 20 oz glass medical flat bottle will yield approximately $2-4 \times 10^7$ cells (HEp2/c, WISH, L929), a 75 cm^2 plastic bottle will yield $1-2 \times 10^7$ cells. Relatively lower numbers of cells will be obtained in the case of contact-inhibited fibroblasts. Therefore, one 20 oz bottle will provide enough cells for up to four microtitre plates, if all 96 wells are used, or up to six plates if 60 wells are used; one 75 cm^2 plastic bottle will provide for half this number of plates.

(viii) Seed the cells into all wells including cell and virus controls. For 1:2 serially diluted samples containing 0.1 ml/well, add cells (appropriately diluted in growth medium) in a further 0.1 ml ($5-10 \times 10^4$ cells)/well so that the final volume is 0.2 ml and the initial dilution of the IFN sample is 1:4 (the series continues 1:8, 1:16, 1:32, etc). For 0.5 $\log_{10}$ serially diluted samples containing 0.055 ml/well, add cells in a further 0.121 ml/well ($\sim$0.120$-$0.125 ml is good enough given the accuracy of micropipettes) so that the final volume is 0.176 ml and the initial dilution of the IFN sample is 1 $\log_{10}$ (1:10).

(ix) Place the completed microtitre plates in a humidified incubator set at 37°C and gassed with 5% CO_2. At this stage it is recommended that the plates are incubated for 16$-$24 h (overnight), although much shorter periods of incubation (4$-$6 h) may be used. Expect some loss of sensitivity if the short route is chosen.

(x) The next day check that the monolayers of cells are complete, show relatively even distribution of cells and are healthy. Remove the growth medium in the wells by flicking out and blotting on a paper towel. There is some risk of contamination here, but experience has shown that with antibiotics in the medium it almost never happens. Dilute virus to around 10$-$30 plaque forming units (p.f.u.)/cell in growth medium, or to whatever m.o.i. is required to effect 100% CPE in unprotected virus controls, and add 0.2 ml to all wells (this should be determined empirically with each new virus stock). Remember to add growth medium, not containing virus, to the cell control wells. Return the plates to the incubator.

(xi) The time required for maximum CPE to develop will depend on the specific cell–virus pairings. The rate of CPE development will also be dependent on the input m.o.i. of the challenge virus. Some experience in these assays will usually have to be acquired before any reliable prediction of when, following virus challenge, the plates will be ready to read. As a rough guide, the HEp2/c–EMCV combination takes 20–24 h at 37°C, the A549–EMCV combination 30–48 h and the L929–EMCV combination 20–30 h depending on the L929 strain. Incubation times for other viruses in the CPER assay will be similar, although this author has no experience with the other popularly used virus, VSV. Suffice it to say that monolayers of the virus control wells should be examined microscopically in order to ascertain when maximum CPE is reached. Ideally, the cells should be seen to be completely lysed.

(xii) End-points for each IFN sample titration may be determined microscopically, but this is a laborious business. It is better to fix and stain the cell monolayers so that they may be directly observed. Several vital strains, that is those which stain only, or preferentially, live cells, are available. The violet dyes such as crystal and gentian violet are suitable; these should be prepared at 1–2% w/v in 20% ethanol–phosphate-buffered saline (PBS), warmed to 70°C and filtered hot. The problem with these is that they tend to be extremely messy. The author would recommend in preference to crystal or gentian violet, amido blue black (also variously called amido black, naphthol blue black, etc). For the latter, the stain solution contains 0.05% amido blue black in 9% acetic acid with 0.1 M sodium acetate and is filtered at room temperature before use. To stain cell monolayers, first remove medium from all wells by flicking out, wash the monolayer once with PBS at 0.2 ml/well (this step is not recommended if the cell monolayers are at all fragile) and add 0.1 ml of the stain solution to all wells. Allow staining to take place for 15–30 min at room temperature. Remove the stain solution by flicking out and fix the cell monolayers with formol–saline (10% formaldehyde solution in normal saline or PBS) or 10% formaldehyde solution in 9% acetate acid with 0.1 M sodium acetate for the amido blue black stain. Use 0.1 ml/well in a fume-hood or well ventilated laboratory and cover the plates as formaldehyde vapour is extremely noxious and irritating to the eyes. The plates may be washed in running tap water in a large beaker after 10–15 min at room temperature with cell-fixing solution. Dry plates at room temperature or in a warm dry incubator (30–40°C).

(xiii) The assay plates may be read by eye at any time after staining and fixing. An example of an amido blue black-stained plate is shown in *Figure 1*. Ideally, the degree of differential staining between cell control and virus control should be large in order that the end-points of IFN titrations are easily discernible. Usually

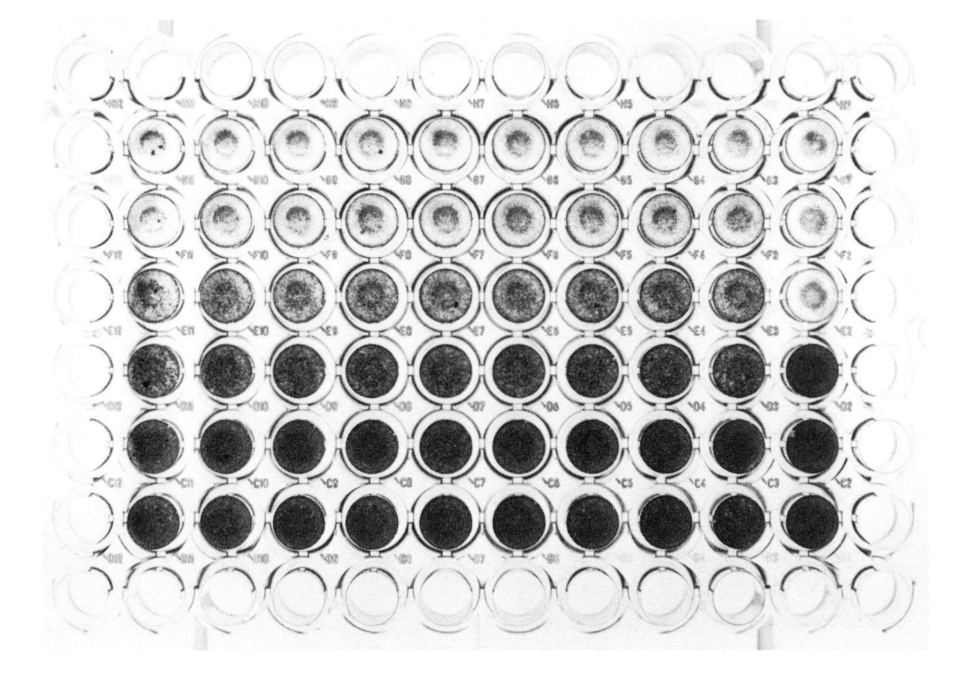

Figure 1. Amido blue black-stained microtitre plate of human HEp2/c cell–EMCV CPER assay. Line 2 (from left), rows B, C, D are cell controls; line 2, rows E, F, G are virus controls. Lines 3–11, titration of a single IFN-α preparation through rows **B–G**, 1/400–1/12 800.

in CPER assays there is but one well which can be considered as the end-point, that is approximately 50% of the staining of the cell control, since the development of CPE occurs over a rather short dilution range. Obviously reading the assays by eye is not all that accurate and is highly subjective, although seasoned 'interferonologists' will swear to the validity of their results derived in this manner. However, the CPER assay can be made objective by dye or stain extraction and measurement of optical densities in a spectrophotometer. Crystal and gentian violet can be extracted in 50% ethanol (0.1–0.15 ml) when the stained monolayers are dry; amido blue black can be extracted in 50 mM sodium hydroxide (0.1–0.15 ml). Make sure that the well contents are uniformly distributed by tapping the sides of the plates before placing the plates in the spectrophotometer. A Titertek Multiskan or other automatic ELISA microplate reader set at 700 nm for the violet dyes or 610 (620) nm for amido blue black are eminently suitable for reading well absorbances. An example of the kind of data produced by this means is given in *Table 4*.

Results of the CPER assay derived as above can be plotted to produce sigmoidal dose–response curves by plotting the IFN concentration (the log of the reciprocal of the IFN dilution) against the dye absorbance or optical density. Only the linear portion

Table 4. Data obtained from the microtitre plate illustrated in *Figure 1* following dye extraction: OD_{620} readings for each well are tabulated below.

	2	*3*	*4*	*5*	*6*	*7*	*8*	*9*	*10*	*11*
B	1.291 cc	1.259	1.260	1.249	1.280	1.265	1.229	1.244	1.246	1.233
C	1.317 cc	1.261	1.279	1.282	1.296	1.281	1.274	1.296	1.266	1.261
D	1.280 cc	1.202	1.232	1.251	1.234	1.243	1.252	1.247	1.254	1.201
E	0.599 vc	1.070	1.112	1.101	1.097	1.063	1.090	1.037	1.043	0.902
F	0.571 vc	0.771	0.801	0.766	0.829	0.696	0.764	0.691	0.689	0.558
G	0.496 vc	0.540	0.539	0.568	0.613	0.581	0.594	0.541	0.531	0.497
Starting dilution					1/400					
Interferon designation					Human IFN-α					

of the curve is used to estimate a titre of the IFN sample. The titre is the IFN concentration interpolated at the arbitrarily defined end-point of the response measured, that is the median absorbance or optical density on the scale of dye uptake measurements that range between those of the cell controls and the virus controls. The IFN titre corresponding to the end-point can be obtained graphically from semi-log plots (*Figure 2*) or by linear regression calculations, using only those points which fall within the linear portion of the curve.

The CPER assay is straightforward and reliably produces useful results, but it has some undesirable features, besides the short dilution range in which the end-point occurs, which are worth mentioning.

Firstly, virtually all plastic microtitre plates are unevenly statically charged, the end result being that the cell monolayers on the outside wells are pulled sideways, often giving a sickle-moon effect. This leads to a more poorly recognizable end-point in the end rows (1 and 12) and usually a lower IFN titre than one derived from the middle rows. A slight to pronounced bowing effect is also sometimes observed across the plate (see *Figure 1*) when the same IFN sample is titrated in rows 1–12 (or 2–11). It is therefore recommended that the outside wells of microtitre plates are excluded from the assays, but this regrettably reduces the number of available wells from 96 to 60.

Secondly, as previously mentioned in (v), if micropipette tips are not changed for every dilution step then there is considerable risk of (uneven) carryover which ultimately leads to a sizeable compounded, and often forgotten, dilution error. Obviously, this error will be small if the end-point occurs in the first, second or third row (A–C or B–D) of the dilution series in the microtitre plate, but will increase progressively with further dilution steps down to row H (or G). Therefore, should financial restraints prevent the assay operator from changing tips, IFN samples containing relatively high IFN

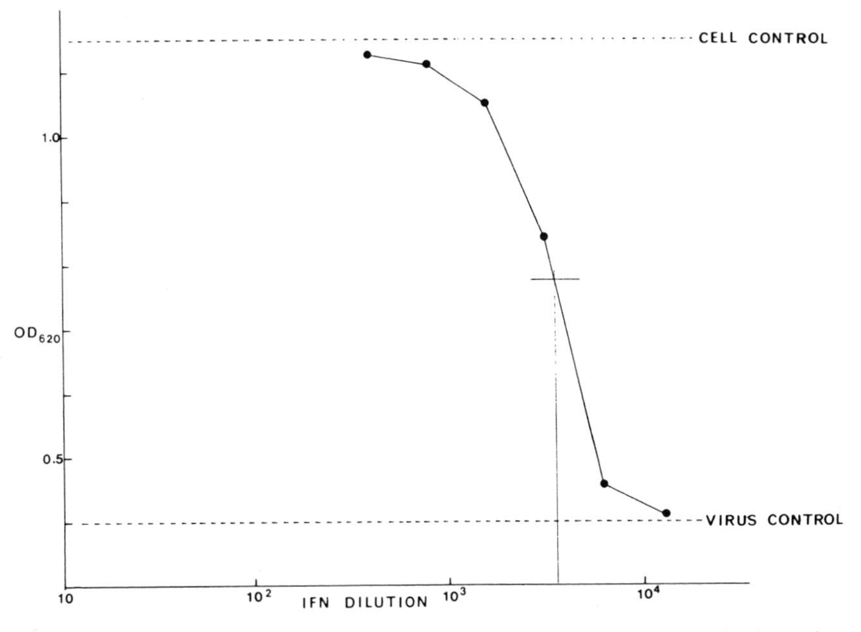

Figure 2. Typical IFN titration curve obtained from a CPER assay. The IFN preparation has a titre of 3600 U/ml.

concentrations should be pre-diluted in order that end-points fall within the top half (rows A−D, or B−D) of the microtitre plate. Ideally, the end-points of the IFN samples should be close to or coincide with that of the IFN LRP, so that the dilution error does not affect normalization of IFN titres to international units (IU).

Thirdly, and lastly, IFN may have a cell growth inhibitory effect on the assay cells which can at worst prevent complete monolayer formation. In addition, synergistic effects with other lymphokines and cytotoxic substances, which are often present in IFN samples, may lead to cell deterioration and death. Such effects can sometimes be observed in the first and second rows of the IFN dilution series where the IFN and toxic substance(s) concentrations are highest. This may mean that the stain intensity of IFN-protected monolayers falls short of that occurring in the cell controls, and that any sigmoidal dose−response curves plotted from such titrations are distorted. Getting around this problem is not always possible, but it is recommended that in such cases the IFN sample is pre-diluted to the point at which it still produces an end-point in the CPER assay, but produces no, or a reduced, cytostatic or cytotoxic effect. A converse effect may be observed when the IFN samples contain high amounts of serum or cell growth factors. In this situation, cells grow more rapidly, form denser monolayers and viral CPE may be sufficiently reduced for an IFN-like effect to be observed. Again dilution, where possible, to overcome such an effect is recommended.

4.1.2 *Calibration of the laboratory IFN reference preparation against the IFN international reference preparation*

Two or three dilutions of both IRP and LRP are prepared in the medium used in the anti-viral assays and IFN anti-viral potencies determined for each in the manner appropriate for the type of assay chosen. The IRP and LRP should be titrated in parallel assays on at least five occasions. The geometric mean titre of each preparation is calculated and the experimentally determined potency for the IRP compared with its assigned value. The correction required to normalize the experimental value, or laboratory unitage, to international unitage is then applied to the geometric mean titre of the LRP. The potency of the LRP may then be expressed in IU. Once this has been done, the LRP should be used in place of the IRP to calibrate all further assays. However, it is advisable to re-check the calibration of the LRP from time to time using the IRP especially if some instability of the laboratory IFN preparation is suspected.

As an example, suppose that the IRP for human leukocyte IFN, 69/19, with an assigned potency of 5000 IU/ampoule, has, because of the sensitivity of the assay on that occasion, an end-point titre of 3.8 $\log_{10}$ laboratory units (LU)/ml (i.e. 6300 LU/ml), and that in the same assay, the LRP of human leukocyte IFN has an end-point titre of 4.1 $\log_{10}$ LU/ml (12 600 LU/ml). The potency of the preparation in IU is then calculated as (12 600/6300) $\times$ 5000 = 10 000 IU/ml. A similar calculation is made when measuring test samples in relation to the LRP in order to express those results in IU.

4.1.3 *Titration of IFN samples of low potency (<10 IU/ml) in CPER assays*

It will have been noticed that for most CPER assays the starting dilutions are 1:4 or 1:10 and that measurement of potencies below 10 IU/ml is rather difficult and subject to large errors. Here the sensitivity of the assay is the determining factor in whether firstly such low levels are detected in the first place and secondly whether good quantitative data can be forthcoming. In the author's experience most CPER assays will titrate IFN-α and IFN-β at or above a 1:1 ratio with the appropriate IRP; that is they are at least as sensitive as the assays used to assign potency values to the IRPs. Thus, most CPER assays will reliably detect down to 10 IU/ml for IFN-α and IFN-β samples. Some increase in sensitivity may be achieved by reducing the m.o.i. of the challenge virus. In some instances, a more sensitive cell line may be found; for example bovine cell lines (EBTR, MDBK) are relatively more sensitive to human IFN-α than human cell lines; human fibroblasts trisomic for chromosome 21, which carries the gene for the IFN-α/β receptor, are intrinsically more sensitive to IFN-α and IFN-β than are human diploid fibroblasts. Thus, in some CPER assays, it is possible to titrate IFN-α and -β samples at or above a 10:1 ratio with the appropriate IRP to give a detection limit of around 1 IU/ml. However, such samples often do not provide sufficient titration points to be plotted graphically, for example the end-point is in the first row of the microtitre plate, and in these circumstances the titre remains very error prone, unless substantiated by data from alternative assays, such as immunoassays. Some samples may give an apparent IFN anti-viral effect without containing any IFN; serum samples are notorious in this respect. For IFN-γ, CPER assays tend to be less sensitive than they are for IFN-α or -β, and therefore there is increased difficulty in measuring IFN-γ of low potency.

4.2 **Plaque-reduction assay**

This is less widely used nowadays. Essentially the procedure is that of the CPER assay steps (i)–(ix) and then in step (x) the cell monolayers are infected with a low m.o.i. of challenge virus, previously found to produce x plaques in untreated cells, for 1 h at 37°C (or room temperature). The virus inoculum is then removed or, since its volume is small (0.025–0.05 ml/well), left in the wells. Cell monolayers are overlaid growth or maintenance medium containing 0.5–0.9% agar (or methyl cellulose) at 39°C, the agar left to set, and the plates returned to a 37°C CO_2 gassed incubator. Plaques may take from 1 to 7 days to develop. They are normally visualized by staining. Buffered neutral red (0.02%) may be used to stain monolayers without removal of the agar layer. If however the agar layer is sloppy it may be carefully removed by tipping it out and then staining may be achieved by using the violet dyes or amido blue black as in step (xii) of the CPER assay procedure.

4.3 **Yield-reduction assay — infectious virus yield inhibition assay**

Essentially the procedure is that of the CPER assay (Section 4.1.1 i–ix) and then in step (x) add virus, for example EMCV, diluted in maintenance medium to give a multiplicity of 5–10 p.f.u./cell in 0.03 ml/well. Virus adsorption may be carried out at room temperature out of the incubator for approximately 45 min. Remove virus inoculum by flicking out and blotting on a paper towel, and wash cells three times with PBS or Earle's balanced salt solution (EBSS) at 0.15–0.2 ml/well. Care should be taken not to dislodge, or make holes in, the cell monolayers at this stage. Add 0.1–0.15 ml of maintenance medium and incubate for 18–24 h at 37°C in a CO_2 gassed incubator. (The time of incubation should be sufficient to allow one growth cycle of the virus.)

The second part of the assay requires infectivity titrations of the virus produced and released into the culture medium of each well of the IFN samples dilution series (*Figure 3*). This will inevitably require the use of a lot more microtitre plates, thus increasing considerably the expense of performing such assays. To minimize the number of titrations, and hence microtitre plates, it is legitimate to pool the contents of replicate wells of IFN dilution series.

The plaque or CPE infectivity titrations may be done in the same cell line as used in the first stage of the assay, but it is often better to use a second cell line in which the virus grows well and has a cytopathic effect. To illustrate the infectivity titration the second step method as employed for an international collaborative reference (1982) bioassay is chosen.

The method is for the titration of EMCV produced in IFN-treated A549 cultures.

(i) Seed microtitre plates with L929 cells at 3–5 $\times$ 10^4 cells/well in growth medium and incubate overnight at 37°C in a CO_2-gassed incubator. Allow at least two plates for each IFN sample or preparation plus the group of virus controls.

(ii) The next day, make serial 10-fold dilutions of each sample (from a single well or a pool of replicate wells) in growth or maintenance medium (*Figure 3*). For dilutions use 0.025 ml of sample plus 0.225 ml of diluent in microtitre trays. As mentioned previously, pipette tips should be changed at each dilution or transfer to avoid carryover; if tips are not changed in both stage 1 and stage 2 expect errors to be compounded.

(iii) Remove the medium from the microtitre trays seeded with L929 cells by flicking out and transfer 0.03 ml of each virus dilution to each of six wells, beginning with the highest dilution. Attach virus for 45 min at room temperature out of the incubator, and then add 0.1 ml of growth medium without decanting the virus

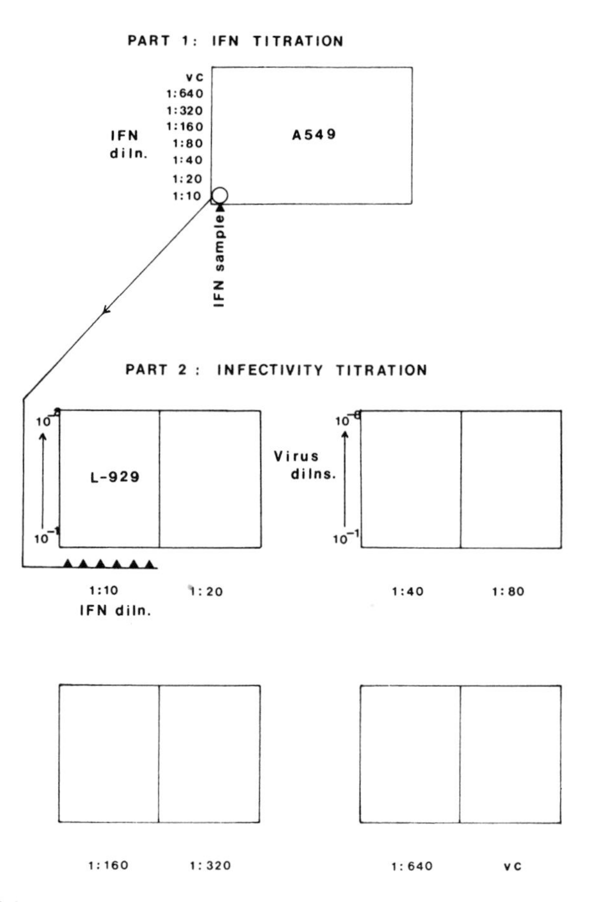

Figure 3. Plan of the procedure for infectious virus yield inhibition assays for IFN estimations.

inoculum or washing cells. Return the plates to a 37°C incubator.

(iv) Examine the cell cultures microscopically for characteristic EMCV CPE after 24 h. Score each well + or 0 for CPE and record the number of wells with CPE (as the numerator) per the total number of wells inoculated (as the denominator). It is recommended that the L929 cell monolayers are stained as per (xii) of the CPER assay procedure when it is judged that CPE is maximum in wells having the highest concentration of virus, for example virus controls. Virus infectivity may be calculated by the Spearman–Karber method (see below) and infectivity titres used to construct dose–response curves to determine IFN titres.

(v) Initial assays will show which virus dilutions from IFN dilution wells are necessary to produce the CPE end-point. Thus, in subsequent assays it should be possible to reduce the number of microtitre plates used for infectivity titrations. As a guide, the figures shown below may be used.

For IFN dilution well: A,B C,D,E F,G,H virus control use infectivity $\log_{10}$ dilutions range:

$-3 \rightarrow -6 \quad -4 \rightarrow -7 \quad -5 \rightarrow -8 \quad -5 \rightarrow -8$

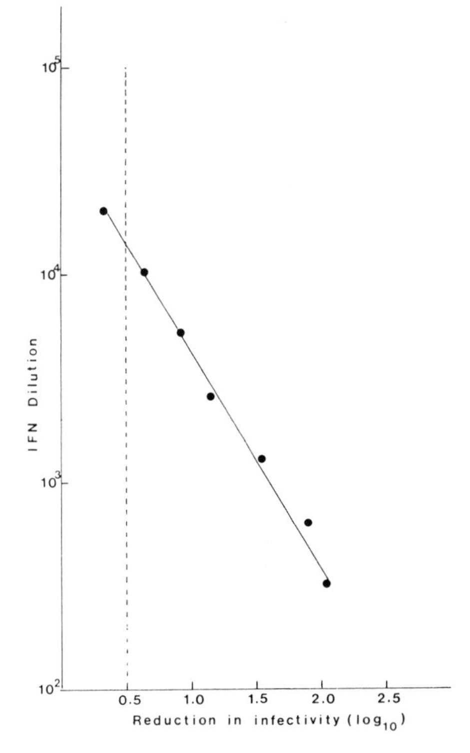

Figure 4. Typical IFN titration curve obtained from an EMCV yield-reduction assay. Data obtained following the second stage titration of infectious virus has been computed by the Spearman–Karber method to give infectivity titres. The IFN titre is the reciprocal of the IFN dilution corresponding to a 0.5 $\log_{10}$ reduction in virus yield, i.e. 14 000 U/ml in this case.

(vi) Calculation of virus titres by the Spearman–Karber method for 10-fold dilution intervals. Transform into a proportion the ratio of CPE positive (+): total wells (+ plus 0) inoculated.

Ratio:	0/6	1/6	2/6	3/6	4/6	5/6	6/6
Proportion:	0	0.17	0.33	0.5	0.67	0.83	1.0

calculate log $TCID_{50}$ from the following equation: log = m – (S – 0.5) where m = log of highest dilution with 100% CPE (proportion = 1), S = sum of proportions infected at dilutions from 'm' and higher.

The titre of virus is the reciprocal of the dilution corresponding to the median infectivity point just calculated, thus the negative of the log.

e.g. dilution: (log)	–3	–4	–5	–6
proportion:	1.0	0.67	0.17	0

log = m – (S – 0.5)

log = – 3 – (1.0 + 0.67 + 0.17 – 0.5) = – 4.34

titre = 4.34

log reduction in virus yield = titre of virus control – titre of virus at each dilution of IFN.

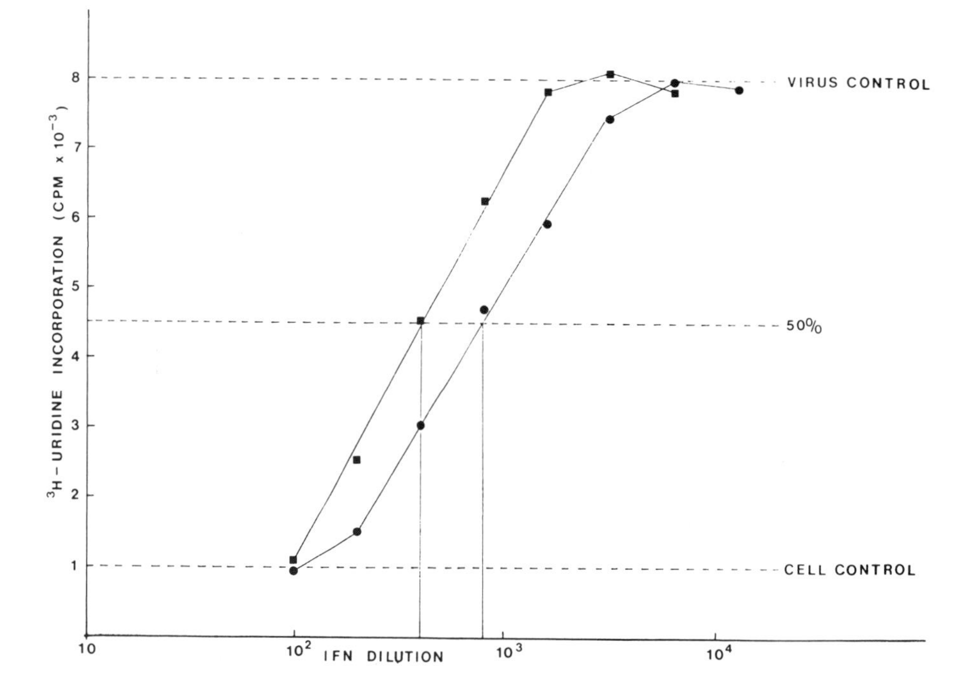

Figure 5. Typical IFN titration curves obtained from inhibition of viral nucleic acid synthesis assays. In this instance, human GM2767 trisomic chromosome 21 fibroblasts were the assay cells and SFV the challenge virus. Note that the titration curve for an IFN-α preparation (●—●) is not parallel to the titration curve for an IFN-β preparation (■—■). The IFN titre is the reciprocal of the IFN dilution corresponding to a 50% reduction in [^{3}H]uridine incorporation, the mid-point on the ordinate between virus and cell control.

Plot: $\log_{10}$ dilution of IFN versus log−reduction in virus yield (see *Figure 4*). The IFN titre is the dilution (reciprocal) of IFN corresponding to a 0.5 $\log_{10}$ reduction in virus yield.

The two stage assay as presented is rather cumbersome, but may appeal to purists. However, virus may be quantitified more readily and simply by alternative means, for example by measuring haemagglutination by released virus (2).

4.4 Inhibition of viral nucleic acid synthesis (IVNAS) method

(i) Use steps (i)−(ix) of the CPER method (Section 4.1.1).

(ii) Remove medium by flicking out, blot on a paper towel and add 0.05 ml of maintenance medium containing actinomycin D (1−3 μg/ml) to cell controls. To all remaining wells, add 0.05 ml of maintenance medium containing actinomycin D (1−3 μg/ml) and the infectious challenge virus. SFV, Sindbis virus (SV) and VSV are suitable as challenge viruses in this particular assay. Use a relatively high m.o.i., for example approximately 25−50 p.f.u./cell, empirically determined to give high levels of [^{3}H]uridine incorporation. Return the plates to the incubator for 2.5 h.

(iii) Without removing the virus inoculum, add a further 0.05 ml of maintenance medium containing tritiated [^{3}H]uridine (10–20 μCi/ml) to each well. Incubate for a further 2.5–3 h at 37°C.
(iv) Remove the medium and flood the plate twice with ice-cold 5% trichloroacetic acid (TCA) and once with ice-cold distilled water.
(v) Dissolve the well contents in 0.1 ml of 2% sodium dodecyl sulphate (SDS) at 37°C for 30 min. Tap the sides of the plate to ensure the contents are mixed and add absorbent filter discs to all wells. Dry the discs in a warm incubator and then transfer them to scintillation counter vials. (Alternatively, following solubilization, the well contents may be transferred using Pasteur or micropipettes to filter discs or direct to scintillation vials.)
(vi) Add appropriate scintillation fluid and count the radioactivity incorporated in a liquid scintillation counter with channels and gain settings appropriate for ^{3}H.
(vii) Plot counts incorporated versus the log of the IFN dilution (see *Figure 5*). The end-point is the dilution of IFN which inhibits the incorporation of [^{3}H]uridine by 50% between the virus control and the cell control.

Alternatively, the whole assay may be performed by seeding cells into glass vials or, better, glass inserts for scintillation vials. In this instance, volumes of medium are adjusted to the size of the glass vial (inserts) and solubilization of TCA-precipitated monolayers is effected by addition of Soluene diluted 1:3 in toluene. Acidified toluene scintillant (0.5% PPO, 0.01% POPOP, 0.01% acetic acid) is added directly into the vials (inserts) and the radioactivity measured.

5. REFERENCES

1. Stewart,W.E., II (1979) *The Interferon System.* Springer-Verlag, Vienna, New York.
2. Grossberg,S.E. and Sedmark,J.J. (1984) In *Interferon. Volume 1: General and Applied Aspects.* Billiau,A. (ed.), Elsevier, Amsterdam and New York, p. 189.
3. Various authors (1981) In *Methods in Enzymology.* Pestka,S. (ed.), Academic Press Inc, London and New York, Vol. 78, p. 339.
4. Mahy,B.W.J., ed. (1985) *Virology – A Practical Approach.* IRL Press, Oxford and Washington D.C.
5. Standardization of Interferons (1983) *Report of a WHO Informal Consultation.* World Health Organization Technical Report Series No 687, p. 35.
6. Standardization of Interferons (1985) *Report of a WHO Informal Consultation.* World Health Organization Technical Report Series No 725, p. 28.

CHAPTER 10

Analysis of anti-viral mechanisms: interferon-regulated 2′5′ oligoadenylate and protein kinase systems

ROBERT H.SILVERMAN and DAVID KRAUSE

1. INTRODUCTION

Interferon treatment of cells results in profound biochemical and physiological alterations leading to the establishment of anti-viral states, inhibition of cell proliferation and immunoregulation (1). The diverse, phenotypic changes seen in interferon-treated cells are the result of specific alterations in gene expression (e.g. ref. 2). The lists of known activities (3) or of unidentified proteins (4) that are either enhanced or diminished after interferon treatment are lengthy and complicated indeed. This technical guide to interferon-induced activities is concerned only with the two best characterized pathways, namely the interferon-regulated 2′5′ oligoadenylate and protein kinase systems (for reviews, see refs 3 – 5). These two pathways mediate some of the effects of interferons in cells by inducing the breakdown of RNA and inhibiting protein synthesis initiation, respectively. The methods described here allow the investigator quantitatively to monitor these activities from intact cells or cell-free systems. These techniques may be applied to determining the interferon responsiveness of cells and to the study of interferon action in general.

2. THE 2′5′ OLIGOADENYLATE SYSTEM

2.1 Background information

2′5′ Oligoadenylates (abbreviated here to 2-5A) constitute an oligomeric series of unusual molecules with the general formula: $p_x(A2'p)_nA$; $x = 2$ or 3, $n \geq 2$. The only well characterized function for 2-5A is in the regulation of RNA stability, although other effects of 2-5A may exist. The 2-5A system, as it is presently known, consists of three types of activities (*Figure 1*).

(i) 2-5A synthetases (3 – 5) are enzymes which require double-stranded (ds)-RNA to catalyse the synthesis of 2-5A from ATP. The synthetases may be the most highly regulated enzymes of the interferon system; in this regard several thousand-fold increases have been observed in certain cell lines after interferon treatment (6). It is for this reason that synthetase activity is frequently used to monitor the biochemical responsiveness of patients to interferon therapy (e.g. refs 7 and 8).

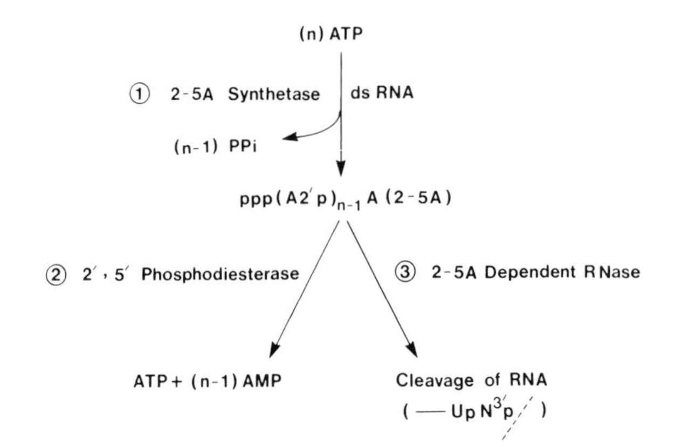

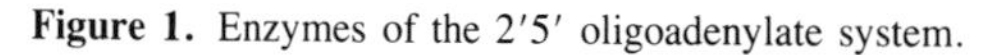

Figure 1. Enzymes of the 2′5′ oligoadenylate system.

(ii) 2-5A oligomers are degraded from their 2′-3′ termini to ATP and AMP by the action of 2′,5′-phosphodiesterase(s). In addition, the 5′-phosphoryl groups of 2-5A may be removed by phosphatases present in cells.

(iii) 2-5A-dependent RNase (RNase L, RNase F) is the enzyme which mediates effects of 2-5A on RNA stability. Subnanomolar levels of 2-5A activate 2-5A-dependent RNase resulting in the cleavage of single-stranded RNA after UpN sequences leaving 3′-terminal phosphoryl groups (UpU and UpA are highly preferred cleavage sites). 2-5A-dependent RNase may be capable of cleaving any RNA molecule, be it viral or cellular, which contains these dinucleotide sequences in a single-stranded conformation. However, a localized activation model has been proposed which describes a mechanism for the preferential degradation of viral RNA under certain circumstances (9). 2-5A-dependent RNase, if indeed there is only one, resides in both the cytoplasm and the nucleus (10). Multiple 2-5A-binding proteins have been described which could represent different segments or forms of the nuclease or other unidentified proteins associated with the 2-5A system (11 – 13).

2-5A is almost certainly responsible for some of the anti-viral activities of interferons. Perhaps the best evidence of an anti-viral role for 2-5A comes from studies on the effect of interferon treatment on the replication of encephalomyocarditis virus (EMCV). The cascade of events resulting in anti-viral activity against EMCV may be described as follows (*Figure 2*).

(i) Interferon binds to its cell surface receptors thus inducing the expression of the 2-5A synthetase and enhancing levels of the nuclease.

(ii) The EMCV then infects the interferon-treated cell and produces dsRNA during its replicative cycle.

(iii) The viral dsRNA in turn activates the pre-existing synthetases to produce 2-5A.

(iv) Finally, the 2-5A activates 2-5A-dependent RNase, cleaving both viral and cellular RNA. The end result is that the virus is either eliminated or greatly inhibited in its ability to produce progeny virus particles.

In addition to its role in the anti-viral effects of interferons the 2-5A system may

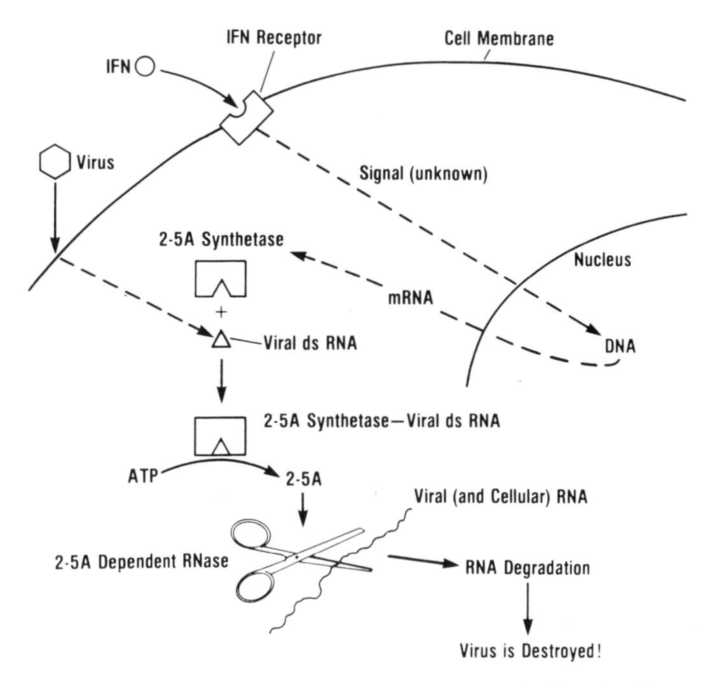

Figure 2. The role of 2′5′ oligoadenylates in the anti-viral response of cells to interferon treatment.

have significance in the regulation of cell growth, in hormone action and in cell differentiation.

The methods described here are those which are performed in our laboratory; accordingly, other methods and related techniques used in some other laboratories may be mentioned or described but space does not allow all of these to be discussed in detail. However, many of these alternate techniques are described in ref. 14.

2.2 Detection of 2′5′ oligoadenylates and related material

In designing methods for detecting 2-5A it is critical that biologically relevant levels of the oligonucleotides be detectable. 2-5A is a potent regulator of 2-5A-dependent RNase with activation occurring in the subnanomolar range. Therefore, it is essential that the limit of detection for 2-5A extend to at least the low nanomolar range. In addition, it is necessary that the assays be relatively specific for 2-5A; for instance, the natural occurrence of molecules with 2-5A-related structure is now well established. There is a variety of methods available for detecting very low levels of 2-5A (for review see ref. 15). Each of these has its advantages and disadvantages. Therefore, it is often necessary to use a combination of methods when attempting to identify naturally occurring 2-5A. In this regard, we strongly recommend that anyone contemplating searching for 2-5A from intact cells first read the excellent article of Reid *et al.* (15). Here we will consider two complementary types of methods: (i) the radiobinding assay (16), and (ii) the core–cellulose assay (17); and, in addition, an rRNA cleavage method will be presented when considering assays of 2-5A-dependent RNase (18). These techniques utilize 2-5A-dependent RNase to measure levels of 2-5A; the first method relies on only the 2-5A binding properties of the enzyme while the latter two monitor the RNase activity mediated by 2-5A. It should be mentioned that several immune assays for detecting 2-5A and related structures have also been developed and are described elsewhere (16, 19–23).

2.2.1 *Assay of 2'5' oligoadenylates by the radiobinding method*

The radiobinding assay developed by Kerr's group (16) is a method in which unlabelled 2-5A and a radioactively labelled 2-5A derivative compete for 2-5A binding activity (most probably 2-5A-dependent RNase) present in cell extracts. The very high affinity of 2-5A-dependent RNase for 2-5A makes this assay possible. There are three major components to the assay: (a) a source of 2-5A binding activity, for example an extract of mouse L cells; (b) a highly radioactive derivative of 2-5A, ppp(A2'p)$_3$A[^{32}P]pCp (2-5A[^{32}P]Cp); and (c) the competitor, namely a known concentration of authentic 2-5A, other oligonucleotide or the unknown material to be tested. Any 2-5A or 2-5A[^{32}P]Cp bound to 2-5A-dependent RNase will also bind to a nitrocellulose filter; and conversely any material not bound to protein will be washed free of the filter.

(i) Synthesis and purification of ppp(A2'p)$_3$A[^{32}P]pCp (2-5A[^{32}P]Cp). Because 2-5A synthetase requires relatively high concentrations of ATP (>100 μM) for the efficient synthesis of 2-5A (24), synthesis of highly radioactive 2-5A by this route is not practical. Instead, 2-5A[^{32}P]Cp is enzymatically synthesized and purified for use in the radiobinding assay (25). This is accomplished by first ligating [5'-^{32}P]pCp to the 3'-OH of tetramer 2-5A using the enzyme T4 RNA ligase; and then purifying the product, 2-5A[^{32}P]Cp, using high performance liquid chromatography (h.p.l.c.). The acceptor, tetramer 2-5A, may by synthesized and purified as described below (Section 2.3.3) or it may be purchased from Pharmacia Inc. There are at least three commercial sources of T4 RNA ligase (New England Biolabs, Pharmacia and Bethesda Research Labs, BRL); in our experience the ligase obtained from BRL gives about 50% more ligation than those from the other two companies.

In the T4 RNA ligase reaction synthesis of 2-5A[^{32}P]Cp involves the following reaction:

$$\text{ppp(A2'p)}_3\text{A} + [5'\text{-}^{32}\text{P}]\text{pCp} + \text{ATP} \xrightleftharpoons{\text{T4 RNA ligase, Mg}^{2+}} \text{ppp(A2'p)}_3\text{A}[^{32}\text{P}]\text{pCp} + \text{AMP} + \text{PPi}$$

dTTP is included to inhibit phosphatases which sometimes contaminate preparations of T4 RNA ligase. However, recent preparations of enzyme were found to be phosphatase-free and therefore addition of dTTP may no longer be required.

(1) Dry 250–500 μCi (9–18 × 10^6 Bq) of [5'-^{32}P]pCp (2000–4000 Ci/mmol, from New England Nuclear or Amersham, in 10 mM Tricine, pH 7.6) *in vacuo* using a Speed Vac (Savant) or similar apparatus. Dissolve the dried [5'-^{32}P]pCp in 25 μl of ligase buffer [100 mM Hepes, pH 7.6; 15 mM $MgCl_2$; 6.6 mM dithiothreitol (DTT); and 20% v/v dimethyl sulphoxide (DMSO) (Mallinckrodt)].

(2) Place the tube containing the pCp solution on ice and add in the following order:

5 μl of (1.0 mM dTTP; 0.1 mM ATP, pH 7.0)
10 μl of 0.5 mM tetramer 2-5A in water
8 μl of water
2 μl of T4 RNA ligase (BRL) (18.0 units)

Mix thoroughly by pipetting up-and-down a few times. The enzyme solution contains 50% glycerol and must be added with mixing.

(3) Fill a small beaker (or the top from the lead container in which the pCp is shipped) with ice. Place the reaction vessel into a refrigerator (2–4°C) for 16–18 h, during which time the ice will melt leading to a gradual acceleration in the rate of the reaction.

(4) Heat the reaction mixture to 90°C for 5 min and then centrifuge at 10 000 *g* in a microfuge for 10 min. Remove the supernatant to a new tube.

The extent of ligation may be easily monitored by thin layer chromatography (t.l.c.). Dilute 1 μl of the reaction in 100 μl of water and spot 2 μl of this to a pre-treated strip of polyethyleneimine (PEI)–cellulose coated plastic. (The method for pre-treatment of the t.l.c. sheet and commercial sources of PEI–cellulose plates are given in *Table 1*). Perform the chromatography in a tank containing 0.25 M ammonium bicarbonate. Remove the t.l.c. strip from the tank when the solvent front reaches the top (20 cm) and then air-dry and expose a sheet of X-ray film (Kodak XAR-5). After about 1–2 h develop the film; an autoradiogram from a t.l.c. separation of a typical preparation is shown in *Figure 3*, lane 1 ([5′-^{32}P]pCp is shown in lane 2). Cut the areas containing the 2-5A[^{32}P]Cp from the plate and determine the radioactivity by scintillation counting. Typically 50–80% of the [5′-^{32}P]pCp becomes ligated to the 2-5A.

Because the unlabelled acceptor molecule, tetramer 2-5A, is present in the reaction mixture in a large molar excess compared with the product, 2-5A[^{32}P]Cp, it is necessary to perform an h.p.l.c. separation to purify the latter. For this purpose we use an h.p.l.c. system from Beckman Instruments which includes a model 421 controller, two model 122 solvent modules (pumps) and a model 160 254 nm spectrophotometer. The column effluent is connected to a model 201 Gilson fraction collector. Separate the reaction mixture on a reverse phase C-18 column (either a μBondapak C18 column, No.

Table 1. Washing of polyethyleneimine (PEI)–cellulose t.l.c. plates.

PEI–cellulose t.l.c. plates may be either Machery Nagel Polygram Cel 300, PEI (Camlab, Cambridge, UK) or Cat. 5504, Pre-coated t.l.c. plastic sheets, PEI–cellulose F (from E.M. Reagents, MC/B Manufacturing Chemists, Inc., 2909 Highland Avenue, Cincinnati, OH 45212). These are prepared for use by performing the following washing procedure (I.M.Kerr, personal communication). Washing can be done using 10 sheets at a time (20 cm × 20 cm) in a glass baking dish.

1. 1 M NaCl, 10 min at room temperature.
2. Three times with water.
3. Pyridine/formic acid (88%) 22 ml/40 ml to 1 litre (in fume hood).
4. Three times with water.
5. Hang to air-dry overnight at room temperature.
6. Store at 2–4°C.

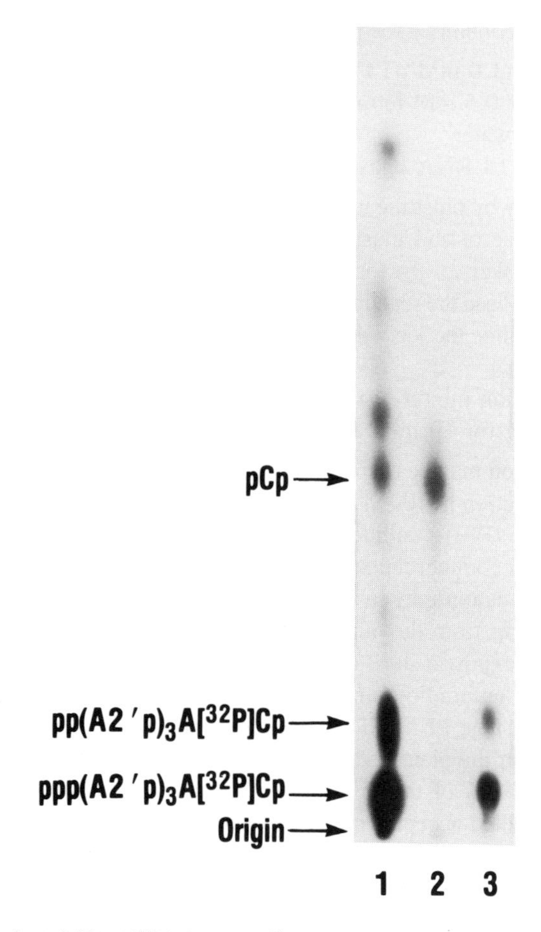

Figure 3. Analysis of ppp(A2′p)$_3$A[^{32}P]pCp (2-5A[^{32}P]Cp) by t.l.c. **(1)** The unresolved reaction products, (2 μl of a 1:100 dilution); **(2)** h.p.l.c. purified [5′-^{32}P]pCp, 2 μl; and **(3)** h.p.l.c. purified ppp(A2′p)$_3$A[^{32}P]pCp (2-5A[^{32}P]Cp), 2 μl were spotted to a sheet of PEI−cellulose coated plastic which was then developed in 0.25 M ammonium bicarbonate. An autoradiogram of the chromatogram is shown.

27324, 3.9 mm × 30 cm, from Waters Co. or a Partisil ODS-3/5 μm, 4.6 mm × 25 cm, column from Ace Scientific Co.) in a linear gradient from 0 to 20% of 1:1 methanol:water in 50 mM ammonium phosphate, pH 7.0 at a flow-rate of 1.0 ml/min in 20 min as described by Brown *et al.* (26).

(1) First standardize the column to determine the times of elution for different oligomers of 2-5A. It is convenient to use a 'crude' preparation of 2-5A obtained in a 2-5A-synthetase reaction, for example using an extract from interferon-treated HeLa cells (Section 2.3.3). However, one may also use a mixture of commercially available oligomers of 2-5A (Pharmacia Inc.) Inject about 5 μl of a millimolar solution of 2-5A onto the column and elute as described above. A typical absorbance profile (at 254 nm) of a separation of the oligonucleotides from a 2-5A synthetase reaction is shown in *Figure 4*.

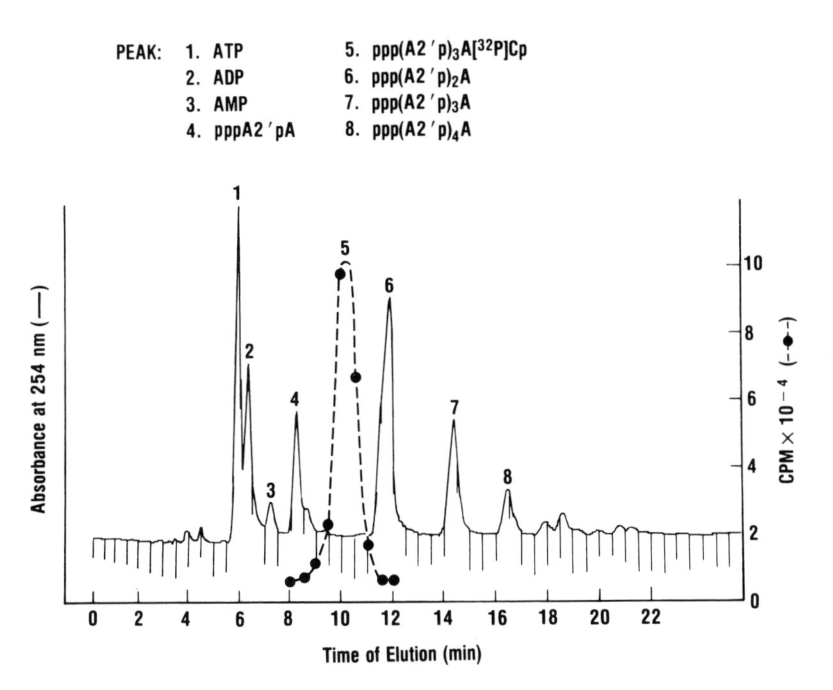

Figure 4. Isolation of 2-5A[^{32}P]Cp by h.p.l.c. A mixture of 2-5A oligomers, ATP, ADP and AMP from a 2-5A synthetase reaction (1 μl) were separated by h.p.l.c. and monitored by absorbance at 254 nm (solid line, see text). In a separate analysis, ppp(A2′p)$_3$A[^{32}P]Cp (2-5A[^{32}P]Cp) was purified from a reaction mixture and the radioactivity was determined from 2 μl of the fractions (broken line).

(2) Once it has been established that the system is optimally separating 2-5A oligomers, wash the injector port thoroughly and then perform a blank run (preferably two blank runs) to remove any traces of 2-5A which could contaminate and thereby reduce the specific activity of the preparation of 2-5A[^{32}P]Cp.

(3) Program the fraction collector to collect fractions every 0.5 min.

(4) Inject the reaction mixture, start the elution program, and collect the fractions.

Locate the peak of 2-5A[^{32}P]Cp by pipetting 2 μl of each fraction into 5 ml of aqueous scintillation counting fluid, and determine the radioactivity. A major radioactive peak occurs at about 10 min of elution (just prior to the time of elution of trimer 2-5A); this peak is the purified 2-5A[^{32}P]Cp (*Figure 4*, peak 5; plotted with the 2-5A standards from a separate analysis). We sometimes observe a second peak of radioactivity of unidentified composition which appears near the time of elution of tetramer 2-5A. Analyse the presumptive 2-5A[^{32}P]Cp by applying 2 μl of the peak fraction to a PEI−cellulose strip and develop in 0.25 M ammonium bicarbonate (the autoradiograph should appear as shown in *Figure 3*, lane 3). The 2-5A[^{32}P]Cp is typically useable in the radiobinding assay for about 6 weeks; after this time radioactive decay results in reduced sensitivity of the assay.

(ii) Preparation of cell extract as a source of radiobinding activity. A variety of different types of cells including rabbit reticulocytes may be used as a source of 2-5A

Table 2. Preparation of cell pellets and cell extracts.

A. Preparation of cell pellet (from one 75 cm^2 plate of cells).

1. Pour out media from monolayer cultures.
2. Add about 20 ml of PBS without calcium and magnesium to cells and wash by rocking a few times; discard the PBS.
3. Add 10 ml of fresh PBS, scrape the cells, pipette the cells up-and-down a few times in order to break up clumps of cells.
4. Transfer the cell suspension to a centrifuge tube, centrifuge at 700 *g* for 10 min at 2°C, and discard the supernatant.
5. Resuspend the cells in 1.0 ml of ice-cold PBS, transfer the cell suspension to a 1.5 ml conical tube, centrifuge at 700 *g* for 5 min.
6. Remove and *discard all of the supernatant* using a micropipette.
7. Quick-freeze cells on dry ice, transfer to a −70°C freezer for storage or use immediately for preparation of cell extract. This method should result in about 100 μl of packed cell volume which will vary depending on the cell type and density of the culture.

B. Preparation of post-mitochondrial fraction from cell pellets (all steps performed at 0−2°C).

1. Prepare NP-40 lysis buffer [0.5% (v/v) NP-40; 90 mM KCl; 1.0 magnesium acetate; 10 mM Hepes pH 7.6; and 2.0 mM 2-mercaptoethanol]
   1.0 ml of NP-40 (Sigma Co.)
   4.5 ml of 4 M KCl
   0.2 ml of 1 M magnesium acetate
   4.0 ml of 0.5 M Hepes pH 7.5
   28 μl of 14.3 M 2-mercaptoethanol
   Adjust to 200 ml with distilled water, store at −20°C.
   Important: on the day on which it is used, a solution of 10 mg/ml of leupeptin (Sigma Co.) is diluted 1:100 in the NP-40 lysis buffer to inhibit proteinase activity. The leupeptin is stored at −20°C as a 10 mg/ml stock solution in water.
2. Lyse the cells by thawing into 1.5 packed cell volumes of NP-40 lysis buffer containing leupeptin. The packed cell volume is estimated by holding the tube up to a 1.5 ml tube with graduated markings on it.
3. Vortex the cell lysate for 10 sec.
4. Centrifuge at 10 000 *g* for 10 min, carefully remove the supernatant to a fresh 1.5 ml tube and centrifuge a second time at 10 000 *g* for 10 min. Remove the supernatant to a fresh 1.5 ml tube.

C. Preparation of complete cell extract (all steps performed at 0−2°C)

Perform cell lysis and vortexing as described in part B steps 1−3 but do not centrifuge (step 4).

D. Determination of protein concentration in cell extracts.

1. Mix one part of Bio-Rad protein assay dye with four parts of distilled water, filter through Whatman filter paper in a funnel.
2. Prepare standards by mixing 2, 5, 10, 20 and 30 μl of a 1 mg/ml solution of bovine serum albumin (BSA) with 2.0 ml of diluted dye.
3. Dilute the cell extracts 1:10 by mixing 5 μl of cell extract in 45 μl of distilled water and add 5 μl of the diluted cell extract to 2.0 ml of diluted dye.
4. Wait 5 min and then determine the A_{595} of the assay mixtures, using diluted dye as a blank.
5. Determine protein concentrations from the standard curve.

binding activity for the radiobinding assay. The major considerations in selecting a source of 2-5A binding activity are:

(1) availability or ease with which the cells are grown;
(2) the basal level of 2-5A binding activity should be relatively high;
(3) the levels of 2-5A degrading activities should be low;

(4) the extract should give a low level of background binding of 2-5A[^{32}P]Cp in assays containing an excess of unlabelled 2-5A.

For instance, some preparations of reticulocyte lysates result in high levels of background binding and HeLa cells are high in 2-5A degrading activities; therefore these extracts are not recommended. Optimal results are obtained using post-mitochondrial fractions of mouse L cells. For this purpose, mouse L cells may be grown in suspension cultures in spinner-modified essential medium supplemented with L-glutamine, non-essential amino acids, vitamins, penicillin – streptomycin and 10% Nu serum (Collaborative Research). Cells are washed twice in phosphate-buffered saline (PBS) and cell extracts are prepared as described in *Table 2B*. These may be stored in small aliquots (100 μl) at −70°C.

(iii) The radiobinding assay for detecting 2'5' oligoadenylates.

(1) Prepare a '2-5A[^{32}P]Cp mixture' by pipetting the following reagents into a 1.5 ml tube on ice:

    5000 c.p.m. of 2-5A[^{32}P]pCp for each assay to be performed;
    1 μl for each assay of '20 × mix': (200 mM Tris-HCl, pH 7.6; 40 mM magnesium acetate; 8.0 mM ATP, pH 7.6; and 40% glycerol);
    Adjust the volume with water to give 10 μl of 2-5A[^{32}P]Cp mixture for each assay to be performed.

(2) Aliquot 10 μl of the 2-5A[^{32}P]Cp mixture [step (1)] to 0.5 ml conical tubes.

(3) Add 5 μl of competitor (i.e. 2-5A, other oligonucleotide, or the sample) to the assay tubes. A standard curve using authentic 2-5A must be performed with each assay by using 5 μl of 1.25, 2.5, 5, 10, 20, 40, 80 and 250 nM of trimer 2-5A.

(4) Add 5 μl (100 – 200 μg) of post-mitochondrial mouse L cell extract (*Table* 2) to each tube.

(5) Incubate the assay tubes on ice for 90 min.

(6) Cut 2.5 cm diameter nitrocellulose discs (Millipore Co., HAWP 025 00, 0.45 μm pores) in half and number each with water-insoluble ink. Place these on a sheet of aluminium foil and spot all of each assay mixture on the filters. The latter must be done quickly so that the last filter is spotted before the first filter becomes dry.

(7) Immediately after spotting the last filter, place the filters in a beaker containing 10 ml of tap distilled water per filter using Millipore forceps, swirl gently for 30 sec, discard the water into an appropriate waste container, replace the same volume of water, swirl again for 30 sec and discard all of the water. (Allowing the filters to remain in the water results in reduced binding.)

(8) Remove the filters, and spread them on a new piece of aluminium foil, dry the filters using a heat lamp for about 15 min.

(9) Place the filters in counting vials, add 5 ml of scintillation fluid and determine the radioactivity.

A typical radiobinding assay is shown in *Figure 5* in which trimer and tetramer 2-5A and some related compounds were tested. The concentration of tetramer 2-5A required to compete 50% of the 2-5A[^{32}P]Cp was 0.5 nM. The level of 2-5A in a sample is

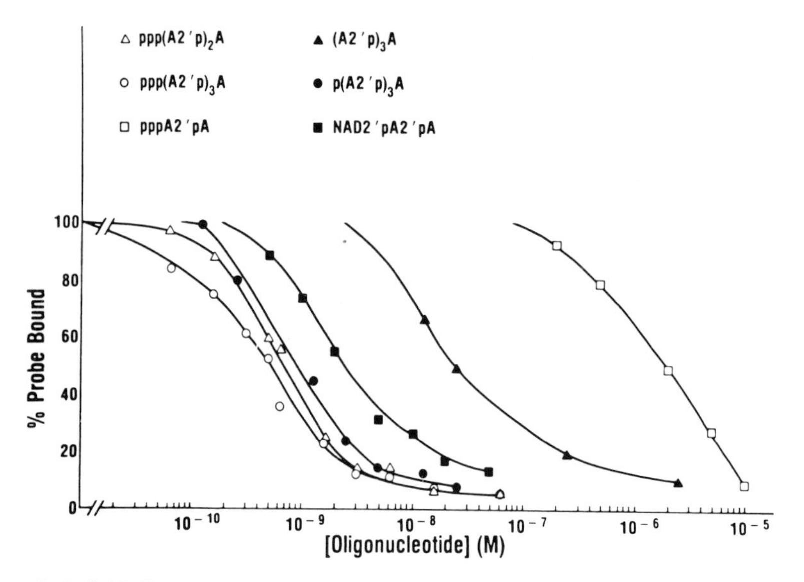

Figure 5. Radiobinding assay for detecting 2′5′ oligoadenylates. The radiobinding assay of Knight *et al.* (16) was performed with 2-5A or analogues of 2-5A as indicated in the figure. (From ref. 17, reprinted with the permission of the publisher.)

estimated by comparing its ability to displace 2-5A[^{32}P]Cp with that found using the authentic 2-5A in the standard curve. The radiobinding assay may be used to determine levels of 2-5A from intact cells (Section 2.2.3) or from 2-5A synthetase reactions (Section 2.3.1). In addition, an adaptation of this assay is used to estimate levels of 2-5A-dependent RNase (Section 2.4.2).

The order in which the components of the radiobinding assay are added is important; adding the 2-5A[^{32}P]Cp to the cell extract prior to adding the competitor results in reduced sensitivity. Conversely, however, pre-incubating 2-5A with cell extract (30 min) and then adding 2-5A[^{32}P]Cp increases the sensitivity of the assay by 3- to 5-fold (P.J.Cayley and I.M.Kerr, personal communication).

2.2.2 *Core – cellulose assay for detecting 2′5′ oligoadenylates*

The radiobinding assay (Section 2.2.1) has become one of the standard methods used in the detection of 2-5A due to its sensitivity and specificity. However, in some circumstances, additional types of analysis may be required because not all molecules with high affinity for 2-5A-dependent RNase are, in fact, authentic 2-5A. This is especially true when analysing material from intact cells (Section 2.2.3). It is often necessary to determine whether the isolated '2-5A' is in fact biologically active; in other words is it capable of activating 2-5A-dependent RNase? The core – cellulose assay (17) measures activation of 2-5A-dependent RNase in the absence of some of the interfering factors present in cells (e.g. endogenous nucleotides and enzymes capable of modifying, degrading or synthesizing 2-5A). This technique consists of immobilizing and partially purifying 2-5A-dependent RNase using core – cellulose [(A2′p)$_3$A-cellulose] (*Figure 6*) and then monitoring the breakdown of poly(U)-3′-[^{32}P]Cp into acid-soluble fragments. 2-5A-dependent RNase has a high affinity for core – cellulose (17) thus per-

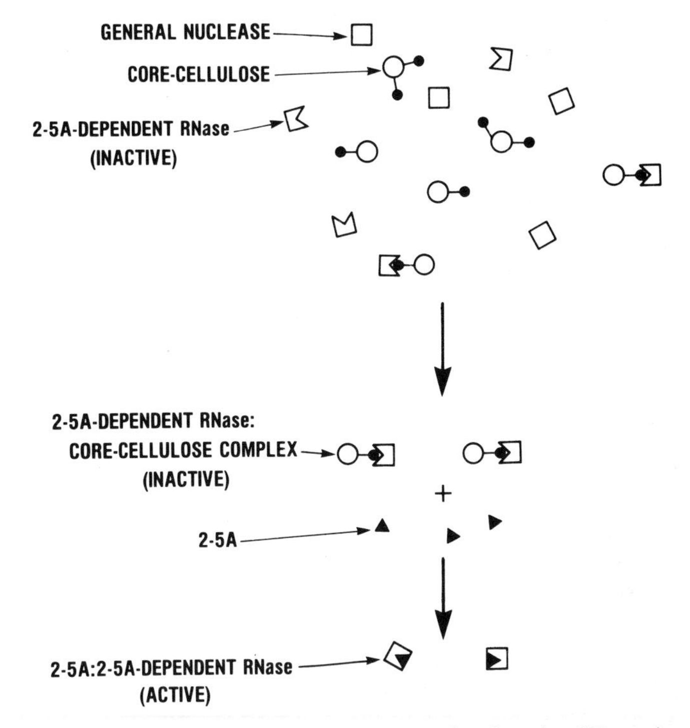

Figure 6. Immobilization, washing and activation of 2′5′ oligoadenylate-dependent RNase using core – cellulose.

mitting it to be separated from general nuclease activities (*Figure 6*). Although core – cellulose binds 2-5A-dependent RNase, it does not activate the enzyme to cleave RNA (17). However, when 2-5A is added to the 2-5A-dependent RNase:core – cellulose complex, it displaces the core – cellulose resulting in the activation of the RNase (*Figure 6*).

(i) Preparation of core – cellulose. The affinity resin, core – cellulose, is prepared by an adaptation of the method developed by George R.Stark (London) and co-workers for the synthesis of poly(I):poly(C) – cellulose (27). The advantages of using finely-divided cellulose as a support are that:

(a) the resin is easily distributed as a suspension using a micropipette (yellow) tip,
(b) the cellulose remains in suspension for several minutes permitting assays to be performed without continuous shaking,
(c) the cellulose particles can be highly derivatized externally, thus permitting assays to be performed using small volumes of cellulose.

For the preparation of 'hydrazide – cellulose' carry out the following steps. Use deionized and distilled water throughout.

(1) Prepare 80 ml of ammoniacal $Cu(OH)_2$ by mixing in a beaker 3.6 g of $Cu(OH)_2$ (ICN Co.), 0.8 g of sucrose, 48 ml of concentrated NH_4OH, and 32 ml of deionized and distilled water. Stir at room temperature overnight. Do not heat!

(2) Centrifuge at 700 *g* for 5 min and collect the supernatant (discard the undissolved material). The solution will be a very deep blue colour. Add 5.0 g of CF-11 cellulose (Whatman Co.) to the ammoniacal $Cu(OH)_2$ solution and stir at room temperature for 1 h.

(3) Centrifuge at 1500 *g* for 10 min, *collect the supernatant* and discard the pellet. There will be a large pellet of undissolved cellulose (perhaps 30–50% of the total volume). This is a deliberate excess but it is desirable to have the maximum amount of cellulose in solution.

(4) Pre-heat 320 ml of de-ionized and distilled water to 70°C in a water bath (make sure the water is up to temperature before proceeding). Add supernatant (from step 3) to the heated water with stirring. Immediately precipitate the cellulose by adding a solution of 25% H_2SO_4 until the pH is about 6 (~35–40 ml is required). The colour will change from dark blue to pale sky blue; monitor the pH with indicator strips.

(5) Centrifuge at 1500 *g* for 10 min, discard the supernatant, and then wash once with 320 ml of water, and then twice with 320 ml of 1.2 M HCl, followed by two washes with 320 ml of water. After the last centrifugation, discard the supernatant. This should result in about 20 ml of packed cellulose which is white in appearance. Each wash is done by centrifuging at 1500 *g* for 10 min, discarding the supernatant, and resuspending. Conical (250 ml capacity) centrifuge tubes are ideal for this purpose. Care must be taken not to pour a little of the cellulose down the drain at each washing step.

(6) Add 20 ml of water and then 40 ml of 2 M Na_2CO_3 to the cellulose pellet and resuspend.

(7) **Proceed to a fume hood, make sure that it is in good working order and that a co-worker is near by in case of accident. Read the warning label on the cyanogen bromide container.** Open a fresh bottle of h.p.l.c. grade acetonitrile (Fisher Scientific Co.). Place a 50 ml beaker with stir bar on a stirring plate in the fume hood. Add to the beaker 2 ml of acetonitrile and then 4 g of cyanogen bromide (Sigma Co.); this must be weighed out on a balance in the hood. The cyanogen bromide should go into solution in 5–10 min; if it does not it is advisable to start again with new reagents.

(8) Place the cellulose suspension in a 500 ml beaker with a stir bar and place on the stir plate in the fume hood. Stir very rapidly while adding the cyanogen bromide solution with a Pasteur pipette. Stir for an additional 5 min. Pipette the cellulose suspension into a screw-capped centrifuge tube and centrifuge at 1500 *g* for 10 min. Discard the supernatant into an appropriate waste container, and then wash the pellet in 250 ml of 0.1 M $NaHCO_3$ pH 9.5. Resuspend the pellet in 80 ml of 0.1 M $NaHCO_3$ pH 9.0.

(9) Dissolve 6 g of ϵ-aminocaproic acid methyl ester HCl as linker (from Chemical Dynamics Corporation, Hadley Road P.O. Box 395, South Plainfield, NJ 07080, USA) into 120 ml of ice-cold 0.1 M $NaHCO_3$, pH 9.0. Add this to the cellulose suspension and stir overnight at 2–4°C.

(10) Centrifuge the cellulose at 1500 *g* for 10 min, wash twice with 250 ml of water, and then resuspend in 20 ml of water. **Return to the fume hood.** Add 28 ml

of hydrazine (Sigma Co.) to the cellulose suspension and then heat it in a water bath in the fume hood at 70°C for 15 min.

(11) Wash the cellulose three times with 250 ml of cold water, wash twice with 250 ml of 0.1 M sodium acetate pH 5.5, and then resuspend in 30 ml of 0.1 M sodium acetate pH 5.5. The volume of the cellulose pellet shrinks by about 30% during steps 10 and 11. This material will be subsequently referred to as 'hydrazide – cellulose'.

For linkage of the tetramer core to hydrazide – cellulose the following procedures should be carried out.

(1) Calculate the amounts of hydrazide – cellulose and core, [(A2′p)$_3$A], you wish to link. Usually 600 nmol of core (in AMP equivalents) are bound per ml of packed cellulose; this concentration or greater is required in order to bind the 2-5A-dependent RNase from a cell extract to a relatively small volume of core – cellulose. This method is for a 5 ml (packed volume) preparation of core – cellulose.

(2) Oxidize 3.0 μmol (46.2 A_{260} units) of tetramer core (e.g. 3.0 ml of a 1.0 mM solution in water) by adding 334 μl of freshly prepared 100 mM sodium metaperiodate; 10 mM final concentration of periodate (Sigma Co.). Incubate at room temperature for 1 h in the dark. The core may be purchased from Pharmacia Inc. or it may be prepared by bacterial alkaline phosphatase digestion of enzymatically synthesized 2-5A (Section 2.3.3).

(3) Quench the excess periodate by adding 1/9th volume (370 μl) of 0.6 M ethylene glycol/50% glycerol. Add the oxidized core solution to a volume of hydrazide – cellulose suspension containing 5 ml of packed cellulose in 0.1 M sodium acetate, pH 5.5 (step 11 above).

(4) Mix briefly and incubate at 25°C for 30 min and then agitate on a motor-driven shaker for 16 h at 2 – 4°C.

(5) Wash the core – cellulose five times by centrifuging and resuspending in 15 volumes (75 ml) of 2 × standard saline citrate (SSC) pH 7.0 (ref. 28: 1 × SSC is 0.15 M NaCl, 0.015 M Na citrate, pH 7.0).

(6) After the final centrifugation discard the supernatant and resuspend the core – cellulose in 1 volume (5 ml) of 2 × SSC and add 2 volumes (10 ml) of ethanol and store at −20°C. (The core – cellulose is stable for at least 1 year under these conditions.)

(7) Determine the concentration of core bound to the cellulose by alkaline hydrolysis. Take 100 μl of the core – cellulose suspension and centrifuge at 10 000 *g* for 5 min, discard the supernatant, suspend in 1.0 ml of 2 × SSC and centrifuge again. Discard the supernatant and suspend to 0.2 ml with 175 μl of 0.3 M KOH. Incubate at 37°C for 18 h, centrifuge at 10 000 *g* for 5 min, remove the supernatant and neutralize it by adding 150 μl of 0.5 M Hepes pH 7.6 and 175 μl of 0.3 M HCl. Measure the absorbance at 254 nm and determine the amount of AMP produced from the hydrolysis of the bound core (15.4 A_{260} units per μmol). Nearly all (>95%) of the core should be covalently attached to the cellulose resulting in a concentration of about 150 μM in the suspension (3.0

μmol of core attached to 5 ml of cellulose diluted to 20 ml in SSC/ethanol).

For the preparation of poly(U)-[5′-^{32}P]Cp the ligation of [5′-^{32}P]pCp to the 3′-OH of polyuridylic acid [poly(U); Pharmacia Inc.] is performed by a modification of the method described above for preparing 2-5A[^{32}P]Cp (Section 2.2.1).

(1) Dry 150 μCi (5.6×10^6 Bq) of [5′-^{32}P]pCp (2000–4000 Ci/mmol) *in vacuo*. Dissolve the [5′-^{32}P]pCp in 25 μl of ligase buffer [100 mM Hepes, pH 7.6; 15 mM $MgCl_2$; 6.6 mM DTT; and 20% v/v DMSO (Mallinckrodt)].

(2) Place the [5′-^{32}P]pCp solution on ice and add in the following order:

  5 μl of (1.0 mM dTTP; 0.1 mM ATP, pH 7.0)
  10 μl of 20 mM poly(U) (concentration is in UMP equivalents)
  8 μl of water
  2 μl of T4 RNA ligase (18 units, BRL)

(3) Mix thoroughly by pipetting, and then place the tube in a small beaker of ice and keep in a refrigerator at 2–4°C for 16–18 h.

(4) Add 100 μl of Tris-HCl, pH 7.6 saturated phenol:chloroform (1:1) (ref. 28), vortex for a few seconds and centrifuge at 10 000 *g* for 5 min. Remove the upper (aqueous) phase to a clean tube.

(5) Repeat step 4, then add 100 μl of chloroform, vortex for a few seconds, centrifuge at 10 000 *g* for 5 min and remove the upper (aqueous) phase to a clean tube.

(6) Add 450 μl of dialysis buffer (20 mM Tris-HCl pH 7.5, 0.1 mM EDTA, 50 mM KCl), place the diluted reaction mixture in a dialysis bag and dialyse against two changes of 1 l of dialysis buffer for 16 h.

(7) Remove the material from the dialysis bag and determine the amount of ligated [5′-^{32}P]pCp. Pipette 2 μl of the poly(U)-[^{32}P]Cp solution (in duplicate) into 1.0 ml of 5% trichloroacetic acid (TCA). 0.2% sodium pyrophosphate, plus 0.1 ml of 5.0 mg/ml yeast RNA (as carrier). Incubate on ice for 30 min. Filter on No. 30 glass-fibre circles, 25 mm diameter (Schleicher and Schuell), wash twice with 10 ml of ice-cold 5% TCA, 0.2% pyrophosphate, and once with 5 ml of 95% ethanol. Also spot 2 μl of the unprecipitated poly(U)-[^{32}P]Cp mixture to two filters without washing. Dry the filters and determine the radioactivity by scintilation counting. This material should be about 50% acid-precipitable. Typically about 25% of the [5′-^{32}P]pCp becomes ligated to the poly(U). The poly(U)-[^{32}P]Cp solution is stored at −20°C. Preparations of poly(C)-[^{32}P]Cp are performed in the same manner by substituting poly(C) (Pharmacia Inc.) for poly(U).

The method for performing the core–cellulose assay for the detection of 2-5A is designed for about 50 determinations.

(1) Prepare cellulose buffer (CB):

  0.92 ml of autoclaved buffer (100 mM Hepes, pH 7.5; 50 mM magnesium acetate; 70 mM 2-mercaptoethanol; and 900 mM potassium chloride). This may be stored at −20°C,

  5 μl 2-mercaptoethanol,
  1.0 ml of 10 mM ATP, pH 7,

80 μl of 10 mg/ml leupeptin (Sigma Co.),
2.0 ml of autoclaved and distilled water.

This gives a total volume of 4.0 ml.

(2) Pipette core – cellulose suspension (1:3 in SSC/ethanol) into a 1.5 ml conical tube and centrifuge at 1250 *g* for 5 min at 2°C (all subsequent centrifugations are identical) and discard the supernatant. Use enough cellulose to give 25 μM bound core in the final assay reaction mixture. Because the core – cellulose suspension is typically about 125 μM and the assay volume is 25 μl it is necessary to take about 5 μl of core – cellulose suspension for each assay to be performed.

(3) Wash the cellulose by resuspending in about 10-times the packed cellulose volume of 0.5 × CB (i.e. a 1:2 dilution of CB in water). Centrifuge and discard the supernatant; repeat the washing step, discarding the supernatant. Place the core – cellulose pellet on ice.

(4) Prepare a post-mitochondrial cell extract from a frozen pellet of mouse L cells (*Table 2*). Determine the volume of cell extract required to perform the desired number of assays; 100 μg of protein are used in each assay (this amount is considered 'one-volume' for step 5).

(5) Resuspend the core – cellulose (step 3) in 1.6 cell extract volumes of 0.8 × CB (i.e. eight parts of CB plus two parts of water). Add the cell extract, mix by vortexing for a few seconds, and then place the tube on ice for 1 h. During this period the 2-5A-dependent RNase in the cell extract will bind to the core – cellulose. Mix by vortexing once or twice during this incubation.

(6) Pellet the cellulose by centrifugation (1250 *g*, 5 min, 2°C) and discard the supernatant. Estimate the volume of cellulose and resuspend it in about 10-fold this volume of 0.5 × CB. Mix by vortexing for a few seconds, centrifuge and discard the supernatant. Repeat this washing procedure twice more. It is very important that the cellulose be kept at 0 – 2°C during this step; therefore make sure the centrifuge is pre-cooled before the centrifugation is performed. After the last centrifugation, discard the supernatant and place the tube on ice.

(7) Prepare the poly(U)-[^{32}P]Cp mixture as follows (this is a recipe for 50 assays):

625 μl of 1 × CB,
25 μl of 0.4 mM poly(U)-[^{32}P]Cp (Section 2.2.2),
350 μl of water.

(8) Resuspend the cellulose pellet from step 6 in enough of the poly(U)-[^{32}P]Cp mixture (step 7) to result in a volume of 20 μl for each assay to be performed. (Example: if there are 50 assays to be performed and the packed cellulose volume is 250 μl, then add 750 μl of poly(U)-[^{32}P]Cp mixture so that there will be a final volume of 1.0 ml.) Mix briefly by vortexing a few seconds.

(9) Aliquot 20 μl of the resuspended cellulose to 0.5 ml conical tubes, add 5 μl of a known concentration of 2-5A or 5 μl of the sample to be tested. Mix gently by vortexing at a low speed. To obtain a standard curve for 2-5A use 5 μl of 1.0, 2.5, 5.0, 10, 50, 100, 250 and 1000 nM tetramer or trimer 2-5A. Include one assay without 2-5A as a control.

(10) Incubate the assay mixtures at 30°C for 2 h to allow cleavage of the po-

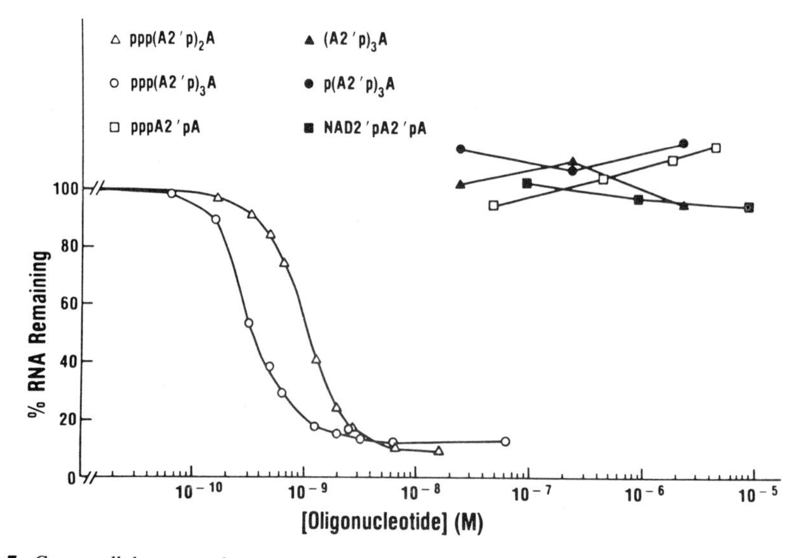

Figure 7. Core – cellulose assay for detecting 2′5′ oligoadenylates. The core – cellulose assay was performed using mouse L cell 2-5A-dependent RNase isolated on core – cellulose. Assays were performed with and without 2-5A or analogues of 2-5A as indicated in the figure. (From ref. 17, reprinted with permission from the publishers.)

ly(U)-[^{32}P]Cp by the 2-5A-dependent RNase.

(11) Prepare fresh tubes for acid precipitation of the RNA. Place conical tubes (1.5 ml) on ice and into each tube pipette 1.0 ml of ice-cold 5% TCA, 0.2% sodium pyrophosphate (PPi) and 0.1 ml of 5 mg/ml yeast RNA as carrier.

(12) After the incubation (step 10) place the assay tubes on ice and pipette all of each reaction into the TCA/PPi/RNA solutions and mix by vortexing for a few seconds. Leave the tubes on ice for at least 30 min to allow the RNA to precipitate.

(13) Pre-soak GF/C glass fibre discs (No. 30, 25 mm diameter, Schleicher and Schuell) in 5% TCA/0.2% PPi, place on a filtering manifold and apply a vacuum. Pipette the precipitated reaction mixtures onto the filters and then wash twice with 10 ml of ice-cold 5% TCA/0.2% PPi and then once with 5 ml of ice-cold 95% ethanol. Dry the filters under a heat lamp for 10 – 15 min and then determine the bound radioactivity by scintillation counting.

As shown in *Figure 7*, the activation of 2-5A-dependent RNase is an indirect measure of the concentration of 2-5A present in the assay mixture. Increasing levels of 2-5A resulted in a gradual enhancement in 2-5A-dependent RNase activity such that there was a 50% breakdown of poly(U)-[^{32}P]Cp with concentrations of 0.3 nM tetramer 2-5A or 1.1 nM of trimer 2-5A (*Figure 7*, from ref. 17). The closely related analogues of 2-5A [pppA2′pA, (A2′p)$_3$A, p(A2′p)$_3$A, and NAD2′pA2′pA] were tested and found to be without activity in the assay, that is these compounds were incapable of activating the 2-5A-dependent RNase (17). By comparison, all of these compounds showed varying levels of activity in the radiobinding assay for 2-5A (*Figure 5*, and ref. 17). Therefore, the radiobinding assay is useful in situations where it is desirable to detect all oligomers which bind to 2-5A-dependent RNase; whereas the core – cellulose assay is useful when monitoring levels of only those oligonucleotides capable of activating 2-5A-dependent RNase. Not all naturally occurring material reactive in the radiobinding assay is capable of activating 2-5A-dependent RNase. The core–cellulose assay has been used to detect 'active' 2-5A in cell extracts from EMCV-infected cells (29). In addition, the core–cellulose assay has been applied to studying the structure–function relationship of analogues of 2-5A (30). Alternate methods for determining 'active' 2-5A usually measure 2-5A-dependent RNase activity in rabbit reticulocyte lysate (31) or in crude cell extracts (e.g. the rRNA cleavage method, 18 and Section 2.4.4). The former method, however, has the disadvantage of not detecting trimer 2-5A.

If there is little or no breakdown of the poly(U)-[^{32}P]Cp in the core–cellulose assay in the presence of 2-5A, then determine the 2-5A binding activity (Section 2.4.2) of the cell extract before and after adsorption to core–cellulose. Typically the core–cellulose should remove between 65 and 75% of the 2-5A binding activity (presumptive 2-5A-dependent RNase); if it does not then there may be a problem with the core–cellulose preparation. Also the level of 2-5A binding activity should be such that 100 μg of L cell protein binds at least 20–30% of the 2-5A[^{32}P]Cp in a radiobinding assay (Section 2.4.2). Levels of radiobinding activity much less than these values indicate that the cell extract does not contain enough 2-5A-dependent RNase to drive the reaction. Also, it is very important to keep the temperature between 0 and 2°C during the washing of the 2-5A-dependent RNase:core–cellulose complex. The poly(U)-[^{32}P]Cp may appear somewhat resistant to breakdown immediately after its synthesis. However, with repeated freeze–thawings or upon storage for a few days the poly(U)-[^{32}P]Cp often gives optimal assay results.

If the poly(U)-[^{32}P]Cp is degraded in incubations without 2-5A then test all of the solutions for RNase contamination by incubating (30°C for 2 h) the individual components with poly(U)-[^{32}P]Cp and then performing the filter assay (steps 11–13). When assaying samples of unknown composition (e.g. extracts from intact cells, Section 2.2.3) it is also necessary to monitor the samples for RNase contamination by incubation with poly(U)-[^{32}P]Cp (i.e. without 2-5A-dependent RNase:core–cellulose complex).

2.2.3 *Extraction of 2'5' oligoadenylates and related material from intact cells*

This method is divided into two stages: (i) the nucleotide pool is extracted from cells using TCA and the TCA is then removed, and (ii) the 2-5A and other nucleotides are further purified using C18 Sep-pak cartridges (Waters Associates).

(i) Extraction of nucleotide pool from intact cells. This technique is a modification of previously described techniques (32,33) and is designed for monolayer cultures of cells but it may be adapted to suspension cultures of cells or to organs (e.g. 34).

(1) Remove all of the culture medium from the monolayer cell surface (75 cm^2 plate) using a pipette for the last few drops.
(2) Place the plate on ice and rinse the cell surface with 5.0 ml of ice-cold PBS, removing the PBS with a pipette.
(3) Add 2.0 ml of 10% TCA and remove the cell extract with a rubber or Teflon cell scraper.

(4) Using a pipette, collect as much of the extract as possible and place in a 4.0 ml centrifuge tube.
(5) Vortex the extract vigorously and then place the tube in an ice bath for 10 min, and then vortex vigorously a second time.
(6) Centrifuge the extract at 750 *g* for 5 min at 2–4°C.
(7) Collect the supernatant into a polypropylene centrifuge tube (if less than 2.0 ml adjust the volume to 2.0 ml with 10% TCA). *Save the pellet for a protein determination.*
(8) In a fume hood, pipette 39 parts of Freon (1,1,2-trichloro-1,2,2-trifluoroethane, Mallinckrodt) and then add 11 parts of tri-*n*-octylamine (Sigma Co.) and mix by vortexing. Add an equal volume of the Freon:tri-*n*-octylamine solution to the supernatant from step 7 and mix by vortexing for 15 sec.
(9) Centrifuge at 3500 *g* for 5 min at 2–4°C.
(10) Transfer the upper aqueous phase to a clean polypropylene tube, add an equal volume of the Freon:tri-*n*-octylamine, vortex, and centrifuge as in step 9.
(11) Collect the upper, neutral aqueous phase into a new tube, being careful not to contaminate it with any of the freon phase.
(12) Check the pH of the extract by removing a few microlitres to a pH indicator strip. The pH should be between 6.5 and 7.5; if not add Hepes, pH 7.0 to 10 mM and if necessary adjust the pH with a solution of 0.1 M KOH.

(ii) Partial purification of 2′5′ oligoadenylates using a C18 Sep-pak cartridge.

(1) Prepare Sep-pak C18 cartridges (Waters Associates) for use by washing with 3.0 ml of 100% methanol and then with 5.0 ml of 50 mM ammonium phosphate (adjusted to pH 7.0 with ammonium hydroxide).
(2) Adjust the sample to 50 mM ammonium phosphate, pH 7.0 and apply to the Sep-pak collecting the effluent; pass this through the cartridge again.
(3) Wash the cartridge first with 3.0 ml of 50 mM ammonium phosphate, pH 7.0 and then with 3.0 ml of water to remove the salt.
(4) Elute the 2-5A from the cartridge with 2.0 ml of 100% methanol.
(5) Lyophilize the sample (or use a Speed-Vac, Savant) until it is dry. Resuspend the sample to the desired volume (usually one-fifth of the estimated packed cell volume) with water.

All cartridge separations may be performed using a Sep-pak cartridge rack (Waters Associates) or using disposable syringes. If syringes are used, however, one must be careful not to create any back pressure.

To monitor the recovery of 2-5A by this method, use 50 μl of 100 nM trimer 2-5A, and measure the 2-5A recovered using the radiobinding assay (Section 2.2.1). Typically there is about 60% recovery of 2-5A by this technique. This method, or the one described below, may be used to de-salt h.p.l.c. fractions of 2-5A [^{32}P]Cp.

Material from cells extracted in this way can be directly assayed for 2-5A by either the radiobinding assay (Section 2.2.1) or the core–cellulose assay (Section 2.2.2) (29). In some situations, however, there may be natural oligonucleotides inhibitory to 2-5A-dependent RNase. Therefore, it is always desirable to perform an h.p.l.c. separation of the Sep-pak purified material prior to assay.

W.G.Hearl and M.I.Johnston (USUHS, Bethesda) have developed an improvement of this method (personal communication). They found that an increased recovery of 2-5A (90%) was obtained by using a buffer with a lower pH. Wash the Sep-pak cartridges with 2 ml of 100% methanol and then with 0.2 M ammonium acetate, pH 5.0. Dissolve the sample and apply in equilibration buffer and wash the cartridge in equilibration buffer containing 2% methanol. Elute the 2-5A with 100% methanol. When analysing material from mouse organs, these researchers first purify the extracts using silica cartridges (Waters Associates) and subsequently with a Sep-pak. The material is then separated using h.p.l.c. and the 2-5A is determined using an enzyme-linked immunoassay (22).

An alternate method of extracting tissue for 2-5A uses phenol followed by ether extraction, Sep-pak purification and de-salting on a Sephadex G-50 column (34). The phenol step results in the removal of contaminating polypeptide material.

2.3 Assays of 2′5′ oligoadenylate synthetase and synthesis and purification of 2′5′ oligoadenylates

As mentioned previouly, 2-5A synthetases may be the most highly regulated activities of the interferon system; therefore, they are often assayed as a biochemical monitor of the interferon-responsiveness of cells, animals and man. When assaying 2-5A synthetase activity in crude cell extracts it is important to keep in mind that there are multiple forms of the enzyme which may vary between cell types. The total level of synthetase in a cell extract, therefore, may consist of different individual activities.

The 2-5A synthetase assays described here involve immobilizing the enzymes from cell extracts on poly(I):poly(C) linked to cellulose (or agarose), washing the complex to remove some of the 2-5A degrading activities (e.g. phosphatases and 2′,5′-phosphodiesterases), and then incubating with ATP to allow the synthesis of 2-5A (*Figure 8*). Some investigators monitor the synthesis of 2-5A in crude cell extracts by incubating in the presence of unbound poly(I):poly(C); however, this measures 2-5A accumulation, that is a net synthesis of 2-5A in the presence of concomitant degradation, making it difficult to calculate synthetase levels in a meaningful way. The method employing poly(I):poly(C)–cellulose is a modification of the technique pioneered by Wells *et al.* (27).

2.3.1 *Poly(I):poly(C)–cellulose method for assaying 2′5′ oligoadenylate synthetase*

(i) Preparation of poly(I):poly(C)–cellulose. The method for covalently attaching poly(I):poly(C) to cellulose is essentially similar to the technique for the synthesis of core–cellulose (Section 2.2.2).

(1) Add 50 ml of 100 mM KCl; 10 mM KPO_4, pH 5.5 to 100 mg of lyophilized poly(I):poly(C) (Pharmacia Inc.) and place in a dialysis bag. Dialyse against several changes of the same buffer for 2–4 days, occasionally inverting the bag, until the RNA is completely dissolved.

(2) Oxidize the poly(I):poly(C) by adding 5.6 ml of freshly prepared 0.1 M sodium periodate (10 mM final concentration) and then incubating for 1 h at room temperature in the dark.

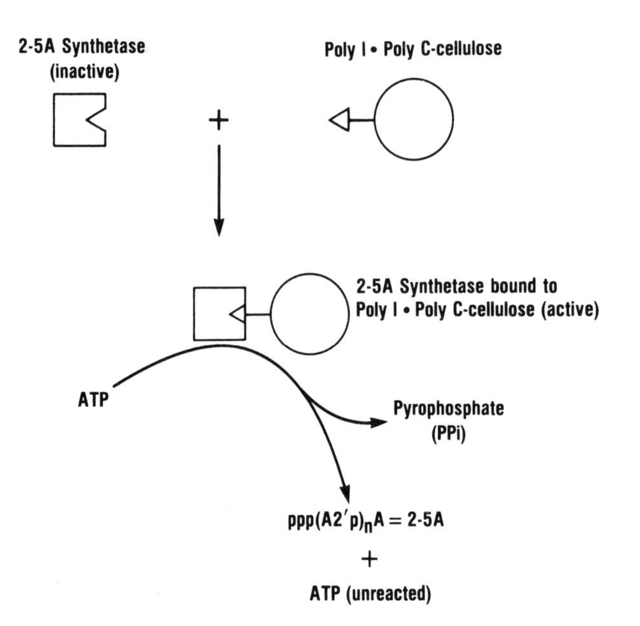

Figure 8. Immobilization, washing and activation of 2′5′ oligoadenylate synthetase using poly(I):poly(C)—cellulose.

(3) Quench the excess periodate by adding 6.4 ml of 600 mM ethylene glycol; 50% glycerol.
(4) To the solution of oxidized poly(I):poly(C) add 50 ml of freshly prepared hydrazide—cellulose suspension (Section 2.2.2). Mix briefly, incubate for 30 min at 25°C, and then agitate at 2–4°C for 16 h.
(5) Wash the poly(I):poly(C)—cellulose five times by centrifuging (1500 *g* for 10 min at 2°C) and resuspending in 15 packed cellulose volumes of 2 × SSC, pH 7.0 (~250 ml) and discarding the supernatant. (1 × SSC is 0.15 M NaCl, 0.015 M sodium citrate, pH 7.0).
(6) After the final centrifugation resuspend the poly(I):poly(C)—cellulose in one packed cellulose volume of 2 × SSC, pH 7.0 and add two packed cellulose volumes of ethanol and store at −20°C.

This preparation is enough for about 10 000 2-5A synthetase assays; this material is also used in the assay of dsRNA-dependent protein kinase (Section 3.3). It is stable at −20°C for at least 1 year. Typically, about 40% of the poly(I):poly(C) becomes linked to the cellulose [~0.9 mg poly(I):poly(C) bound per ml of cellulose suspension] as determined by alkaline hydrolysis and absorbance at 254 nm (Section 2.2.2).

(ii) Immobilization and isolation of 2′5′ oligoadenylate synthetase on poly(I): poly(C)—cellulose.

(1) Pipette 5 μl of poly(I):poly(C)—cellulose suspension for each assay to be per-

Table 3. Reagents for 2′5′ oligoadenylate synthetase assays.

1. 2 × DBG (20 mM Hepes pH 7.5; 100 mM KCl; 3.0 mM magnesium acetate; 14 mM mercaptoethanol; and 40% v/v glycerol)

   4.0 ml of 0.5 M Hepes pH 7.5
   2.5 ml of 4 M KCl
   0.3 ml of 1 M magnesium acetate
   98 μl of 14.3 M 2-mercaptoethanol
   40 ml of glycerol

   Adjust to 100 ml with distilled water. To obtain DBG (1×) dilute 1:2 with distilled water.
2. 'Unlabeled ATP assay mixture' (10 mM Hepes pH 7.5; 50 mM KCl; 16.5 mM magnesium acetate; 7 mM mercaptoethanol; 20% v/v glycerol; and 8.0 mM ATP)

   5.0 ml 2 × DBG (20 mM Hepes pH 7.5; 100 mM KCl; 3.0 mM magnesium acetate; 14 mM mercaptoethanol; and 40% v/v glycerol),
   0.8 ml of 0.1 M ATP *adjusted to pH 7.0,*
   0.15 ml of 1 M magnesium acetate,
   4.05 ml of distilled water.

   Store −20°C.
3. '[^{3}H]ATP assay mixture' (10 mM Hepes pH 7.5; 50 mM KCl; 9.5 mM magnesium acetate; 7 mM mercaptoethanol; 20% v/v glycerol; 4 mM ATP containing 200 μCi/ml (7.4 × 10^6 Bq/ml) [2,8,5′-^{3}H]ATP. Place 0.4 mCi (1.48 × 10^7 Bq) of [2,8,5′-^{3}H]ATP (New England Nuclear) in a 4.0 ml plastic tube and evaporate to dryness *in vacuo*. Add the following to the tube containing the dried [^{3}H]ATP.

   16 μl of 1 M magnesium acetate,
   80 μl of 100 mM ATP, pH 7,
   1.0 ml of 2 × DBG,
   0.9 ml of water

   Store at −70°C (−20°C may be sufficient).
4. BAP mixture. Centrifuge 30 units of BAP suspension in ammonium sulphate (P-4252, Sigma Co.) at 10 000 *g* for 10 min, discard the supernatant, and resuspend in 100 μl of DBG. Use immediately or quick freeze on dry ice and store at −70°C.

---

formed all into one tube (the amount of cellulose suspension used = one volume).

(2) Add 5 volumes of DBG (*Table 3*, ref. 24) (10 mM Hepes pH 7.5; 50 mM KCl; 1.5 mM magnesium acetate; 7 mM mercaptoethanol; and 20% v/v glycerol), centrifuge at 1500 *g* for 5 min (room temperature), and discard most of the supernatant using a micropipette, being very careful not to disturb the cellulose pellet.

(3) Resuspend in 5 volumes of DBG, vortex briefly, centrifuge and discard the supernatant.

(4) Resuspend the cellulose to 3 volumes with DBG, (i.e. 15 μl for every assay performed).

(5) Aliquot 200 μg (protein) of post-mitochondrial cell extract (*Table 2*) into 0.5 ml conical tubes on ice, adjust the volume to 15 μl by adding DBG, and then add 15 μl of washed cellulose (step 4), and vortex briefly. It is advisable to include an assay with an extract containing a known amount of synthetase as a control. Cytoplasmic extract of α-interferon-treated (100 unit/ml for 20 h) HeLa cells is ideal for this purpose.

(6) Incubate for 1 h at room temperature [this step is to bind the 2-5A synthetase in the cell extract to the poly(I):poly(C) – cellulose], vortexing a few times dur-

ing the incubation. Improved stability of the poly(I):poly(C)—cellulose may be achieved by addition of NAD during this step (27).

(7) Add 250 μl of DBG, vortex briefly, centrifuge at 1500 *g* for 5 min (2–4°C) and very carefully remove almost all of the supernatant, making sure to leave the cellulose pellet undisturbed (leave a small amount, < 15 μl, of the supernatant behind).

(8) Repeat step 7 twice more.

(9) After the last wash adjust the volume to 25 μl using DBG.

(iii) Synthesis of unlabelled 2′5′ oligoadenylates and determination of 2′5′ oligoadenylate levels by the radiobinding assay.

(1) Add 25 μl of 'unlabelled ATP assay mixture' to the assay tubes (DBG plus 15 mM magnesium acetate; 8.0 mM ATP pH 7, see *Table 3*).

(2) Incubate for 2 h at 37°C, vortex a few times during incubation, then centrifuge at 1500 *g* for 5 min at 2°C and transfer the supernatant containing the 2-5A to a fresh 0.5 ml tube.

(3) In a microtitre dish dilute the supernatant from step 2 in water. Add 5 μl of supernatant to 45 μl of water and then make 10-fold serial dilutions to a final dilution of 10^{-5}. Perform the radiobinding assay for the detection of 2-5A (Section 2.2.1) using 5 μl of each dilution of sample and include a standard curve with trimer 2-5A.

(4) Determine the level of 2-5A made in each assay using a standard curve obtained with trimer 2-5A (see *Figure 5*). Find a point in the standard curve from each set of radiobinding assays which gives a displacement of 2-5A[^{32}P]Cp in the 50% range. Calculate the amount of synthetase in the assay in terms of units of activity (one unit = 1 pmol of 2-5A synthesized/mg of protein/h at 37°C).

Example: suppose a sample gives 50% displacement of probe using a 10^{-4} dilution corresponding to 1 nM 2-5A in the standard curve (*Figure 5*). There would be $10^4 \times$ 1 nM × 50 μl (volume of synthetase assay) = 500 pmol of 2-5A synthesized. Therefore, there would be 500 pmol ÷ 0.2 mg protein ÷ 2 h = 1250 units of synthetase activity in the sample.

This method is probably the most sensitive technique available for measuring 2-5A synthetase activity; the lower limit of detection is about 0.1 units of activity. By comparison, α-interferon treatment (100 units/ml for 20 h) of HeLa cells results in levels of about 50 000–100 000 synthetase units. An interferon dose–response curve for induction of 2-5A synthetase in murine NIH 3T3, clone 1 cells is shown in *Figure 9*. The cells were incubated for 18 h with and without varying doses of murine α + β interferons (7 × 10^7 units/mg). The levels of cytoplasmic 2-5A synthetase activity were enhanced from 0.4 units in the control cells to 3660 units in the interferon-treated cells, representing greater than 9000-fold induction. The extent of induction, however, varies with the cell type, dose and type of interferon, and the growth state of the cells. For instance, it is well known that synthetase levels usually increase during growth arrest of cells (35). Therefore, in general, the degree of induction of synthetase by interferon will be less in growth-arrested cells as compared with actively-growing cells. Also, some cell lines, such as HeLa cells grown in spinner culture (32), have very high basal

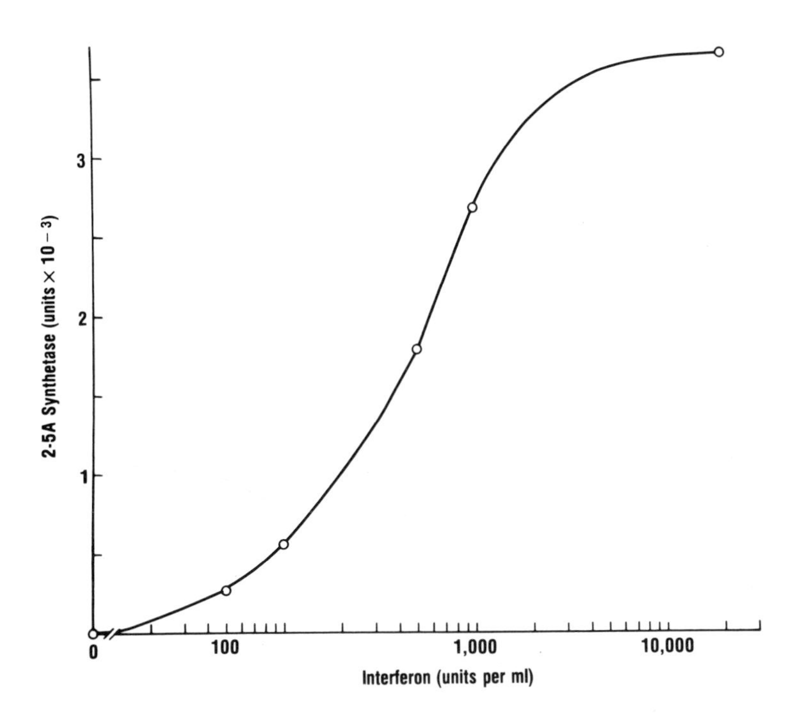

Figure 9. Induction of 2′5′ oligoadenylate synthetase by interferon treatment of NIH 3T3, clone 1 cells. NIH 3T3, clone 1 cells were incubated for 18 h without or with varying doses of murine $\alpha + \beta$ interferons as indicated in the figure. The levels of 2-5A synthetase activity were determined from post-mitochondrial fractions of cell extract using the poly(I):poly(C)−cellulose technique and the radiobinding assay to determine levels of 2-5A. Synthetase amounts are given in units, 1 unit = 1 pmol 2-5A/mg cell protein/h at 37°C.

levels of synthetase and show only about a 3- to 10-fold induction of synthetase levels by interferon.

(iv) Determining 2′5′ oligoadenylate synthetase levels using [^{3}H]ATP. The 2-5A synthetase assay employing [^{3}H]ATP is at least 100-fold less sensitive than the method based on the radiobinding assay. However, when there are large numbers of samples it is easier to perform because there is only one 2-5A determination per synthetase assay. In addition, the radiochemical, 2,8-[^{3}H]ATP is commercially available whereas, presently, 2-5A[^{32}P]Cp must be synthesized (Section 2.2.1).

The first part of the assay in which the synthetase activities are bound to poly (I): poly (C)−cellulose and then washed is identical to the method just described (Section 2.3.1 ii). However, here [^{3}H]ATP rather than unlabelled ATP is polymerized into 2-5A. The amount of [^{3}H]2-5A synthesized is determined by digesting with bacterial alkaline phosphatase (BAP), spotting on PEI−cellulose squares, and differentially eluting the BAP-resistant core of 2-5A (*Figure 10*). This technique is a modification of the method developed by Ian Kerr's group (31).

(1) To the 2-5A synthetase bound to poly(I):poly(C)−cellulose, suspended to 25 μl in DBG (from Section 2.3.1.ii. step 9), add 25 μl of [^{3}H]ATP assay mixture

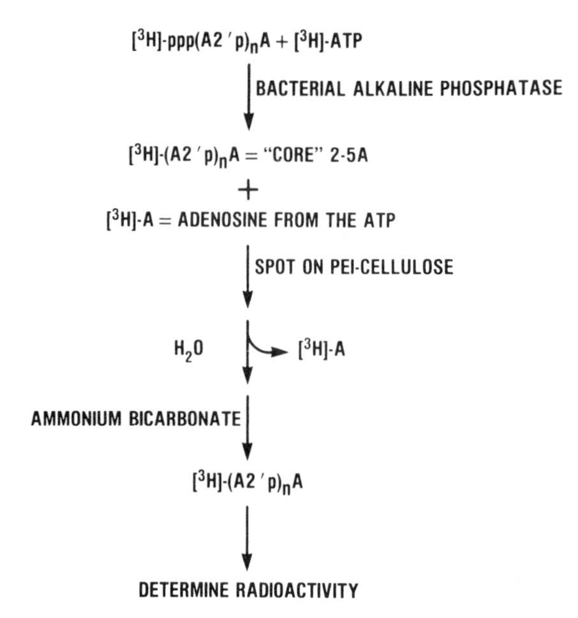

Figure 10. Detection of the BAP – core of 2′5′ oligoadenylates using PEI – cellulose and differential elution.

[DBG plus 8.0 mM magnesium acetate; 4.0 mM ATP, pH 7; and 200 μCi/ml (7.4 × 10^6 Bq/ml) [^{3}H]ATP, see *Table 3*].

(2) Incubate for 2 h at 37°C, vortex a few times during incubation, centrifuge at 1500 *g* for 5 min and transfer the supernatant containing the 2-5A to a fresh 0.5 ml tube.

(3) Mix 10 μl of sample (supernatant from step 2) with 2 μl of BAP mixture (*Table 3*). Set up two additional tubes for 0 and 100% controls.

Zero percent (background) control: a tube containing 5 μl of [^{3}]ATP assay mixture, 5 μl of DBG and 2 μl of BAP mixture;
100 percent (input) control: a tube containing 5 μl of [^{3}H]ATP assay mixture, 5 μl of DBG and 2 μl of water.

Be sure to use a fresh tip each time when aliquoting BAP. Incubate at 37°C for 16 – 20 h.

(4) Cut out 1.5 cm^2 pieces of PEI – cellulose t.l.c. plastic-coated plates, pre-washed as described in *Table 1*.

(5) Spot 4 μl of BAP-treated reaction product on each square and air-dry for 5 min.

(6) Place PEI squares cellulose-side down in scintillation vials containing 1.0 ml of water, incubate for 5 min, swirl occasionally for a few seconds, discard the water using a disposable pipette tip, wash four more times in the same manner. This step removes [^{3}H]adenosine from the squares while retaining the [^{3}H]core from the 2-5A.

(7) Elute the BAP core of 2-5A from the squares by adding 0.5 ml of 0.5 M NH_4HCO_3, incubating for 5 min at room temperature with occasional shaking.

(8) Add 15 ml of scintillation counting fluid (for aqueous samples) and determine the radioactivity.

(9) Determine the percent conversion of ATP to 2-5A using the 100% (input) control and the 0% (background) control values.
(10) Express the results in units of synthetase activity, where one unit = 1 pmol of 2-5A synthesized/mg of protein/h at 37°C.

Example: the background is 500 c.p.m. (from the control in which [^{3}H]ATP was incubated with BAP); the input level of [^{3}H]ATP is 250 000 c.p.m.(from the control in which [^{3}H]ATP was incubated with water); the synthetase assay resulted in 10 000 c.p.m.

Calculation: percent conversion of ATP to 2-5A = 10 000 − 500 ÷ 250 000 = 0.038 (or 3.8%); the initial assay conditions include 2 mM ATP; therefore at the end of the incubation with synthetase there would be 2 mM × 0.038 = 7.6×10^{-5} M 2-5A, and 7.6×10^{-5} M × 50 μl (the volume of the assay) = 3.8 nmol of 2-5A were synthesized; 3800 pmol ÷ 0.2 mg (the amount of cell protein used in this assay) ÷ 2 h (the length of the incubation) = 9500 pmol 2-5A/mg protein/h or 9500 units of synthetase.

2.3.2 *Alternative methods for assaying 2′5′ oligoadenylate synthetase*

(i) Substituting poly(I):poly(C)−agarose for poly(I):poly(C)−cellulose. Because poly(I):poly(C)−agarose is commercially available [Pharmacia Inc.; Ag poly(I):poly (C) Type 6] whereas poly(I):poly(C)−cellulose must be prepared (Section 2.3.1.i) it may be more convenient to substitute the agarose-linked polynucleotide for cellulose-bound material. In our experience about 25 μl of the commercial poly(I):poly (C)−agarose suspension results in similar levels of synthetase activity to about 5 μl of the poly(I):poly(C)−cellulose suspension. However, the agarose particles are much larger than the finely-divided cellulose, and are difficult to equally distribute to the assay tubes; for example it is necessary to cut the ends from yellow micropipette tips in order to pipette the agarose. In addition, the agarose rapidly settles whereas the cellulose may remain in suspension for several minutes. The agarose-bound poly(I):poly (C) may be used in synthetase assays employing either the radiobinding assay (Section 2.3.1.iii) or the [^{3}H]ATP method (Section 2.3.1.iv).

(ii) Substituting 2′,5′-ADP−Sepharose for poly(I):poly(C)−cellulose. In this method (36), 2-5A synthetase is first bound to 2′,5′-ADP−Sepharose, and is then washed and incubated with ATP and dsRNA. This technique allows for the testing of different dsRNAs for their ability to activate synthetase from various sources. It is also used when assaying synthetase in extracts with high levels of dsRNA-degrading activity (M.I.Johnston, personal communication).

(iii) Substituting DE 81 paper circles for PEI−cellulose squares. Circles of DE 81 DEAE−cellulose paper (Whatman, 25 mm diameter) may be substituted for the PEI−cellulose squares in differential elution of the BAP core of 2-5A from adenosine (31). Following the synthetase reaction using [^{3}H]ATP and the BAP digestion (Section 2.3.1.iv., step 3), 6 μl of the reaction mixtures are spotted onto DE 81 circles. The adenosine may be removed by gently washing five times with water and the core (2-5A) is then eluted into 5 ml of 350 mM KCl; 20 mM Hepes, pH 7.5; 5 mM magnesium

acetate (B.R.G.Williams, personal communication) and counted in scintillation fluid for aqueous samples. This method may result in slightly higher background counts than using PEI – cellulose.

(iv) Automating the radiobinding method for large numbers of samples. A.Yeh and O.Preble, (USUHS, Bethesda) have adapted the radiobinding assay to facilitate determination of synthetase activities from large numbers of samples. The radiobinding assay is performed in round-bottom 96-well microtiter plates and reagents are added by using a multichannel pipette. After the incubation for binding of the 2-5A[^{32}P]Cp, the reaction mixtures are transferred with a multichannel pipette and filtered through a 96-well Millititer-HA filtration plate in a Millititer vacuum holder (Millipore Co.). The original wells (in which the reactions were performed) are rinsed with 50 μl of water which is transferred to the filtration plate. The filtration plate is then washed twice with about 300 μl of water per well. After the filters dry, individual wells from the plate are punched into vials using a special Millititer filter punch, and the radioactivity is determined in scintillation fluid. 2-5A synthetase levels are calculated as described above (Section 2.3.1.iii).

(v) Measuring 2'5' oligoadenylate synthetase levels using an enzyme-linked immunoassay. An enzyme-linked immunoassay for 2-5A has been described which may be applied to measuring 2-5A produced from large numbers of synthetase assays (21,22). This method has the advantages of being largely automatable and does not require any isotopes.

(vi) Additional assays for 2'5' oligoadenylate synthetase. Finally, it should be mentioned that other techniques for assaying 2-5A synthetase levels have been described (14). For instance, one method polymerizes [^{32}P]2-5A from [α-^{32}P]ATP and then uses alumina columns to separate the BAP core of 2-5A from [^{32}P]inorganic phosphate (37).

2.3.3 *Preparative synthesis and purification of 2'5' oligoadenylate*

This method describes a technique for the synthesis and purification of relatively large amounts (in the range of 500 A_{260} units) of 2-5A.

(i) Grow a 1 litre culture of HeLa S3 cells in suspension culture using spinner-modified minimal essential medium containing 10% Nu serum (Collaborative Research), sodium pyruvate, non-essential amino acids, vitamin mixture, L-glutamine, and penicillin – streptomycin and/or gentamicin. When the cell density reaches 10^6 cells/ml add human α-interferon (Interferon Sciences, Inc.) to 100 units/ml and incubate at 37°C for 20 h.
(ii) Harvest the cells by centrifugation, wash twice in PBS, and prepare a post-mitochondrial supernatant fraction as described in *Table 2*.
(iii) Pipette 1.2 ml of poly(I):poly(C) – cellulose [0.9 mg poly(I):poly(C) bound per ml] into a 50 ml conical tube and centrifuge at 1500 *g* for 5 min.
(iv) Discard the supernatant from step iii, resuspend the poly(I):poly(C) – cellulose in 50 ml of 1 × DBG (10 mM Hepes pH 7.5; 50 mM KCl; 1.5 mM magnesium

acetate; 7 mM mercaptoethanol; and 20% v/v glycerol, see *Table 3*) and centrifuge again at 1500 *g* for 5 min.

(v) Repeat step iv.

(vi) Resuspend the poly(I):poly(C) – cellulose in 5.0 ml of DBG and then add 2.0 ml of post-mitochondrial supernatant fraction of the interferon-treated HeLa cells (step ii).

(vii) Incubate for 60 min at room temperature with occasional vortex mixing.

(viii) Centrifuge at 1500 *g* for 5 min (2 – 4°C) and decant and discard the supernatant.

(ix) Resuspend in 15 ml of DBG and centrifuge again at 1500 *g* for 5 min and then discard the supernatant.

(x) Repeat step ix twice more.

(xi) Resuspend the pellet in 3.0 ml of DBG and add 3.0 ml of unlabelled ATP assay mixute (*Table 3*). Mix briefly by vortexing.

(xii) Incubate for 20 h at 37°C, centrifuge at 1500 *g* for 5 min and remove the supernatant containing the 2-5A to a new tube and store at −20°C (this is the first round of 2-5A production).

(xiii) Resuspend the cellulose in 3.0 ml of DBG and add 3.0 ml of fresh unlabelled ATP assay mixture and incubate at 37°C for 72 h. Centrifuge at 1500 *g* for 5 min and collect the supernatant (second round of 2-5A production).

(xiv) Heat the supernatant fractions containing the 2-5A (steps xii and xiii) in a boiling water bath for 5 min and centrifuge at 1500 *g*. Remove the supernatants to new tubes. Filter each supernatant through three successive 0.22 μm filter units (Millex-GS, Millipore Co.).

(xv) Analyse the products of the synthetase reaction (3 μl) using h.p.l.c. with the conditions described in Section 2.2.1 (detector range = 0.1 at A_{254} nm). The products of a typical large-scale synthesis of 2-5A are indicated in *Table 4*.

(xvi) Purify the individual oligomers of 2-5A by performing several separations on preparative C_{18} h.p.l.c. columns as described by Imai and Torrence (38).

Table 4. Distribution of 2′5′ oligoadenylate oligomers from preparative synthesis as determined by h.p.l.c. analysis.

Oligomer	*Percent conversion of ATP (4 mM) to different oligomers of 2-5A*	
	1st round of synthesis[a]	*2nd round of synthesis*
$ppp(A2'p)_2A$	28.4	27.1
$ppp(A2'p)_3A$	18.2	16.3
$ppp(A2'p)_4A$	8.7	9.5
$ppp(A2'p)_5A$	4.1	6.5
$ppp(A2'p)_6A$	2.7	6.7
$ppp(A2'p)_7A$	1.8	3.2
$ppp(A2'p)_8A$	1.6	2.8
$ppp(A2'p)_9A$	1.0	2.1
Sum	66.5	74.2

[a]Synthesis of 2-5A was in two rounds; the first was for 24 h and the second was for 72 h using the same preparation of 2-5A synthetase bound to poly(I):poly(C) – cellulose as described in the text (Section 2.3.3).

(xvii) De-salt the h.p.l.c. purified oligomers by a modification of the method described by Imai and Torrence (38) (P.F.Torrence, personal communication). Prepare 1 M triethylammonium bicarbonate (TEAB) by adding 140 ml of triethylamine (h.p.l.c. grade, from Pierce Chemicals) to 500 ml of ice-cold water, dropwise, under the continuous bubbling in of CO_2 gas through a dispersion tube for 1 h. Continue to add CO_2 until the pH reaches about 7.6–7.8. Adjust the volume to 1 litre with water.

(xviii) Prepare a slurry of DEAE–Sephadex A-25 in 1 M TEAB and leave overnight. Pour a column (1.0 × 10.0 cm) of DEAE–Sephadex A-25 (7.8 ml) and wash with about 30 ml of 1 M TEAB. Wash the column for 16–18 h with water (~100 ml). Load about 100 A_{260} units of an h.p.l.c. purified 2-5A oligomer. There should be little or no material absorbing at 254 nm in the pass-through fraction. Elute the 2-5A either in a gradient of 0.25–0.5 M TEAB (150 ml each) or in a stepwise manner with increasing concentrations of TEAB.

(xix) Lyophilize the pooled fractions containing the 2-5A (from step xviii), resuspended in a few ml of water and lyophilize again. Repeat this process twice more and store 2-5A at −70°C.

Alternatively, the 2–5A oligomers are purified on a 100 ml DEAE-Sephadex A-25 column using a 2-l gradient of 0.1 M-to-1.0 M TEAB (pH 7.4). The sample is loaded in 0.05 M TEAB and the column is washed with 100 ml of 0.05 M TEAB prior to starting the gradient (C.W.Dieffenbach,USUHS, Bethesda).

2.4 Assays for 2′5′ oligoadenylate-dependent RNase

2-5A-dependent RNase may mediate most, if not all, of the activities of the 2-5A system in cells. The regulation of 2-5A-dependent RNase is somewhat complex. Although there are relatively high basal levels of 2-5A-dependent RNase in many cell lines and in organs, several cell lines contain either very low (39) or no detectable levels (13) of the enzyme. Most cell lines show some (2- to 4-fold) inducation in levels of 2-5A-dependent RNase by interferon treatment whereas others show much greater (as much as 10- to 20-fold) induction (e.g. ref. 40).

2-5A-dependent RNase is monitored by either its 2-5A binding activity or by its RNase function. In the former methods, 2-5A[^{32}P]Cp is bound to 2-5A-dependent RNase and the analysis is either by gel electrophoresis and autoradiography or by a filter binding assay. 2-5A-dependent RNase has no measurable nuclease activity in the absence of 2-5A and the activation of 2-5A-dependent RNase by 2-5A can be monitored in a variety of ways (for a description of some of the earlier methods see ref. 14). Here we describe methods which measure 2-5A-mediated RNase activity from either partially purified enzyme (17) or in crude cell-free systems (18) and intact cells (41). When monitoring levels of 2-5A-dependent RNase it is important to recognize that the enzyme is present in both the nucleus and cytoplasm (10).

2.4.1 *Affinity labelling of 2′5′ oligoadenylate-dependent RNase*

Wreschner *et al.* (11) and Floyd-Smith *et al.* (42) have developed methods in which a radioactive derivative of 2-5A is covalently attached to 2-5A-dependent RNase either chemically or by u.v. irradiation, respectively; here we will describe only the former method. After the covalent attachment of the probe, the labelled 2-5A-dependent RNase is analysed by denaturing gel electrophoresis and autoradiography. Typically, the enzyme appears as a band of 77 000 – 85 000 daltons. It is critical that these assays be performed in the presence and absence of excess unlabelled 2-5A. There are many proteins in the cell which bind 2-5A[^{32}P]Cp weakly in a manner which is not competed with a 250-fold excess of unlabelled 2-5A. Presumably these reflect relatively non-specific interactions between 2-5A[^{32}P]Cp and various proteins in cell extracts, for example ATP-binding proteins. In contrast, 2-5A-dependent RNase displays a high specific affinity for 2-5A (16) and consequently 2-5A[^{32}P]Cp binding to the nuclease is competed with unlabelled 2-5A. The following method is for 14 assays (for instance seven samples incubated with and without unlabelled 2-5A).

Using 100 000 c.p.m. of 2-5A[^{32}P]Cp per assay results in a final concentration of about 1 nM; using more than about 2 μl of 2-5A[^{32}P]Cp per assay may result in decreased binding to the enzyme due to the presence of ammonium phosphate in the solution of 2-5A[^{32}P]Cp. Therefore, it may be necessary to de-salt and concentrate the 2-5A[^{32}P]Cp using a C_{18} Sep-pak cartridge (Waters Associates) as described above (Section 2.2.3) prior to labelling the 2-5A-dependent RNase.

(i) Pipette into a 0.5 ml conical tube 1.5 × 10^6 c.p.m. of 2-5A[^{32}P]Cp (Section 2.2.1), evaporate to dryness *in vacuo*, re-dissolve in 25 μl of distilled water and place on ice.
(ii) Pipette 2 units of a BAP suspension in ammonium sulphate (Sigma Co., Cat. No. P-4252), into a 0.5 ml tube, centrifuge at 10 000 *g* for 10 min at 2°C, discard the supernatant and re-dissolve the BAP pellet in 18 μl of 100 mM Hepes, pH 7.6.
(iii) Add 1.3 μl of the BAP preparation (step ii) to the 25 μl of 2-5A[^{32}P]Cp from step i, incubate on ice for 30 min, and then at 4°C for an additional 30 min. This mild BAP digestion step is designed to remove the 3′-terminal phosphoryl group while leaving 1–2 5′ phosphoryl groups (11).
(iv) Add 1.0 μl of 50 mM EDTA, pH 7 to the material from step iii, place the tube in a 90°C bath for 5 min, centrifuge at 10 000 *g* for 5 min and transfer the supernatant to a new tube.
(v) Add 3 μl of 100 mM sodium metaperiodate (prepared fresh), and incubate at room temperature in the dark for 60 min.
(vi) Prepare assay reactions.
    (a) Add samples of cell extracts — 200 μg of protein diluted to 10 μl with Nonidet P-40 (NP-40) lysis buffer containing leupeptin (*Table 2*). These may be either post-mitochondrial fractions or complete cell lysates prepared by lysing cell pellets in NP-40 lysis buffer (plus leupeptin) with vortexing but without subsequent centrifugation (*Table 2*).
    (b) Add 1.3 μl of 25 mM magnesium acetate; 10 mM ATP, pH 7.
    (c) Add 1.4 μl of water or competitor (2.5 μM unlabelled trimer 2-5A).
    (d) Add 2 μl of BAP digested and oxidized probe from step v (~100 000 c.p.m.).
(vii) Incubate the reactions for 2 h on ice to permit binding of the probe to the enzyme.
(viii) Add 1.5 μl of 100 mM sodium cyanoborohydride (prepared fresh) to each assay, and incubate for 1 h at 0°C.

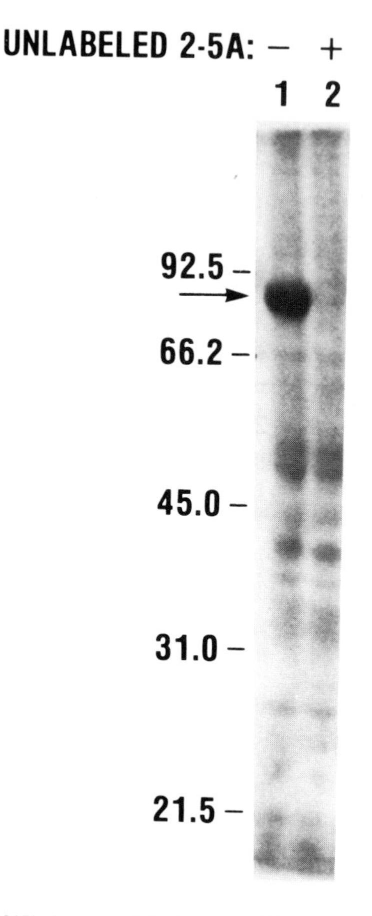

Figure 11. Affinity labelling of 2′5′ oligoadenylate-dependent RNase from complete cell lysates of murine Ehrlich ascites tumour cells. 2-5A-dependent RNase was affinity labelled from a complete cell lysate using a ^{32}P-labelled 2-5A derivative {$p_{1-3}(A2'p)_3A[^{32}P]C$} as described in the text. The incubations contained 200 μg of cell protein and 2 nM (200 000 c.p.m.) of the probe, and were performed without (**lane 1**) or with (**lane 2**) 250 nM of unlabelled 2-5A as competitor. The positions and the molecular masses (in kd) of the protein markers and the 2-5A-dependent RNase (arrow) are indicated. An autoradiogram of the dried gel is shown.

(ix) Add an equal volume of 2 times gel sample buffer and heat at 90°C for 5 min.

(x) Separate the proteins in the samples by electrophoresis on a slab 10% polyacrylamide/SDS gel, dry the gel *in vacuo*, and prepare an autoradiogram by exposing X-ray film (XAR-5, Kodak) for at least 16 h.

It is also possible to reduce substantially the incubation times in this assay; step iii to 15 min (ice)/15 min (4°C), step vii to 1 h, and step viii to 30 min.

Results of a typical affinity labelling experiment are shown in *Figure 11*. This analysis was performed using a complete cell lysate of murine Ehrlich ascites tumour (EAT) cells incubated without or with unlabelled 2-5A (*Figure 11*, lanes 1 and 2, respectively). In this experiment a higher than normal amount of 2-5A[^{32}P]Cp was used (2 nM or 200 000 c.p.m.) in order to highlight the non-specifically labelled bands which may appear. Although there were many bands in the autoradiogram, the labelling of only one band, presumptive 2-5A-dependent RNase (lane 1, see arrow), was specifically prevented by the addition of excess unlabelled 2-5A (lane 2). It is also apparent that the band of 2-5A-dependent RNase represents the most heavily labelled protein. Addition of leupeptin to the cell lysis buffer is critical in order to prevent the occurrence of proteinase digestion products of 2-5A-dependent RNase (13).

2.4.2 *Radiobinding assay for 2′5′ oligoadenylate-dependent RNase*

In addition to labelling covalently 2-5A-dependent RNase, it is also possible to measure levels of the nuclease by a simple filter binding assay (11,16,43,44). The method described here is adapted from the radiobinding assay for detecting 2-5A (Section 2.2.1). This technique is a nitrocellulose filter binding method which measures the specific binding of 2-5A[^{32}P]Cp to 2-5A-dependent RNase.

(i) Prepare a '2-5A[^{32}P]Cp mix':
   (a) enough 2-5A[^{32}P]Cp for 5000 c.p.m. per assay;
   (b) 1 μl per assay of '20 × RB mix', (200 mM Tris-HCl pH 7.6; 40 mM magnesium acetate; 8.0 mM ATP pH 7.6; and 40% glycerol);
   (c) make up to 5 μl per assay with water.

(ii) Aliquot in duplicate 50, 100, 200 and 300 μg of either post-mitochondrial fraction (*Table 2*) or complete cell lysate (*Table 2*) to 0.5 ml conical tubes. Also prepare one tube with no cell extract to determine the background radioactivity.

(iii) To one set of each of the cell extracts add competitor in the form of 5 μl of 400 nM trimer 2-5A (unlabelled).

(iv) Adjust the cell extract volume (with and without added 2-5A) to 15 μl with NP-40 lysis buffer containing leupeptin (*Table 2*).

(v) Add 5 μl of 2-5A[^{32}P]Cp mix to each assay tube, mix briefly by vortexing, and incubate for 90 min on ice to allow the probe to bind to the 2-5A-dependent RNase.

(vi) Cut 25 mm nitrocellulose discs (Millipore, HAWP 025 00, 0.45 μM pores) in half, place on aluminium foil and number with water-insoluble ink.

(vii) At the end of the incubation period, spot all of each reaction mixture on the filters (it is important to do this quickly so that the last filter is spotted before the first filter dries out). Place all of the filters into a beaker containing 10 ml of tap distilled water per filter. Gently swirl by hand for 30 sec, discard the water and wash once more in the same way. Discard the water, spread the filters on a fresh piece of aluminum foil, dry for 15 min with a heat lamp and determine the radioactivity by liquid scintillation counting.

A typical radiobinding assay for 2-5A-dependent RNase is shown in *Figure 12*. Levels of 2-5A-dependent RNase were determined in post-mitochondrial fractions of untreated or interferon-treated murine JLS-V9R cells (*Figure 12, A* and *B*, respectively, performed in collaboration with H.Jacobsen, Heidelberg). These assays were done with varying levels of competitor 2-5A. The binding of 2-5A[^{32}P]Cp (present at 36 pM) was almost completely prevented by 10 nM 2-5A (*Figure 12*). The binding is specific for 2-5A or closely-related analogues of 2-5A (16 and *Figure 5*). The assays showed a large

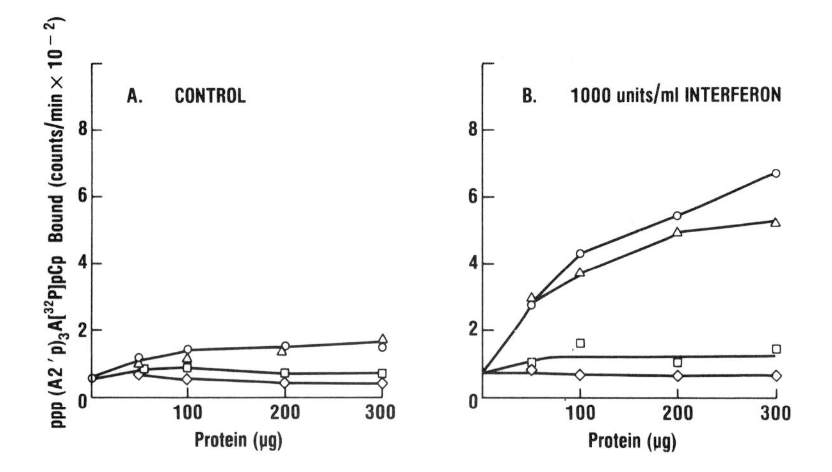

Figure 12. Radiobinding method for measuring 2′5′ oligoadenylate-dependent RNase levels: induction of the enzyme by interferon treatment of JLS-V9R cells. Subconfluent cultures were incubated without **(A)** or with **(B)** 1000 units/ml of murine $\alpha + \beta$ interferons (7×10^7 units/mg) for 16 h. The post-mitochondrial cell extracts were prepared and labelled with 2-5A[^{32}P]Cp in the absence (○) or presence of 0.5 (△), 10 (□), and 100 (◇) nM trimer 2-5A.

induction in levels of 2-5A-dependent RNase by interferon (40). Levels of 2-5A binding activity usually reach a plateau value with about 200 μg of protein.

2.4.3 *Core–cellulose assay for 2′5′ oligoadenylate-dependent RNase activity*

In contrast to the previous two methods, the core–cellulose assay measures 2-5A-mediated RNA cleavage rather than 2-5A binding activity. Because this assay involves immobilization and a partial purification of 2-5A-dependent RNase it avoids some of the problems associated with assays in crude cell-free systems or in intact cells, such as modification, synthesis or degradation of 2-5A during the assay and the presence of non-2-5A-dependent RNases (17,30). The following is an adaptation of the core–cellulose assay for detecting levels of 2-5A (Section 2.2.2).

(i) Prepare post-mitochondrial fractions of cell extracts and determine the protein concentrations (*Table 2*).

(ii) Determine the amount of core–cellulose required to perform the desired number of assays. First, calculate the final assay volume (250 μl $\times$ number of cell extracts to be tested). Take enough core–cellulose to give a final concentration of 25 μM bound core in the assays. For instance, if there are six extracts to be tested, and the core–cellulose stock suspension is 125 μM, then 6×250 μl $= 1.5$ ml final volume; a 5-fold dilution of the core–cellulose will give 25 μM; therefore, 1.5 ml $\div$ 5 $=$ 300 μl of core–cellulose suspension is required. Place the core–cellulose suspension (Section 2.2.2) in a conical centrifuge tube and centrifuge at 1500 *g* for 5 min at 2°C.

(iii) Discard the supernatant and resuspend the core–cellulose in 10 times the packed cellulose volume of $0.5 \times$ CB (section 2.2.2) with vortexing. Centrifuge at 1500 *g* for 5 min at 2°C.

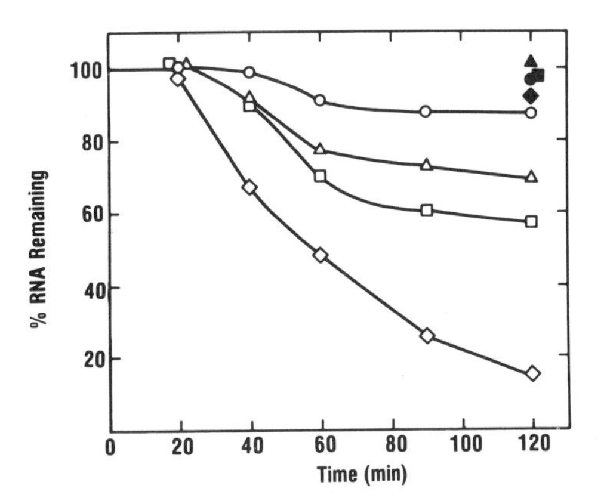

Figure 13. Core–cellulose assay for measuring 2′5′ oligoadenylate-dependent RNase levels: induction of the enzyme in NIH 3T3, clone 1 cells by interferon treatment and growth arrest. 2-5A-dependent RNase was immobilized and partially purified from NIH 3T3, clone 1 cells using core–cellulose. 2-5A-dependent RNase was measured by the breakdown of poly(U)-[^{32}P]Cp in the presence (open symbols) or absence (closed symbols) of 2-5A. The cells were subconfluent, untreated (○, ●); subconfluent and interferon treated (△, ▲); late-confluent, untreated (□, ■); or late-confluent, and interferon treated (◇, ◆). Interferon treatment was with 5000 units/ml (murine α + β interferons, 2 × 10^8 units/mg) for 20 h. (Figure from ref. 39, reprinted with permission of the publisher.)

(iv) Discard the supernatant, resuspend the core–cellulose in 10 times the packed cellulose volume of 0.5 × CB and vortex until the cellulose pellet is completely suspended. Distribute the core–cellulose suspension equally into 0.5 ml conical tubes (one for each cell extract to be tested).

(v) Centrifuge the distributed core–cellulose at 1500 *g* for 5 min, and discard the supernatants. To each tube add 160 μl of 0.8 × CB, vortex briefly, and then add the cell extracts [1.9 mg of protein; each adjusted to 100 μl with NP-40 lysis buffer containing leupeptin, (*Table 2*)]. This results in 150 μg of cell extract protein used for each assay time point.

(vi) Incubate on ice for 60 min, vortexing briefly every 20 min.

(vii) Centrifuge at 1500 *g* for 5 min at 2°C.

(viii) Discard the supernatants, and resuspend each pellet in 200 μl of 0.5 × CB. Centrifuge at 1500 *g* for 5 min at 2°C.

(ix) Repeat step viii twice more.

(x) Discard the supernatants from the last centrifugation.

(xi) Prepare the assay mixture (volumes given are for six cell extracts):
   (a) 800 μl of 1 × CB;
   (b) 32 μl of 0.4 mM poly(U)-[^{32}P]Cp, 125 Ci/mol (Section 2.2.2);
   (c) 768 μl of autoclaved, distilled water.

(xii) Resuspend cellulose pellets from step x to a final volume of 250 μl with assay mixture (step xi). Vortex briefly and then remove 50 μl of the resuspended cellulose from each tube to a new tube for a minus 2-5A control. To the tubes containing 200 μl of resuspended cellulose add 5 μl of 10 μM trimer 2-5A.

(xiii) Take 20 μl zero time points from all tubes (including minus 2-5A controls), into

1.0 ml TCA/PPi/carrier RNA mixtures exactly as described above (Section 2.2.2).

(xiv) Place assays at 30°C, taking 20 μl aliquots into 1.0 ml TCA/PPi/RNA mixtures at 20, 40, 60, 90 and 120 min for the plus 2-5A incubations and at only the 120 min point for the minus 2-5A controls.

(xv) Determine the acid-insoluble radioactivity by filtering on glass fibre filters (Section 2.2.2).

Results of a typical core–cellulose assay for 2-5A-dependent RNase are shown in *Figure 13* (from ref. 39). There was no breakdown of the poly(U)-[^{32}P]Cp in the minus 2-5A controls (solid symbols) indicating an absence of non-2-5A-dependent RNase in the assays. In contrast, the assays containing 2-5A resulted in RNA breakdown. This particular experiment was with extracts of NIH 3T3, clone 1 cells and demonstrated that there was little 2-5A-dependent RNase in control, actively-growing cells but that the nuclease could be induced by either interferon treatment or by growth arrest during confluency (39). In addition, these effects on 2-5A-dependent RNase levels appeared to be additive in cells which were first growth arrested during confuency and then interferon treated (*Figure 13*).

It is also possible to assay 2-5A-dependent RNase using 2-5A–cellulose instead of core–cellulose (17). Because 2-5A–cellulose binds 2-5A-dependent RNase at a much higher affinity than core–cellulose, it is possible to immobilize the nuclease using a much smaller volume of cellulose. However, this assay has the disadvantgae of lacking a minus 2-5A control. The 2-5A-dependent RNase is activated by merely binding the 2-5A–cellulose although there is an enhancement in nuclease activity upon the addition of free 2-5A (17). One can monitor general nuclease activity in such assays by incubating with poly(C)-[^{32}P]Cp which is not a substrate for the 2-5A-dependent RNase (17).

It should be mentioned that some cell types, for example EAT cells, have high levels of general nuclease activity which cannot be completely removed, making the assay of 2-5A-dependent RNase by the core–cellulose assay extremely difficult. Therefore, in such instances it is necessary to measure 2-5A-dependent RNase activity by other methods, for example the rRNA cleavage assay (Section 2.4.4).

2.4.4 *2′5′ Oligoadenylate-mediated cleavage of rRNA*

Activated 2-5A-dependent RNase cleaves rRNA in ribosomes in cell-free systems and in intact cells into discrete sets of characteristic cleavage products (18,32,41,45) which are then analysed by denaturing gel electrophoresis (46). Different RNases result in different patterns of rRNA cleavage (18), and the cleavage patterns produced by 2-5A-dependent RNases appear to be unique and species specific (18,45, and R.H.S. and D.K., unpublished). The cleavage of rRNA is, therefore, an index of 2-5A activity (45) and as such has been very valuable in establishing whether 2-5A-dependent RNase is switched on in interferon-treated, virus-infected cells (e.g. 18,32,45,47). In addition, the cell-free system assay is used as a method for detecting authentic 2-5A (e.g. 18,48). Here we describe the analysis of rRNA breakdown as a measure of 2-5A-dependent RNase activity in both cell-free systems and intact cells.

Table 5. Reagents for rRNA cleavage assay.

1.	Preparation of buffer A: 300 μl of [100 mM Hepes pH 7.5; 0.9 M KCl; 30 mM magnesium acetate; and 70 mM 2-mercaptoethanol]. This may be stored at −20°C. 16 μl of 25 mM spermidine, 2 μl of fresh 14.3 M 2-mercaptoethanol, 325 μl of 10 mM ATP; 1 mM GTP; 6 mM CTP; and 100 mM creatine phosphate adjusted to pH 7.0 with NaOH, 660 μl of autoclaved, distilled water.
2.	Buffer B: 50 mM sodium acetate pH 5.0 (with acetic acid); 10 mM EDTA; and 0.5% SDS.

(i) Cell-free system for producing rRNA breakdown. This assay uses modified protein synthesis conditions of Kerr *et al.* (49). However, similar results are obtained by merely incubating the cell extracts with and without added 2-5A (I.M.Kerr, personal communication). Typically these assays are performed with and without added 2-5A and the appearance of 2-5A-mediated breakdown of rRNA is monitored.

(1) Add the following components to 0.5 ml conical tubes on ice in this order:
   (a) 125 μl of buffer A (*Table 5*);
   (b) enough autoclaved distilled water to give a final volume of 250 μl;
   (c) 25 μl of 10 μM trimer or tetramer 2-5A (to give a final concentration of 1 μM) (or without 2-5A for a negative control);
   (d) post-mitochondrial fraction of cell extract containing 650 μg of protein (prepared as described in *Table 2*).
(2) Mix briefly by vortexing and incubate for 2 h at 30°C.
(3) Add 0.5 ml of buffer B (*Table 5*) to each assay tube and then add 0.75 ml of Tris-HCl, pH 7.5-saturated phenol/chloroform/isoamyl alcohol (25:24:1) (28), vortex vigorously for 30 sec and centrifuge at 10 000 *g* for 10 min at 4−6°C.
(4) Remove the upper, aqueous phase to a new tube and re-extract with 0.75 ml of phenol/chloroform/isoamyl alcohol, vortex and then centrifuge at 10 000 *g* for 10 min. Extract once with chloroform in the same manner.
(5) Remove the upper, aqueous phase (~0.75 ml) to a new tube and add 1/30th volume of 3 M sodium acetate pH 5.5 (made with acetic acid) to give a final concentration of 0.1 M and then add 2.5 volumes of ethanol. Place tubes on dry ice for at least 1 h or at −70°C for 16−18 h.
(6) Collect the RNA by centrifuging at 10 000 *g* for 20 min and carefully decant the supernatant, being careful not to dislodge the pellet. Invert the tubes on clean, absorbant paper for a few minutes, then dry *in vacuo* in a Speed-vac (Savant).
(7) Resuspend the RNA in 6 μl of autoclaved, distilled water. The RNA is now ready for analysis by denaturing agarose gel electrophoresis (Section 2.2.4).

(ii) Production of 2′5′ oligoadenylate-mediated rRNA cleavages in intact cells: the calcium phosphate co-precipitation method. Due to its negative charge and size, 2-5A does not normally penetrate the cell membrane. However, Hovanessian *et al.* (41) showed that 2-5A could be introduced into intact cells by an adaptation of the calcium

Table 6. Reagents for introduction of 2′5′ oligoadenylates into cells by calcium phosphate co-precipitation.

1.	2.5 M $CaCl_2$. Weigh out 36.9 g of $CaCl_2$ dihydrate, make up to 100 ml with water, adjusting the pH to 7.0 with HCl, and filter through a 0.45 μm Nalgene filter and store in plastic tubes.
2.	2 × Hepes-buffered saline (2 × HBS) (42 mM Hepes, pH 7.0; 274 mM NaCl; 10 mM KCl; 1.4 mM Na_2HPO_4; and 11 mM glucose). Weigh out 8.0 g of NaCl; 0.37 g of KCl; 0.125 g of Na_2HPO_4 $2H_2O$, 1.0 g of glucose; and 5.0 g of Hepes. Add 400 ml of distilled, de-ionized water and adjust the pH to 7.0 with NaOH. Adjust the volume to 500 ml with water and filter sterilize through a 0.2 μm Nalgene filter.

phosphate co-precipitation method of Graham and van der Eb (50). The activation of 2-5A-dependent RNase in the cells is then monitored by analysis of the rRNA. The following is a modification of this method designed for monolayer cultures of cells in 150 cm^2 flasks.

(1) Prepare solutions of 2.5 M $CaCl_2$ and 2 × Hepes-buffered saline (2 × HBS) as indicated in *Table 6*.
(2) For each plate of cells to be analysed prepare 10 ml of 125 mM $CaCl_2$ in 1 × HBS using the solutions from step 1 and autoclaved distilled water.
(3) To the solution from step 2 add trimer or tetramer 2-5A to 1 μM (for instance by making a 1/1000 dilution of 1 mM 2-5A). Place the mixture on ice for 30 min to allow the precipitate to form.
(4) Remove the medium from the monolayer cultures by aspiration.
(5) Wash the cells twice with 10 ml of pre-warmed (37°C) 1 × HBS, being careful to remove as much of the HBS as possible after the second wash.
(6) Mix the 2-5A solution from step 3 by vortexing vigorously and add 10 ml to each plate of cells.
(7) Allow the precipitate to settle on the cells for 15 min at room temperature.
(8) Add 15 ml of pre-warmed (37°C) Earle's minimal essential medium supplemented only with ITS pre-mix (Collaborative Research) and L-glutamine.
(9) Place the cells in a 37°C incubator with 5% CO_2 for 15 min.
(10) Remove the precipitation solution from the cells by aspiration.
(11) Wash the cells twice with 10 ml of pre-warmed (37°C) 1 × HBS.
(12) Add 20 ml of pre-warmed (37°C) normal growth medium containing serum and incubate the cells at 37°C for 90 min.
(13) Remove the growth medium by aspiration and wash the cells with 10 ml of ice-cold 1 × HBS. **Note:** be very careful not to dislodge the cells from the plate during this procedure because the cells become less adherent. If the monolayers you are working with can tolerate 2 h incubations, you may wish to make this adjustment.
(14) Scrape the cells in 10 ml of fresh ice-cold 1 × HBS and place in a 15 ml centrifuge tube.
(15) Centrifuge the cells very gently at 60 *g* for 5 min and decant and discard the supernatant.
(16) Resuspend the cells in 1.0 ml of 1 × HBS and pipette into a 1.5 ml conical tube.
(17) Centrifuge at 60 *g* for 5 min and aspirate the supernatant. The dry cell pellet may be frozen on dry ice and stored at −70°C.

Table 7. Reagents for glyoxal/agarose gels.

1. 1 M sodium phosphate pH 7.0: mix solutions of 1 M dibasic sodium phosphate and 1 M monobasic sodium phosphate until a solution which is pH 7.0 is obtained.
2. 10 M urea: make fresh by mixing 1.1 g of urea with 1.0 ml of water.
3. Glyoxal: the glyoxal must be extensively de-ionized prior to use. De-ionization is achieved by repeated swirling in mixed bed resin [(AG501-X8(D) Bio-Rad]. Add the resin to about 50% of the volume of the glyoxal, swirl, remove glyoxal to fresh resin, repeat about 10 more times. Then prepare aliquots (200 μl) and store at −70°C. Upon removing the glyoxal aliquots for use add about 100 μl of the same resin and vortex. The beads should sink within a few seconds if the glyoxal is de-ionized.

(18) Prepare a post-mitochondrial cell extract as described in *Table 2* by thawing the cell pellet in NP-40 lysis buffer, and then centrifuging at 10 000 *g* for 10 min at 2°C.

(19) Immediately extract the RNA from the supernatant by diluting the supernatant with 20 volumes of buffer B (*Table 5*) and adding an equal volume of phenol/chloroform/isoamyl alcohol (25:24:1).

(20) Re-extract the upper aqueous phase with phenol/chloroform/isoamyl alcohol mixture and then with chloroform, make the final upper phase 0.1 M in sodium acetate, pH 5.5 and add to this 2.5 volumes of ethanol. Precipitate the RNA at −70°C and collect by centrifugation at 10 000 *g* for 20 min. Decant and then dry *in vacuo* in a Speed-vac (Savant) and resuspend the RNA in autoclaved, distilled water.

Note: Total cell RNA may also be used in this type of assay (32,48).

(iii) Analysis of the rRNA by glyoxal/agarose gel electrophoresis (46).

(1) Prepare the necessary reagents as described in *Table 7.*

(2) Prepare the RNA for electrophoresis by mixing the following in 0.5 ml conical tubes:

   (a) 5 μl of 10 M urea;
   (b) 15 μl of DMSO;
   (c) 10 μg of RNA in 6 μl of water (the concentration of RNA is determined by diluting 1 or 2 μl in water and determining its absorbance at 260 nm);
   (c) 5 μl of glyoxal mix [144 μl of de-ionized glyoxal (see *Table 7*) plus 10 μl of 1 M sodium phosphate pH 7.0].

(3) Heat the RNA mixture from step 2 at 50°C for 1 h and then add 1 μl of 0.1% bromophenol blue.

(4) During the incubation (step 3) prepare the gel. Mix in a 500 ml flask: 3.6 g of agarose, 2 ml of 1 M sodium phosphate pH 7.0 and 200 ml of water. Heat in a microwave oven for 5 or 6 min until the agarose is completely in solution, swirl to mix, replace any water which may have evaporated and pour the gel (20 cm^2 horizontal gel system). Insert a 1 mm thick sample comb. Let the gel set at room temperature for about 15 min.

(5) Place the gel in the electrophoresis apparatus, pour the running buffer (10 mM sodium phosphate, pH 7.0) about half way up the gel (*do not flood the gel*).

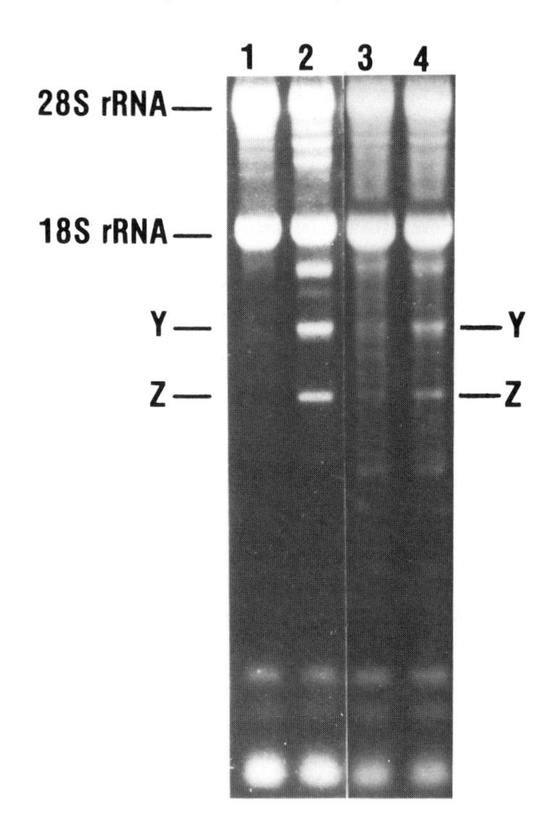

Figure 14. Analysis of 2′5′ oligoadenylate-mediated rRNA cleavage products as a measure of 2-5A-dependent RNase activity. **Lanes 1** and **2**: cell-free system analysis. Post-mitochondrial cell extracts from untreated murine L cells were incubated for 2 h without (**lane 1**) and with (**lane 2**) 1.0 μM 2-5A. **Lanes 3** and **4**: intact cell analysis. Murine L cells were pre-incubated for 20 h without (**lane 3**) and with (**lane 4**) 1000 units/ml of murine $\alpha + \beta$ interferons (7×10^7 units/mg) and then both were treated with calcium chloride-precipitated trimer 2-5A at 1.0 μM. The RNA was isolated, denatured and analysed on a glyoxal/agarose gel. The position of the rRNAs and the 2-5A-specific cleavage products of the 18S rRNA (Y and Z) are indicated. A photograph of the ethidium bromide-stained gel is shown.

Add the RNA samples (from step 3) to the wells. Apply 100 mA until the dye has run into the gel. Add running buffer until the gel is just covered. Circulate the running buffer between the chambers using an electric pump (*this is very important*). Apply 150 mA of current to the gel for about 3 h.

(6) Turn off the current and remove the gel to a dish containing 500 ml of 50 mM NaOH, and shake for 40 min.

(7) Remove the NaOH solution, replace with 500 ml of 50 mM sodium phosphate, pH 7.0 and shake for 20 min.

(8) Repeat step 7.

(9) Replace the buffer with 500 ml of 50 mM sodium phosphate, pH 7.0 containing 100 μl of 5 mg/ml ethidium bromide and shake for 10 min.

(10) Place the stained gel on a 302 nm u.v. light source box and photograph the gel using Type 55 positive/negative film (Polaroid). Expose the film for 20 sec with a wide-open aperture of 4.7.

Note: the best results are obtained by using a narrow sample comb (1 mm). Also, take care not to overload the wells because using greater than about 10 μg of RNA per lane results in diffuse bands. Good circulation of the running buffer during the electrophoresis is required to maintain the pH of the buffer. An alternative buffer system for analysis of rRNA cleavage products may be used (48).

A typical analysis of 2-5A-mediated cleavage of rRNA from a mouse L cell-free system and from the intact cells is presented in *Figure 14*. Lanes 1 and 2 were from the cell-free system incubated in the absence and presence of 2-5A, respectively. The most characteristic cleavage products are the 18S rRNA breakdown products (45) labelled 'Y' and 'Z' (*Figure 14*). In addition, 2-5A was introduced into control and interferon-treated cells, lanes 3 and 4 respectively, by calcium phosphate co-precipitation. As shown, there was a similar cleavage pattern with the production of bands 'Y' and 'Z' in both the intact cells (lanes 3 and 4) and in the cell-free system (lane 2). 2-5A-mediated cleavage of rRNA is, therefore, an excellent index of 2-5A activity (45).

2.5 Assay for 2′,5′-phosphodiesterase activity

The levels of 2′,5′-phosphodiesterase activity in cell extracts may be estimated by using the radiobinding assay (Section 2.2.1) to monitor the breakdown of 2-5A added to cell-free systems (25). These assays are performed under modified protein synthesis conditions containing 1.25 mM ATP which protects the 5′-phosphoryl groups of 2-5A from endogenous phosphatase activity. However, to verify the stability of the 5′-phosphates of 2-5A it is desirable to perform h.p.l.c. analysis using material obtained from the incubations (Section 2.2.1).

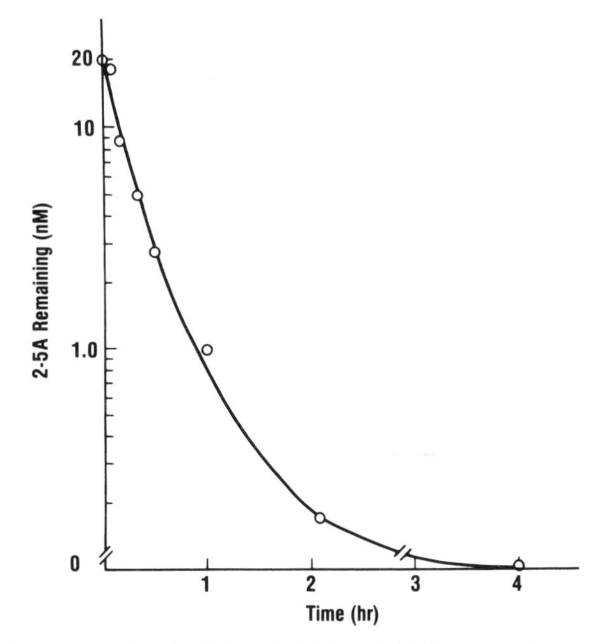

Figure 15. Breakdown of 2′5′ oligoadenylates by 2′,5′-phosphodiesterase in reticulocyte lysate. At various times after the addition of 25 nM tetramer 2-5A to a reticulocyte lysate system, aliquots were removed and the levels of 2-5A were determined by the radiobinding assay (Section 2.2.1).

(i) Set-up the incubations by mixing the following:
 (a) 50 μl of buffer A (see *Table 5*);
 (b) 20 μl of water;
 (c) 20 μl of cell extract (see *Table 2*); (these should be normalized to contain the same amount of cell protein, typically ~400 μg);
 (d) 10 μl of 200 nM trimer or tetramer 2-5A.

(ii) Incubate at 30°C and at 0, 5, 20, 30, 60, 120 and 240 min remove 10 μl aliquots to 0.5 ml conical tubes containing 15 μl of water and immediately heat to 90°C for 5 min.

(iii) Centrifuge the heat-treated aliquots at 10 000 *g* for 5 min and remove 5 μl of the supernatants to new 0.5 ml tubes for 2-5A determination.

(iv) Perform the radiobinding assay for detecting 2-5A as described in Section 2.2.1. Determine the amounts of 2-5A remaining from a standard curve of 2-5A.

A typical 2-5A degradation curve from a rabbit reticulocyte lysate system is shown in *Figure 15*. There was greater than 50% breakdown of the added 2-5A in about 10 min. Using EAT cells, the added 2-5A had a half-life of 30 min (25).

3. THE INTERFERON-REGULATED PROTEIN KINASE SYSTEM

3.1 **Background information**

The protein kinase, like 2-5A synthetase, is both induced by interferon treatment and is dependent on dsRNA for its activity (see refs. 3–5 for review). Activation of the protein kinase results in the phosphorylation of the α subunit of initiation factor eIF2 leading to an inhibition of protein synthesis by an indirect mechanism in which the recycling of eIF2 is prevented (reviewed in ref. 51). In addition, the protein kinase phosphorylates histones and an endogenous, interferon-inducible 65 000–67 000 molecular weight protein in mouse cells or a 68 000–72 000 molecular weight protein in human cells (these are also called protein P1). Recent evidence indicates that the 68 000-dalton protein is a kinase that must first be phosphorylated in order to phosphorylate exogenous substrates (52). The 68 000-dalton kinase has been extensively purified using either monoclonal antibody (52) or conventional enzyme purification methods (53).

Here we present two methods for monitoring levels of the protein kinase. The first technique measures the phosphorylation of the P1 protein (65 000- or 68 000-dalton kinase) in solution; and the second method involves the immobilization and partial purification of the kinase on poly(I):poly(C)–cellulose prior to the phosphorylation reaction. An elegant two-dimensional immunoblot method for determining phosphorylation of the α subunit of eIF2 from intact cells has been described but will not be further considered here (47,54,55).

3.2 **'In-solution' assay of the protein kinase**

This is a modification of the method developed in Kerr's laboratory (56). Addition of excess dsRNA is inhibitory to the kinase; therefore, when altering the conditions of the assay it is advisable to determine the concentration of dsRNA which results in maximal kinase activity.

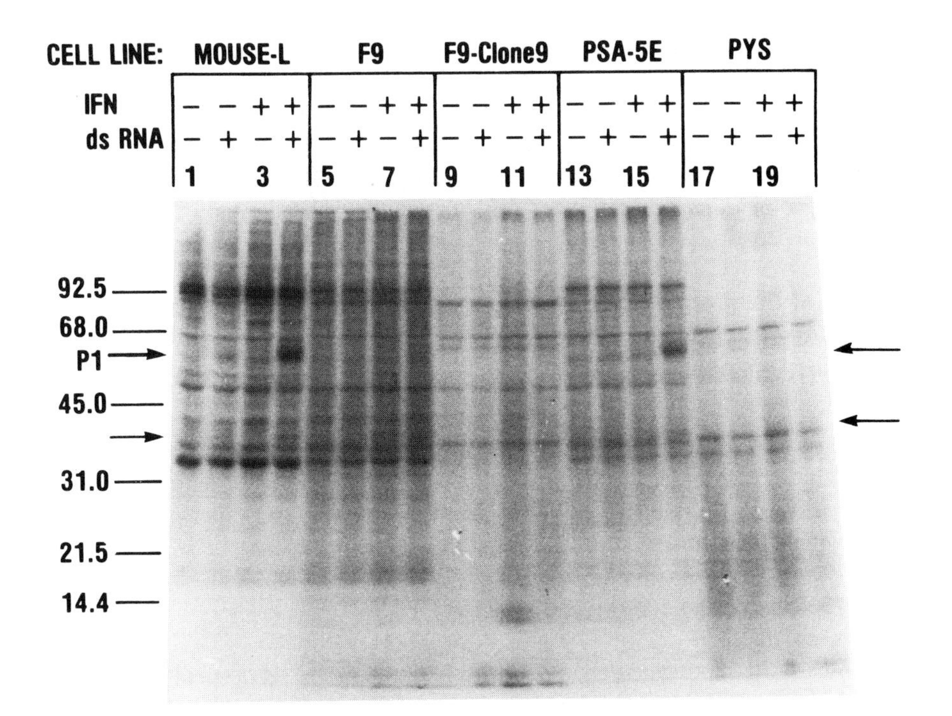

Figure 16. Assay of the protein kinase system by the 'in-solution' method using cell extracts from murine L and teratocarcinoma cells. Cells were incubated for 16 h in the absence or presence of 1000 units/ml of murine $\alpha + \beta$ interferons ($>7 \times 10^7$ units/mg) as indicated. Post-mitochondrial cell extracts were prepared and incubated with [γ-^{32}P]ATP in the presence or absence of 1.0 μg/ml poly(I):poly(C) ('dsRNA' in the figure). The labelled proteins were analysed by electrophoresis on an SDS/8–20% gradient polyacrylamide gel. The positions and the molecular masses (in kd) of the protein markers, the protein P1 and the α subunit of eIF2 (lower arrow) are indicated. An autoradiogram of the dried gel is shown.

(i) Mix the following in 0.5 ml conical tubes (on ice):
   (a) 75 μg of post-mitochondrial cell extract protein (see *Table 2*) adjusted to 7 μl with NP-40 lysis buffer containing leupeptin;
   (b) 1 μl of 10 μg/ml poly(I):poly(C) or 1 μl of water for the controls;
   (c) 2 μl of protein kinase assay mix [5 μl of 1 M $MgCl_2$; 10 μl of 5 mM ATP, pH 7.0; 10 μl of 5.0 mCi/ml (1.85×10^8 Bq/ml) [γ-^{32}P]ATP (Amersham, >1000 Ci/mmol); and 75 μl of water].
(ii) Mix the samples briefly by vortexing and then incubate at 30°C for 15 min.
(iii) Add 10 μl of two times gel sample buffer containing SDS, heat at 90°C for 5 min and apply to an SDS/8–20% gradient polyacrylamide gel.
(iv) After electrophoresis dry the gel *in vacuo* and expose to X-ray film for 16–20 h or longer as required.

A typical 'in-solution' protein kinase assay for levels of dsRNA-dependent phosphorylations in extracts of control and interferon-treated cells is shown in *Figure 16* (performed in collaboration with H.Jacobsen, Heidelberg and R.M.Friedman, Bethesda). This particular experiment compares the levels of the kinase in murine L cells and in four murine teratocarcinoma cell lines; [the F9 cells are an undifferentiated line and the others are differentiated (13,57)]. The phosphorylation of protein P1 in the L cells and in the PSA-5E cells was dependent on both prior treatment of the cells with interferon and on the presence of dsRNA in the incubation mixtures (*Figure 16*). In the case of the L cells one can also visualize the phosphorylation of the α subunit of eIF2 (*Figure 16*, lane 4, see lower arrow). In contrast, there was no detectable dsRNA-dependent protein kinase in the other cell lines, only a background of non-dsRNA-dependent phosphorylations.

3.3 Assay of the protein kinase system using poly(I):poly(C) – cellulose

This is a modification of the technique developed by Hovanessian and Kerr (56) in which poly(I):poly(C) – cellulose (Section 2.3.1) is substituted for poly(I):poly(C) – Sepharose.

(i) Steps i to ix are at 0 – 4°C. Into a 1.5 ml conical tube, pipette enough poly(I):poly(C) – cellulose suspension (0.9 mg poly(I):poly(C)/ml) (Section 2.3.1) to have 15 μl for each assay (15 μl × the number of assays to be performed = one volume). Centrifuge at 1250 *g* for 5 min.

(ii) Discard the supernatant, resuspend the cellulose in 5 volumes of 1 × DBG buffer (*Table 3*) and centrifuge at 1250 *g* for 5 min.

(iii) Repeat step ii.

(iv) Discard the supernatant and resuspend the cellulose to its original volume (i.e. 15 μl times the number of assays) in 1 × DBG.

(v) Pipette 150 μg of post-mitochondrial cell extract protein (see *Table 2*) into 0.5 ml conical tubes on ice and adjust volumes to 15 μl with DBG. Add 15 μl of washed poly(I):poly(C) – cellulose (from step iv) after briefly vortexing.

(vi) Incubate on ice for 60 min with occasional vortexing to allow the protein kinase to bind to the poly(I):poly(C). Add 250 μl of NP-40 buffer containing leupeptin (*Table 2*), vortex briefly, centrifuge at 1250 *g* for 5 min at 2 – 4°C and then *very carefully* remove the supernatants making sure to leave the cellulose pellets undisturbed (leave a small amount, <15 μl, of the supernatant behind).

(vii) Wash the protein kinase:poly(I):poly(C) – cellulose complex twice more in 250 μl of NP-40 buffer plus leupeptin as in step vi.

(viii) Adjust the volumes to 15 μl by adding NP-40 buffer plus leupeptin.

(ix) Add 15 μl of assay mixture prepared as follows (for 20 assays):

- (a) 3 μl of 1 M magnesium acetate;
- (b) 30 μl of 1 mM ATP pH 7;
- (c) 264 μl of NP-40 buffer containing leupeptin (*Table 2*);
- (d) 3 μl of 10 μCi/μl (3.7×10^5 Bq/μl) [γ-^{32}P]ATP, 1000 Ci/mmol.

(x) Incubate for 30 min at 30°C, vortexing gently every 10 min. Add 30 μl of two times gel sample buffer containing SDS and heat at 90°C for 5 min. Centrifuge at 10 000 *g* for 10 min and remove the supernatant to new tubes. Apply 30 μl of the supernatants to an SDS/10% polyacrylamide gel.

An analysis of levels of phosphorylation of protein P1 is shown in *Figure 17* for three undifferentiated and three differentiated lines of murine teratocarcinoma cells in-

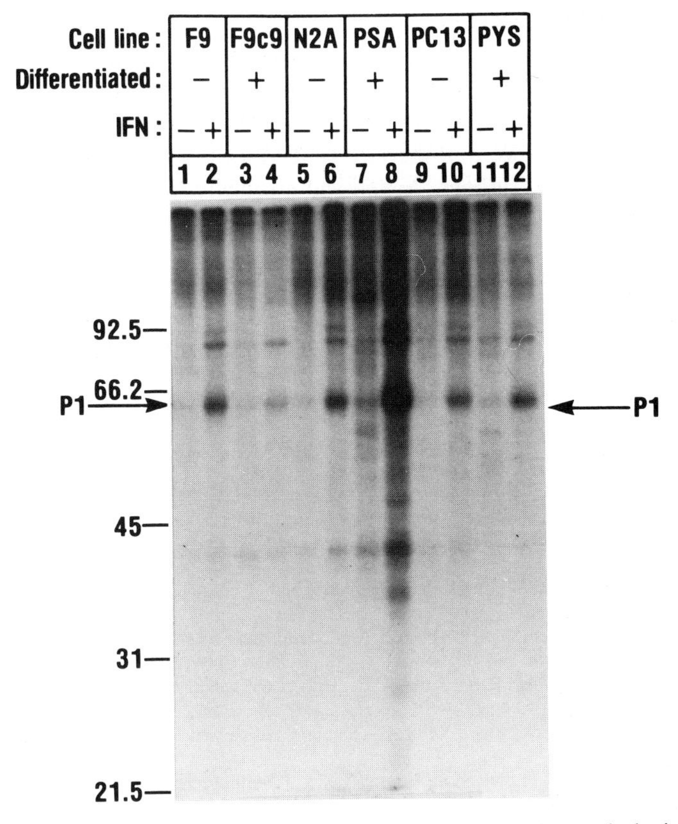

Figure 17. Assay of the protein kinase system by the poly(I):poly(C) – cellulose method using cell extracts from murine teratocarcinoma cells. The dsRNA-dependent protein kinase from post-mitochondrial cell extracts was assayed following adsorption to poly(I):poly(C) – cellulose. The labelled proteins were analysed on an SDS/10% polyacrylamide gel. The cell lines and presence or absence of murine $\alpha + \beta$ interferons (1000 units/ml for 16 h, $>7 \times 10^7$ units/mg) are indicated. The arrows indicate the position of protein P1 and the numbers to the left indicate the positions and molecular masses (in kd) of the protein markers. An autoradiograph of the dried gel is shown. Certain of the cell lines are abbreviated; Nulli 2A (N2A), F9 clone 9 (F9c9), PSA 5E (PSA). The figure is from ref. 13 and is reprinted with permission of the publisher.

cubated in the presence and absence of interferon (from ref. 13). The background is reduced as a result of the immobilization and partial purification of the kinase on poly(I):poly(C) – cellulose (compare *Figures 16* and *17*). In addition, the sensitivity of the assay is greatly improved. For instance, there was no detectable phosphorylation of protein P1 in the F9, F9-clone 9 and PYS cells using the 'in-solution' assay whereas phosphorylated protein P1 was clearly detectable in extracts of these same cell lines using the poly(I):poly(C) – cellulose method (*Figure 17*). This could be due in part to the removal of protein phosphatases in the latter method. The additional interferon-inducible proteins (i.e. other than the P1 protein) have not been identified. The level of protein P1 phosphorylation can be determined either by densitometer tracing or by measuring the radioactivity in the excised gel bands (13).

4. ACKNOWLEDGEMENTS

We would like to thank all those who contributed to the development of these methods including; R.E.Brown, P.J.Cayley, C.W.Dieffenbach, R.M.Friedman, C.S.Gilbert, W.G.Hearl, A.G.Hovanessian, J.Imai, H.Jacobsen, T.C.James, M.I.Johnston, M.Knight, O.Preble, W.K.Roberts, G.R.Stark, P.F.Torrence, J.A.Wells, B.R.G.Williams, D.H.Wreshner, A.Yeh and especially to I.M.Kerr. We also thank C.W.Dieffenbach for valuable discussions.

5. REFERENCES

1. Stewart,W.E.,II (1981) *The Interferon System.* 2nd edn, Springer-Verlag, New York.
2. Freidman,R.L., Manly,S.P., McMahon,M., Kerr,I.M. and Stark,G.R. (1984) *Cell,* **38**, 745.
3. Williams,B.R.G. (1983) In *Interferon and Cancer.* Sikora,K. (ed.), Plenum Publishing Corp., p. 33.
4. Johnston,M.I. and Torrence,P.F. (1984) In *Interferon, vol. 3: Mechanisms of Production and Action.* Friedman,R.M. (ed.), Elsevier Science Publishers, B.V., New York, p. 189.
5. Lengyel,P. (1982) *Annu. Rev. Biochem.*, **51**, 251.
6. Ball,L.A. (1979) *Virology*, **94**, 282.
7. Williams,B.R.G., Read,S.E., Freedman,M.H., Carver,D.H. and Gelfand,E.W. (1982) In *Chemistry and Biology of Interferons: Relationship to Therapeutics, UCLA Symposia on Molecular Biology and Cellular Biology.* Merrigan,T., Friedman,R.M. and Fox,C.F. (eds), Academic Press, New York, Vol. **25**, p. 253.
8. Shattner,A., Merlin,G., Wallach,D., Rosenberg,H., Bino,T., Hahn,T., Levin,S. and Revel,M. (1981) *J. Interferon Res.*, **1**, 587.
9. Nilsen,T.W. and Baglioni,C. (1979) *Proc. Natl. Acad. Sci. USA*, **76**, 2600.
10. Nilsen,T.W., Wood,D.L. and Baglioni,C. (1982) *J. Biol. Chem.*, **257**, 1602.
11. Wreschner,D.H., Silverman,R.H., James,T.C., Gilbert,C.S. and Kerr,I.M. (1982) *Eur. J. Biochem.*, **124**, 261.
12. St. Laurent,G., Yoshie,O., Floyd-Smith,G., Samanta,H., Sehgal,P.B. and Lengyel,P. (1983) *Cell*, **33**, 95.
13. Krause,D., Silverman,R.H., Jacobsen,H., Leisy,S.A., Dieffenbach,C.W. and Friedman,R.M. (1985) *Eur. J. Biochem.*, **146**, 611.
14. Pestka,S., ed. (1981) *Methods in Enzymology.* Vol. **79**, Part B, Academic Press. Inc., New York.
15. Reid,T.R., Hersh,C.L., Kerr,I.M. and Stark,G.R. (1984) *Anal. Biochem.*, **136**, 136.
16. Knight,M., Cayley,P.J., Silverman,R.H., Wreschner,D.H., Gilbert,C.S., Brown,R.E. and Kerr,I.M. (1980) *Nature*, **288**, 189.
17. Silverman,R.H. (1985) *Anal. Biochem.*, **144**, 450.
18. Wreschner,D.H., James,T.C., Silverman,R.H. and Kerr,I.M. (1981) *Nucleic Acids Res.*, **9**, 1571.
19. Hersh,C.L., Reid,T.R., Freidman,R. and Stark,G.R. (1984) *J. Biol. Chem.*, **259**, 1727.
20. Cailla,H., LeBorne DeKaouel,C., Roux,D., Delaage,M. and Marti,J. (1982) *Proc. Natl. Acad. Sci. USA*, **79**, 4742.
21. Johnston,M.I., Imai,J., Lesiak,K. and Torrence,P.F. (1983) *Biochemistry*, **22**, 3453.
22. Johnston,M.I., Imai,J., Lesiak,K., Jacobsen,H., Sawai,H. and Torrence,P.F. (1985) *Biochemistry*, **24**, 4710.
23. Sawai,H., Ishibashi,K., Itoh,M. and Watanabe,S. (1984) *Biochem. Biophys. Res. Commun.*, **125**, 1061.
24. Hovanessian,A.G., Brown,R.E., Martin,E.M., Roberts,W.K., Knight,M. and Kerr,I.M. (1981) In *Methods in Enzymology.* Pestka,S. (ed.), Academic Press, New York, Vol. **79**, Part B, p. 184.
25. Silverman,R.H., Wreschner,D.H., Gilbert,C.S. and Kerr,I.M. (1981) *Eur. J. Biochem.*, **115**, 79.
26. Brown,R.E., Cayley,P.J. and Kerr,I.M. (1981) In *Methods in Enzymology.* Pestka,S. (ed.), Academic Press, New York, Vol. **79**, Part B, p. 208.
27. Wells,J.A., Swyryd,E.A. and Stark,G.R. (1984) *J. Biol. Chem.*, **259**, 1363.
28. Maniatis,T., Fritsch,E.F. and Sambrook,J. (1982) *Molecular Cloning: A Laboratory Manual.* Cold Spring Harbor Laboratory Press, New York.
29. Meurs,E., Krause,D., Robert,N., Silverman,R.H. and Hovanessian,A.G. (1985) In *The 2-5A System: Molecular and Clinical Aspects of the Interferon-Regulated Pathway.* Williams,B.R.G. and Silverman,R.H. (eds.), A.R.Liss Inc., New York, p. 307.
30. Krause,D., Lesiak,K., Imai,J., Sawai,H., Torrence,P.F. and Silverman,R.H. (1986) *J. Biol. Chem.*, **261**, 6836.

31. Williams,B.R.G., Brown,R.E., Gilbert,C.S., Golgher,R.R., Wreschner,D.H., Roberts,W.K., Silverman,R.H. and Kerr,I.M. (1981) In *Methods in Enzymology*. Pestka,S. (ed.), Academic Press, New York, Vol. **79**, Part B, p. 199.
32. Silverman,R.H., Cayley,P.J., Knight,M., Gilbert,C.S. and Kerr,I.M. (1982) *Eur. J. Biochem.*, **124**, 131.
33. Hersh,C.L., Brown,R.E., Roberts,W.K., Swyryd,E.A., Kerr,I.M. and Stark,G.R. (1984) *J. Biol. Chem.*, **259**, 1731.
34. Brown,R.E. and Kerr,I.M. (1985) In *The 2-5A System: Molecular and Clinical Aspects of the Interferon-Regulated Pathway*. Williams,B.R.G. and Silverman,R.H. (eds), A.R.Liss Inc., New York, p. 3.
35. Stark,G.R., Dower,W.J., Schimke,R.T., Brown,R.E. and Kerr,I.M. (1979) *Nature*, **278**, 471.
36. Johnston,M.I., Friedman,R.M. and Torrence,P.F. (1981) In *Methods in Enzymology*. Pestka,S. (ed.), Academic Press, New York, Vol. **79**, Part B, p. 228.
37. Revel,M., Wallach,D., Merlin,G., Schattner,A., Schmidt,A., Wolf,D., Shulman,L. and Kimchi,A. (1981) In *Methods in Enzymology*. Pestka,S. (ed.), Academic Press, New York, Vol. **79**, Part B, p. 149.
38. Imai,J. and Torrence,P.F. (1984) *Biochemistry*, **23**, 766.
39. Krause,D., Panet,A., Arad,G., Dieffenbach,C.W. and Silverman,R.H. (1985) *J. Biol. Chem.*, **260**, 9501.
40. Jacobsen,H., Czarniecki,C.W., Krause,D., Friedman,R.M. and Silverman,R.H. (1983) *Virology*, **125**, 496.
41. Hovanessian,A.G., Wood,J., Meurs,E. and Montagnier,L. (1979) *Proc. Natl. Acad. Sci. USA*, **76**, 3261.
42. Floyd-Smith,G., Yoshie,O. and Lengyel,P. (1982) *J. Biol. Chem.*, **257**, 8584.
43. Silverman,R.H., Watling,D., Balkwill,F.R., Trowsdale,J. and Kerr,I.M. (1982) *Eur. J. Biochem.*, **126**, 333.
44. Nilsen,T.W., Wood,D.L. and Baglioni,C. (1981) *J. Biol. Chem.*, **256**, 10751.
45. Silverman,R.H., Skehel,J.J., James,T.C., Wreschner,D.H. and Kerr,I.M. (1983) *J. Virol.*, **46**, 1051.
46. McMaster,G.K. and Carmichael,G.G. (1977) *Biochemistry*, **74**, 4835.
47. Rice,A.P., Duncan,R., Hershey,J.W.B. and Kerr,I.M. (1985) *J. Virol.*, **54**, 894.
48. Rice,A.P., Roberts,W.K. and Kerr,I.M.(1984) *J. Virol.*, **50**, 220.
49. Kerr,I.M., Olshevsky,U., Lodish,H.F. and Baltimore,D. (1976) *J. Virol.*, **18**, 627.
50. Graham,F.L. and van der Eb,A.J. (1973) *Virology*, **52**, 456.
51. Safer,B. (1983) *Cell*, **33**, 7.
52. Galabru,J. and Hovanessian,A.G. (1985) *Cell*, **43**, 685.
53. Berry,M.J., Knutson,G.S., Lasky,S.R., Munemitsu,S.M. and Samuel,C.E. (1985) *J. Biol. Chem.*, **260**, 11240.
54. Duncan,R. and Hershey,J.W.B. (1983) *J. Biol. Chem.*, **258**, 7228.
55. Samuel,C.E., Duncan,R., Knutson,G.S. and Hershey,J.W.B. (1984) *J. Biol. Chem.*, **259**, 13451.
56. Hovanessian,A.G. and Kerr,I.M. (1979) *Eur. J. Biochem.*, **93**, 515.
57. Silverman,R.H., Krause,D., Jacobsen,H., Leisy,S.A., Barlow,D.P. and Friedman,R.M. (1983) In *The Biology of the Interferon System 1983*. DeMaeyer,E. and Schellekens,H. (eds), Elsevier Science Publishers B.V., p. 189.

CHAPTER 11

Cell growth inhibition by interferons and tumour necrosis factor

G.B.DEALTRY and F.R.BALKWILL

1. INTRODUCTION

Interferons (IFNs) were discovered as anti-viral agents (1), but they are now recognized as cell regulators, and one of their most important regulatory activities is the inhibition of cell growth (2). The extent of this growth inhibition varies with cell type, the IFN used and the dosage and duration of treatment. Generally, short-term exposure leads to cytostasis, whilst prolonged treatment may be cytotoxic. Both normal and transformed cells are sensitive to IFN. However, although the most sensitive cell lines are tumour derived, there is no direct correlation between the degree of sensitivity and transformation.

Cell cycle-related responses to IFN-α and IFN-γ have been identified (3,4) and recent work has indicated that IFN-β may be involved in the normal regulation of cell growth and differentiation as an autocrine negative growth factor (5). Thus analysis of the mechanisms of IFN-mediated cell growth inhibition is valuable to the understanding of IFN function and to the study of growth regulation. Furthermore, since the use of IFNs in the treatment of cancer is being investigated in a large number of clinical trials, it is important to have detailed knowledge of their action on normal and transformed cell growth *in vitro* and *in vivo*.

Tumour necrosis factor (TNF) is a cytotoxic factor with potential as an anti-cancer agent. Tumour necrosis factor was first identified in 1975 in the serum of BCG-infected mice injected with lipopolysaccharides (LPS) (6). Unlike IFNs, TNF shows rapid cytotoxicity for some transformed cell lines (6). It is not toxic, as a single agent, for normal cells. In fact, high doses of TNF can be mitogenic to normal fibroblasts (7).

The production and commercial availability of pure recombinant IFNs and TNF of both human and murine origin allows analysis of their individual functions and the study of their synergy. Indeed early studies using serum purified TNF may have been affected by trace contaminants of IFN and other serum factors. Similarly, the impure virus or mitogen-induced IFNs used in the first experiments probably contained contaminating TNF.

In this chapter the major techniques for assessing the direct effects of these biological agents on cell growth *in vitro* and on human tumours in the nude mouse *in vivo* are described.

2. EFFECTS ON CELL GROWTH IN TISSUE CULTURE

The techniques outlined in the next four sections of this chapter all require some basic knowledge of tissue culture. We describe here the methods currently in use in our laboratory. However the conditions required for successful cell culture vary with the different cell lines and primary cell cultures used. More general texts are available that deal in detail with all aspects of equipping and running a tissue culture laboratory, and with specific culture environments required by certain cell lines (8).

Cells may be grown as a monolayer on plastic, as colonies within a semi-solid medium, such as agar, or in suspension. In all the *in vitro* techniques described in this chapter, cells are grown at a constant 37°C in a humidified incubator in an atmosphere of 7% carbon dioxide in air.

All procedures described must be carried out under sterile conditions and using aseptic technique.

2.1 Growth in a monolayer

Most mammalian cells can be cultured *in vitro* as a monolayer in tissue culture grade plastic vessels. The cells attach to the plastic and spread out forming a layer one cell deep over the base of the vessel. The exceptions to this are haemopoietic cells and ascitic tumour cells.

Tumour-derived cells or virally transformed cells will readily form cell lines which can be passaged many times. However lines from normal cells generally have a more restricted life-span, usually between 20 and 80 population doublings. Some normal cell types such as the mammary epithelial (HumE) cells described in Section 2.3, cannot easily be cultured beyond a single primary culture.

The procedure for producng a cell line will not be described here since the more commonly used cell lines are available from commercial sources or other laboratories. Stocks of cell lines are generally stored frozen in liquid nitrogen. The procedures for preparing and thawing out such stocks are shown in *Table 1*.

Table 1. Frozen cell stocks.

A. *Preparation*

1. Prepare a single cell suspension and centrifuge at 200 *g* for 5 min.
2. Resuspend the cell pellet in 2–5 ml of medium containing 10% FCS (depending on cell concentration).
3. Adjust with medium + 10% FCS to give 2×10^6 cells/ml.
4. Prepare an equal volume of medium + 10% FCS + 20% dimethyl sulphoxide (DMSO).
5. Mix 3. and 4. and immediately aliquot 1 ml volumes into 1.8 ml sterile vials (Nunc, Kamstrup, Denmark).
6. Lay the sealed vial on two thicknesses of paper towel on top of solid carbon dioxide for 90 min (this allows the cells to freeze at ~1°C/min).
7. Transfer the vial to a tank of liquid nitrogen for storage. This procedure gives a final concentration of 1×10^6 cells in medium + 10% FCS and 10% DMSO.

B. *Thawing*

1. Rapidly thaw frozen vial of cells in a 37°C water bath.
2. Transfer the vial contents to a sterile universal.
3. Add cold medium + 10% FCS dropwise to cells (10 ml over 10 min).
4. Spin the cells down at 200 *g* for 5 min at room temperature.
5. Remove the supernatant (containing DMSO).
6. Adjust the cells to the number and volume required for culture and add to a culture vessel.

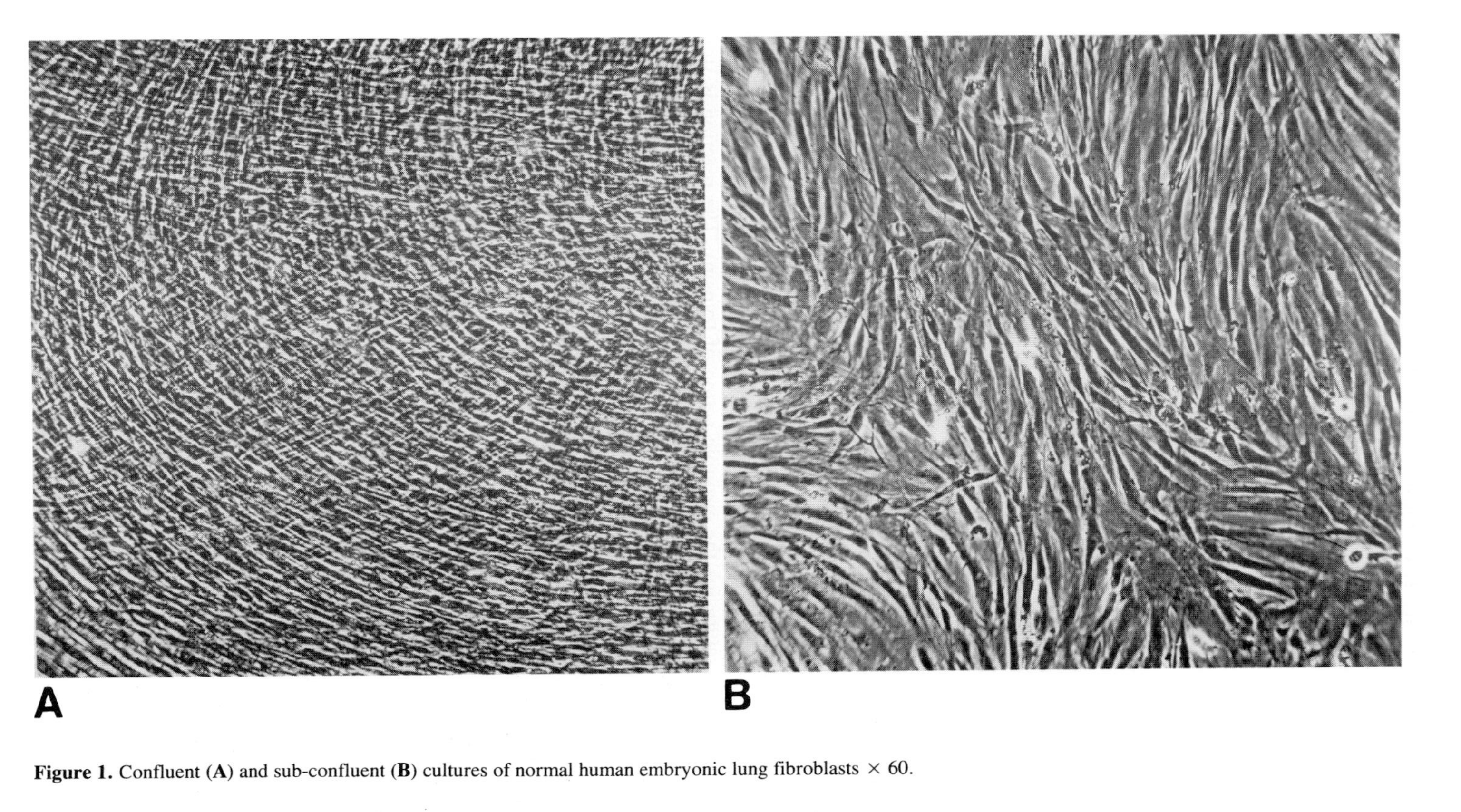

Figure 1. Confluent **(A)** and sub-confluent **(B)** cultures of normal human embryonic lung fibroblasts × 60.

Once a cell culture is established, the medium must be regularly changed as the nutrients in the medium are used up and toxic waste products from the growing cells increase. The growth of normal cells such as diploid fibroblasts is contact inhibited, that is once the cells become confluent they cease dividing and arrest in the G1/G0 phase of the cell cycle (see *Figure 1*). These cells can be left as confluent cultures for a few days provided the medium is changed regularly. Transformed cells and permanent cell lines however are not contact inhibited and deteriorate rapidly at high cell density and must be immediately passaged. The intervals between medium changes and passages vary with the cell line depending on its growth and metabolism rates.

In our laboratory normal diploid human embryo lung fibroblasts and a variety of human tumour cell lines such as MCF7, BT20 and HT29 are routinely used (9–11). All are cultured in RPMI 1640 medium containing 3.7% bicarbonate (Gibco Biocult, Paisley, UK) and 10% fetal calf serum (FCS) (Sera Labs, Crawley Down, Sussex). Stock cultures are seeded at a concentration of 1×10^6 cells in 30 ml in a 75 cm^2 tissue culture flask. The fibroblasts are passaged twice weekly and the tumour lines once a week.

A mixture of trypsin and versene (EDTA) prepared as described in *Table 2* is used to detach and disperse the cells. Trypsin enzymically digests surface molecules involved in cellular adhesion, whilst versene is a chelator of Ca^{2+} and Mg^{2+} ions which are required for attachment and aggregation.

Serum inactivates trypsin, therefore the culture medium must be removed and the cells washed briefly with a small volume of trypsin/versene to ensure rapid detachment. The cells are released from the culture vessel using fresh trypsin/versene at 37°C (optimal temperature for the enzymic action of trypsin). The reaction is stopped by

Table 2. Stock solutions for detaching monolayer cell cultures.

A. *Trypsin* 0.25% in Tris saline pH 7.7 (filter sterilized and stored at −20°C)	
For 1 litre	
Trypsin (Difco 1:250, Difco)	2.5 g
NaCl	8 g
KCl (19% solution w/v in distilled water)	2 ml
Na_2HPO_4	0.1 g
Dextrose	1 g
Trizma base [Tris (hydroxy methyl) amino methane]	3 g
Phenol red (1% solution w/v in distilled water)	1.5 ml
Antibiotics: Penicillin (sodium) (Gibco)	1×10^5 units
Streptomycin (Gibco)	0.1 g
Adjust pH to 7.7. with 1 M HCl	
B. *Versene* in PBS pH 7.2 (autoclave and store at 4°C)	
For 1 litre	
Diaminoethane tetraacetic acid disodium salt (versene-BDH quality)	0.2 g
NaCl	8 g
KCl	0.2 g
Na_2HPO_4	1.15 g
KH_2PO_4	0.2 g
Phenol red (1% solution w/v in distilled water)	1.5 ml
For use add 1 volume of trypsin stock to 4 volumes of versene.	

the addition of fresh medium containing 10% FCS as soon as the cells detach. Treatment should not normally exceed 15 min duration as prolonged exposure to trypsin can damage the cells. Dispersal of the cells is aided by gently tapping the culture vessel. The cells are washed by centrifugation in culture medium, the pellet resuspended in a small volume of fresh medium and dispersed using a syringe and 19 gauge needle to produce a single cell suspension. The suspension is counted using a haemocytometer and adjusted to the required seeding density.

2.2 Viable cell counts

The simplest method to assess the growth of cells in the presence of mediators such as IFN is to do a series of cell counts at set times during the experiment.

Two methods are available for cell counting. The simplest and quickest procedure for counting one sample is to prepare a single cell suspension, and count the number of viable cells in a small sample on a haemocytometer slide. However, if many samples are counted at once, as in a cell growth experiment, then it is quicker to use an electronic cell counting device, the Coulter counter (Coulter Electronics Ltd., Dunstable, UK), although this method does not distinguish viable from dead cells. However in monolayer cultures dead cells detach and float in the culture medium and the washing procedure prior to trypsinization will remove the majority of these cells.

2.2.1 *Cell counts using an improved Neubauer haemocytometer*

A haemocytometer is a microscope slide with a calibrated counting grid which, with a coverslip attached, forms a chamber 0.1 mm deep in which a single cell suspension may be counted on a microscope under a low power objective (see *Figure 2*). The cen-

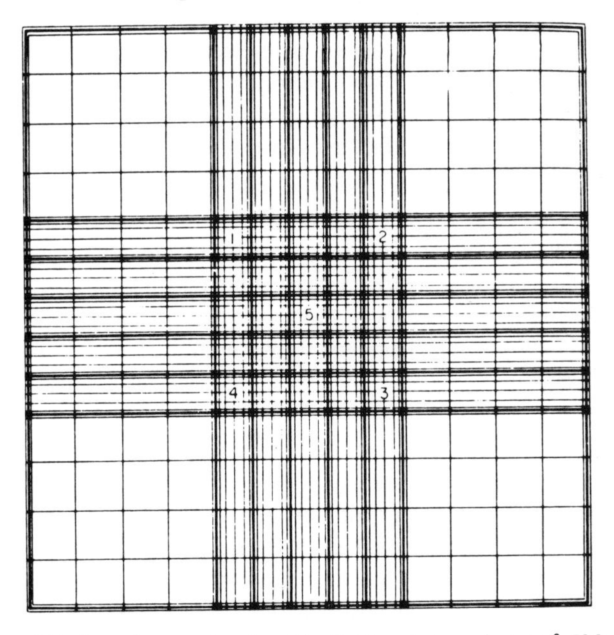

Figure 2. Improved Neubauer haemocytometer slide. Total central area is 1 mm^2. If few cells are present count all the cells in this area; if there are many cells, count the five numbered small squares (at least 100 cells should be counted).

tral 25 squares of the grid form a 1 mm^2 area, giving a total volume counted of 0.1 mm^3 or 1×10^{-4} ml. Ideally all the cells within this volume should be counted, although five of the squares may be counted and the result multiplied up. For greatest accuracy the mean of two sample counts should be taken and at least 100 cells recorded at each count. Under phase contrast unstained viable cells appear bright and refractive, whilst dead cells are dull and dark coloured. Non-electrolyte dyes such as nigrosin or trypan blue may be used. These are absorbed solely by the dead cells, and so these cells are stained and the viable cells remain unstained.

The cell count per ml is calculated as follows:

$$\text{cell count/ml} = \text{number in central } 1\ mm^2 \text{ area} \times 10^4 \times \text{dilution factor}$$

2.2.2 *Cell counts using a Coulter counter*

This apparatus is produced by Coulter Electronics. An electrical current is passed through a 100 μm aperture in a probe, which is placed into a single cell suspension in 10 ml of electrolyte such as phosphate-buffered saline (PBS) or isoton (Coulter Electronics Ltd). The cells are drawn through the probe by a mercury manometer controlled pump altering the resistance to the current flow and producing a series of pulses which are counted and displayed on a digital read out. The change in resistance produced is proportional to the volume of the cell or particle passing through the probe. Thus a threshold can be set so that background counts from cell debris are excluded, and the final reading corresponds to the number of cells in 0.5 ml of the sample.

(i) Treat cultures to be counted with trypsin/versene as described in Section 2.1.
(ii) Stop the reaction with 5% FCS in electrolyte, disperse the cells with a syringe and 19 gauge needle, and wash the culture dishes three times with electrolyte to give a final 10 ml cell suspension in electrolyte.
(iii) Flush the Coulter counter with electrolyte prior to use.
(iv) Shake each sample vigorously to disperse the cells evenly and then count the sample following the manufacturers instructions.

Since the counter counts 0.5 ml, the total number of cells in the sample is found by multiplying the readings by 20. Readings below 10 000 are accurate and can be used directly. Above 10 000 the accuracy decreases due to coincident passage of cells through the aperture, and a conversion chart supplied by the manufacturers must be used to correct the numbers. The mean counts for replicate cultures at each time point and for each treatment are calculated and drawn up to give a growth curve similar to that shown in *Figure 3*.

2.2.3 *Experiment design*

Cultures of cells to be examined are usually seeded at a density of 1×10^5 cells per 30 mm Petri dish in 2 ml of medium, although the seeding density varies with cell type. The aim is to obtain a complete growth cycle through the initial lag phase, logarithmic or exponential growth and finally a plateau at confluence. This gives a characteristic growth curve for each cell type against which treated cultures can be compared. The cells are generally counted on days 3, 5 and 7 after seeding.

The cultures may be treated with IFN or TNF continuously through the experiment,

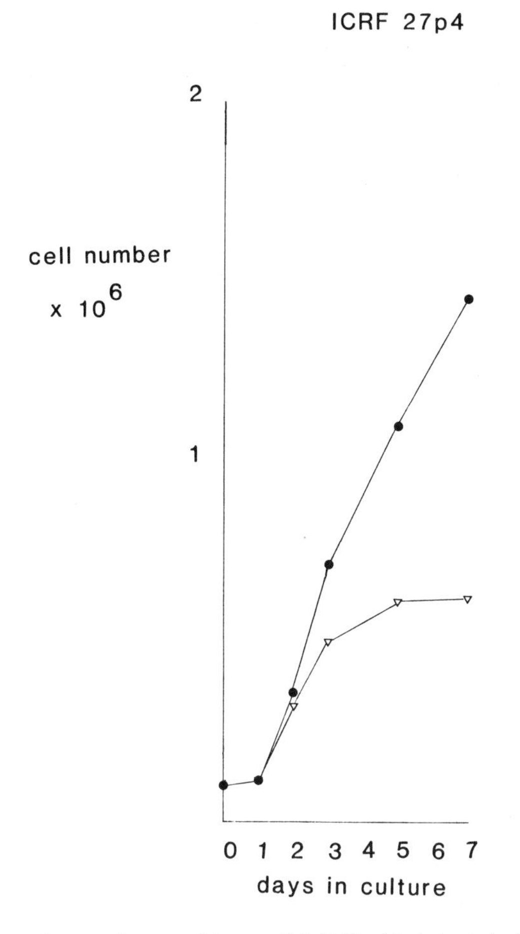

Figure 3. A typical growth curve for normal human diploid fibroblasts treated with IFNγ. -●- control; -▽- 10^3 U/ml IFN-γ. Medium was removed from all cultures and fresh medium ± IFN-γ added twice weekly.

or for brief time periods as required. The IFN and TNF are made up in stock solutions in PBS containing 3 mg/ml bovine serum albumin (BSA) fraction V (Sigma Chemical Co., Dorset, UK), stored at −70°C, and as small a volume as possible added to give the required final concentration. Parallel control cultures are treated with PBS and BSA. If IFN or TNF is added for a brief period of time during the experiment, the treated and parallel control cultures must be washed four times with medium, then fresh culture medium added. Alternatively an excess of blocking antibodies to the IFN or TNF may be added. Combinations of IFNs, TNF and other factors may be used to study their synergy or antagonism. Examples of cell growth experiments using normal human fibroblasts and MCF7 cells are shown in *Figures 3* and *4*.

2.3 **Colony assays**

Fibroblasts are used for most studies on the action of IFN and TNF on normal cell growth. However, many of the strongly IFN-sensitive and TNF-sensitive tumour lines

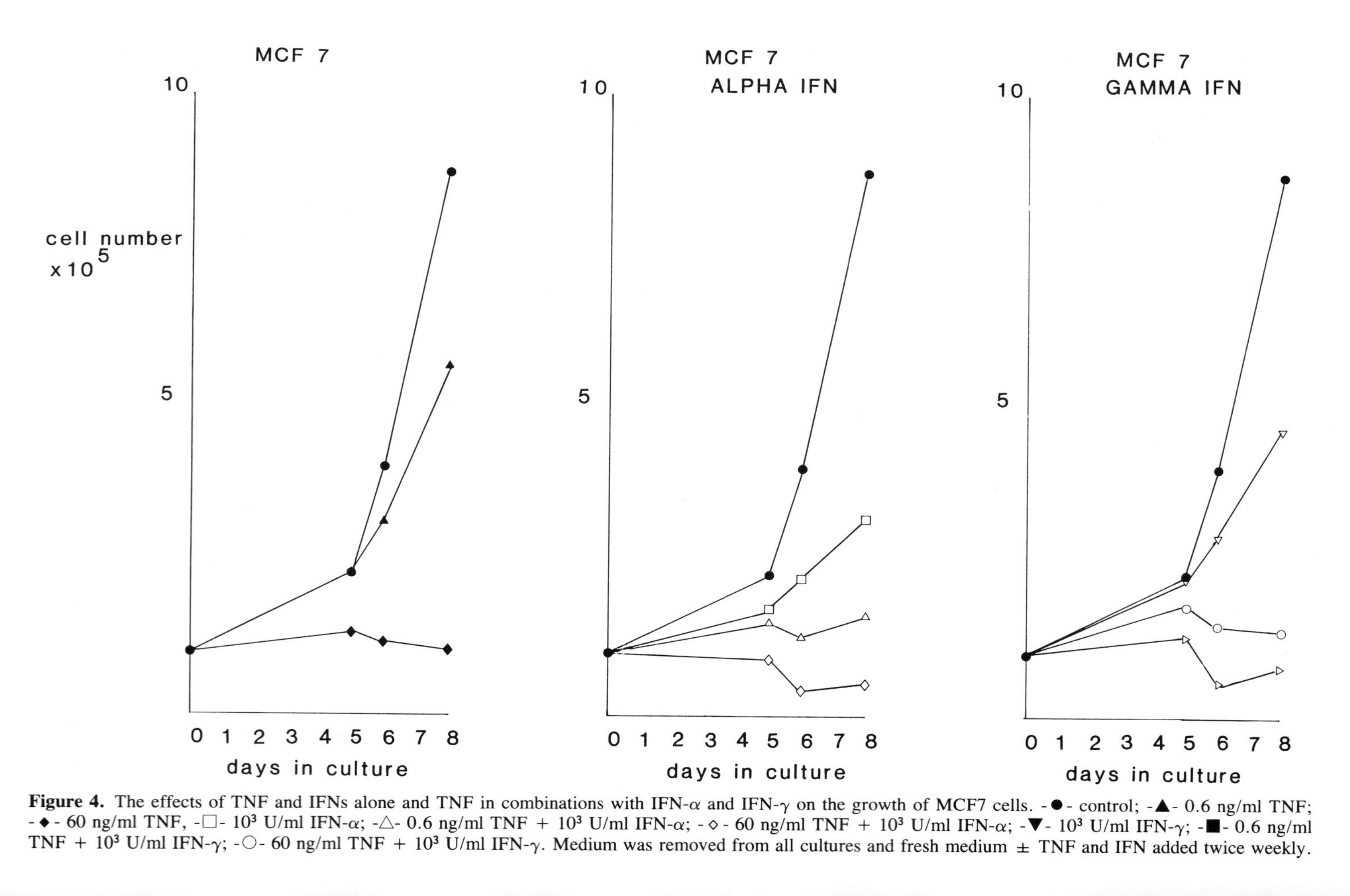

Figure 4. The effects of TNF and IFNs alone and TNF in combinations with IFN-α and IFN-γ on the growth of MCF7 cells. -●- control; -▲- 0.6 ng/ml TNF; -◆- 60 ng/ml TNF, -□- 10^3 U/ml IFN-α; -△- 0.6 ng/ml TNF + 10^3 U/ml IFN-α; -◇- 60 ng/ml TNF + 10^3 U/ml IFN-α; -▼- 10^3 U/ml IFN-γ; -■- 0.6 ng/ml TNF + 10^3 U/ml IFN-γ; -○- 60 ng/ml TNF + 10^3 U/ml IFN-γ. Medium was removed from all cultures and fresh medium ± TNF and IFN added twice weekly.

are epithelial in origin. It is important, therefore, to study the effects of IFN and TNF on normal epithelial cell growth.

It is difficult routinely to obtain primary cultures of normal epithelial cells, but HumE cells derived from early lactation milk produce discrete colonies in liquid cultures that can be used for growth studies. These cells, however, do not survive repeated passaging and require highly supplemented medium containing extra FCS, human serum, insulin and hydrocortisone to stimulate growth (see *Table 3*).

The HumE cells are shed with the colostrum into breast milk during the first few days of lactation. Small samples of milk (up to 20 ml) are collected from women in maternity wards up to 1 week after they have given birth. The cells are cultured as soon as possible after collection following the protocol developed by Taylor-Papadimitriou (12) outlined in *Table 3*. Both the number and size of colonies in treated and control cultures are quantitated and compared (see *Figure 5*).

Colony formation by some tumour lines can also be studied by culturing the cells in soft agar or by plating at very low density in liquid medium. There are several methods for promoting colony growth in soft agar. In our experience the modified Courtenay and Mills method is suitable for most cell lines and primary preparations. This is outlined in *Table 4*.

Fewer cell lines will grow as discrete colonies in liquid medium. A seeding density

Table 3. Preparation of human mammary epithelial (HumE) cells.

1. Collect samples of breast milk in sterile universals.
2. Centrifuge milk samples at 200 *g* for 7 min.
3. Pour off the milk into fresh universals, pooling any small volumes. Retain the cell pellets.
4. Re-centrifuge the milk at 200 *g* for 7 min to collect any remaining cells.
5. Resuspend the cell pellets in RPMI 1640 medium containing 3.7% sodium bicarbonate and 5% FCS. Pool all samples into two universals. Re-centrifuge, and repeat the washing process three or four times until all the milk fat has been washed out and the medium supernatant is clear.
6. After the last wash pool the cells in 10 ml of the following medium:

   30 ml of FCS
   20 ml of human serum
   2 ml of insulin (1 mg/ml stock solution)
   2 ml of antibiotics [gentamycin (2.5×10^4 units/ml), fungizone (50% v/v), penicillin (5×10^4 units/ml) and streptomycin (0.05 g/ml)] in PBS
   2 ml of hydrocortisone C (500 μg/ml stock solution)
   2 ml of cholera toxin (5 μg/ml stock) may be required in the initial seeding medium to encourage formation of epithelial cell colonies.
   The medium is made up to 200 ml with RPMI 1640 + 3.7% sodium bicarbonate, and sterilized by passage through a 0.22 μm pore filter.

7. Plate out the cells at a seeding density of 3×10^5 cells per 30 mm Petri dish. Culture in a humidified incubator at 37°C containing 5–10% CO_2.
8. Replace the medium twice weekly. The cells will continue to grow for approximately 1 month.
9. After 1 week of culture add IFN or TNF and replace at each feeding.
10. After 2–4 weeks, fix the cells with formal saline, and stain for 10 min with Giemsa.
11. Examine under a low power microscope and count the colonies. Measure the colony diameter at the widest part using a slide with 1 mm divisions. Take a second measurement at right angles to the first, and record the mean diameter.

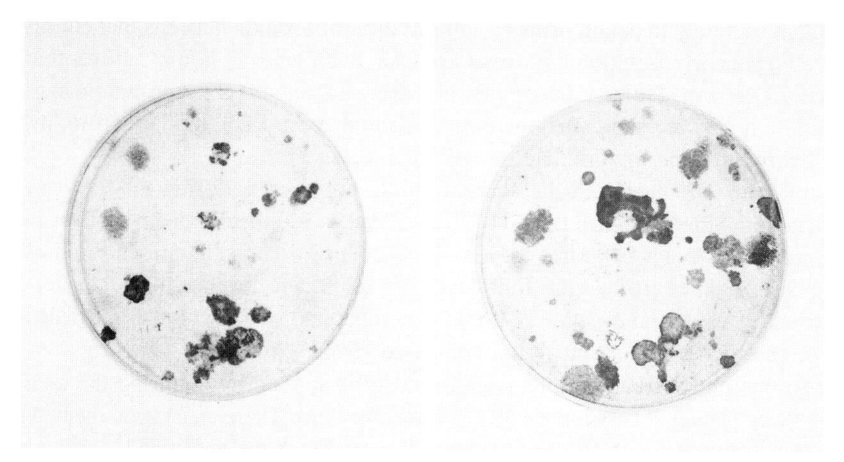

Figure 5. Control and TNF-treated cultures of HumE cells cultured as described in *Table 3*.

Table 4. Modified Courtenay and Mills soft agar colony assay.

This assay was adapted by Hill and Whelan (14) from an original soft-agar assay developed by Courtenay and Mills (15).

1. Prepare soft agar: 2.5% (w/v) low gelling temperature agarose (Uniscience Labs) in distilled water. Boil to dissolve the agarose, and allow to cool slightly.
2. Prepare appropriate medium + 10% FCS and warm to 37°C.
3. Mix 2 ml of agar with 8 ml of medium (to give a final concentration of 0.5% agarose). Keep at 37°C until required.
4. Dilute packed and washed August rat red blood cells 1 in 8 with medium + 10% FCS.
5. Trypsinize a stock culture of cells as described in Section 2.1.
6. Resuspend the cells in medium + 10% FCS to 6 × 10^4 cells in 1.8 ml.
7. Mix the cell suspension with 0.6 ml of diluted rat red blood cells, and 3.6 ml of medium + 10% FCS + 0.5% agarose. This gives a final volume of 6 ml.
8. Add the required concentration of IFN in a small volume.
9. Dispense 1 ml volumes of the mix into 205 l Falcon tubes, and push on caps to create a gas-permeable seal.
10. Plunge into ice to set the agar.
11. Incubate at 37°C in a humidified incubator.
12. Feed every 6 days by adding 1 ml of medium + 10% FCS + the appropriate IFNs to the top of the agar layer.
13. After approximately 1 month examine the agar for colonies by tipping into a gridded Petri dish and covering with the inverted lid to compress the agar. Examine by low power microscopy. Count and measure the size of the colonies formed.

of $10^2 - 10^3$ cells per 30 mm culture dish in 2 ml of medium will establish individual colonies after 5 days' culture. The colonies are cultured in the presence of factors such as IFN for a further 2 weeks with twice weekly medium changes, then fixed, stained with Giemsa and the colonies counted and measured as for HumE (13).

Table 5. Production of resistant cells.

1.	Titrate *N*-methyl-*N*′-nitro-*N*-nitrosoguanidine (Sigma) against stock cultures of cells from 0.1 μg/ml to 100 μg/ml. Select the concentration which gives approximately 50% cell survival.
2.	Add the appropriate concentration of *N*-methyl-*N*′-nitro-*N*-nitrosoguanidine. Treat for 18 h at 37°C.
3.	Wash the cells by centrifugation at 200 *g* for 5 min.
4.	Resuspend in RPMI + 10% heat-inactivated FCS, and culture for 5 days at 37°C.
5.	Add a low dose of IFN, TNF or other factor.
6.	Maintain this concentration of factor until the cells grow at the same rate as untreated cells, then double the concentration. If the cells appear unhealthy the concentration should be dropped until they recover.
7.	Continue this process over several weeks until the cells will grow at normal rates in doses of IFN or TNF that are cytostatic or cytotoxic for the parent cell line.
8.	Resistant cells can be stored in liquid nitrogen until required, then thawed and cultured immediately in RPMI + 10% heat-inactivated FCS containing the IFN or TNF dose to which they are resistant.

2.4 Preparation of resistant cell lines

Resistant cell lines derived from IFN- or TNF-sensitive cells provide a useful tool for studying the effects of these agents on cell growth. By comparing the response of sensitive and resistant cells, important mechanisms mediating cytostasis or cytotoxicity can be identified and the role of induced enzymes, such as 2′5′-oligoadenylate (2-5A) synthetase and protein kinase, in the IFN system investigated. Silverman *et al.* (16) reported the selection of stable IFN-resistant Daudi cells using the mutagen *N*-methyl-*N*′-nitro-*N*-nitrosoguanidine to aid the production of mutant forms of this cell line. The procedure is outlined in *Table 5*, and can be adapted to apply to any cell line. For lines that form monolayers, cells are passaged as required and IFN concentration increased 24 h after passaging to allow the cells time to plate down.

3. CELL CYCLE STUDIES

The cell cycle is defined as the period between completion of mitosis in the parent cell and completion of the next mitotic division in one of the daughter cells (17). It follows a regulated sequence of DNA synthesis and cell growth leading to a doubling of cellular components and cell division resulting in two daughter cells. The cycle has been subdivided into four phases:

(i) the pre-synthetic or post-mitotic gap, G1;
(ii) DNA synthesis (S) phase;
(iii) post-synthetic or pre-mitotic gap, G2;
(iv) mitosis (M).

Quiescent non-cycling cells are said to be in G0.

As seen in Section 2, overall changes in cell viability and mean generation time (i.e. the time taken for one population doubling) can be identified from simple growth curve experiments. However, growth inhibitory agents may be cytotoxic or cytostatic to cells at all stages of the cell cycle, or alternatively may selectively affect cells at a particular stage. In the latter case cells may be killed or arrested in one stage of the cycle, or the passage of the cells through that stage may be delayed. IFNs α and γ have been shown to lengthen the cycle time of normal human synchronized fibroblasts by extending the G1 and S + G2 phases in the case of IFN-α, and G1 only in the case of IFN-γ (18,4).

Cell cycle length varies with cell type, nutrient availability and other environmental factors. Actively growing cultures of homogeneous cell lines have characteristic mean generation times, but individual cells within a population have variable cycle lengths distributed around the population mean. The action of growth inhibitors on the cell cycle of non-synchronous cells can be studied using fluorescence-activated cell sorting (FACS) analysis of DNA replication (see Section 3.2.3), or alternatively a synchronous cycle can be imposed upon the population.

3.1 Methods for synchronization

Many techniques have been developed for arresting cells at a particular stage of the cycle, such that on removal of the block the cells will re-enter the cycle in unison. The cycle inhibitors used are generally DNA synthesis inhibitors such as 5-fluoro-2′-deoxyuridine, aphidicolin, hydroxyurea and high thymidine concentration; or metaphase inhibitors such as colcemid, colchicine, vinblastine and nocodazole (reviewed in ref. 19). Less disruptive methods can be used to synchronize normal cells, but cannot easily be applied to transformed cells, which are not subject to the same growth control. However, the selective detachment of rounded up mitotic cells from monolayer cultures will apply for normal and transformed cells. Alternatively, resting 'G0' cells can be selected by culture to confluence, or culture at low density in limited serum. The latter techniques, which can only be used with normal diploid cells, are routinely employed in our laboratory since these provide good synchrony without disruption to the cells or the need for subculturing prior to growth stimulation. Two procedures for synchronizing normal diploid human fibroblasts (18) and mouse 3T3 cells (20) are outlined in *Table 6*. Both are based on the requirement of normal cells for

Table 6. Cell synchronization.

A. *Synchronization of normal human diploid fibroblasts*

1. Remove medium from stock fibroblasts — 'spent' medium.
2. Dilute 'spent' medium 1 in 20 with fresh RPMI 1640 medium to give a final concentration of 0.5% FCS.
3. Passage the cells, wash by centrifugation at 200 *g* for 5 min in RPMI 1640 and resuspend in the RPMI 1640 containing 5% spent medium.
4. Count and plate out 1×10^5 cells in a 30 mm dish in 2 ml of RPMI containing 5% 'spent' medium, and culture for 7 days.
5. Remove medium and replace with fresh RPMI 1640 + 10% FCS or defined growth factors. This is stimulation time 0. For the normal fibroblast cell lines ICRF23 and ICRF27 the cell cycle length is approximately 30 h; therefore samples must be taken during this time period.

B. *Synchronization of BALB/c 3T3 and Swiss 3T3K cells*

1. On passaging stocks retain the old 'spent' medium as above and use it to make RPMI 1640 containing 5% 'spent' medium.
2. Plate out 1×10^5 cells in a 30 mm Petri dish in 2 ml of RPMI 1640 + 10% FCS, and culture to confluence (1 week).
3. Replace the medium with RPMI 1640 containing 'spent' medium and culture for 24 h.
4. Replace with fresh RPMI 1640 + 10% FCS or defined growth factors to stimulate the cells.

growth factors (present in serum) to stimulate cell growth. On stimulation with serum or growth factors the cells in G0 enter the cycle with a high degree of synchrony. One drawback with all methods for synchronizing cells is that the close synchrony only lasts for one cell cycle, and by the third cycle the distribution is almost random again. Therefore all experiments must be performed during the cycle after stimulation.

3.2 DNA synthesis

The synthesis of DNA can be quantitated by autoradiographic measurement of the percentage of nuclei incorporating [^{3}H]thymidine (labelling index), by measuring the total amount of [^{3}H]thymidine incorporated over a defined time span, or by FACS analysis of the DNA content. Comparison of treated cells with parallel control cultures gives information on the effects of IFN on the rate and amount of DNA synthesis.

3.2.1 *Labelling index*

The labelling index is a measure of the percentage of nuclei incorporating [^{3}H]thymidine. An outline of the procedure is given in *Tables 7* and *8*. Nuclei synthesizing DNA incorporate [^{3}H]thymidine which is detected autoradiographically *in situ* at set times following growth stimulation, thus giving information on the rate of passage through the early cell cycle to S phase (see *Figure 6*).

Table 7. Stock solutions for determination of labelling index.

1. Tris-saline (filter sterilized and stored at 4°C).
   For 1 litre

NaCl	8 g
KCl (19% solution w/v in distilled water)	2 ml
Na_2HPO_4	0.1 g
Dextrose	1 g
Trizma base	3 g
Phenol red (1% solution w/v in distilled water)	1.5 ml
Adjust pH to 7.4 with 1 M HCl	
Add Penicillin	5×10^4 units
Streptomycin	0.5 g

2. 5% TCA in distilled water (prepare fresh).
3. Chrome alum

Solution A	*Solution B*
5 g gelatin	0.05 g chromic potassium sulphate
400 ml distilled water	400 ml distilled water

   Heat until the gelatin dissolves
   Allow to cool.
   Mix the two solutions and make up to 1 litre with distilled water.
   Dispense into 200 ml bottles, and add 0.5 ml of formaldehyde to each bottle.
   Store at 4°C.
4. Kodak D19 developer (Eastman Kodak Co., Rochester, New York). Make up according to the manufacturer's instructions. Store at room temperature in the dark.
5. Ilford fixative (Ilford Ltd, Cheshire, UK) diluted 1:4 in distilled water. Prepare fresh.

Table 8. Determination of labelling index in cell monolayers.

1. Stimulate the synchronized cells as required. Add 5 μCi/ml (1.85×10^5 Bq/ml) [^{3}H]thymidine (sp. act. 25 Ci/mmol, 925 GBq/mmol) (Amersham International PLC, Aylesbury, UK) to each dish.
2. At required times post-stimulation place the cultures on ice to stop DNA synthesis.
3. Wash twice with Tris-saline buffer and once with 5% TCA; then add fresh 5% TCA and leave for 20 min on ice to extract the acid-soluble DNA precursors.
4. Wash twice with absolute alcohol to fix the cells.
5. Dry overnight at room temperature.

Applying film

All the work is done in the dark room using only a photographic red 'safe' light. Treat 10 dishes at a time.

6. Add prepared cooled molten chrome alum to one dish, leave for a few seconds, then pour the chrome alum from this into a second dish. Continue the cascade system to treat 10 dishes. Pour away remaining chrome alum and drain the dishes onto tissues.
7. Prepare Kodak fine grain stripping film AR10 (stored at 4°C until just prior to use), by scoring film into squares approximately 2 cm² — just large enough to cover the base of a 30 mm Petri dish.
8. Fill dishes with distilled water.
9. With forceps lift the film square from the plate, invert and place a square emulsion side up in each dish.
10. Leave a few seconds to allow the film to absorb water and flatten. Then carefully pour off the remaining water and drain the dishes to leave the film covering the base of the dish.
11. Store the dishes in foil-wrapped dark boxes at room temperature for 1 week.

Developing film

Work in the dark room using the red 'safe' light.

12. Pipette 1 ml of developer onto each dish, leave for 5 min then rinse thoroughly with tap water.
13. Pipette 1 ml of fixative onto each dish and leave for 5 min. Ensure that the film clears.
14. Rinse under the tap for 15 min.
15. Stain for 10 min in Giemsa stain, then wash with distilled water and air-dry.
16. Examine under low power magnification. Labelled nuclei appear dark on film. Unlabelled nuclei are stained with the Giemsa. Count approximately 1000 nuclei in five different fields to give the labelling index (see *Figure 6*).

3.2.2 *Measurement of trichloroacetic acid-precipitable radioactivity*

In this procedure, developed by Rozengurt and Hepel (21), [^{3}H]thymidine incorporation into trichloroacetic acid (TCA)-precipitable material is quantitated at intervals post-stimulation (see *Table 9*). This method is a more quantitative and less subjective measurement of DNA synthesis by the whole cell population than the labelling index, but the percentage of the population entering DNA synthesis at a given time post-stimulation cannot be calculated.

3.2.3 *Flow cytometric analysis of the cell cycle*

Several fluorochromes are available to label DNA including fluorescent antibiotics such as mithramycin, chromomycin and ovimycin, fluorescently labelled antibodies to DNA, Feulgen Schiff reagents such as acriflavine and auramine O, and the phenanthridinium dyes, ethidium bromide and propidium iodide (22). The most commonly used are the phenanthridinium dyes, which are simple, efficient and quick to use, propidium iodide being the best for unfixed cells. The dye selectively and quantitatively intercalates into double-stranded nucleic acid. Thus the amount of dye taken up is proportional to the

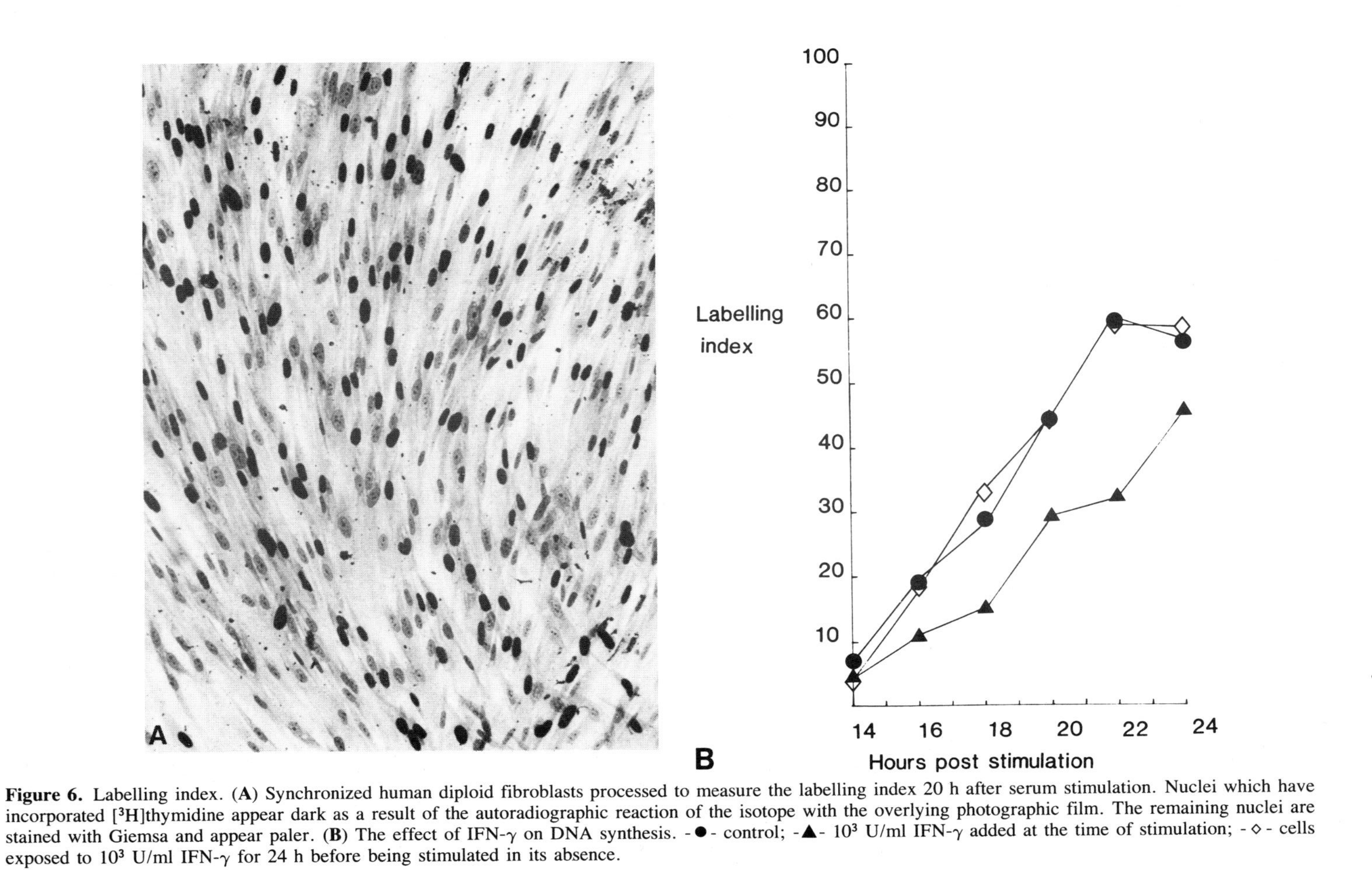

Figure 6. Labelling index. (**A**) Synchronized human diploid fibroblasts processed to measure the labelling index 20 h after serum stimulation. Nuclei which have incorporated [^{3}H]thymidine appear dark as a result of the autoradiographic reaction of the isotope with the overlying photographic film. The remaining nuclei are stained with Giemsa and appear paler. (**B**) The effect of IFN-γ on DNA synthesis. -●- control; -▲- 10^3 U/ml IFN-γ added at the time of stimulation; -◇- cells exposed to 10^3 U/ml IFN-γ for 24 h before being stimulated in its absence.

Table 9. Measurement of incorporation of [^{3}H]thymidine into DNA.

1. Stimulate the cells with fresh medium, + 10% FCS or growth factors, + 10 μCi/ml (3.7 × 10^5 Bq/ml) [^{3}H]thymidine (sp. act. 25 Ci/mmol).
2. Incubate the cultures at 37°C for 28 or 40 h to allow incorporation of the [^{3}H]thymidine[a].
3. Prepare solution buffer:

   For 500 ml:
   NaOH 2 g (final concentration 0.1 M)
   Na_2CO_3 10 g (final concentration 2%)
   Dissolve in 500 ml of distilled water.
   Just prior to use add 1% SDS (10 ml of a 10% stock/100 ml medium)

4. To stop the incorporation:

   Wash twice with cold PBS.
   Add 5% cold TCA.
   Leave at 4°C for 20 min.
   Wash once with 5% TCA.

5. Wash thoroughly with ethanol to remove all the TCA.
6. Solubilize the cells with 2 ml of solubilization buffer.
7. Incubate at 37°C for 1 min.
8. Determine the radioactivity of a 0.2 ml sample in 4 ml of picofluor (Packard Instrument Co. Inc., IL) (or other aqueous scintillant) in a liquid scintillation counter.

[a]This procedure gives results which are influenced by both the rate of DNA synthesis and the specific activities of the intracellular thymidine nucleotide pools. To reduce the effect of variation in the latter it is necessary to expand the pool size by inclusion of 0.1 – 1 mM unlabelled thymidine in the incubation medium. The radioactivity of a 0.5 ml sample in 10 ml of picofluor is then determined in step 8.

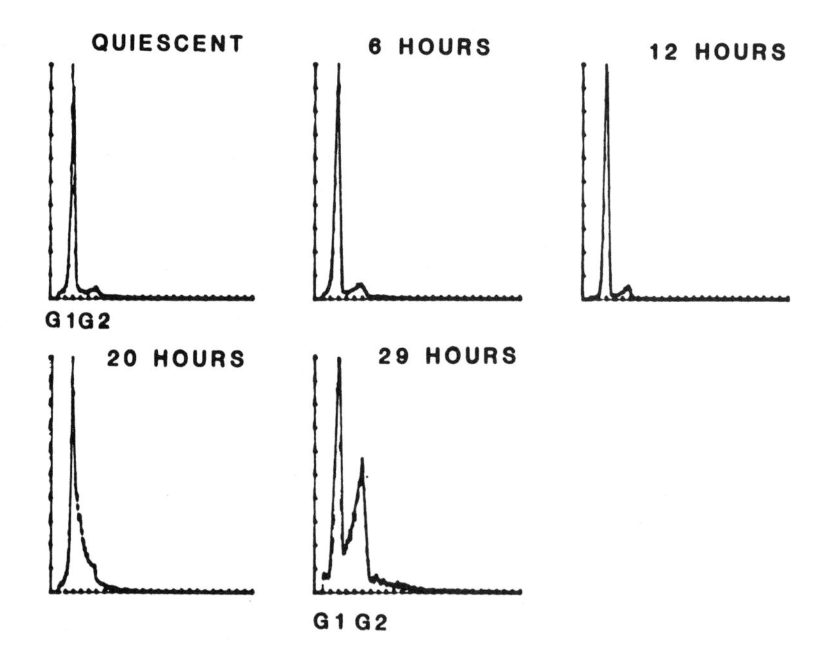

Figure 7. DNA histograms of synchronized fibroblasts at intervals post-serum stimulation. The large G1 peak shown for quiescent cells was normalized to channel 30. As the cells progress through the cycle the size of the G2 peak at channel 60 increases — as shown in the 29 h post-stimulation histogram.

Table 10. Stock solutions for FACS analysis of propidium iodide-stained cells.

1.	*1 M Tris-HCl buffer stock pH 7.5*	
	6.05 g of Tris base	
	50 ml of distilled water	
	Dissolve Tris base in 2/3 volume of distilled water.	
	Chill on ice, and adjust the pH to 7.5 with concentrated HCl.	
	Make up to 50 ml with remaining distilled water.	
2.	*SMT buffer* (0.25 M sucrose, 5 mM $MgCl_2$, 20 mM Tris-HCl, pH 7.5)	
	Sucrose	42.7 g
	Magnesium chloride	0.5 g
	1 M Tris-HCl pH 7.5	10 ml
	Make up to 500 ml with distilled water. Store at 4°C.	
3.	*Ribonuclease A* (Sigma) made fresh at 0.5 mg/ml in SMT buffer.	
4.	*Lysis buffer* (10 mM Tris-HCl, pH 7.5, 0.6 M sucrose, 4 mM $MgCl_2$, 0.5% Triton X-100)	
	Triton X-100	1 ml
	Magnesium chloride	0.17 g
	Sucrose	41 g
	1 M Tris-HCl pH 7.5	2 ml
	Make up to 200 ml with distilled water. Filter and store at 4°C.	
5.	*Propidium iodide* (Sigma) made fresh and stored at 4°C in the dark. 0.05 mg/ml in 0.1% tri-sodium citrate (w/v in distilled water).	

Table 11. FACS analysis of propidium iodide-stained cells.

1. Trypsinize the cells and spin down at 200 *g* for 5 min.
2. Resuspend the cells in serum-free RPMI and transfer to the tubes used for FACS analysis (e.g. for Becton Dickinson FACS1-LP3 tubes).
3. Spin down at 200 *g* for 5 min, remove most of the supernatant, and resuspend the cell pellet in the remaining supernatant ($\sim 100\ \mu l$).
4. Add 1.5 ml of lysis buffer, mix thoroughly, leave for 5 min at room temperature, then spin down at 200 *g* for 4 min.
5. Remove the supernatant, resuspending the pellet of lysed cells in a little residue supernatant; then add 1.5 ml of RNase, mix thoroughly and pass the suspension through a 25-gauge needle to break up any clumps.
6. Incubate at 37°C for 30 min.
7. Pass the suspension through a 25-gauge needle again and check under the microscope that it consists of single nuclei.
8. Spin down at 200 *g* for 4 min, remove the supernatant and resuspend the nuclei in a little of the residue.
9. Add 1.5 ml of propidium iodide, mix thoroughly, then leave on ice in the dark for 3 min.
10. Spin down at 200 *g* for 4 min at 4°C (the nuclear pellet should appear pink after taking up the propidium iodide).
11. Remove the supernatant and resuspend the nuclei in 0.5 ml of cold PBS, place on ice and analyse immediately on the FACS (do not keep for longer than 30 min).

amount of nucleic acid present, and a nucleus in G2 will be stained twice as much as one in G1, while during S phase the level will fall between the two extremes. Since RNA binds some dye the isolated nuclei are treated with RNase prior to staining.

Flow cytometric analysis is carried out by FACS (e.g. Becton Dickinson Modified FACS 1, Becton Dickinson, California, USA). An aqueous suspension of fluorochrome-stained cells or sub-cellular components, such as nuclei, is passed at uniform high speed through a sensing region where each particle is exposed briefly to uniform illumination

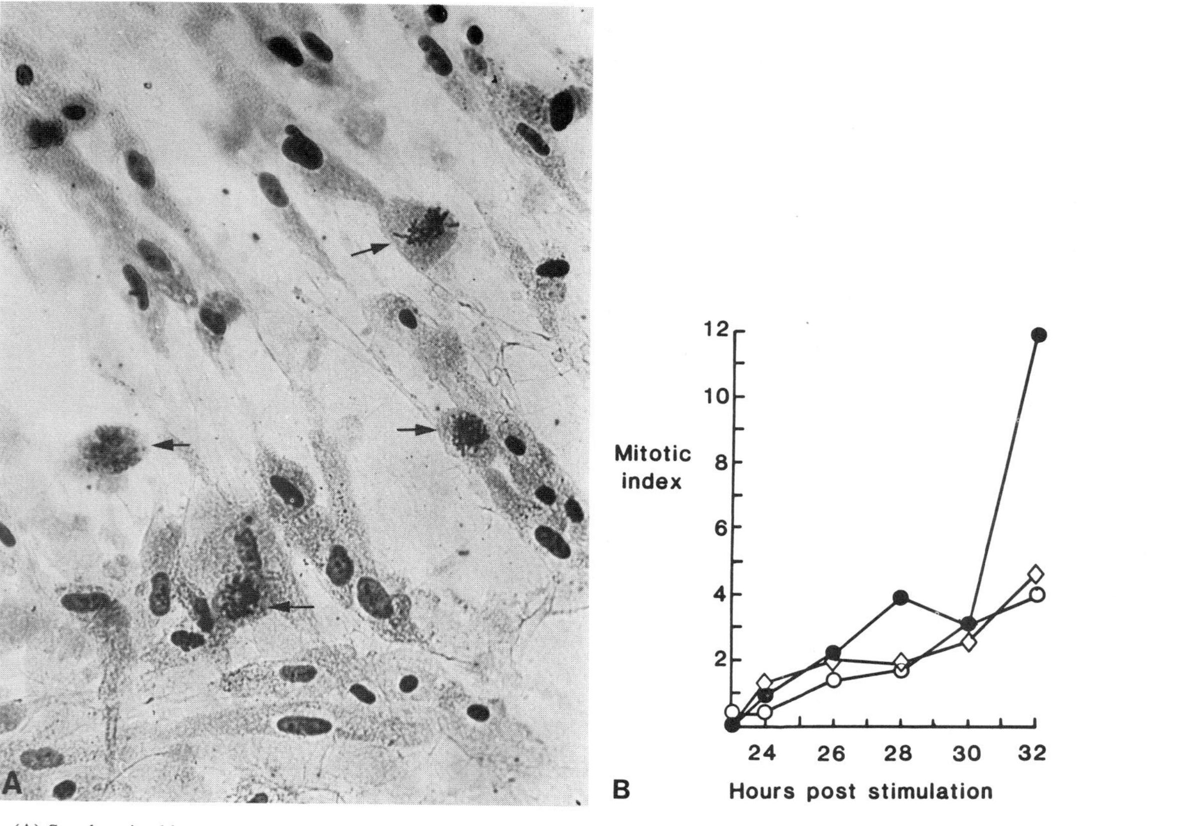

Figure 8. Mitotic index. **(A)** Synchronized human diploid fibroblasts 32 h after serum stimulation stained with Giemsa, showing mitotic spreads (→) and non-mitotic nuclei. **(B)** The effect of IFN-γ on mitosis. -●- control; -○- cells exposed to 10^3 U/ml IFN-γ for 24 h prior to, and during stimulation; -◇- cells exposed to 10^3 U/ml IFN-γ for 24 h before being stimulated in its absence.

from an argon ion laser beam of correct wavelength to excite the bound fluorochrome. Sensors in this region measure the intensity of fluorescence emitted by excited fluorochrome. The signals for a sample are accumulated and an analysis of the fluorochrome-binding profile of the whole sample population obtained. The peak of staining for G1 is normalized to channel 30 or 60 on the FACS. A DNA histogram is produced from which the percentage of nuclei at each stage of the cycle can be calculated on a computer (see *Figure 7*). The positions of the G1 and G2/M peaks are defined, and the area under the curve for each section calculated and expressed as a percentage of the total histogram. A control culture is used to establish the positions of the peaks and all following experimental cultures are analysed using the same parameters. More complex computer analyses of the cycle may be performed in which a control culture is used to derive the positions and shapes of the G1 and G2/M peaks. These are then fitted to the peaks on the test sample and the difference between the two analysed.

The propidium iodide staining procedure is outlined in *Tables 10* and *11*. This technique can be used with synchronized cultures to obtain detailed information on the timing of the cell cycle, or with unsynchronized cultures to assess overall effects on cycle progression.

3.3 **Mitotic index**

The time and rate of DNA synthesis by stimulated synchronized cells can be analysed by the preceding methods. For a complete picture of the cell cycle, rate of entry into mitosis should be studied. The procedure involves gentle lysis of the cell, fixation of mitotic spreads and staining of chromatin with orcein acetate at intervals following growth stimulation (see *Table 12*). The condensed chromosomes of mitotic nuclei are clearly visible against a background of lightly stained nuclei (see *Figure 8*). The percentage of mitotic nuclei is calculated at each time point to assess the effect of IFN treatment on the rate of progress to mitosis (*Figure 8*).

Table 12. Determination of mitotic index.

Solutions

1. Tri-sodium citrate: 0.684% in distilled water (keep at 4°C).
2. Carnoy's fixative: 10 ml of glacial acetic acid
   (make fresh) 30 ml of absolute alcohol.

Technique

1. Remove medium, then add 1.9 ml of tri-sodium citrate to each 30 mm dish.
2. Immediately add 0.4 ml of distilled water dropwise. Mix by very gently swirling the plate.
3. Leave at room temperature for 10 min.
4. Add 2.3 ml of fresh Carnoy's fixative, pour off and add the same volume of fixative.
5. Leave at room temperature for 10 min.
6. Pour off fixative and allow the dishes to air dry.
7. Stain when absolutely dry with filtered orcein acetate (LaCoeur) for 10 min.
8. Wash once with absolute alcohol, then dry.
9. Examine under low power magnification to see mitotic nuclei. Count approximately 1000 cell nuclei (from five different fields) per dish to give the percentage of mitoses.

Table 13. Assay for cytotoxic effects of TNF.

1. Prepare target cells in their growth medium at a concentration of $1.5-2 \times 10^5$ cells/ml.
2. In 96-well microtitre plates prepare serial dilutions of TNF using medium, so that the final volume per well is 100 μl.
3. Add 100 μl of cell suspension to each well (giving a final cell number of $1.5-2 \times 10^4$ cells per well).
4. Incubate for 48 h at 37°C in a humid incubator with 7% CO_2 in air.
5. Fix in formal saline.
6. Stain for 10 min in 0.5% crystal violet, 8% (v/v) formalin (40% formaldehyde), 0.17% NaCl, 22.3% (v/v) ethanol.
7. Wash thoroughly in tap water.
8. Record the titre of TNF causing lysis of 50% of the cells.
9. For more accurate quantitation, dissolve the dye staining the cells with 33% acetic acid (100 μl per well) and measure the optical density of the released dye in a spectrophotometer at 577 nm. Plot the change in OD with increased TNF titre.

   Alternatively the target cells can be seeded in the 96-well plate 24 h prior to TNF treatment. On the day of treatment the growth medium is aspirated and replaced with serial dilutions of TNF made directly into the plates. By this method the action of TNF on cell growth only, and not on seeding efficiency, is measured.

Table 14. ^{51}Chromium release cytotoxicity assay.

1. Prepare a suspension of $1-2 \times 10^7$ cells in a 15 ml sterile centrifuge tube.
2. Spin at 200 *g* for 5 min and resuspend the pellet in 0.2 ml of medium + 10% FCS.
3. Add 200 μCi (7.4×10^6 Bq) of ^{51}Cr in 200 μl (obtained as Na $^{51}CrO_4$ in sterile isotonic saline from Amersham). Use ^{51}Cr no more than 1 month old, as the decay products of the isotope are toxic to cells.
4. Incubate for 1 h in a lead shielded beaker in a 37°C incubator, gently shaking the tube every 15 min to ensure all cells are labelled.
5. Wash three times with medium + 10% FCS to remove non-incorporated ^{51}Cr and any ^{51}Cr released by dead or injured cells.
6. Meanwhile serially dilute the TNF with PBS + 3 mg/ml BSA in a 96 well v-bottomed plate to give final volumes of 100 μl per well. Include the following control wells:
   (a) 100 μl of PBS to measure spontaneous release;
   (b) 100 μl of distilled water to measure total release.
7. Resuspend the cells in medium + 10% FCS to give $1-2 \times 10^5$ cells/ml. Disperse 100 μl to each well giving a final concentration of 1×10^4 cells per well.
8. Incubate at 37°C for 4 h.
9. Centrifuge the plate at 200 *g* for 5 min to pellet all the cells and cell debris, in a specially designed plate carrier, e.g. for Beckman JS-4.2 and JS-3.0 rotors use microplate carrier cat. no. 270-341978.
10. Take 100 μl samples from each well, and count on a gamma counter.
11. Calculate ^{51}Cr release as follows,

$$\% \text{ specific } ^{51}\text{Cr release} = \frac{\text{experimental release} - \text{spontaneous release}}{\text{total release} - \text{spontaneous release}} \times 100$$

4. MEASUREMENT OF CYTOTOXIC EFFECTS

Since cytotoxic agents such as TNF rapidly lyse sensitive cells, a cell-lysis assay has been developed in which the effect of TNF is measured by the decrease in viable cells stained by crystal violet (23). This is outlined in *Table 13*. An alternative cytotoxicity assay involves the release of ^{51}chromium (^{51}Cr) from pre-labelled cells, and is outlined in *Table 14*. The assay must be performed over a short time period because the background level of ^{51}Cr release is relatively high, and after 24 h a large proportion

Table 15. Measurement of total cellular RNA synthesis by [^{3}H]uridine incorporation.

1. Culture the cells for 24 h at 1×10^5 cells/well in 2 ml of medium + 10% FCS in 24 well culture plates; then add IFNs or TNF.
2. At set times after treatment add 10 μCi (3.7×10^5 Bq) [^{3}H]uridine (sp. act. 5 Ci/mmol, 185 GBq/mmol) to plates. Other plates should be unlabelled and cell counts made at relevant time points such that RNA synthesis can be related to cell number.
3. Incubate at 37°C for 1 h.
4. To stop incorporation and harvest TCA-precipitable material:
   Wash once with PBS on ice
   Wash three times with cold 5% TCA
   Add 1 ml of cold 5% TCA
5. Heat the culture plate for 1 h in an 80°C water bath to hydrolyse the RNA.
6. Place 200 μl samples from each well into 4 ml of scintillation fluid and determine the radioactivity in a scintillation counter.
7. Using the following equation calculate the effect on RNA synthesis corrected for changes in cell number (TNF causes rapid cell death in tumour cell lines).

$$\text{Incorporation Index} = \frac{\text{c.p.m. TNF-treated cells}}{\text{c.p.m. control}} \times \frac{\text{cell number control}}{\text{cell number TNF-treated cells}}$$

This gives an incorporation index in which no difference between treated and control cells is recorded as 1, an increase in synthesis above that of control cells is recorded as >1 and a decrease as <1.

Measurement of total protein synthesis by [^{3}H]leucine incorporation follows essentially the same procedure, except that the 1 h incubation period at 80°C is omitted. Instead, after 20 min TCA treatment (step 4) wash the precipitates with ethanol and solubilize them as described in *Table 9* for determination of radioactivity (see *Figure 9*).

Measurement of DNA synthesis by [^{3}H]thymidine incorporation is described in *Table 9*, and can be used for non-synchronized cells as well as synchronized cells.

of the ^{51}Cr will leak from the cells.

The involvement of protein synthesis and RNA synthesis in mediating the cytotoxic effect or protective cellular repair mechanisms can be studied by adding cycloheximide (5 μg/ml) to inhibit protein synthesis or actinomycin D (1 μg/ml) to inhibit RNA synthesis.

5. MEASUREMENT OF MACROMOLECULAR SYNTHESIS

The effects of IFNs, TNF and other agents on cellular metabolism are studied by the following procedures. Overall changes in the rates of synthesis of protein and RNA can be assessed by the incorporation of radiolabelled precursors, [^{3}H]uridine for RNA and [^{3}H]leucine for protein, over defined time periods. The degradation of RNA and protein is measured by pulse labelling with the radiolabelled precursors prior to the addition of IFNs or TNF. The methodology for assessing effects on RNA and protein synthesis is described in *Table 15*; that for measuring degradation is essentially similar except that after a 1 h pulse label, the radioisotope is washed off the cells prior to the addition of IFNs or TNF.

6. THE GROWTH OF HUMAN TUMOURS IN NUDE MICE

The genetically immunosuppressed nude mouse mutant is congenitally athymic. This animal lacks T cells and therefore cannot reject xenografts of normal or neoplastic tissue.

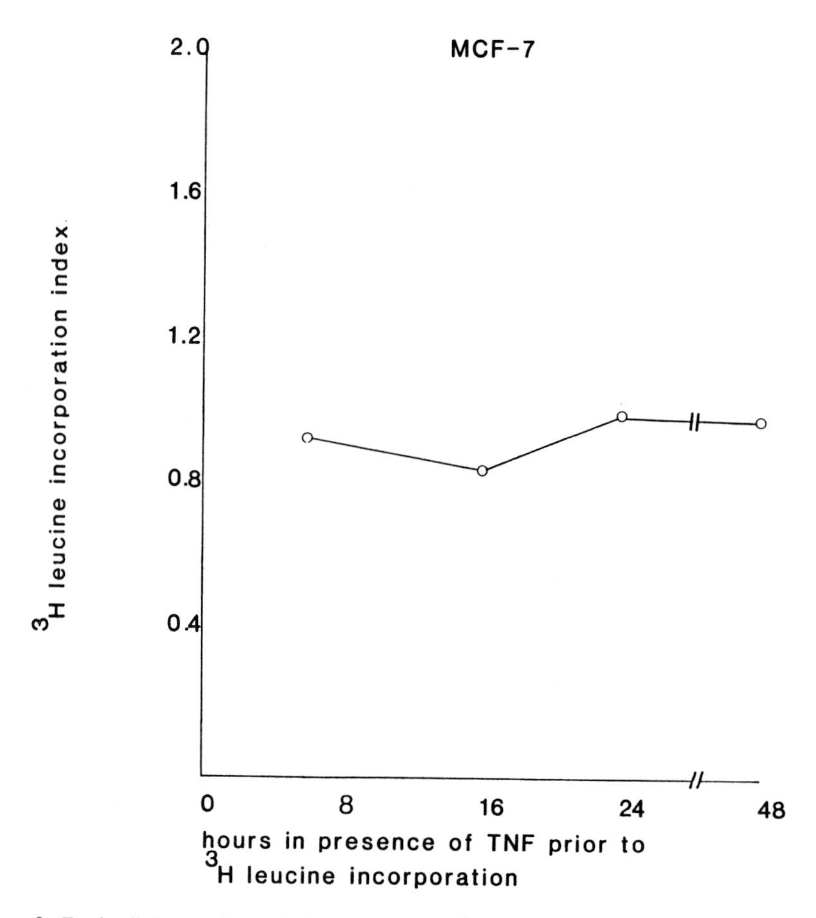

Figure 9. Total cellular protein synthesis shown by the [^{3}H]leucine incorporation index for MCF7 cells treated with 60 mg/ml TNF. The incorporation index over the 48 h time periods remains approximately 1, indicating no change in the rate of protein synthesis.

Human tumours grown in mice are extremely useful for studies of tumour cell biology, biochemistry and for experimental therapeutics (24). Extensive studies on panels of several different tumour types support the validity of this model as a predictive system for testing new anti-cancer agents and for determining optimal treatment schedules and drug combinations (25). Biological therapies such as IFNs are thought to exert at least some of their anti-tumour activity via the host immune system. The nude mouse is not an appropriate model for investigating such actions, but is useful for studying direct effects of IFNs on human tumours. Interferons show a degree of species specificity and we have shown that human IFNs do not affect the nude mouse host in any measurable way (26). Thus we can exclusively investigate direct effects of IFNs on the human tumour xenografts. In contrast, human TNF is active on mouse and human cells. Therefore in this model system we can study direct effects on the human tumours and non-T cell-dependent effects on the host immune system.

6.1 Nude mice

6.1.1 *Source of mice*

Nude mice should be obtained from a reliable and established breeding colony and should be either specific pathogen free (SPF) or gnotobiotic. The nude gene has been introduced into a variety of different genetic backgrounds, for example BALB/c, CBA, NSF, NIH and Swiss mice but while some studies show that some human tumours behave differently in these different mouse strains, other studies have shown similar growth (24).

6.1.2 *Maintenance of tumour-bearing nude mice*

Ideally these mice should be housed and manipulated in negative pressure isolators (~2 atmospheres) with high efficiency air filtration (Vickers Medical, Basingstoke, UK or La Calhene, Cambridge, UK). Under such conditions mice are protected from environmental pathogens and staff protected from the theoretical risk of xenotropic viruses arising from the human tumours and their surrounding mouse stroma. All routine isolator supplies are sterilized by heat, 2.5 Mrad irradiation or ethylene oxide fumigation. Items entering or leaving the isolators are sprayed with a surface-active disinfectant such as Alcide. Under such conditions the life span of nude mice is comparable with that of normal SPF mice.

If isolators are not available, mice can be kept for several weeks in filter cages preferably located within laminar flow hoods.

6.2 Human tumour xenografts

These should be established and maintained in young adult (6–12 weeks) mice which are strong enough to withstand implantation techniques, but still have low levels of natural killer cells.

6.2.1 *Establishment from surgical specimens*

Tumours from untreated individuals are obtained at the time of surgery and should be implanted into mice within 4 h of removal from the patient as described in *Table 16*.

The take rate of human tumour material transplanted directly from the patient to the nude mouse is dependent on the quality of the original surgical specimen, whether it is derived from primary, recurrent or metastatic tumours (the latter two grow more readily), the degree of malignancy, the hormone dependency of the tumour and the histological type (24).

Table 16. Implantation of human tumour material.

All manoeuvres should be carried out aseptically and mice treated with antibiotics for 5 days post-implantation.

1. Wash the tumour four times in tissue culture medium containing antibiotics.
2. Dissect away any obvious areas of necrotic or connective tissue.
3. Slice the tumour with crossed scalpels into millimetre cubes.
4. Load a Bashford 16-gauge needle (Holborn Surgical, London, UK) with four pieces of tumour.
5. Insert the needle s.c. and through one ventral incision, manoeuvre the needle to implant the tumour cubes in four well spaced sites.
6. Inject mice s.c. with 0.05 ml of neobiotic (Upjohn Ltd., 50 mg/ml) and add 0.1 ml of neobiotic to each 200 ml of drinking water daily for 5 days.

6.2.2 *Establishment from human cell lines*

Cells in log phase of growth are harvested by trypsinization (if growing as a monolayer), centrifuged then resuspended so that an appropriate number of cells can be injected in 0.05 ml. Successful tumours can be obtained from injections of $10^4 - 10^7$ cells, depending on the tumourigenicity of the cell line. Cells can be injected subcutaneously (s.c.) in a ventral or dorsal site, intramuscularly (i.m.) in the muscle of the leg or intraperitoneally (i.p.).

6.2.3 *Transplanting established tumours*

Once tumours are established they can be passaged from mouse to mouse when they have reached a size of 1 – 1.5 cm. Tumours larger than this tend to show central necrosis.

6.3 **Experimental therapy of human tumour xenografts growing subcutaneously in nude mice**

Therapy of established tumours is best initiated 2 – 3 weeks after transplantation when tumours are approximately 0.5 cm in diameter. In order to obtain significant results it is important that tumours borne by mice on experimental therapy should be of a uniform size. If tumours are obtained from existing tumours in the mice this is often difficult to control. It is useful to set upper and lower limits for tumour size and exclude all tumours outside those limits. If tumours are of uniform size 4 – 10 mice each bearing a single tumour are required per therapy group.

6.3.1 *Measuring effects of therapy on subcutaneous tumour growth*

Effects of therapy on such tumours are easily quantitated because of their site and the lack of body hair in the nude mice. Calipers are used to measure the longest (D) and shortest (d) diameter once or twice weekly. These two measurements can be used in the following ways.

(i) Tumour size index $= d \times D$

(ii) Volume $= \dfrac{\Pi D d^2}{6}$

(iii) From the volume measurements, response to therapy can be assessed as partial response (>50% reduction in original tumour volume) or complete response (total regression).

(iv) Response to therapy can be compared between tumours of different growth rates by calculating the specific growth delay, SGD (26). The median tumour volume doubling time, T^D, is calculated from data obtained from volume calculations — see equation (ii) above.

$$\text{SGD} = \frac{T^D\ (\text{treated}) - T^D\ (\text{control})}{T^D\ (\text{control})}$$

Therapeutic efficacy of an agent can also be measured by excising and weighing tumours at the end of the experiment. An example of the different ways of expressing data is shown in *Figure 10*.

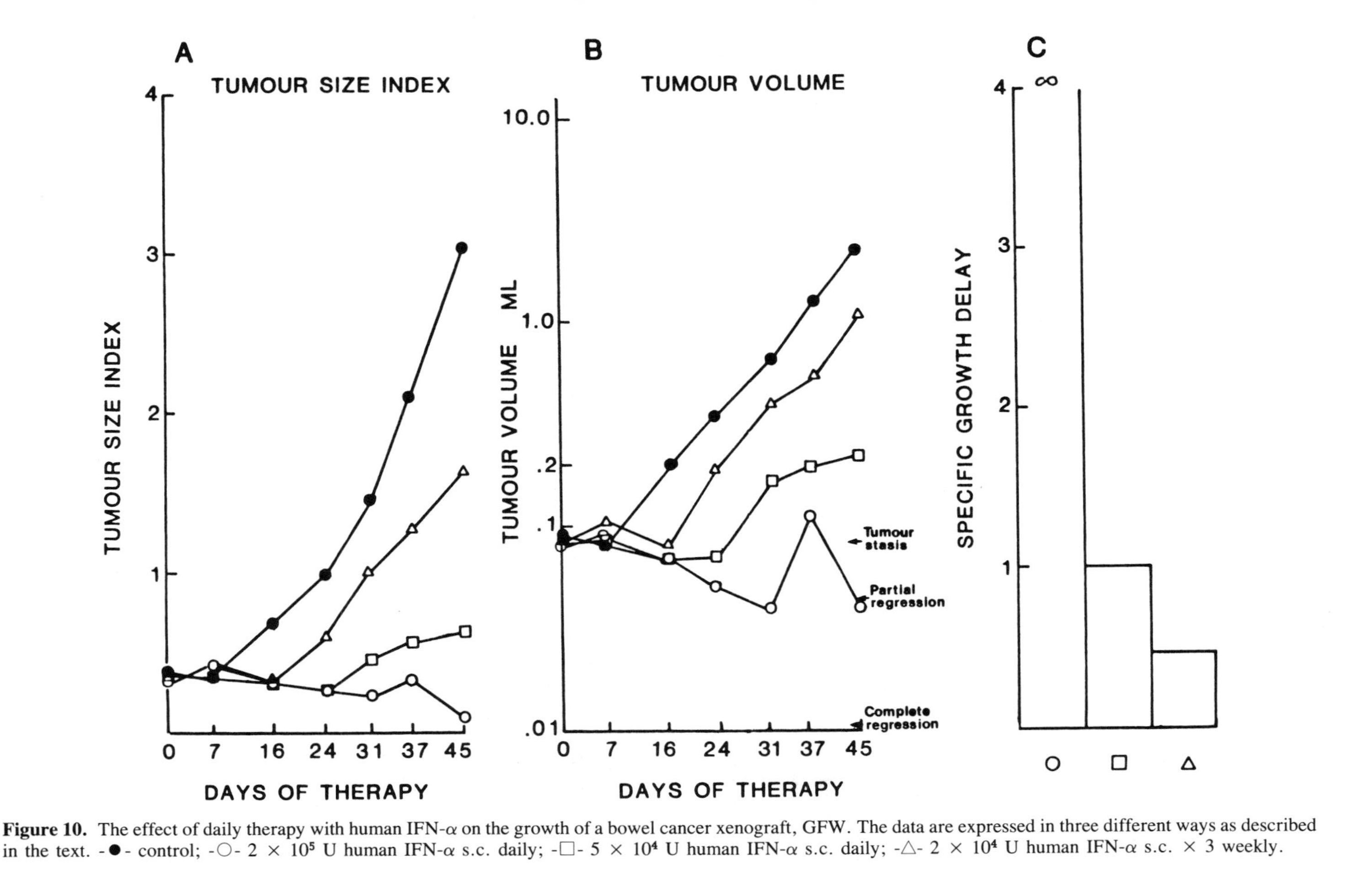

Figure 10. The effect of daily therapy with human IFN-α on the growth of a bowel cancer xenograft, GFW. The data are expressed in three different ways as described in the text. -●- control; -○- 2 × 10^5 U human IFN-α s.c. daily; -□- 5 × 10^4 U human IFN-α s.c. daily; -△- 2 × 10^4 U human IFN-α s.c. × 3 weekly.

6.3.2 *Measuring experimental therapy of human tumour xenografts growing intraperitoneally in nude mice*

Some primary tumours and cell lines will successfully grow intraperitoneally in nude mice. Ovarian cancer xenografts in particular can closely mimic the clinical picture in man producing solid tumours and ascites in the peritoneum (B.G.Ward, personal communication). In such models survival, and the increase in lifespan caused by therapy, can be measured.

7. REFERENCES

1. Isaacs,A. and Lindenmann,J. (1957) *Proc. R. Soc. B.*, **147**, 258.
2. Taylor-Papadimitriou,J. and Rozengurt,E. (1985) In *Interferons — Their Impact in Biology and Medicine.* Taylor-Papadimitriou,J. (ed.), Oxford University Press, Oxford, p. 81.
3. Taylor-Papadimitriou,J., Balkwill,F.R., Ebsworth,N. and Rozengurt,E. (1985) *Virology*, **147**, 405.
4. Balkwill,F.R. and Bokhon'ko,A.I. (1984) *Exp. Cell Res.*, **155**, 190.
5. Yarden,A., Shure-Gottlieb,H., Chebath,J., Revel,M. and Kimchi,A. (1984) *EMBO J.*, **3**, 969.
6. Old,L.J. (1985) *Science*, **230**, 630.
7. Kohase,M., Henriksen-Destefano,D., May,L.T., Vilcek,J. and Sehgal,P.B. (1986) *Cell*, **45**, 659.
8. Freshney,R., ed. (1986) *Animal Cell Culture — A Practical Approach.* IRL Press, Oxford and Washington, DC.
9. Soule,H.D., Cozquez,J., Long,A., Albert,S. and Brennan,M. (1973) *J. Natl. Cancer Inst.*, **51**, 1409.
10. Lasfargues,E.Y. and Ozello,L. (1958) *J. Natl. Cancer Inst.*, **21**, 1131.
11. Fogh,J. and Trempe,G. (1975) *Human Tumour Cells In Vitro.* Plenum Press, New York.
12. Taylor-Papadimitriou,J., Shearer,M. and Stoker,M. (1977) *Int. J. Cancer*, **20**, 903.
13. Balkwill,F.R., Watling,D. and Taylor-Papadimitriou,J. (1978) *Int. J. Cancer*, **22**, 258.
14. Hill,B.T. and Whelan,R.D.H. (1983) *Cell Biol. Int. Rep.*, **7**, 617.
15. Courtenay,V.D. and Mills,J. (1978) *Br. J. Cancer*, **37**, 261.
16. Silverman,R.H., Watling,D., Balkwill,F.R. and Kerr,I.M. (1982) *Eur. J. Biochem.*, **126**, 333.
17. Baserga,R. (1985) *The Biology of Cell Reproduction.* Harvard University Press, Cambridge, MA.
18. Balkwill,F.R. and Taylor-Papadimitriou,J. (1978) *Nature*, **274**, 798.
19. Lloyd,D., Poole,R.K. and Edwards,S.H. (1982) *The Cell Division Cycle.* Academic Press, London.
20. Rozengurt,E., Legg,A. and Pettican,P. (1979) *Proc. Natl. Acad. Sci. USA*, **76**, 1284.
21. Rozengurt,E. and Heppel,L.A. (1975) *Proc. Natl. Acad. Sci. USA*, **72**, 4492.
22. Crissman,H.A., Stevenson,A.P., Kissane,R.J. and Tobey,R.A. (1979) In *Flow Cytometry and Sorting.* Melamed,M.R., Mullaney,P.F. and Mendelsohn,M.L. (eds), J.Wiley and Son, New York, p. 243.
23. Ruff,M.R. and Gifford,G.E. (1981) In *Lymphokines 2.* Pick,E. (ed.), Academic Press, New York, p. 235.
24. Giovenella,B.C. and Fogh,J. (1985) *Adv. Cancer Res.*, **44**, 69.
25. Giovenella,B.C., Stehlin,J.S., Shepard,R.C. and Williams,L.J. (1983) *Cancer Res.*, **52**, 1146.
26. Balkwill,F.R., Moodie,E.M., Freedman,V. and Fantes,K.H. (1982) *Int. J. Cancer*, **30**, 231.
27. Kopper,L. and Steel,G.G. (1975) *Cancer Res.*, **35**, 2704.

CHAPTER 12

Cytotoxicity assays for tumour necrosis factor and lymphotoxin

N.MATTHEWS and M.L.NEALE

1. INTRODUCTION

Tumour necrosis factor (TNF) and lymphotoxin (LT) have recently been cloned and sequenced and found to have over 40% homology at the DNA level. Not surprisingly they have a similar spectrum of growth inhibitory and cytotoxic activity against tumour cell lines *in vitro*. Although the recent upsurge in interest in these molecules is due to their potential as anti-cancer agents, the resultant availability of highly purified recombinant material has revealed that they have a role in many other biological processes other than killing tumour cells. Despite these other properties and the potential of immunoassay, for the present, TNF and LT are usually assayed by *in vitro* bioassays based on killing tumour cells.

In our laboratory, TNF/LT samples are incubated for 1–3 days with monolayers of L929 cells, a plastic-adherent, murine tumour cell line. Any cells that are killed detach from the plastic and, at the end of the assay, the remaining adherent, viable cells are stained with crystal violet. The amount of dye taken up is proportional to the number of residual cells and can be quantitated spectrophotometrically with an enzyme-linked immunosorbent assay (ELISA) reader. We employ two variants of this technique — a '1 day' and a '3 day' assay. These are described in detail below with their differing applications being considered in Section 5.

2. TARGET CELLS

Our L929 cells were originally obtained from Gibco. There is great variation in TNF/LT susceptibility of L929 cells from different suppliers and even in cells of different passage number from the same source. The best policy is to test a validated preparation of TNF or LT against L929 cells from different sources and select the most susceptible line. In this laboratory we use a subline of L929 cells derived as follows. L929 cells were cloned by limiting dilution, 20 clones were tested and the most susceptible clone was re-cloned. The most susceptible secondary clone was then selected.

After prolonged passage cells may develop increased or decreased susceptibility. For standardization purposes, freeze a large number of ampoules and initiate new stock cultures every 1–2 months. Grow the cells in RPMI medium with 5% fetal calf serum (FCS), glutamine and penicillin/streptomycin (culture medium). When confluent, trypsinize the cells (see below) and split 1:10 in fresh culture medium; this is necessary twice per week. Best results are obtained with healthy cells in the log phase of growth.

3. SETTING UP THE ASSAY

3.1 Three day assay

(i) Tip off the culture medium from the culture flasks and wash the cells once with Dulbecco's phosphate-buffered saline (PBS).
(ii) Add 2 ml of trypsin/EDTA solution (Flow) diluted 1 in 4 with PBS. If used at the supplier's recommended dilution, the trypsin is toxic to the cells.
(iii) Leave at 37°C for the minimum time necessary to detach more than 95% of the cells (usually 1–2 min).
(iv) Dilute the suspension with culture medium to neutralize the trypsin and adjust the volume to 10 ml.
(v) Take a small amount for counting in a Neubauer chamber and whilst counting centrifuge the cells at 100 *g* for 5 min, then resuspend at 10^5/ml in culture medium.
(vi) By means of a micropipette (Oxford, Finn or similar), dispense 75 μl volumes of this cell suspension into the wells of a flat-bottomed, 96-well microtitre plate. Leave one row of eight wells without cells, as a blank for the ELISA reader (Section 4.2). When dispensing the cells, push the plunger of the micropipette to the first position only and ignore the last few microlitres of cell suspension in the pipette tip. This prevents introduction of air bubbles into the plate; air bubbles cause uneven monolayers.
(vii) Shake the plate for 5–10 sec on a microshaker (Dynatech) to distribute the cells evenly. At this stage it is preferable to leave the plate at 37°C for 2-3 h in a gassed incubator (5% CO_2 in air) to permit the cells to adhere. However if time is short this step can be omitted and test samples are added to the cells before they have adhered. In this case defer shaking until after addition of the test material.
(vii) If the test material is not sterile then filter through a 0.22 μm filter.
(ix) Make a series of dilutions of the test sample (4-fold is suitable) in culture medium. For each dilution add triplicate 75 μl volumes to the L929 cells. The exact range of dilutions depends upon the nature of the test sample and the sensitivity of the target cells but as a rough guide we would use dilutions in the range 50 000–1 000 000 for high titre, TNF-containing serum from rodents with endotoxic shock and 500–10 000 for unconcentrated supernatants from tissue culture cells.
(x) On each plate include a negative control (i.e. tissue culture medium alone) and a positive control. As yet there are no universally acceptable standards for TNF and LT and as a positive control we use our own laboratory standards — partially purified preparations of either serum-derived rabbit TNF or lymphoblastoid-derived human LT as appropriate. An interim reference reagent for TNF is now available from NIBSC, South Mimms, Herts.
(xi) Incubate the microtitre plate at 37°C for 3 days in a gassed incubator (5% CO_2 in air). The assay is terminated as described in Section 4.

3.2 One day assay

This is set up in essentially the same manner as the '3 day' assay but with two modifications.

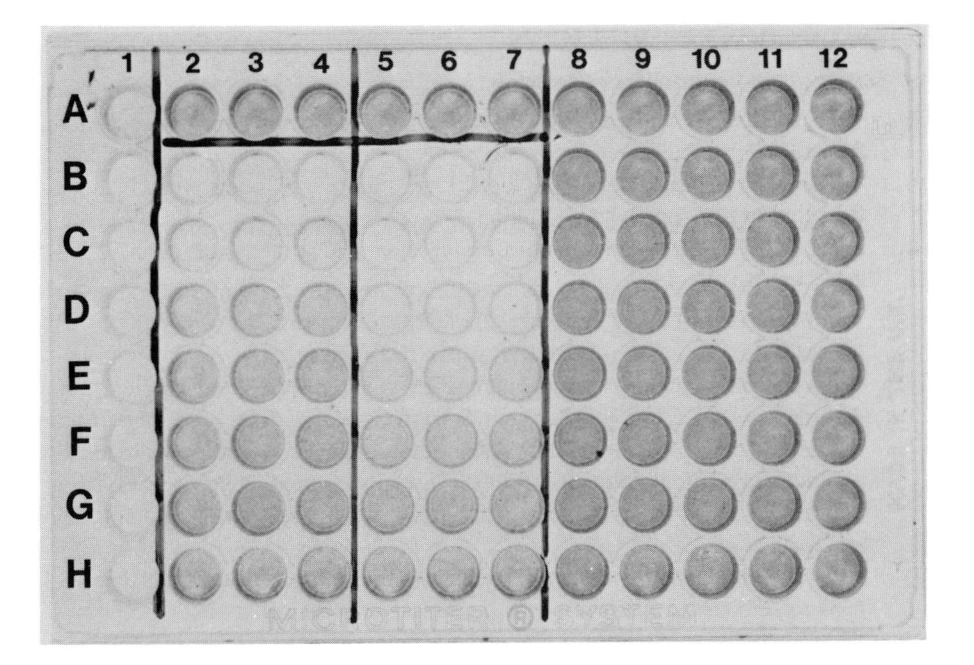

Figure 1. '1 day' assay of rabbit TNF. L929 cells were plated in all wells of the plate except those in column 1. A crude TNF-containing serum was titrated in triplicate in columns 2–4 and a partially purified preparation in triplicate in columns 5–7. In both cases, A = medium alone (negative control); B = sample dilution of 1/10 000; C = 1/40 000; D = 1/160 000; E = 1/640 000, etc.

(i) Firstly, plate out the cells at 5×10^5/ml. For this assay, the cells must be allowed to adhere before adding the test material — a minimum of 4 h is required, but it is preferable to plate them out a day before required at 3×10^5/ml.

(ii) Secondly, dilute the test samples in culture medium containing 2 μg/ml actinomycin D to give a final culture concentration of 1 μg/ml actinomycin D when added to the cells. Dissolve actinomycin D (Sigma) in PBS at 20 μg/ml (warming at 37°C if necessary) and store at −20°C in 5 ml portions. Thaw out as required and dilute to 2 μg/ml with culture medium — the thawed concentrated solution is stable for up to 2 weeks at 4°C.

This assay is about 10 times more sensitive than the '3 day' assay and samples should therefore be assayed at higher dilutions. After addition of the test samples incubate the microtitre plate at 37°C for 16–24 h in a gassed incubator (see *Figure 1*).

Because of the short duration of this assay and the anti-microbial activity of actinomycin D it is often not necessary to filter test samples, for example if screening large numbers of column fractions. Emetine can be used in place of actinomycin D (at the same concentration). Others have found that cycloheximide and mitomycin C are further alternatives although with our strain of L929 cells they are much less effective than actinomycin D or emetine. The increased susceptibility of L929 cells in the presence of these inhibitors of protein and RNA synthesis is presumably because of their interference with repair mechanisms although at present this is still speculation.

4. TERMINATION OF THE ASSAYS

4.1 Staining the plate with crystal violet

Wear disposable gloves and carry out the following procedures in a fume cupboard, preferably one with a sink.

(i) After incubation invert the plate over the sink and flick out the culture medium. Rinse the sink with copious amounts of water, especially if actinomycin D has been used.

(ii) With the plate face upwards, immerse in 5% formaldehyde (in PBS) in a sandwich box, allowing one row of wells to fill at a time.

(iii) Leave for 5 min to allow the cells to fix to the plastic.

(iv) Wash the plate three times with running tap water, flick off any residual water and immerse for 5 min in 1% aqueous crystal violet in a sandwich box to stain the cells. When preparing crystal violet filter through No. 1 filter paper to remove undissolved particulate matter. After a few days a precipitate may reform on the surface and it is advisable to filter again, otherwise spurious staining may result.

(v) After staining, exhaustively wash the plate with running tap water and then flick hard to remove as much water as possible from the wells.

(vi) Wipe the under surface of the plate with a paper tissue to remove residual water and any streaks of dried dye. To ensure the plate is completely dry, either invert over a paper towel and leave for several minutes or dry with a hot air dryer, taking care not to melt the plate!

4.2 Quantitation

Normally the stained plate is read 'dry' in the ELISA reader. However, uneven monolayers of cells can give erroneous readings. To obtain an even distribution of dye, add 100 μl of 33% aqueous glacial acetic acid to each of the wells, including the blank row, shake for 10 sec on a microshaker and then read the plate 'wet'. The absorbance maximum for crystal violet of approximately 580 nm is not a standard filter on some ELISA readers but we obtain satisfactory results with the 540 nm setting on the Titertek Multiskan.

Blank the machine against the row of eight wells on the plate which lacks cells and then determine the absorbance of the other wells at 580 nm. For each dilution of the test sample calculate the percentage cytotoxicicity from the formula $100\,(a-b)/a$ where a and b are the mean absorbances of triplicate wells with culture medium alone and test sample, respectively. Either graphically or by means of a computer or programmable calculator, determine the titre — the highest dilution giving 50% cytotoxicity. In our laboratory the ELISA reader is interfaced to a microcomputer which is programmed to perform the appropriate calculations. Whilst this is an advantage it is by no means an essential and often sufficient information can be gleaned by simple visual inspection of the plate, without resort to the ELISA reader.

5. GENERAL COMMENTS

The assays described here are sensitive, robust, relatively cheap and correlate well with isotopic assays. The L929 target cells also grow rapidly and are undemanding to culture. At least *in vitro* TNF and LT are not species specific and L929 cells are suitable targets for TNF/LT from many species.

The '1 day' assay with actinomycin D is more sensitive than the '3 day' assay, obviously produces results more quickly and has the advantage that it is often possible to assay samples without membrane filtration, for example column fractions. Until we understand the exact mechanism by which actinomycin D and other inhibitors of protein or RNA synthesis increase TNF susceptibility, it is wiser to use the '3 day' assay when testing agents for modulation of TNF/LT cytotoxicity.

The '1 day' assay is sensitive enough for most purposes but sensitivity can be increased by lowering the serum concentration in the culture medium, raising the incubation temperature to 40°C or using actinomycin D and emetine together.

6. ACKNOWLEDGEMENT

The financial support of the Cancer Research Campaign is gratefully acknowledged.

CHAPTER 13

Anti-microbial and hydrogen peroxide assays for MAFs

DOUGLAS B.LOWRIE and MAGDY FAHMY

1. INTRODUCTION

Interferon-γ (IFN-γ) and other lymphokines with the capacity to activate macrophage anti-microbial activities can be readily detected in supernatants from stimulated lymphocytes. These macrophage activating factors (MAFs) show species specificity, working best on macrophages from the same animal species. Therefore murine MAFs are assayed with murine macrophages for example and human MAFs with human.

Macrophages, unfortunately, are a very variable commodity. Not only do macrophages differ in form and function from tissue to tissue but also cells from a single human donor or from an inbred strain of mouse can vary widely in responses to MAFs from one experiment to another. The underlying causes of this variability have not been properly defined but undetected infections in the donors of the cells are probably implicated. Minor variations in the way the macrophages are handled can also be significant and rigorous standardization of procedures is essential.

1.1 **Anti-microbial assays**

How many distinct MAFs there are in any species and whether they have different biological functions are unresolved issues. Part of the problem is that macrophage activation means different things to different investigators, but even when the definition is restricted to enhancement of anti-microbial and anti-tumour activity there is clear evidence for multiple MAFs in murine and human lymphokines (1,2). This is shown, for example, by the chromatographic separation from one another of MAFs that are effective against different microorganisms (1). Thus the MAFs revealed depend on the assays used.

IFN-γ is certainly one MAF and in the murine MAFs especially there is also the question whether the non-IFN-γ activities might actually be due to contaminating bacterial endotoxin or other microbial products. These issues will only be resolved when the chemical structures are fully defined, probably following gene cloning. However, it is no small task to set up assays of macrophage anti-microbial activity against a range of target microorganisms in any one laboratory. This is partly because the target microorganisms that are of most interest are frequently potentially pathogenic in man and therefore require special facilities, but more importantly the different microorganisms present widely different technical problems in handling and assay. Furthermore there are, perhaps, as many different assays as there are laboratories working in this field. Apart from differences in microbial targets, the assays may measure MAF effects on phagocytic uptake, killing during uptake, intracellular multiplication rate and intracellular death rate, but discrimination between these phenomena, particularly the latter two (3), is not always straightforward.

Described in detail here are three assays based on counting bacterial colony-forming units (c.f.u.) and stained bacteria which permit various degrees of discrimination and which the authors' experience suggests could be of general interest.

1.2 Peroxide assays

Because of the difficulties associated with measuring MAF activity against actual micro-organisms there is often a need for easier and faster indirect assays. This is possible only where the alternative parameter to be measured has been shown to be causally related to the anti-microbial activity of interest. There is strong evidence that for a number of microorganisms MAFs work at least partly by increasing the release of 'toxic oxygen' as superoxide and hydrogen peroxide during phagocytosis (4); enhancement of this function has proved to be amenable to assay and three procedures using mouse peritoneal macrophages and human U937 cells are described in detail here.

2. MOUSE PERITONEAL MACROPHAGE ANTI-STAPHYLOCOCCUS AND ANTI-MYCOBACTERIUM ASSAYS

2.1 Introduction

In order to measure macrophage anti-bacterial activity and its enhancement by MAFs it is not strictly necessary to distinguish intracellular activity (after phagocytosis) from extracellular activity (before, during or in the absence of phagocytosis). However, the fate of the bacteria in the intracellular and extracellular compartments is likely to be dramatically different: most bacteria of interest multiply rapidly in macrophage maintenance media but not intracellularly; *Legionella* do the opposite and die rapidly extracellularly but not intracellularly. Therefore, in practice, macrophage anti-bacterial activity can be obscured unless steps are taken to remove, or restrict and control for, bacteria that have not come into contact with the macrophages.

There are those who believe that antibiotics can be used under carefully controlled conditions to selectively kill extracellular bacteria on the grounds that the drugs penetrate poorly into the intracellular environment. This practice is not recommended since direct tests show penetration can be far from insignificant (5) and MAF is likely to alter both penetration and intracellular activity. The main alternative is thorough rinsing of the macrophages to remove extracellular bacteria. Mouse peritoneal macrophages lend themselves to this approach particularly well as they form firmly adherent, contiguous monolayers.

The murine peritoneal cavity provides a convenient source of about 5×10^6 macrophages per animal. Although the numbers can be greatly increased by injecting irritants into the cavity before harvest, this yields cells that are different and introduces another variable and is not recommended as a rule. Inbred strains of mice with inherited differences in cell-mediated immunity present the possibility of macrophages with genetically determined different responses in MAF assays (6) whereas the few mouse macrophage cell lines that we have tested have given disappointingly poor responses in MAF assays. Although it is likely that most inbred and outbred strains of mice can provide macrophages that respond well in MAF assays, it is our experience that mouse colonies can vary widely in this respect with time and the cause of this variation has not been established.

2.2 **Macrophage preparation**

2.2.1 *Reagents and equipment*

Medium 199.
Heparin, preservative-free (Weddel Pharmaceuticals Ltd., Wrexham, Clwyd, UK).
Fetal calf serum (FCS).
Sharp-pointed scissors, dog-toothed forceps, dissecting pins, cork board.
5-ml disposable syringes (Gillette Ltd., Reading, Berks, UK) fitted with 1.5″ × 21 gauge needle and containing 3.0 ml of medium 199 containing 10 IU heparin/ml. Note that failure of macrophages to form good monolayers can be due to occasional batches of syringes having toxic lubricant on the rubber plungers: this seems more of a problem with some other manufacturers.
Tissue culture flask on ice.
Capped centrifuge tubes or 30-ml Universal bottles.
Refrigerated centrifuge.
Haemocytometer and phase-contrast microscope.

2.2.2 *Procedure*

(i) Obtain specific pathogen-free BALB/c mice weighing 18–20 g from a reliable accredited source and use them within 2 days of receipt to avoid fluctuation in microbiological infection status.
(ii) Kill a mouse by cervical dislocation and pin it on its back through each limb.
(iii) Slit the abdominal skin from above the vent to the chest wall and pull back the skin on the right half of the body and pin it under tension to the board. The peeling-back of the skin exposes the sterile surface of the abdominal wall.
(iv) Through this sterile surface and close to the skin/body wall junction inject medium containing heparin, taking care to avoid piercing internal organs with the needle. Do not withdraw the needle.
(v) Agitate the contents of the peritoneal cavity by massage to encourage macrophages to become suspended in the medium then withdraw the medium into the syringe. About 2.5 ml containing about 2×10^6 cells/ml should be recovered from each mouse. Discard cell harvests that are visibly contaminated with blood.
(vi) Pool the lavage fluids from the mice in a sterile tissue culture flask on ice and transfer them to sterile capped centrifuge tubes.
(vii) Centrifuge the cells at 150 *g* for 10 min at 4°C and gently resuspend the loose pellet in medium 199 containing 10% FCS.
(viii) Transfer a drop using a Pasteur pipette to a Neubauer blood cell counting chamber and count the number of cells per unit volume using a phase-contrast microscope.
(ix) Dilute the cell suspension in the same medium to a density of 10^6 cells/ml.

2.3 **Anti-*Staphylococcus aureus* assay**

2.3.1 *Advantages and disadvantages*

Results are available next day; lysostaphin permits effective selective elimination of extracellular bacteria; the bacteria are not hazardous. Disadvantages are that dead bacteria disappear rapidly inside macrophages so that distinction between MAF effects on intracellular division and death rates by comparison of total and viable counts is not simple; the relationship of the results to what would happen with other, more pathogenic, microorganisms is not necessarily a close one.

2.3.2 *Reagents and equipment*

S. aureus Oxford strain.
Nutrient broth (Oxoid Ltd.).
Nutrient agar plates (Oxoid Ltd.).
Hanks' balanced salt solution (HBSS) supplemented with 25 mM Hepes buffer, pH 7.3.
Heat-inactivated (56°C for 30 min) horse serum.
Newborn calf serum.
Medium 199.
Plastic tissue culture dishes (3 cm diameter) or multidishes (16 mm well, Linbro, Flow Laboratories, Woodcock Hill, Herts, UK) and/or 8-chambered tissue culture slides (Lab-Tek, Miles Scientific Labs., Slough, UK).
Lysostaphin (10 μg/ml in HBSS; Sigma Ltd.).
Trypsin (0.125% in HBSS).
Saponin (1% w/v; Sigma).
Ultrasonicator with a roving micro-probe for use within a safety cabinet (Rinco Ultrasonics UK Ltd.).
Phenolic-based disinfectant.
Formol saline.
Bacteria counting chamber for phase-contrast microscopy.
Graduated pipettes (1 ml and 10 ml) or micropipettors (e.g. Gilson) and suitable tubes to make serial 10-fold dilutions.
Pasteur pipettes.
Crushed ice in an insulated tray.
37°C water bath on a time-switch in a cold (4°C) room.

2.3.3 *Procedures*

(i) *Activation of macrophages.*

(1) Swirl the ice-cold macrophage suspension and dispense it into either 16-mm Linbro plates or 3-cm Petri dishes (0.5 or 1.0 ml, respectively). Each experimental condition to be tested requires six or nine monolayers for duplicate or triplicate assays, respectively.
(2) Allow the cells to adhere to the surface during 2 h at 37°C in 5% CO_2.
(3) Remove the medium and rinse the monolayers four times by adding 1 ml of HBSS (37°C), swirling it vigorously and removing it.

(4) Culture the monolayers in maintenance medium, comprising 75% v/v medium 199, 20% v/v heat-inactivated horse serum, 4% v/v bovine embryo extract (Flow Labs) and 1% v/v of 1 mg/ml liver fraction L (United States Biochemical Corporation, Cleveland, Ohio), or in medium supplemented with dilutions of MAF or sham MAF preparations. Lymphocyte supernatant concentrations of 1, 3, 10 and 30% are usually satisfactory.

(5) Remove and replace the medium and supplements daily for 3 days.

Although activation begins to be detectable in hydrogen peroxide assays after 24 h, consistent anti-bacterial effects are not detected without 3 days of repeated exposure to MAF. Where economy of MAF consumption is of paramount importance, monolayers can be established in the multichamber glass slides, using 2×10^5 cells and 100– 200 μl per chamber. An additional advantage of the glass slides is the superior optical properties for microscopic analyses, however they are somewhat fiddly to handle and require vigorous pipetting with either a Pasteur pipette or a micropipettor to rinse adequately.

(ii) *Preparation of bacterial inoculum.*

(1) Make a stock of bacteria by growing a culture in nutrient broth to the logarithmic phase and storing 1-ml aliquots in ampoules in liquid nitrogen.

(2) For each experiment thaw an ampoule, transfer the contents into a bottle containing 100 ml of nutrient broth and place this at 4°C in a water bath which is timed to switch to 37°C at about 5 am.

(3) Prepare the inoculum for the experiment from the resultant fresh log phase culture at about 9 am.

(4) Centrifuge the culture at 2000 *g*, wash it by resuspension in HBSS, centrifuge it once more and resuspend it in HBSS containing 2.5% newborn calf serum (to opsonize the bacteria).

(5) Briefly ultrasonicate the suspension to disperse clumps. The ultrasonicator probe is conveniently disinfected before and after use by immersion in 74 o.p. ethanol which is then flamed or evaporated off. Duration and intensity of ultrasonication depend on the instrument in use and should be optimized in preliminary tests to the point where the increase in viable bacterial counts (c.f.u.) due to unclumping begins to be offset by loss of viability due to bacterial cell disruption.

(6) Count the total bacterial number per ml by phase-contrast microscopy of a sample diluted 2-fold in formol saline and adjust the suspension to 1×10^7/ml.

(iii) *Assay of anti-Staphylococcus activity.*

(1) Remove the medium from the control and activated monolayers with a Pasteur pipette and rinse the monolayers once with warm (37°C) HBSS.

(2) Put 1.0 ml of bacterial suspension onto each monolayer then incubate for 30 min at 37°C to allow phagocytosis. Save a portion of the inoculum on ice for viable counts.

(3) Remove the supernatant and discard it into disinfectant.

(4) Rinse the monolayers once with HBSS, the rinses also going into disinfectant.

(5) Put 1 ml of lysostaphin on each monolayer and incubate for 15 min. Remove this, replace with 1 ml of trypsin then incubate for 10 min.

(6) Rinse the monolayers twice with HBSS.
(7) Put some of the monolayers on ice for subsequent viable (c.f.u.) counts, allow some to dry at room temperature for subsequent fixing, staining and total bacterial counts.
(8) To the remainder of the monolayers add 1 ml of medium 199 containing 10% heat-inactivated horse serum and incubate for 1, 2 or 3 h.
(9) Rinse the monolayers twice with HBSS and put them on ice for subsequent viable counts.

(iv) *Total counts (estimation of phagocytic uptake of bacteria).*

(1) Fix the air-dried monolayers by flooding with methanol, drain them and allow them to dry again.
(2) Stain the fixed monolayers with Gram stain, decolourizing with ethanol instead of acetone.
(3) Examine at least 100 cells per monolayer with an oil-immersion microscope objective and score the cells as containing 0, 1–2, 3–5 or more than 5 bacteria.
(4) Calculate the average number of bacteria per macrophage. The percentages in the different categories can also be calculated.

(v) *Viable counts (estimation of intracellular fate).*

(1) Dry the agar plates at 37°C overnight with their lids on or for 1 h upside down with their lids off.
(2) Add 1 ml of saponin to the monolayers and incubate for 10 min at 37°C.
(3) Vigorously pipette the well or dish contents (or briefly ultrasonicate) to disperse the dissolved cells and break up any bacterial clumps, then serially dilute in 10-fold dilutions in saline to 10^{-4} and pipette 0.1 ml volumes of the dilutions onto 1/3rd segments of Dubos oleic base agar plates; three different dilutions should go onto one plate. Prepare duplicate plates.
(4) Allow the inocula to dry into the agar before transferring the plates to an incubator.
(5) Count the colonies after incubation at 37°C for 18 h.

(vi) *Estimates of monolayer macrophage number (DNA).* The number of macrophages in monolayers can be conveniently estimated by measuring the DNA content of the monolayer lysates with the Hoechst compound 33258 fluorimetric method of Labarca and Paigen (7). Prepare the standards in 1% saponin. In our experience with calf thymus DNA standards, 7.5 μg DNA $\equiv 10^6$ mouse macrophages.

(vii) *Results.*

(1) Calculate the effects of MAF on phagocytosis from the total counts at the end of phagocytosis (T_0).
(2) Calculate the viability of the phagocytosed bacteria from the viable count per monolayer, the DNA content of the monolayer and the average number of stained bacteria per macrophage at T_0.
(3) Compare the ratios of total counts to viable counts (viability) of the inoculum

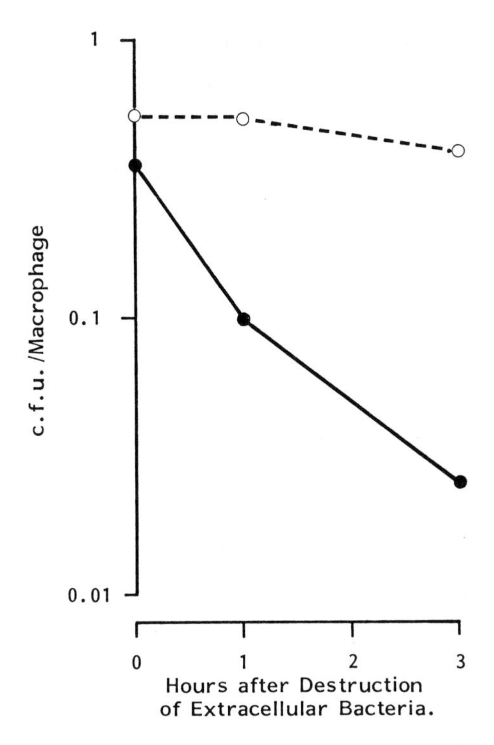

Figure 1. MAF enhancement of killing of *Staphylococcus aureus* in mouse peritoneal macrophages. Normal mouse peritoneal macrophages were maintained for 3 days, with daily medium changes, in either medium alone (○) or medium supplemented with 50% v/v supernatant from PPD-stimulated spleen cells from BCG vaccinated mice (●). Extracellular bacteria were destroyed with lysostaphin after 30 min of phagocytosis.

and in the monolayers at T_0 to see the effect of phagocytosis on bacterial viability.

MAF can affect phagocytosis and can affect killing concomitant with phagocytosis. The viable counts in the control (no MAF or sham MAF) macrophages may increase, remain unchanged or decline in the period after phagocytosis, and it is against this that the effect of MAF on intracellular bacteria is measured. Representative results showing the effect of MAF on intracellular anti-*Staphylococcus* activity are shown in *Figure 1*.

2.4 Anti-*Mycobacterium microti* assay

2.4.1 *Advantages and disadvantages*

M. microti in the mouse causes a disease resembling tuberculosis and hence the organism in mouse macrophages *in vitro* provides a model for testing MAF that is likely to be relevant to tuberculosis. *M. microti* presents only modest hazard to man, being used as a vaccine similar to BCG (which is in hazard category 2), unlike *M. tuberculosis* which, being in category 3V, requires special facilities (8). The bacteria multiply at a relatively leisurely pace (doubling time >12 h), both extracellularly and intracellularly, so that thorough rinsing and medium changes can be easily used to control extracellular bacteria without causing excessive disturbance to the monolayers (or the experimenter). Dead bacteria persist for long periods so MAF effects on intracellular multiplication and death rates can in principle be assessed by comparisons of changes in total and viable counts.

Disadvantages are that results are available only after 4–5 weeks; there is no equivalent to lysostaphin for selective total removal of extracellular bacteria.

2.4.2 *Reagents and equipment*

Mycobacterium microti OV254 strain.
Dubos broth base (double strength but without supplements; Difco Laboratories Ltd., West Molesey, Surrey, UK).
Dubos oleic base agar (Difco).
Mouse peritoneal macrophages, prepared as described above (Section 2.2.2).
16-mm Linbro multidishes (Flow Laboratories).
HBSS supplemented with 25 mM Hepes buffer.
Macrophage maintenance medium (as Section 2.3.3).
Heat-inactivated (56°C for 30 min) FCS.
Saponin (1% w/v).
Ultrasonicator (as Section 2.3.2).
Phenolic disinfectant.
Formol saline.
Bacteria counting chamber and phase-contrast microscope.
Pipettes and tubes for serial 10-fold dilutions.

2.4.3 *Procedures*

(i) *Activation of macrophages.* Set up macrophage monolayers and treat them with MAF as described above (Section 2.3.3).

(ii) *Preparation of bacterial inoculum.*

(1) Maintain *M. microti* in Dubos broth base at 37°C by weekly subculture of 0.2 ml into 10 ml in static Universal (25 ml) bottles.
(2) To prepare an inoculum wash a 6-day culture twice by centrifugation at 1000 *g* for 15 min and resuspension in HBSS then sonicate the suspension for 15 sec to break up clumps.
(3) Count the bacteria in a sample diluted 2-fold in formol saline and finally dilute in medium 199 containing 10% heat-inactivated FCS.
(4) Select a dilution such that when a convenient volume is added to a monolayer the bacterium:macrophage ratio in the dish or well is 1:1.

(iii) *Assay of anti-mycobacterial activity.*

(1) Rinse the monolayers, add the bacterial suspension and incubate at 37°C for 2 h for phagocytosis. Retain a sample of the inoculum suspension on ice for viable counts.
(2) Wash the monolayers four times with warm HBSS to remove unattached bacteria,

discarding supernatants and rinses into disinfectant. Place duplicate monolayers on ice for subsequent viable counts.

(3) Add fresh maintenance medium or medium containing MAF to the remainder and incubate them at 37°C for up to 6 days.

(4) Remove the medium and MAF and replace with fresh every day; save representative supernatants for viable counts.

(5) Every 2 days wash duplicate monolayers twice with ice-cold HBSS, saving the pooled washings and washed monolayers on ice for viable counts.

(iv) *Viable counts.*

(1) Before use, dry the agar plates at 37°C overnight with their lids on or for 1 h upside down with their lids off.

(2) Process the inoculum sample, washed monolayers, supernatants and washings as soon as possible after placing them on ice.

(3) Add 1 ml of saponin to monolayers and equal volumes to the inoculum sample, supernatants and washings, then incubate for 10 min at 37°C and briefly sonicate to lyse the macrophages and disperse any bacterial clumps.

(4) Make serial 10-fold dilutions in distilled water and plate out as described (Section 2.3.3) on Dubos oleic base agar plates.

(5) After the inocula have dried into the agar, seal the plates with Parafilm, pack them in polythene bags and incubate them for 4–5 weeks at 37°C before counting the colonies.

(v) *Macrophage number (DNA).* Assay the DNA content of macrophage lysates at the end of phagocytosis and after 6 days as described for the anti-*Staphylococcus* assay [Section 2.3.3 (iii)].

(vi) *Total counts.* The ratio of 1:1 bacteria presented to macrophages is suitable for following changes of bacterial viable count in monolayers, but at this ratio there are too few bacteria in macrophages for microscopically determining total bacterial counts. The effect of MAF on phagocytosis is therefore best examined separately using a 10-fold bigger inoculum. Then fix the rinsed monolayers in methanol for 5 min at the end of phagocytosis, let them air-dry and stain them with cold Ziehl–Neelsen acid-fast stain. Counter-stain with 2% methylene blue and count as described for *S. aureus* (Section 2.3.3).

(vii) *Results.* Killing during phagocytosis is revealed by decrease in viability, as described for the anti-*Staphylococcus* assay, and MAF-enhanced killing is seen as lower viable counts relative to the control (no MAF or sham MAF) without proportionately lower total counts. MAF-enhanced killing of *M. microti* during phagocytosis is shown in the results of Khor *et al.* (9). With healthy monolayers there is no perceptible loss of macrophages (DNA) during subsequent maintenance for at least 6 days and the sum of the viable bacteria in supernatants and washings at any one time does not exceed 5% of the number in the monolayers. Under these circumstances the post-phagocytic change in viable count in the monolayers is taken as a reflection solely of intracellular

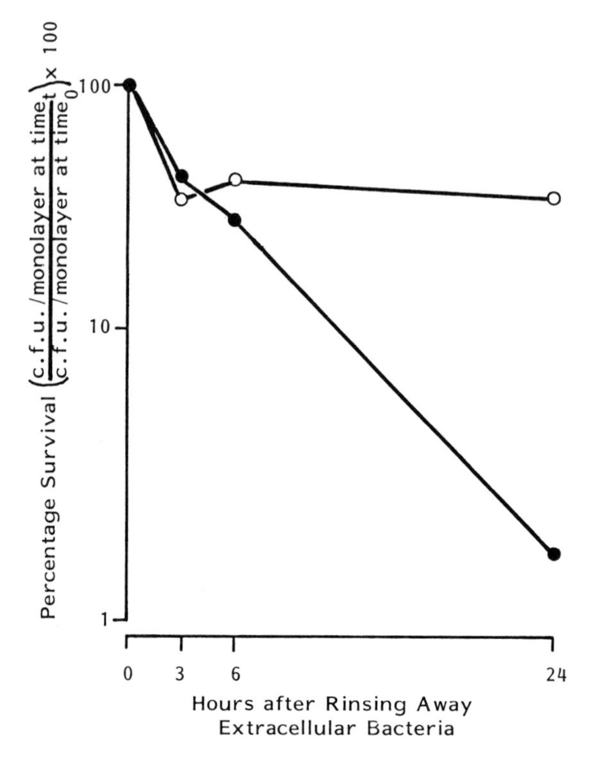

Figure 2. MAF enhancement of killing of *Mycobacterium microti* in mouse peritoneal macrophages. Symbols and methods are as for *Figure 1* except that extracellular bacteria were removed by rinsing after 2 h of phagocytosis.

events, unconfounded by loss of infected cells or extracellular multiplication and continued phagocytosis (*Figure 2*).

3. HUMAN U937 CELL LINE ANTI-LISTERIA ASSAY

3.1 Introduction

Continuous macrophage cell lines have the attractions of being in principle less variable than fresh macrophages and available in essentially unlimited quantity. However, there are few macrophage-like cell lines with substantial anti-microbial activity. The human macrophage-like tumour cell U937 has modest activity although it is essentially non-adherent and is rather undifferentiated (10). Variant sub-lines have arisen during maintenance in different laboratories and mycoplasma contamination is common. We use a strain which is strongly activated by human IFN-γ and MAF to release hydrogen peroxide in response to phagocytic stimuli and which has been cured of mycoplasma contamination by antibiotic treatment.

3.2 Maintenance of the cell line

(i) Grow the cells in upright 250-ml plastic culture flasks containing 50 ml of antibiotic-free medium RPMI 1640 containing 10% FCS. The absence of antibiotics is essential for subsequent anti-bacterial assays and has the advantage of quickly revealing microbial contamination.

(ii) Every 3–4 days count the cell density in a haemocytometer and dilute in fresh medium to 2.5×10^5/ml.

(iii) Assay for mycoplasma every 2 months.

3.3 **Anti-*Listeria* assay**

3.3.1 *Advantages and disadvantages*

The bacteria grow fast on agar plates and so give rapid results. They are intracellular pathogens and hence relevant target organisms; they are also only hazard category 2. The major disadvantage is that U937 cells are only moderately phagocytic and poorly adherent so that the usual washing procedures for removing non-phagocytosed bacteria from monolayers are not satisfactory. One solution is to expose the cells to MAF and the control conditions only after phagocytosis has been completed and the infected cells have been washed in bulk by centrifugation. However, any effect of MAF on events during phagocytosis is not then seen. The alternative adopted here is to activate before infection but to use a combination of a very low bacteria:macrophage ratio and centrifugation to maximize the proportion of bacteria phagocytosed. Too few bacteria are involved for making total counts. In consequence little can be concluded regarding the mechanism of any enhancement of anti-microbial action, whether it is by promoting phagocytosis, killing during phagocytosis, intracellular killing or intracellular stasis. Since a substantial proportion of the inoculum may remain extracellular it is necessary to run controls with extracellular bacteria only (i.e. no U937 cells) in order to exclude direct effects of MAF preparations on the bacteria.

3.3.2 *Materials and equipment*

Listeria monocytogenes NCTC 9373.
Tryptic soy broth (TSB; Difco).
Tryptic soy agar (Difco).
U937 cells.
FCS.
Medium RPMI 1460.
Flat-bottom 96-well microtitre trays.
Centrifuge with microtitre tray carriers.
Crushed ice in an insulated tray.
Cryotube liquid nitrogen storage system.
Water bath on a time-switch in a cold-room.

3.3.3 *Procedures*

(i) *Activation of U937.*

(1) Split the U937 cell culture to 2.5×10^5/ml 24 h before the cells are needed.

(2) Count the cells in an aliquot of the rapidly dividing 24 h culture.

(3) Centrifuge the culture at 1000 *g* for 10 min and resuspend the cells in fresh medium at room temperature to 1×10^6/ml.

(4) Dispense 0.1 ml aliquots into microtitre plate wells and for each well with cells in set up another well with 0.1 ml of medium only.
(5) Add 0.1 ml of medium or medium containing MAF at 2–60% to the pairs of U937-containing and cell-free wells. Set up all wells in triplicate.
(6) Incubate the plates at 37°C for 3 days, each day carefully removing 0.1 ml of supernatant and replacing it with 0.1 ml of medium or medium containing 1–30% MAF.

(ii) *Preparation of the bacterial inoculum.*
(1) Grow *L. monocytogenes* to logarithmic phase in TSB and store 200 μl aliquots directly in cryotubes in liquid nitrogen.
(2) Thaw a tube and add the contents to 10 ml of TSB in a plastic 30 ml Universal bottle.
(3) Place the bottle in a water bath at 4°C which is set to switch to 37°C about 8 h before the inoculum is needed; for example 1 am for a 9 am start. The time needed to give a suitable log phase culture must be determined in advance.
(4) Centrifuge the culture at 3000 *g* for 15 min at 4°C.
(5) Wash the bacteria by resuspension in 10 ml of HBSS and re-centrifugation.
(6) Resuspend the bacteria in 200 μl of 50% AB serum and incubate at 37°C for 30 min to opsonize.
(7) Dilute with about 40 ml of RPMI 1640, mix a 100 μl sample of this with 100 μl of formol saline and make a total count by phase-contrast microscopy using a bacteria counting chamber.
(8) Further dilute the bacterial suspension to 5×10^3/ml. Make a viable count of this inoculum suspension by mixing 50 μl with 50 μl of RPMI 1640 and 100 μl of saponin and plating 100 μl on tryptic soy agar.

(iii) *Assay of anti-Listeria activity.*
(1) Inoculate each microtitre well with 20 μl of bacterial suspension containing on average 10–200 bacteria.
(2) Centrifuge the microtitre trays at 260 *g* for 10 min at room temperature then incubate the trays at 37°C.
(3) After 6 h remove 120 μl of supernatant from each well and replace it with 100 μl of saponin.
(4) Incubate the plate for a further 10 min to allow lysis then make serial 10-fold dilutions in saline to 10^{-2} and pipette 100 μl volumes of the lysate and dilutions onto 1/3rd segments of tryptic soy agar plates. Prepare duplicate plates.
(5) Incubate the plates for about 40 h and count the colonies.

(vi) *Results.* The number of viable bacteria per well increases over 100-fold in 6 h in wells without U937 cells, less in wells with U937 and least when the U937 cells have activated in response to MAF. Apparent MAF activity, expressed as percentage decrease of bacterial numbers in MAF-treated relative to untreated U937 cells, should be corrected by subtraction, in proportion, of any effect of MAF on U937-free controls,

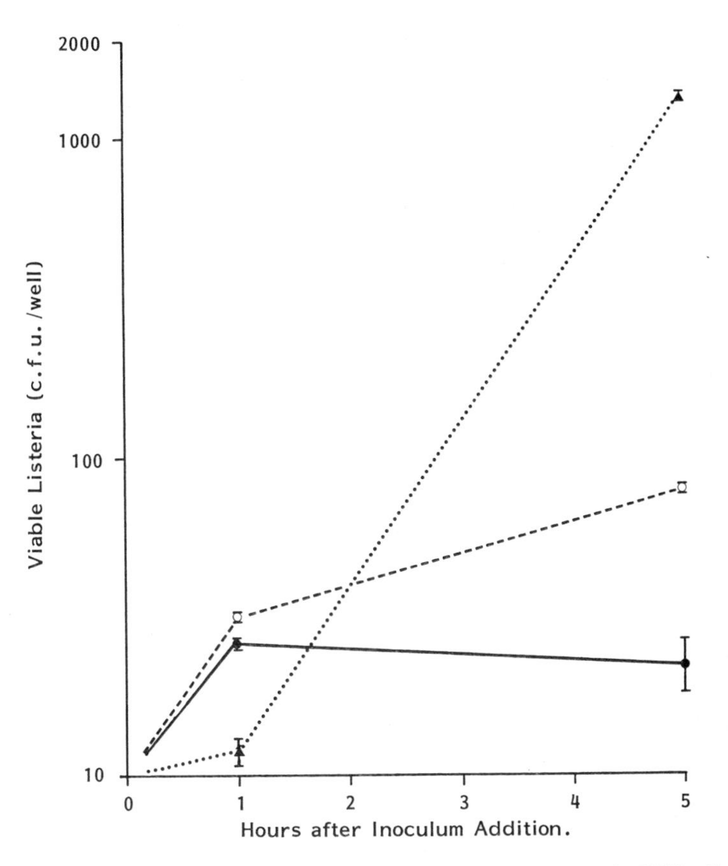

Figure 3. MAF-enhanced suppression of multiplication of *Listeria monocytogenes* by U937 cells. U937 cells were maintained in microtitre tray wells for 3 days, with daily medium changes, in either medium alone (○) or medium containing IFN-γ at 50 U/ml (●). Wells were then inoculated with an average of 10 bacteria per well and incubated without rinsing. Control wells that did not contain U937 cells were also inoculated (▲).

although this is likely to be an over-correction. Representative results are given in *Figure 3*.

4. FLUORIMETRIC PEROXIDE RELEASE ASSAYS

4.1 Introduction

There are a number of colorimetric assays for peroxide release, of which the most notable is that based on peroxidation of phenol red (11). We have opted for fluorescence-based assays because of their potential for greater sensitivity. Like the colorimetric assays, they depend on peroxidation of an indicator in the presence of peroxidase and this is arranged to take place continuously during peroxide release. What is measured is essentially the balance of the rates of synthesis, destruction and release; and it is worth remembering that the process of measurement is certain to alter one or more of those rates. Two assays are described: one measures peroxide as a loss of fluorescence (scopoletin) and the other as a gain in fluorescence (*p*-hydroxyphenylacetic acid, pOHA).

The scopoletin assay is done on a microtitre tray and is convenient for large numbers of samples but utilizes an expensive microtitre tray fluorescence reader; the other assay uses a spectrofluorimeter with a micro-flow cell. The choice of assay will doubtless be determined by equipment availability.

Rate of peroxide release can be expressed as specific activity in relation to cell number or cell substance such as protein. Since the cell maintenance media usually contain serum we prefer to estimate cell numbers from DNA content. The DNA assay utilizes the compound Hoechst 33258 (2-[2-(4-hydroxyphenyl)-6-benzydimidozyl]-6-[1-methyl-4-piperazyl]benzimidazole) whose fluorescence is greatly enhanced from a negligible background on binding to DNA (7).

4.2 Microtitre tray scopoletin-based peroxide assay with U937 cells

Scopoletin has been successfully used in a number of peroxide assays, most recently by De la Harpe and Nathan (12) as part of an adherent monocyte or macrophage system. The fact that scopoletin is an umbelliferone-based fluorophore has been turned to advantage since the new microtitre tray fluorescence readers have been designed to operate with umbelliferone derivatives. We have further exploited the capacity of the fluorescence reader by assaying DNA *in situ* using the fluorophore Hoechst 33258 after assaying peroxide release. The excitation and emission maxima of the DNA-bound stain (356 and 458 nm, respectively) are sufficiently close to those of scopoletin to obviate filter changes. Link steps were devised which enabled the scopoletin and Hoechst 33258-based quantitations to be carried out successively, without mutual interference, in the same plate as the preceding cellular activation. The assay has been found suitable both for adherent and non-adherent cells.

4.2.1 *Macrophage preparation and activation*

Obtain U937 cells (Section 3.2) and activate them in microtitre trays (10^5 cells/well) as described above for anti-*Listeria* assay (Section 3.3.3). Prepare cell-free controls, containing 200 μl of medium, and test each sample in triplicate.

4.2.2 *Advantages and disadvantages*

This assay has the benefit of requiring small volumes of the supernatant under test, and is suitable for processing many samples simultaneously. Advantage can be taken of the availability of multichannel micropipettors and other labour- or time-saving devices designed around the microtitre plate. Only one analytical instrument is required for all stages, and readings are all taken at the same pre-set excitation and emission wavelengths so minimizing confusion. However, since it is loss of fluorescence that is measured, presence of peroxide much in excess of that anticipated will register as out of range.

4.2.3 *Reagents and equipment*

U937 cells.
RPMI medium 1640.
FCS, heat-inactivated at 56°C for 30 min.
Hydrogen peroxide solution of known molarity (E_{230} = 81 l/mol/cm).

DNA, calf thymus, highly polymerized (BDH Chemicals Ltd., Poole, Dorset, UK).
Catalase (EC 1.11.1.6; Sigma Ltd., Cat. no. C40).
Scopoletin (Sigma Ltd.).
Horseradish peroxidase (HRP) Type I (EC 1.11.1.7; Sigma Ltd., Cat. no. P8125).
4β-Phorbol 12β-myristate 13α-acetate (PMA; Sigma Ltd.).
HBSS buffered with 20 mM Hepes, pH 7.3.
Phosphate buffer, 0.125 M, pH 7.4, prepared in 5 M NaCl containing 1 mM ethylenediaminetetracetic acid (EDTA).
Hoechst 33258 (Sigma Ltd.).
Dynatech Microfluor MR600 (or similar microtitre plate fluorescence reader).

Prepare reagents as follows.

(i) Cell maintenance medium: 10% FCS in RPMI 1640.
(ii) DNA solution: 200 μg/ml in 5 mM NaOH.
(iii) Peroxide solution: 5 mM in HBSS.
(iv) Catalase stock: 10^5 U/ml in HBSS; diluted to 100 U/ml working solution.
(v) Scopoletin stock: 1 mM in HBSS.
(vi) Peroxidase stock: 100 U/ml in HBSS.
(vii) PMA stock: 1 mg/ml in dimethyl sulphoxide (DMSO), diluted to 10 μg/ml working solution in HBSS.
(viii) Scopoletin reagent: prepare from the respective stocks a solution in HBSS containing a final concentration of 60 nmol/ml scopoletin, 2 U/ml peroxidase and 200 ng/ml PMA.
(ix) Hoechst 33258 stock: 2.5 mg/ml in DMSO, diluted to 12.5 μg/ml working solution in 0.125 M phosphate buffer.

4.2.4 *Assay procedures*

(i) After the 3-day MAF treatment, remove 150 μl of supernatant from each well and replace it with an equal volume of HBSS.
(ii) Centrifuge the plate at 260 *g* at room temperature for 5 min and again remove 150 μl of supernatant.
(iii) Repeat the washing, leaving the cells in a volume of 50 μl.
(iv) Prepare a separate microtitre tray for a DNA calibration curve by adding DNA dilutions to triplicate wells, covering the range 0–10 μg/well and making up the volume to 50 μl as necessary with HBSS. From this stage, treat these wells in the same way as the MAF sample test wells.
(v) Allow the microtitre tray fluorimeter to stabilize by switching it on 10 min before use.
(vi) Add 50 μl of scopoletin reagent to each well and read the fluorescence immediately.
(vii) Incubate the plates for 1 h at 37°C and again take fluorescence readings. For time course studies, readings can be taken at successive intervals during incubation.
(viii) After the decrease in fluorescence due to peroxide release has been recorded, the residual fluorescence must be eliminated by total oxidation of scopoletin.

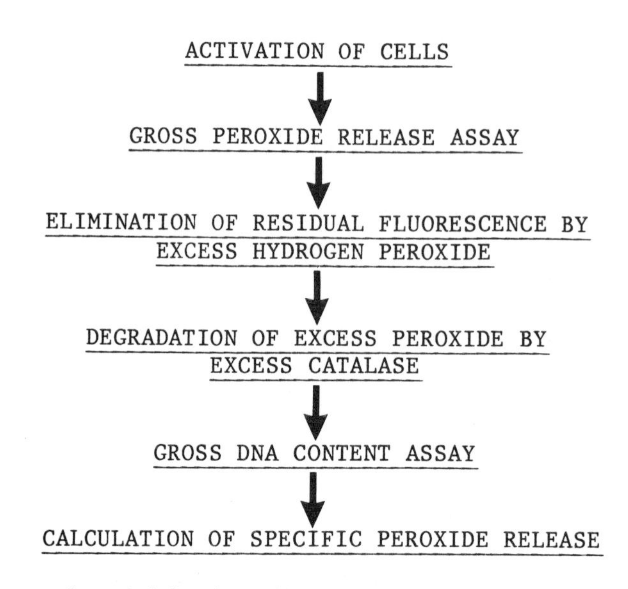

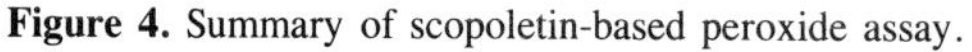
Figure 4. Summary of scopoletin-based peroxide assay.

Add 10 μl of hydrogen peroxide reagent to each well and incubate until fluorescence readings are at background levels (usually ~5 min).

(ix) Decompose the excess peroxide by adding 10 μl of catalase solution to each well and incubating for 30 min at 37°C. At the end of this period, check that readings of all wells are at background levels.

(x) Add 100 μl of Hoechst 33258 reagent to each well and leave the plates in the dark for 3 h at room temperature before again reading the fluorescence, which is now due to binding of Hoechst 33258 to DNA.

4.2.5 *Data handling*

Calculate the amount of peroxide released in each well from the following formula:

$$\text{Peroxide release (nmol/well)} = \frac{F_o}{C_o} - \frac{F_t}{C_t} \times 3.0$$

where F_o = sample fluorescence reading at the start, C_o = fluorescence of cell-free control at the start, F_t = sample fluorescence after incubating for time *t* (usually 1 h), and C_t = fluorescence of cell-free control after time *t*.

The proportional correction relative to a cell-free control is necessary to compensate for non-oxidative loss of fluorescence due to temperature or other factors (12).

Calculate the number of cells per well from the DNA detected, and hence the peroxide release per million cells.

Figure 4 summarizes the main steps in the analyses.

If a BBC Microcomputer is available, the accompanying programs (see Appendix) may be found useful for data capture via the Microfluor RS232 interface and BBC user port and for subsequent calculations and data analysis. The processing of numerous

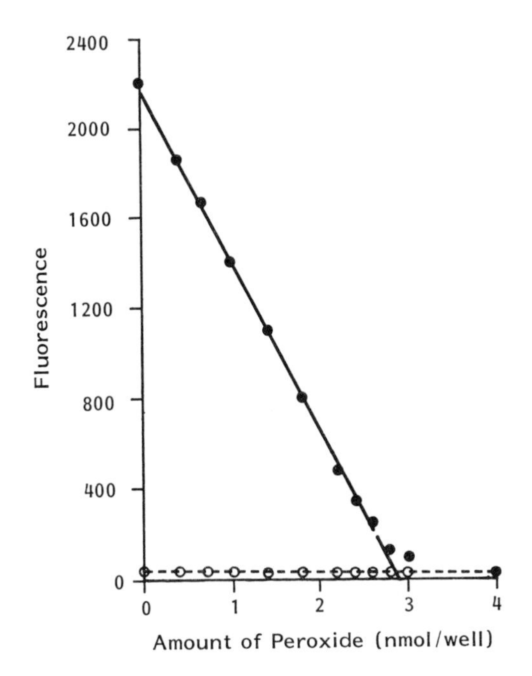

Figure 5. Reaction of scopoletin with hydrogen peroxide. Scopoletin, peroxide and catalase reagents were added sequentially to wells containing known amounts of hydrogen peroxide. Conditions were as described in Section 4.2.4. Closed circles (●): fluorescence after addition of scopoletin reagent. Open circles (○): fluorescence after quenching with excess peroxide followed by addition of catalase reagent.

samples is greatly eased by this means. Programs modified to run on microcomputers using 059 are under development.

4.2.6 *Results*

The results of a calibration experiment, using known amounts of reagent hydrogen peroxide in wells in place of cells, are shown in *Figure 5*. This confirms the direct proportionality of scopoletin fluorescence decrease to the amount of peroxide present. Unimolar stoichiometry is maintained for more than 93% of the scopoletin present. Addition of an excess of peroxide results in reduction of residual fluorescence to background; this baseline is not significantly altered by the incubation with catalase.

Figure 6 shows the results of a separate experiment exploring the feasibility of Hoechst 33258-based DNA estimation on wells that had been through the peroxide assay. While a slight depression of fluorescence in treated relative to untreated control wells is apparent, direct proportionality of fluorescence to DNA present is maintained. The necessity of the catalase-mediated degradation step is also apparent from the same figure: when the assay was performed in the presence of hydrogen peroxide fluorescence was drastically reduced.

Figure 7 shows that direct proportionality was also observed for detection of DNA and was maintained with up to 1×10^6 cells per well. A comparison of fluorescences detected from reagent and cellular DNA (*Figure 7a* and *b* respectively) enabled calculation of a direct conversion factor between number of cells and amounts of DNA,

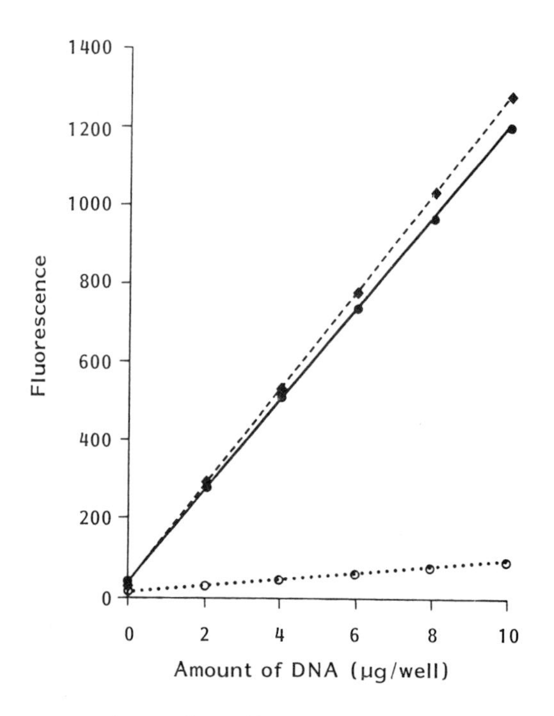

Figure 6. Effect of presence or absence of peroxide and catalase on Hoechst 33258–DNA fluorescence. Fluorescence readings were taken after adding Hoechst 33258 reagent to wells containing: DNA and Hoechst 33258 reagent alone (◆); DNA in the presence of 50 nmol peroxide (○); or DNA after the addition of 50 nmol peroxide followed by decomposition by catalase (10 U/well for 30 min at 37°C) (●).

and hence between the number of cells and fluorescence. The data yield a value of 8.06 μg DNA/10^6 U937 cells.

Typical results obtained using this MAF assay are given in *Figure 8*.

4.2.7 *Assays with other macrophages*

The assay appears suitable for use with almost any cell type. We frequently use it with fresh human monocytes.

(i) Separate mononuclear cells from a fresh peripheral blood sample by differential centrifugation (e.g. on Lymphoprep; Flow Ltd., Rickmansworth, UK) and re-suspend in RPMI 1640.

(ii) Place 10^6 cells into each well, incubate for 2 h and wash off non-adherent cells. This leaves approximately 10^5 monocytes adhering to each well.

(iii) Carry out the rest of the assay as described previously, but all liquid can be removed when changing medium or removing supernatant.

4.3 Assay with *p*-hydroxyphenylacetic acid and U937 cells

The method is based on that developed by P.W.Andrew (13) and is capable of sensitivity similar to the scopoletin assay (Section 4.2). A similar method has been described by Ruch *et al.* (14).

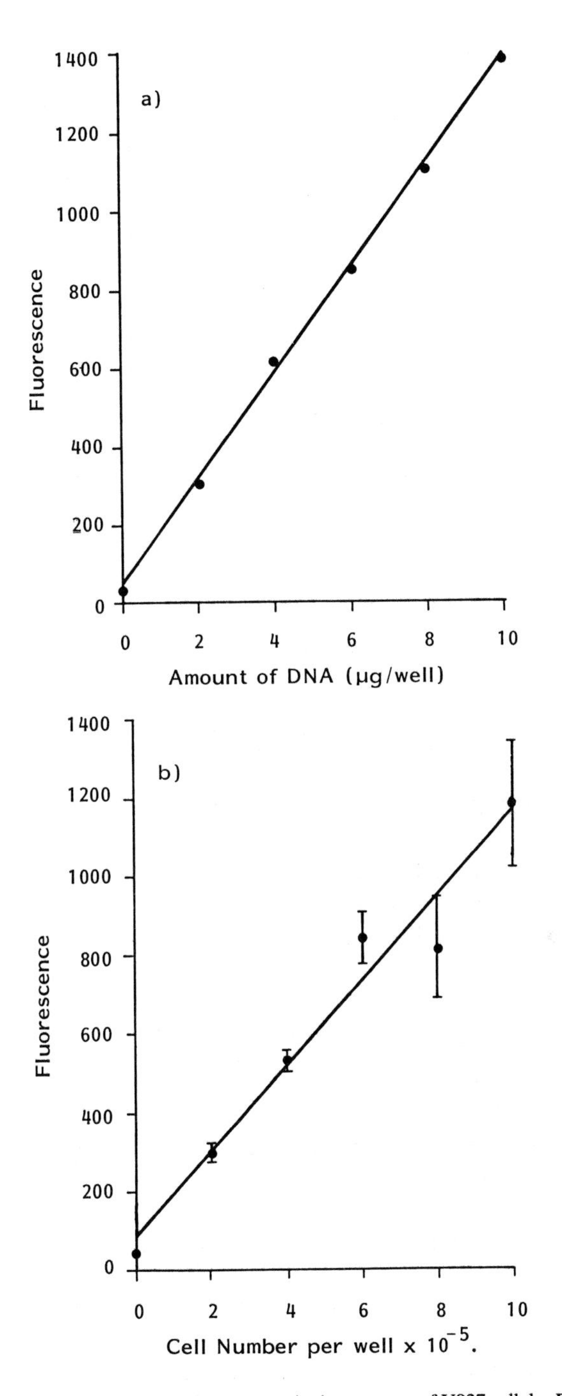

Figure 7. Determination of Hoechst 33258 fluorescence in the presence of U937 cellular DNA (**b**) and standard DNA (**a**). DNA assay was performed as described in the text in wells containing a known amount of DNA or a known number of U937 cells. All wells had been previously treated sequentially with scopoletin, peroxide and catalase reagents.

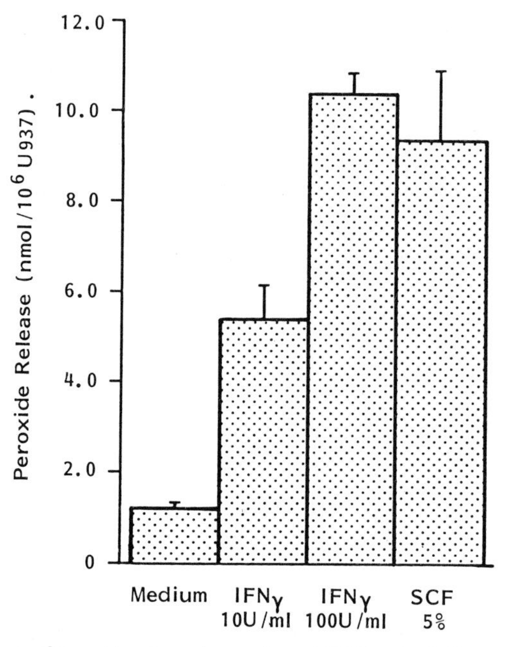

Figure 8. Scopoletin assay of peroxide release from activated U937 cells. Cells were activated as described in the text with IFN-γ, h.p.l.c.-fractionated human spleen cell supernatant (2) (peak activity fraction at 5% final concentration), or maintenance medium control prior to assay. Values are means of triplicates ± standard errors.

4.3.1 *Advantages and disadvantages*

Peroxide release is detected as an increase in fluorescence. The method uses culture tubes for activation and assay and does not require the expensive microtitre tray fluorimeter but, as described, uses more cells and larger amounts of MAF. Precise results are obtained after technical proficiency is achieved, but the assay is comparatively time-consuming if many samples are to be tested.

4.3.2 *Macrophage preparation and activation*

(i) Maintain the U937 cells as previously described (Section 3.2).
(ii) 24 h after splitting, aliquot 2.5×10^5 cells into a Falcon tube 2058, centrifuge for 10 min at 260 *g* and carefully remove the supernatant with a Pasteur pipette.
(iii) Gently resuspend the pellet in 500 μl of the test medium for activation and incubate the tubes at 37°C.
(iv) Change the solutions by centrifugation and gentle resuspension after 24 and 48 h and assay after 72 h.

4.3.3 *Reagents and equipment*

Perkin Elmer 1000M (or similar fluorimeter), fitted with a micro-flow cell (50 μl) and with excitation and emission wavelengths set to 314 and 414 nm, respectively. Sampling pump with timed peristaltic action (Ilacon Ltd., Gilbert House, River Walk, Tonbridge, UK). This is an economical means of reproducibly transferring samples into the fluorimeter flow-cell.

Prepare working reagents as follows when needed.

(i) HBSS buffered with 2.5% Hepes, pH 7.3.
(ii) pOHPA: 7.4 mg/ml in HBSS.
(iii) PMA: 10 μg/ml in HBSS.
(iv) HRP: 10 U/ml in HBSS.
(v) pOHPA reagent: mix 800 μl of HBSS; 50 μl of pOHPA; 100 μl of PMA and 50 μl of HRP. 1 ml of this mixture is required per assay tube.
(vi) 0.2 M borate buffer (pH 10.4): prepare from boric acid and adjust the pH with sodium hydroxide solution (10 M).
(vii) 0.08 M phosphate buffer (pH 7.4): prepare in 3.2 M NaCl containing 1 mM EDTA.
(viii) Hoechst 33258 reagent: dilute the stock (described in Section 4.2.3) to 5 μg/ml in 0.08 M phosphate buffer.

All other materials are the same as for the scopoletin-based assay (Section 4.2.3); scopoletin and catalase are not needed.

4.3.4 *Procedures*

(i) *Peroxide reaction.*

(1) Place five or six different known amounts of hydrogen peroxide from 0 to 30 nmol in final volumes of 100 μl in duplicate tubes to prepare a peroxide standard curve.
(2) Pre-label one tube for each test sample, and place the borate buffer at 4°C to cool.
(3) Prepare the pOHPA reagent and keep it in the dark until needed.
(4) Centrifuge the tubes and wash the cells in 1.0 ml of HBSS. Discard the wash solution and add 1.0 ml of pOHPA reagent to all tubes (including the hydrogen peroxide standards). Incubate the tubes at 37°C for 30 min.
(5) Stop the reaction by rapid addition of 900 μl of cooled borate buffer.
(6) Centrifuge the tubes and pipette the supernatants into pre-labelled tubes. These can be stored in the dark for several hours without significant change in fluorescence.
(7) Add 500 μl of HBSS to the cell pellets, cap the tubes, resuspend the pellets and store at −20°C for subsequent DNA assay.

(ii) *Peroxide fluorimetry.*

(1) Switch on the fluorimeter at least 10 min before it is needed to allow stabilization. Check that excitation and emission wavelengths are 314 and 414 nm. respectively.
(2) Flush the flow-cell with borate buffer.
(3) Zero the meter on the 0 nmol sample and set the instrument scale to read 1000 units with the 30 nmol sample then read the remainder of the samples.

(iii) *DNA assay.*

(1) Thaw the DNA samples and resuspend them by vigorous agitation.
(2) Prepare a range of DNA standards (0−50 μg DNA/ml in 500 μl of HBSS).
(3) Add 750 μl of Hoechst 33258 reagent to each tube, mix and leave for 3−16 h at room temperature and in the dark.

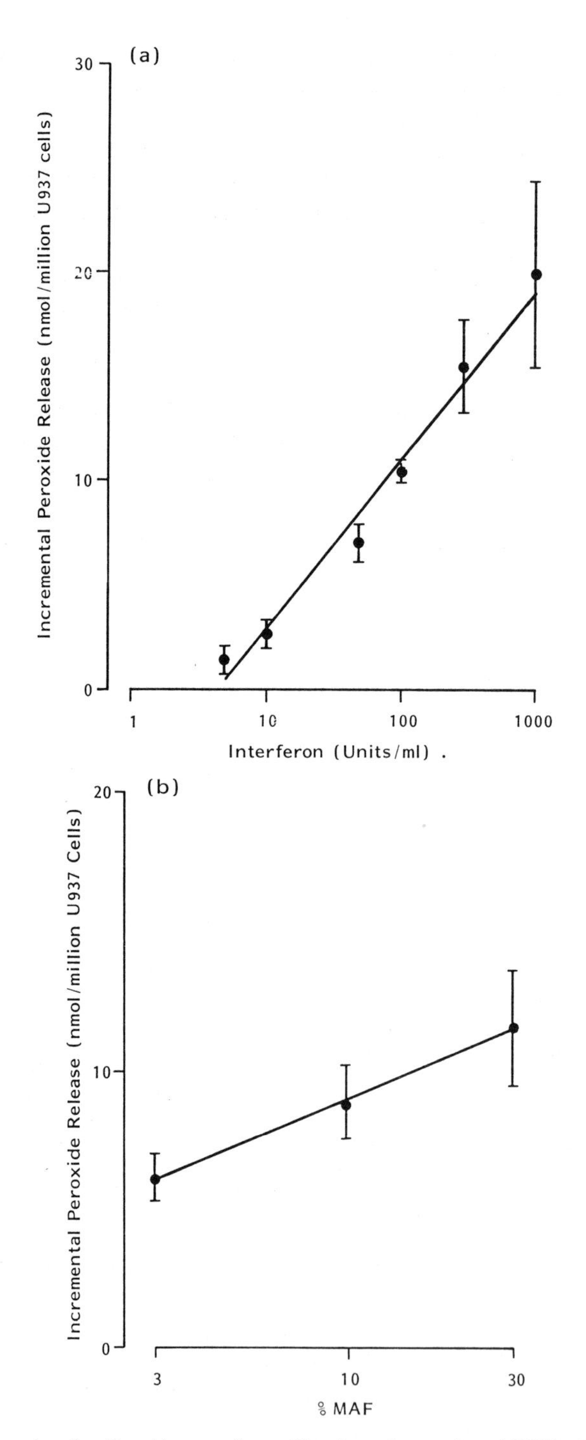

Figure 9. *p*-Hydroxyphenylacetic acid assay of peroxide release from activated U937 cells. Cells were activated as described in the text with recombinant DNA-derived human IFN-γ (**a**) or human T cell clone supernatant (**b**). Values are means ± standard errors of three separate assays.

(4) Read the fluorescence of samples and standards with an excitation wavelength of 365 nm and emission of 456 nm.
(5) Construct a standard curve and calculate the DNA in samples by reference to this.
(6) On at least one occasion estimate the amount of DNA present in reference suspensions of washed U937 cells, the concentrations of which have been accurately determined by direct phase-contrast microscopy. We currently find 7.1 μg of DNA $\equiv 10^6$ U937 cells.
(7) Convert the amount of DNA detected in the experimental samples to number of cells per tube and express the results as nmol peroxide/10^6 cells.

4.3.5 *Results*

Figure 9 illustrates the results obtained with this assay. Peroxide release increases with the log of the concentration of MAF and the relationship approximates to a straight line (although probit transformation is likely to be more appropriate for analysis). The same relationship is seen with recombinant DNA-derived IFN-γ so that, by using this material as a standard in each MAF assay, MAF activity can be expressed as IFN-γ equivalent units.

4.3.6 *Alternative cells*

This assay has proved satisfactory in our laboratory not only with non-adherent U937 cells as described, but also with a wide range of adherent cells, for which purpose the cells are activated and assayed as monolayers in plastic Petri dishes or Linbro multidishes. The cells assayed include human, rabbit, guinea pig and mouse alveolar macrophages, human peripheral blood-derived monocytes and mouse peritoneal macrophages. For each of these cell types MAF preparations have been found to enhance peroxide release.

5. REFERENCES

1. Nacy,C.A., Oster,C.N., James,S.L. and Meltzer,M.S. (1984) In *Macrophage Activation.* Adams,D.O. and Hanna,M.G. (eds), Plenum Press, New York and London, p. 147.
2. Andrew,P.W., Rees,A.D.M., Scoging,A., Dobson,N., Matthews,R., Whittall,J.T., Coates,A.R.M. and Lowrie,D.B. (1984) *Eur. J. Immunol.*, **14**, 962.
3. Lowrie,D.B., Aber,V.R. and Carrol,M.E.W. (1979) *J. Gen. Microbiol.*, **110**, 409.
4. Andrew,P.W., Jackett,P.S. and Lowrie,D.B. (1985) In *Mononuclear Phagocytes: Physiology and Pathology.* Dean,R.T. and Jessup,W. (eds), Elsevier Science Publishers B.V., Amsterdam, p. 311.
5. Lowrie,D.B., Peters,T.J. and Scoging,A. (1982) *Biochem. Pharmacol.*, **27**, 415.
6. Buchmuller-Rouiller,Y. and Mauel,J. (1986) *J. Immunol.*, **136**, 3884.
7. Labarca,C. and Paigen,K. (1980) *Anal. Biochem.*, **102**, 344.
8. Advisory Committee on Dangerous Pathogens (1984) *Categorisation of Pathogens According to Hazard and Categories of Containment.* HMSO, London.
9. Khor,M., Lowrie,D.B. and Mitchison,D.A. (1986) *Br. J. Exp. Pathol.*, **67**, 707.
10. Wing,E.J., Koren,H.S., Fischer,D.G. and Kelley,V. (1981) *J. Reticuloendothel. Soc.*, **29**, 321.
11. Pick,E. and Keisari,Y. (1980) *J. Immunol. Methods*, **38**, 161.
12. De la Harpe,J. and Nathan,C.F. (1985) *J. Immunol. Methods*, **78**, 323.
13. Jackett,P.S., Andrew,P.W., Aber,V.R. and Lowrie,D.B. (1981) *Br. J. Exp. Pathol.*, **62**, 419.
14. Ruch,W., Cooper,P.H. and Baggiolini,M. (1983) *J. Immunol. Methods*, **63**, 347.

6. APPENDIX

Programs for the BBC microcomputer for data capture via the microfluor RS232 interface and BBC user port are given below.

```
   10REM  MicroFlour Data Collection Program
   20REM
   30REM  23/March/1987
   40REM
   50REM   Version 1:50
   60REM
   70CLS
   80PRINT "MicroFlour Data Collection Program"
   90PRINT
  100PRINT "     Version 1:50"
  110PRINT
  120PRINT
  130PRINT
  140REM
  150REM
  160  calavail$="N"
  170  dnadata$="N"
  180  reading%=0
  190  cellp%=60
  200  flaga%=0
  210REM
  220REM  ********************
  230REM
  240REM   Set drive to 1
  250REM
  260    *DRIVE 1
  270REM
  280REM  ********************
  290REM
  300REM   Set RS423 input port to :
  310REM
  320REM       8 Bit
  330REM       Even Parity
  340REM       One Stop Bit
  350REM       9600 Baud
  360REM
  370  *FX156,24,227
  380  *FX7,7
  390  *FX8,7
  400REM  ********************
  410REM
  420REM   Query New run or another reading
  430REM
  440  CLS
  450  PRINT "Do you wish to start a new run (Y/N)"
  460  PRINT
  470  INPUT "",ans$
  480
  490  IF ans$="Y" OR ans$="y" THEN PROCStartNew
  500  IF ans$="N" OR ans$="n" THEN PROCStartOld
  510  IF flaga%=0 THEN 450
  520REM
  530REM ************************
  540REM
  550  cont=OPENOUT("C."+ident$+"0")
  560  IF dnadata$="N" THEN reading%=reading%+1
  570  PRINT#cont,reading%,plate%,samp%,lsamp%,cellp%,cells%,all$,calavail$
  580  CLOSE #cont
  590REM
  600REM ************************
  610REM
  620REM Is it a DNA calibration plate ?
  630REM
  640  IF reading%>3 THEN:
  650  PRINT  "Current version (1.50) of the analysis program":
  660  PRINT  "does not support other than plates at time intervals":
  670  PRINT  "of   i)   T = 0mins":
  680  PRINT  "     ii)  T = 60mins":
```

```
690  PRINT  "      iii) T = 180mins   (DNA)":
700  PRINT:
710  PRINT  "though collection program will accept upto 10 and set":
720  PRINT  "the pointers correctly."
730  IF reading%<3 THEN GOTO 870
740  IF calavail$="Y" THEN 870
750  CLS
760  PRINT
770  PRINT
780  INPUT "Is this a DNA calibration plate Y?N :- ",ans$
790  IF ans$="y" OR ans$="Y" THEN calavail$="Y":dnadata$="Y":GOTO 830
800  IF ans$="n" OR ans$="N" THEN 830
810  GOTO 750
820REM
830REM ************************
840  temp%=reading%
850  IF dnadata$="Y" THEN temp%=0
860  data=OPENOUT("D."+ident$+STR$(temp%))
870REM ************************
880REM
890REM Collect Data
900REM
910REM First flush RS423 buffer then switch
920REM to input from RS423 port
930REM
940  *FX21,1
950  *FX2,1
960REM
970  IF dnadata$="Y" THEN plate%=1
980  FOR X% = 1 TO plate%
990REM
1000  CLS
1010  PRINT
1020  PRINT
1030  IF all$="Y" THEN PRINT "Cell Free sample on ALL PLATES"
1040  IF all$="N" THEN PRINT "Cell Free sample on FIRST PLATE ONLY"
1050  PRINT
1060  PRINT
1070REM Prompt user to start plate reader
1080REM
1090  PRINT "Start MicroFLOUR reading plate number ";X%
1100REM
1110REM Then read data
1120REM
1130  INPUT "",A$
1140  INPUT "",B$
1150  INPUT "",C$
1160  INPUT "",D$
1170  INPUT "",E$
1180  INPUT "",F$
1190  INPUT "",G$
1200  INPUT "",H$
1210REM
1220REM ****************************
1230REM
1240REM Save data on disk
1250REM
1260REM
1270  PRINT#data,A$,B$,C$,D$,E$,F$,G$,H$
1280  NEXT X%
1290REM
1300REM  ****************************
1310REM
1320REM  Close i/o channels and reset
1330REM  compputer before terminating
1340REM  the program.
1350REM
1360  CLOSE#data
1370  *DRIVE 0
1380  *FX2,0
1390  END
1400REM
1410REM  ****************************
1420REM  ****************************
1430REM
```

```
1440REM  PROCEDURE  StartNew
1450REM
1460REM   Get  "RUN" identification, open
1470REM  control file and first data file.
1480REM
1490REM
1500  DEF PROCStartNew
1510  PRINT "Please enter identification for"
1520  PRINT "this run  (Maximum 6 Characters)"
1530  PRINT
1540  INPUT "Identification :",ident$
1550  CLS
1560  INPUT "Number of plates ",plate%
1570  CLS
1580  INPUT "Number of samples on last plate ",lsamp%
1590  CLS
1600REM  INPUT "Number of cells per sample ",cells%
1610  cells%=3
1620  PRINT
1630  PRINT "Number of wells per sample set at 3 in this "
1640  PRINT "version (1.50) of the programs."
1650  PRINT
1660  CLS
1670  INPUT "Cell Free sample on every plate (Y/N) ? ",all$
1680  IF all$="y" THEN all$="Y"
1690  IF all$="n" THEN all$="N"
1700  samp%=(cellp%/cells%*plate%)+lsamp%
1710  flaga%=1
1720  ENDPROC
1730REM
1740REM  ************************
1750REM  PROCEDURE StartOld
1760REM
1770REM    Read control file and get reading
1780REM    number, then open appropriate data file
1790REM
1800REM
1810  DEF PROCStartOld
1820  PRINT "Please enter identification used for"
1830  PRINT "Previous readings "
1840  INPUT "Identification :",ident$
1850  cont=OPENIN("C."+ident$+"0")
1860  INPUT# cont,reading%,plate%,samp%,lsamp%,cellp%,cells%,all$,calavail$
1870  CLOSE# cont
1880  flaga%=1
1890  ENDPROC
1900REM
1910REM  **************************
1920
>
 10REM MicroFlour Data Analysis Program
 20REM
 30REM Development version 1:50
 40REM
 50REM Tested for single plate use only  25/03/1978
 60REM
 70REM
 80CLS
 90PRINT "MicroFlour Data Analysis Program"
100PRINT
110PRINT "    Version  1:50"
120PRINT
130PRINT
140PRINT
150REM
160REM
170REM *********************************
180REM
190REM Variables used :
200REM   background      : Background reading
210REM   cell0           : Cell free control , time 0 min
220REM   cell60          : Cell free control , time 60 min
230REM   dnagrad         : Gradient of dna calibration curve
240REM   dnaincpt        : Intercept of dna calibration curve
250REM   dna             : Dna per 1e6. cells
260REM
```

```
270REM   cal%(12,8)      : Calibration values (one plate)
280REM   data%(3,200)    where :-
290REM   data%(1,xxx)    : Data values , time 0 min
300REM   data%(2,xxx)    : Data values , time 60 min
310REM   data%(3,xxx)    : Data values , time 3 hours
320REM   tdata%(8,12)    : Temporary array for data manipulation
330REM
340REM   plate%          : Number of plates in run
350REM   samp%           : Total number of samples in run
360REM   lsamp%          : Number of samples on last plate
370REM   psamp%          : Number of samples per full plate
380REM   cells%          : Number of cells per sample
390REM   tcells%         : Total number of cells in run
400REM   reading%        : Reading number ie. 1=0min, 2=60min etc
410REM   ident$          : Run identification string
420REM   value$          : Reading value in ASCII
430REM   all$            : Cell free control on all plates Y/N
440REM   calplate$       : Calibration plate (DNA) file name
450REM   calavail$       : Calibration plate (DNA) available
460REM   data            : Channel opened to current data file
470REM   cont            : Channel opened to current control file
480REM   calib           : channel opened to current dna calibration file
490REM
500REM
510REM   first%          : Pointer to string in a string
520REM   length%         : Length of string in a string
530REM   count%          : Pointer to a position in a string
540REM   T$              : Temp string
550REM
560REM ********************************
570REM
580    *DRIVE 1
590    MODE 0
600REM   Initialise variables etc
610REM
620    DIM cal%(8,12)
630    DIM data%(3,200)
640    DIM tdata%(8,12)
650    DIM A%(205)
660    DIM B%(205)
670    DIM D(205)
680    DIM H(205)
690    DIM NC(205)
700    DIM PROD(205)
710    DIM horiz1(10)
720    DIM vert1(10)
730REM
740REM
760REM ********************************
770REM Get name of data files to analyse
780REM
790REM
800    INPUT "Identification used for collection :",ident$
810REM
820REM
830REM ********************************
840REM
850REM Retrieve information from control file
860REM
870REM
880    cont=OPENIN("C."+ident$+"0")
890    PRINT "C."+ident$+"0"
900    INPUT# cont,reading%,plate%,samp%,lsamp%,cellp%,cells%,all$,calavail$
910    CLOSE# cont
920REM
930REM
931    IF plate%=1 THEN tcells%=60:GOTO 950
940    tcells%=(cellp%*(plate%-1))+(lsamp%*cells%)
941REM
950    background%=26
960    INPUT "Background Reading = 26 -- Change Background Reading Y/N :- "ans
$
970    IF ans$="Y" OR ans$="y" THEN  INPUT "Background Reading = ",background%
980REM
990REM
1000REM ********************************
```

```
1010REM
1020REM Get DNA Calibration data from file
1030REM
1040REM
1050    IF calavail$="Y" THEN 1120
1060    CLS
1070    PRINT "A DNA calibration file does not exist for ";ident$
1080    PRINT
1090    PRINT "Please enter the identity of the DNA calibration"
1100    PRINT "file to be used for this analysis ";
1110    INPUT ":- ",calplate$
1120    calplate$=ident$
1130    calib=OPENIN("D."+ident$+"0")
1140    INPUT#calib,T$
1150    FOR y%= 0 TO 6
1160      INPUT#calib,T$
1170      count%=1
1180      PROCIgnoreSpace
1190      PRINT "Row       ";MID$(T$,count%,1)
1200      IF MID$(T$,count%,1)<>CHR$(66+y%) THEN PROCErrorExit
1210      count%=count%+1
1220      FOR x% = 1 TO 12
1230        PROCIgnoreSpace
1240        first%=count%
1250        PROCFindSpace
1260        length%=count%-first%
1270        value$=MID$(T$,first%,length%)
1280        cal%(y%+1,x%)=VAL(value$)-background%
1290REM     cal%(y%+1,x%)=VAL(value$)
1300      NEXT x%
1310    NEXT y%
1320    CLOSE#calib
1330REM
1340
1350REM
1360    PROCCalcSlope
1370REM
1380REM ********************************
1390REM
1400REM Retrieve data from file and place
1410REM in data array.
1420REM
1430REM
1440REM Read data files
1450REM       reading%=1  :-  Data for time = 0
1460REM       reading%=2  :-  Data for time = 60 minutes
1470REM       reading%=3  :-  Data for time = 3 hours  (DNA readings)
1480REM
1490REM    Note :  flag% = 0    :-  No errors
1500REM            flag% = 1    :-  Error 1st cell of sample
1510REM            flag% = 2    :-  Error 2nd cell of sample
1520REM            flag% = 3    :-  Error 3rd cell of sample
1530REM          and
1540REM            flag% = 13   :-  Error in 1st and 3rd cells of sample
1550REM          etc.
1560REM
1570   FOR reading% = 1 TO 3
1580     flag%=0
1590     cellposition%=0
1600REM
1610REM
1620     position%=4
1630     data=OPENIN("D."+ident$+STR$(reading%))
1640     REPEAT
1650REM
1660REM  Ignore First row of samples
1670REM
1680       INPUT#data,T$
1690REM
1700REM  Input values for half a tray
1710REM
1720       FOR halftray%=1 TO 2
1730         FOR y% = 1 TO 3
1740           INPUT#data,T$
1750           count%=1
```

```
1760          PROCIgnoreSpace
1770          temp1%=y%
1780          IF halftray%=2 THEN temp1%=y%+3
1790          PRINT"Row         ";MID$(T$,count%,1)
1800          IF MID$(T$,count%,1)<>CHR$(65+temp1%) THEN PROCErrorExit
1810          count%=count%+1
1820          FOR x% = 1 TO 12
1830            PROCIgnoreSpace
1840            first%=count%
1850            PROCFindSpace
1860            length%=count%-first%
1870            value$=MID$(T$,first%,length%)
1880            tdata%(y%,x%)=VAL(value$)
1890          NEXT x%
1900        NEXT y%
1910REM
1920REM  Copy results of half a plate to data array
1930REM
1940        FOR m%= 2 TO 11
1950          FOR n%=1 TO 3
1960            data%(reading%,position%)=tdata%(n%,m%)-background%
1970            position%=position%+1
1980          NEXT n%
1990        NEXT m%
2000      NEXT halftray%
2010REM   IF halftray%=1 THEN GOTO @@@@
2020      IF position%>64 AND all$="N" THEN GOTO 2090
2030      FOR m%=1 TO 3
2040        data%(reading%,position%-(60+m%))=data%(reading%,position%-m%)
2050      NEXT m%
2060      IF position%=64 AND all$="N" THEN position%=61
2070REM
2080REM Ignore last row of samples
2090REM
2100      INPUT#data,T$
2110REM
2120    UNTIL EOF#data
2130    CLOSE#data
2140  NEXT reading%
2150REM
2160  GOTO 2170
2170REM ********************************
2180REM
2190REM Analyse data
2200REM
2210REM
2220REM
2230  INPUT "Date : ",date$
2240  PRINT "Cells = ";tcells%
2250  INPUT "DNA Conversion factor = 8.06 change Y/N  ",ans$
2260  dnaquan=8.06
2270  IF ans$="Y" OR ans$="y" THEN:INPUT "Conversion factor (dna) : ",dnaquan
2280REM
2290REM  data%(1,xxx)   :-  A(c)
2300REM
2310REM
2320REM  data%(2,xxx)   :-  B(c)
2330REM
2340  position%=1
2350  c%=1
2360  start%=4
2370  end%=tcells%
2380REM
2390  REPEAT
2400  IF start%=4 THEN GOTO 2480
2410  IF start%<>4 AND all$="N" THEN:
2420           start%=end%+1:
2430           end%=end%+60:
2440           GOTO 2480
2450  IF start%<>4 AND all$="Y" THEN:
2460           start%=end%+4:
2470           end%=end%+60
2480  temp1%=position%
2490  IF temp1%=1 OR all$="Y" THEN:
2500           cellfree0%=0:
```

```
2510            cellfree60%=0:
2520            FOR m%=0 TO 2
2530              cellfree0%=data%(1,templ%+m%)+cellfree0%
2540              cellfree60%=data%(2,templ%+m%)+cellfree60%
2550            NEXT m%:
2560            cellfree0%=cellfree0%/3
2570            cellfree60%=cellfree60%/3
2580            position%=position%+3
2590REM
2600    FOR x%=start% TO end%
2610        H(c%)=3*((data%(1,x%)/cellfree0%)-(data%(2,x%)/cellfree60%))
2620REM Should background be subtracted from dna readings ?
2630        D(c%)=(data%(3,x%)-dnaintcept)/dnagrad
2640        NC(c%)=D(c%)/dnaquan
2650        PROD(c%)=H(c%)/NC(c%)
2660        c%=c%+1
2670     NEXT x%
2680REM
2690REM  Out of range well values flag routines
2700REM
2710REM  H(c%)=*** ERROR Flag ****
2720REM  c%=c%+1
2730   UNTIL end% >= tcells%
2740REM
2750REM
2760REM ************************************
2770REM
2780REM Print results
2790REM
2800   *FX6
2810   VDU 2
2820   PRINT
2830   PRINT TAB(25);"Hydrogen Peroxide Production."
2840   PRINT TAB(25);"-----------------------------"
2850   PRINT
2860   PRINT
2870   PRINT TAB(55);"Date  : ";date$
2880   PRINT
2890   PRINT TAB(55);"Title : ";ident$
2900   PRINT
2910   PRINT
2920REM PRINT "Plate ";plate
2930   PRINT
2940   PRINT
2950   PRINT TAB(9);"
nmolH202"
2960   PRINT TAB(9);"Point         nmolh202     ugDNA          nCELLS           p
er 1E6 cells"
2970   PRINT
2980   @%=&90E
2990   FOR x% = 1 TO tcells%
3000     PRINT x%;
3010     @%=&2030E
3020     PRINT H(x%),D(x%),NC(x%),PROD(x%)
3030     @%=&90E
3040   NEXT x%
3050   PRINT
3060   PRINT
3070   PRINT " DNA Calib. : y=";dnagrad;"X + ";dnaintcept;"    Based on ";dnaq
uan;" ugDNA/Million cells."
3080   PRINT
3090   PRINT "CFC Time 0 min = ";cellfree0%+background%
3100   PRINT "CFC Time 60 min = ";cellfree60%+background%
3110   PRINT
3120   PRINT
3130   FOR x% = 1 TO tcells% STEP 3
3140     PRINT x%;
3150     @%=&2030E
3160     PRINT (PROD(x%)+PROD(x%+1)+PROD(x%+2))/3
3170     @%=&90E
3180   NEXT x%
3190 REM
3200 REM   TEST ROUTINE
3210 GOTO 3320
3220 REM
```

```
3230   PRINT
3240   FOR x%= 4 TO tcells%
3250     PRINT x%;"  ";
3260     PRINT data%(1,x%)+background%;"   ";
3270     PRINT data%(2,x%)+background%;"   ";
3280     PRINT data%(3,x%)+background%;"   ";
3290     PRINT PROD(x%)
3300   NEXT x%
3310   PRINT
3320   VDU3
3330   *DRIVE 0
3340END
3350REM ********************************
3360REM
3370REM Procedure IgnoreSpace
3380REM
3390REM Entry  T$     = Temp string to be scanned
3400REM        count% = Position in string to search from
3410REM
3420REM Exit   count% = Position of next non space character
3430REM
3440REM
3450   DEF PROCIgnoreSpace
3460   count% = count%-1
3470   REPEAT
3480     count% = count%+1
3490   UNTIL MID$(T$,count%,1)<>CHR$(&20)
3500   ENDPROC
3510REM
3520REM ********************************
3530REM
3540REM Procedure FindSpace
3550REM
3560REM Entry  T$     = Temp string to be scanned
3570REM        count% = Position in string to search from
3580REM
3590REM Exit   count% = Position of next space character
3600REM
3610REM
3620   DEF PROCFindSpace
3630   count% = count%-1
3640   REPEAT
3650     count% = count%+1
3660   UNTIL MID$(T$,count%,1)=CHR$(&20)
3670   ENDPROC
3680REM
3690REM ********************************
3700REM
3710   DEF PROCErrorExit
3720   PRINT
3730   PRINT "ERROR "; MID$(T$,count%,1)
3740   ENDPROC
3750REM
3760REM ********************************
3770REM
3780REM Procedure CalcSlope
3790REM
3800REM
3810   DEF PROCCalcSlope
3820   N=9
3830   N9=9
3840REM  Get DNA Readings - Vert or Y values
3850   FOR x%=2 TO 10
3860     sum=cal%(1,x%)+cal%(2,x%)
3870     ave=sum/2
3880     vert1(x%-1)=ave
3890   NEXT x%
3900REM
3910REM   Get Horiz. or Y values
3920REM
3930   FOR N%=1 TO 7
3940     horiz1(N%)= N%-1
3950   NEXT N%
3960   horiz1(8)= 8
3970   horiz1(9)= 10
```

```
 3980    numpoints=0
 3990    sumx=0
 4000    sumsqx=0
 4010    sumy=0
 4020    sumsqy=0
 4030    sumxy=0
 4040    FOR N1=1 TO N9
 4050REM  IF F(N1)= 0 THEN zzz
 4060      X=horiz1(N1)
 4070      Y=vert1(N1)
 4080      numpoints = numpoints+1
 4090      sumx=sumx+X
 4100      sumy=sumy+Y
 4110      sumsqx=sumsqx+X*X
 4120      sumsqy=sumsqy+Y*Y
 4130      sumxy=sumxy+X*Y
 4140    NEXT N1
 4150REM RETURN
 4160REM IF numpoints<3 THEN PRINT "Not enough valid results "
 4170    sumxdivn=sumx/numpoints
 4180    sumydivn=sumy/numpoints
 4190    calc1=sumsqx-sumx*sumxdivn
 4200    calc2=sumsqy-sumy*sumydivn
 4210    calc3=sumxy-sumx*sumydivn
 4220    grad=calc3/calc1
 4230    yintcept=sumydivn-grad*sumxdivn
 4240    xintcept=-yintcept/grad
 4250    dnagrad=grad
 4260    dnaintcept=yintcept
 4270    PRINT "Gradient = ";dnagrad
 4280    PRINT "Intercept = ";dnaintcept
 4290    ENDPROC
 4300
>
```

CHAPTER 14

In vitro tumour cytotoxicity assays for macrophage activation

L.VODINELICH and E.S.LENNOX

1. INTRODUCTION

Macrophage activation is a term which refers to a series of physiological and phenotypic changes that occur in the macrophage after contact with various humoral mediators — lymphokines, and also with bacteria, parasites and some chemical compounds. *In vivo*, these phenomena are relevant for macrophage-mediated control of inflammation, destruction of bacteria, parasites and possibly tumour cells. Killing of microbes and tumour cells by activated macrophages can also be observed *in vitro* and this has provided a practical basis for the *in vitro* assays of macrophage activation.

This chapter describes the assays which use the ability of activated macrophages to kill and lyse tumour cells *in vitro*. In contact with some tumour cells, macrophages effect cell stasis without the ensuing cytolysis. In these cases the decrease in the number of dividing tumour cells is measured in the assay. Theoretical considerations about the mechanism of macrophage activation and tumour cell destruction or stasis will not be discussed here, and readers are referred to several articles which deal with those aspects of macrophage activation (1–3). The main purpose of this chapter is to describe practical details of the macrophage-mediated tumour cytotoxicity assays as performed in our laboratory and to draw attention to major possible variations of the assay adopted by different investigators. Variations are found, for example, in the choice of the source of macrophages, which can be obtained from different organs. Recently some macrophage cell lines have been shown to kill tumour cells after activation (4,5). As a source of macrophages, cells resident in the peritoneal cavity can be harvested by peritoneal lavage. Alternatively, macrophages can be attracted from the circulation into the peritoneal cavity by introducing substances such as proteose-peptone, thioglycollate, casein or mineral oil intraperitoneally. Such macrophages are known as 'elicited'. Further variations are found in the type of the tumour target cell and the various radioactive and non-radioactive compounds used to label them. Investigators use different ways to harvest macrophages, the type of assay vessels used (test tubes, 96-well, 48-well, 24-well plates, etc.) and the length of incubation time of the assay mixture. Often these variations are not significant for the final result of the experiment. However, it is worth noting that activating agents may act selectively depending on the source of macrophages and that activated macrophages display different cytolytic or cytostatic activities for different neoplastic cells. Also, selection of a particular radioisotope for optimal cell labelling will vary depending on the target cell type. It is important to emphasize these points to ensure that users of methods described here should spend some time to determine which set of conditions works best for their assay system.

2. PRACTICAL METHODS

2.1 General principles and points of caution

2.1.1 *Assay parameters*

Activation of macrophages can be achieved *in vivo* (bacterial and parasitic infections, inflammation) or *in vitro* by addition of activating agents to macrophages in culture. Activated macrophages lyse a proportion of radioactively labelled tumour cells which then release the intracellular label into the assay medium. The degree of macrophage activation is quantified by the amount of radioactive label released and the potency of activating agents is proportional to the extent of the dilution at which they can still activate macrophages. The dilution at which an activating agent stimulates macrophages to achieve tumour cell kill equivalent to that of 50% of the maximum cytotoxicity is taken to represent 1 unit of that activating agent.

(i) *Maximum release.* Maximum release of the label from a target cell line is usually determined by adding a small volume of detergent into a set of wells containing only labelled tumour cells. 1% sodium dodecylsulphate (SDS) is added to the cell suspension at the end of the assay and the mixture vigorously pipetted several times. Care should be taken that none of the detergent contaminates any of the other assay wells. An alternative way of determining maximum release is repetitive freeze/thawing of tumour cell targets. The amount of released label in the medium by these procedures is generally equal to or slightly less than the total amount of label taken up by the cell. Hence some workers prefer to measure the total radioactivity of target cell aliquots containing the same number of targets as is present in assay wells.

(ii) *Spontaneous release.* Tumour cells release some radioactivity spontaneously during the period of the assay and that proportion of the released label is commonly known as spontaneous release. It is essential that the spontaneous release is low and we recommend that it should not be greater than 10 – 15% of the total releasable label in a 24-h assay. In our laboratory we usually disregard the experiments in which spontaneous release exceeds that level. The extent of spontaneous release is affected by the type of radioactive compound used for any given tumour cell, by the physiological state of the cell and by the length of the assay incubation time. Tumour cells should be in their log phase of growth and 95 – 100% viable at the time of labelling. If the labelling technique has been chosen correctly for the tumour cell line, the spontaneous release will be 5 – 15% over a period of 24 h.

(iii) *Background cytotoxicity.* Since the purpose of the experiment is to measure macrophage activation, the degree of tumour cytotoxicity given by activated macrophages is compared with that carried out by non-activated macrophages in each assay. Non-activated macrophages should not kill tumour cells or should do so to a negligible degree. Consequently the radioactivity released by the action of non-activated macrophages should be close to that of spontaneous release. We term this parameter ‘background cytotoxicity’ of the assay. Experiments in which this background value exceeds spontaneous release by more than 30% are not easily interpreted and should be discarded. For example, if spontaneous release is 15%, the background should not be greater than 20% of the total releasable label.

This is an arbitrary cut off point and is determined by the stringency that the investigator wishes to impose on the assay. It must be understood however, that killing of the tumour cells by 'non-activated' macrophages means that they may be contaminated by non-macrophage cell populations, for example NK cells, or that there is an activated macrophage component in the control cell preparation. In both cases misleading results may be obtained.

2.1.2 *Calculation of the specific cytotoxicity*

Specific cellular cytotoxicity mediated by activated macrophages is calculated as follows:

$$\% \text{ cytotoxicity} = \frac{\text{c.p.m. (test} - \text{SR)} - \text{c.p.m. (B} - \text{SR)}}{\text{c.p.m. (MR} - \text{SR)}} \times 100$$

where SR = spontaneous release; B = background cytotoxicity; MR = maximum release.

2.1.3 *Preparation of macrophages*

Most strains of mice can be used for cytotoxicity assays. Mice should be 8–12 weeks of age. It is essential that the animals are free of pathogens, in particular hepatitis virus, murine pneumonia virus, Sendai virus, CMV and also parasites.

(i) *Resident murine peritoneal macrophages.* These are obtained by harvesting the native population of cells from the peritoneal cavity of mice (see Section 2.2.1). Resident peritoneal macrophages are normally present at about $1-2 \times 10^6$ per mouse and a significantly higher yield of these cells usually implies that mice are in a state of inflammation or infection. In that case it is not advisable to use the cells in the assay. However, macrophages could be activated *in vivo* even without an obvious total cell increase in the peritoneal cavity and appropriate experimental controls will show if that has happened.

(ii) *Elicited murine peritoneal macrophages.* These can be recruited from the circulation into the peritoneal activity after intraperitoneal inoculation of substances such as thioglycollate and proteose peptone (*Table 1*) (6,7). The so-called 'elicited' macrophage population consists of younger macrophage cells and is different from the resident cells which are older, more mature macrophages. Elicitation gives up to 5-, or at the most 10-fold increase in macrophage yield. If the increase in the number of elicited cells surpasses that, it can mean that the macrophages of the animal may be in an activated state due to infection or inflammation.

Further details about harvesting and preparation of murine peritoneal macrophages for the assay are mentioned in Sections 2.2.1 and 2.2.2 and in Chapter 13.

(iii) *Murine bone marrow macrophages.* Bone marrow macrophages are prepared by releasing cells from punctured femoral bones (8).

(1) Dip the bone into 70% alcohol and withdraw the cells from the cavity with a

Table 1. Methods for the elicitation of murine macrophages

Thioglycollate (3%)

(i) Preparation

1. Dissolve 30 g of thioglycollate powder (Difco) in 1 litre of cold distilled water. Pyrogen-free water (Gibco) and glassware or plasticware are preferable.
2. Gradually heat to boiling and stir until the powder is completely dissolved.
3. Dispense the 3% thioglycollate into 20-ml glass Universal bottles and autoclave (15 p.s.i., 121°C for 15 min).
4. Cool, wrap in aluminium foil and store at room temperature in the dark for 2–4 months before use.

(ii) Application

Inject 2–3 ml of aged 3% thioglycollate solution intraperitoneally into mice and harvest peritoneal exudate (PE) 3–4 days later.

Proteose peptone (3%)

(i) Preparation

1. Dissolve 30 g of proteose peptone (Difco) in 1 litre of pyrogen-free phosphate-buffered saline (PBS).
2. Dispense into 20-ml glass Universal bottles and autoclave (15 p.s.i., 121°C, 15 min).
3. Cool and store at room temperature. It can be used immediately.

(ii) Application

Inject 2–3 ml of 3% proteose peptone intraperitoneally and harvest PE cells 2–3 days later.

syringe by squirting in and withdrawing the medium (Iscoves, Gibco) containing 5% fetal calf serum (FCS).

(2) Wash the cells once and resuspend them at 10^6/ml in Iscoves medium supplemented with 10% FCS and 20% conditioned medium from L929 murine fibroblasts. (The conditioned medium is the culture supernatant of confluent L929 cells.).

(3) Incubate the cells at 37°C, 5% CO_2 in T25 cm^2 tissue culture flasks (Falcon) laid down flat, for 2–5 days.

(4) Take off floating cells with a pipette, and replace fresh medium into the flask.

(5) Incubate for a further 3–5 days and then harvest.

(iv) *Murine macrophage cell lines.* Macrophage cell lines are generally not used as effector cells. Macrophage cell lines PU5 1.8 (4) and RAW-264 (5) have recently been shown to kill some tumour cells, in cytotoxicity assays. We have used them successfully with target cells P815 and TU-5.

(v) *Human monocytes.* Fresh human peripheral blood is used as a source of monocytes (9). First, mononuclear cells are separated from granulocytes and erythrocytes using density gradient centrifugation. Commercial density gradient media for preparation of mononuclear cells are available from Nyegaard (Norway), Pharmacia (Sweden), Flow (UK) and other suppliers, but can be prepared in individual research laboratories (10). See Section 2.2.3 for details.

2.1.4 *Labelling of tumour cells used as targets in macrophage cytotoxicity assays*

Murine mastocytoma cell line P815 (4) and an SV40-transformed murine kidney line TU-5 (11) are frequently used as target cells for murine macrophage cytotoxicity assays.

Table 2. Labelling procedures

Labelling of P815 with indium-oxine

1. Wash cells once in RPMI medium with 5% FCS.
2. Adjust the cell concentration to 10^7/ml.
3. Add 50 μCi of ^{111}indium-oxine (medical grade, Amersham).
4. Incubate at 37°C, 5% CO_2 for 15 min (surround the incubation vessel with a lead shield).
5. Wash three times with 50 ml of medium containing 5% FCS, removing radioactive liquid into a special waste container.
6. Adjust to required cell concentration prior to adding to the assay wells.

Labelling of TU-5 with [^{3}H]thymidine

1. Use TU-5 cells when 50% confluent.
2. Pour off the culture medium and wash the cells in the flask twice by adding 10 ml of PBS, swirling around the flask.
3. Add 2 ml of trypsin–EDTA (Gibco) and incubate the cells for 5–10 min at 37°C.
4. Dislodge the trypsinized cells by *shaking* and add to 10 ml of RPMI with 5% FCS. Spin the cells down once and resuspend them at 5 × 10^5/ml.
5. Replace 10 ml of the cell suspension into a fresh T75 flask (Falcon).
6. Add [^{3}H]thymidine (sp. act. 2 μCi/mmol) at 1 μCi/ml of cells and incubate for 16–18 h (overnight) at 37°C, 5% CO_2.
7. Remove the radioactive supernatant carefully into a radioactive waste container.
8. Wash twice with medium warmed to room temperature, containing 5% FCS.
9. Incubate for a further 2 h at 37°C, 5% CO_2.
10. Remove the supernatant and wash the cells twice in PBS.
11. Add 2 ml of trypsin–EDTA to the flask.
12. Leave for 5 min at 37°C.
13. Remove the cells into a sterile centrifuge tube, wash twice with RPMI–5% FCS.
14. Adjust the cells to give the required concentration prior to adding them to the assay wells.

Labelling of SW620 with ^{125}iodine

1. Use SW620 cells when 30–50% confluent in a T75 flask.
2. Pour off the culture medium and wash the cells twice *in situ* with warm RPMI–5% FCS.
3. Replace with 20 ml of RMPI–5% FCS and add 15 μCi of [^{125}]iododeoxyuridine (100 mCi/M) (Amersham) and 10^{-5} M 5-fluorodeoxyuridine (Sigma).
4. Incubate for 24 h at 37°C, 5% CO_2.
5. Remove the cells by trypsin–EDTA and wash (see above).
6. Adjust the cells to the required concentration prior to adding to the assay wells.

Human monocyte cytotoxicity assays predominantly use human colon carcinoma cell line SW620 or human adenocarcinoma HT-29 and others (9), but activated human monocytes can kill murine cell TU-5. These cell lines can be obtained from the American Tissue Culture Collection, Rockville, MD and can be cultured simply, in the routine cell culture media (RPMI, DMEM), supplemented with 5–10% FCS. P815 cells grow as a single cell suspension and can be labelled with tritiated thymidine (11) or indium-oxine (12). In our laboratory labelling P815 with indium gave better results due to lower spontaneous release of the label.

Cells growing in suspension should be spun down and resuspended at 10^6–10^7 cells in 1 ml of tissue culture medium, supplemented with 5% FCS. Radioactive label is then added using a procedure defined for each particular type of label (see below). Cells such as TU-5 and SW620 are adherent, and attach quite firmly to the walls of the culture flask. They can be labelled *in situ* or, after detachment from the flask, by means of trypsin or EDTA (See Sections 2.2.2 and 2.2.3 for details).

Cells should be growing actively when used for labelling. We label cells from culture flasks which are only 50% confluent and growing in fresh medium. We have obtained best results with TU-5 by labelling with thymidine. SW620 can be labelled with indium or iododeoxyuridine (13).

Chromium (^{51}CrNa-chromate) has also been used to label target cells (5). It is, however, a label which is more useful for short-term (4 h) assays, since spontaneous release significantly increases at longer incubation times. Labelling procedures for the cell lines mentioned in this section are given in *Table 2*.

2.2 **Basic experimental protocols for activated macrophage-mediated tumour cytotoxicity assays**

2.2.1 *Assay using interferon-γ and lipopolysaccharide-activated resident BALB/c peritoneal macrophages and P815 tumour cells labelled with indium*

(i) Harvest resident macrophages by peritoneal lavage. Kill the mouse and wash the abdomen with 70% ethanol. Inject 5–10 ml of sterile PBS, Hanks balanced salt solution or culture medium – each containing 2% FCS into the peritoneal cavity with a 24-gauge needle. Some workers add 1–10 U/ml of heparin to the lavage medium, but this is not obligatory. Gently massage the abdomen and withdraw the fluid by syringe into a sterile universal container kept on ice.

(ii) Pool the extracted cells from all the mice required and wash once in the medium (low endotoxin grade Hyclone, USA or Gibco) containing 5% FCS.

(iii) Adjust the cell concentration to (2–4) $\times$ 10^6/ml and place 5 ml of the suspension into a 90-mm plastic Petri dish. It is important that the medium contains FCS at this stage. Incubate for 1–2 h at 37°C, 5% CO_2.

(iv) Remove all medium from the Petri dish. Add about 3 ml of fresh medium, squirting vigorously with a Pasteur pipette several times to wash the base of the Petri dish, to remove the non-macrophage non-adherent cells. Repeat washing twice.

(v) Scrape off the adherent cells (macrophages) gently in 1–2 ml of medium with serum, using a sterile plastic scraper (Costar). Transfer into a 20 ml Universal container kept on ice. Repeat this process if there are remaining adherent cells on the plate (check under a microscope). This population of cells contains more than 95% macrophages and more than 90% cells should be viable. Check the viability, using 0.2% (final concentration) trypan blue dye (Gibco). If the viability is less than 90%, the cells should not be used in the assay.

(vi) Adjust the cells in fresh medium to 1 $\times$ 10^6/ml. Dispense into a 96-well plate (Falcon 3072) in 100 μl volumes. Add murine interferon-γ (IFN-γ) at 1–5 U/ml and lipopolysaccharide (LPS, Sigma L-4130 *E. coli* 0111 B4) at 0–100 ng/ml, to the macrophages in the assay plate, in 100 μl volume. Incubate for 16–24 h at 37°C, 5% CO_2 in a moist atmosphere.

(vii) Aspirate all medium from the wells and wash twice by adding 200 μl of fresh medium and aspirating it off.

(viii) Add 200 μl of indium-labelled P815 cells at a concentration of 1 $\times$ 10^5/ml or 5 $\times$ 10^4/ml to the assay wells.

(ix) Incubate for 16–24 h at 37°C, 5% CO_2.

(x) Spin the plates in a centrifuge at 4°C for 30 sec at 200 *g*.

(xi) Take out 100 μl of the assay medium from each well and transfer to a counting tube.

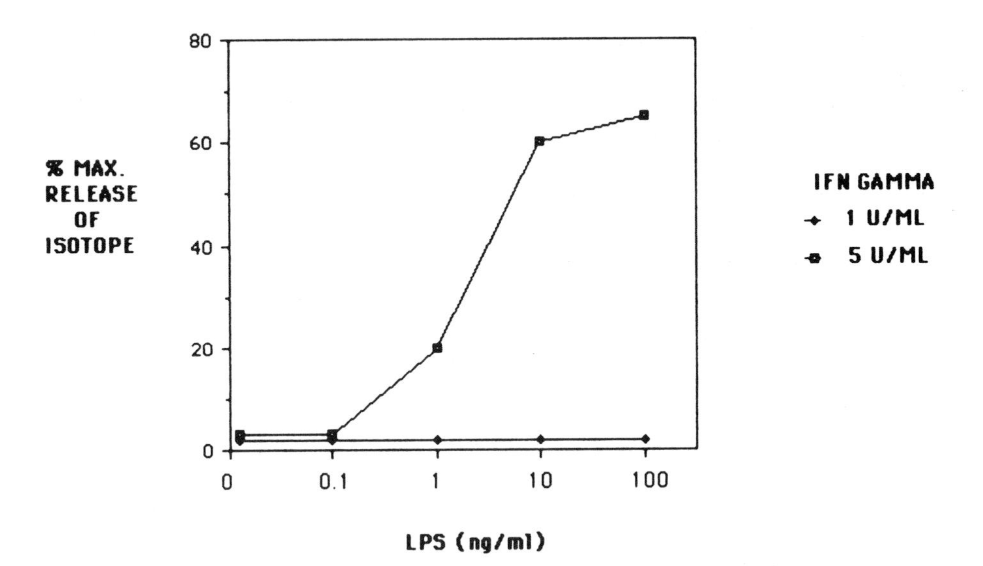

Figure 1. Killing of P815 tumour cells by murine macrophages activated with murine interferon-γ and LPS. Killing is expressed as a % of maximum release of ^{111}indium-oxine from the tumour cells.

(xii) Count in a gamma counter.

An example of the results of such an assay is presented in *Figure 1*. Note that LPS did not activate macrophages in the absence of IFN-γ.

2.2.2 *Assay using activated elicited C_3H/He peritoneal macrophages and TU-5 tumour cells labelled with [^{3}H]thymidine*

(i) Elicit macrophages by injecting 3 ml of sterile proteose peptone (Difco) intraperitoneally 3 days before the lavage.
(ii) Harvest macrophages by peritoneal lavage (see Section 2.2.1)
(iii) Pool PE cells from all the mice required and wash three times in the medium containing 5% FCS.
(iv) Adjust the cell concentration to 1.2×10^6/ml.
(v) Add 100 μl of cell suspension to the 96-well plate.
(vi) Add dilutions of lymphokine (IFN-γ, concanavalin-A stimulated spleen cell supernatant) to the cells in 50 μl volumes.
(vii) Add 50 μl of [^{3}H]thymidine-labelled TU-5 cells at a concentration of 1.2×10^5 per ml.
(viii) Incubate for 48 h at 37°C, 5% CO_2.
(ix) Take out 100 μl of assay medium into a scintillation counting vial, adding 1 ml of aqueous miscible scintillant (Beckman EP).
(x) Count in a scintillation counter.

2.2.3 *Assay using activated human monocytes and SW620 human colon carcinoma cells labelled with ^{125}iodine*

(i) Collect 10–50 ml of fresh human blood into a 20–50 ml sterile syringe containing 50 U of heparin (5000 U/ml) (Flow).

(ii) Dilute 1 volume of blood in 1–2 volumes of the tissue culture medium (RPMI).
(iii) Layer 10 ml of diluted blood onto 5 ml of the density gradient medium (Lymphoprep, Nyegaard). This can be done using a Pasteur pipette, or a 10 ml pipette, by slowly releasing a thin stream of blood along the wall of the centrifuge tube, so that the blood sample sits on a cushion of Lymphoprep.
(iv) Centrifuge at room temperature at 200 *g* for 15–30 min. After centrifugation a white band of peripheral blood mononuclear cells (PBMC) is visible at the plasma/Lymphoprep interface.
(v) Remove the top layer of plasma carefully and aspirate the PBMC immediately into a tube containing fresh medium.
(vi) Wash the PBMC three times in medium supplemented with 2–5% FCS, by centrifuging cells, at 1800 r.p.m. for 10 min in a bench centrifuge. Resuspend the cell pellet in fresh medium and wash at 1200 r.p.m. for 5 min. Repeat this once more.
(vii) Count the cells. The yield of PBMC should be approximately 1×10^6/ml of whole blood. Adjust to $(1-3) \times 10^6$/ml and plate out this cell suspension in a 90 mm plastic Petri dish, at 5 ml per dish. Incubate for 1–2 h at 37°C, 5% CO_2.
(viii) Remove the non-adherent cells and harvest the adherent monocytes (see Section 2.2.1).
(ix) After counting and determining the viability (must be >95%), the concentration of cells should be adjusted to 10^6/ml.
(x) Dispense 100 μl volume of these cells into the wells of a 96-well microtitre plate.
(xi) Add 100 μl of IFN-γ or another test or control sample to the monocytes, and incubate for 12–24 h at 37°C, 5% CO_2.
(xii) Remove the supernatant from the wells and wash.
(xiii) Add 200 μl of the target cell line SW620 at 5×10^4/ml, labelled with [^{125}I]iododeoxyuridine (Amersham).
(xiv) Incubate the assay mixture for 72 h at 37°C, 5% CO_2.
(xv) Withdraw a 100 μl volume from each well for counting in a gamma counter.

4. ACKNOWLEDGEMENTS

We are grateful to Trevor Whittall, Paul Depledge, Lesley Steers, Lindsay Bawden, Claire Moore, Sharon Redmond and Veronica Newton for their careful work on the assays described in this chapter and for help in the preparation of the manuscript. We are also grateful to Jacquie Butler for typing of the manuscript.

5. REFERENCES

1. Annales De L'Institut Pasteur, Immunology, Forum: Macrophage Activation. (1986) Elsevier, Paris.
2. Adams,D.O. and Hamilton,T.A. (1984) *Annu. Rev. Immunol.*, **2**, 283.
3. Adams,D.O. and Nathan,C.F. (1983) *Immunol. Today*, **4**, 166.
4. Hamann,U. and Krammer,P.H. (1985) *Eur. J. Immunol.*, **15**, 18.
5. Gorecka-Tisera,A.M., Snowdowne,K.W. and Borle,A.B. (1986) *Cell. Immunol.*, **100**, 441.

6. Johnston,W.J., Marino,P.A., Schreiber,R.D. and Adams,D.O. (1983) *J. Immunol.*, **131**, 1038.
7. Crapper,M., Vairo,G., Hamilton,J.A., Clark-Lewis,I. and Schrader,J.W. (1985) *Blood*, **66**, 859.
8. Lohmann-Matthes,L.H., Lang,H. and Sun,D. (1982) *Immunobiology*, **161**, 401.
9. Le,J., Prensky,W., Yip,Y.K., Chang,Z., Hoffman,T., Stevenson,H.C., Balazs,I., Sadlik,J.R. and Vilcek,J. (1983) *J. Immunol.*, **131**, 2821.
10. Hudson,L. and Hay,F.C. (1980) *Practical Immunology*, Vol. 17, Blackwell Scientific Publications.
11. Meltzer,M.S., Benjamin,W.R. and Farrar,J.J. (1982) *J. Immunol.*, **129**, 2802.
12. Wiltrout,R.H., Taramelli,D. and Holden,H.T. (1981) *J. Immunol. Methods*, **43**, 319.
13. Svedersky,L.P., Benton,C., Berger,W.H., Rinderknecht,E., Harkins,R.N. and Palladino,M.A. (1984) *J. Exp. Med.*, **159**, 812.

CHAPTER 15

Measurement of interleukin-1 activity

JULIAN A.SYMONS, ELISABETH M.DICKENS, FRANCESCO DI GIOVINE and GORDON W.DUFF

1. INTRODUCTION

Interleukin-1 (IL-1) was the name given to a peptide factor derived from monocytes and macrophages that had a co-mitogenic effect on T lymphocytes (in conjunction with specific antigen or mitogenic lectins). Originally this factor was described in 1972 by Gery *et al.* who studied its effects on thymocytes and called it lymphocyte activating factor (LAF) (1).

Outside the field of immunology many other soluble 'factors' from activated macrophages had been described by biological activity. The best studied of these were: 'endogenous pyrogen' or EP, the host mediator of fever; leukocytic endogenous mediator (LEM), the mediator of acute phase inflammatory responses; and mononuclear cell factor (MCF), an inducer of fibroblast and synoviocyte production of prostaglandins and procollagenase. Biochemical characterization of these different 'factors' suggested increasingly that they might all be closely related or even identical peptide(s) capable of mediating a surprisingly large range of biological effects both *in vivo* and *in vitro*.

As biological and biochemical characterization increased it also became clear that cell types other than macrophages could produce peptides that were very related to IL-1 (e.g. epidermal cell-derived T cell activating factor or ETAF). Thus shortly after the term 'interleukin 1' was introduced, investigators realized that the name was inappropriate since IL-1 was not made exclusively by leukocytes and it was certainly active on a wide variety of non-leukocytic target cells. The early history of IL-1 research including previous terminology is well reviewed in reference (2).

In recent years IL-1 has been characterized at the molecular level (3,4). In most species studied (including human) two distinct IL-1 genes have been identified and cloned. The proteins encoded by these genes have little overall homology (26%) but there are subregions of greater homology which may represent active sites (receptor binding sites) of IL-1.

In healthy humans the most abundant form of monocyte IL-1 is a 17 000 molecular weight peptide with an isoelectric focusing point of pI 7. This has been called IL-1 β while the other gene product (mol wt 17 000, pI 5) is called IL-1 α. Both are believed to be processed from larger (31-K) propeptides and both seem to mediate identical ranges of biological activity. This is supported by competitive-binding studies of radiolabelled IL-1 α and β to target cells which suggest that both bind to the same receptor on lymphocytes and fibroblasts (5). The physiological significance of the two forms of IL-1 characterized so far is unknown since they appear to perform the same roles *in vitro* and *in vivo*. It is possible that access to particular target cells or expression of bioactivity in certain biochemical environments might be different. However, only antibody-dependent assays are at present capable of distinguishing between the two forms and it can be assumed that all of the biological assays described below apply equally to IL-1 α and IL-1 β.

1.1 The clinical measurement of IL-1

Most of the available evidence suggests that IL-1 genes are not constitutively expressed but are induced in response to infection or tissue damage. Thus although low levels of IL-1 have been reported in the blood of healthy individuals, elevated levels probably imply a pathological process and indeed raised IL-1 levels have been reported in a number of infectious diseases and also in sterile inflammatory conditions such as Crohn's disease. IL-1 has also been found in the synovial fluids of patients with arthritic diseases, in pleural and peritoneal fluids during bacterial infection and in the brain and cerebrospinal fluid of animals with neurological damage or inflammation. Some diseases have been associated with apparently reduced IL-1 production when monocytes were tested *in vitro* (eg. systemic lupus erythematosus, advanced neoplasia and malnutrition).

It has sometimes been difficult to evaluate data on IL-1 levels in clinical specimens since many body fluids contain factors that interfere with IL-1 bioassays. The situation is more confusing since it seems that interference may occur in some IL-1 bioassays and not others (e.g. we have found some synovial fluid factors will suppress the T cell response to IL-1 but not the T cell response to IL-2 or fibroblast responses to IL-1). The identity and significance of these factors is not clear at the moment but their existence argues for the availability of a range of IL-1 assays (including bioassays and immunoassays).

2. MATERIALS

The majority of IL-1 assays utilize *in vitro* cell or tissue culture techniques, it is therefore necessary to have access to clean air facilities. Additionally human material such as synovium (Section 4.2.3) should be handled in class II grade facilities.

The media used in the assays are commercially available and unless otherwise stated contain 10% fetal calf serum (FCS; heat inactivated; 56°C, 30 min), 2 mM L-glutamine and antibiotics penicillin (100 IU/ml) and streptomycin (100 μg/ml). All cultures are incubated at 37°C in a 5% CO_2/95% air atmosphere, preferably humidified. Most of the cell lines described are available from the American Type Culture Collection (ATCC), Rockville, Maryland, USA or the European Collection of Animal Cell Cultures (ECACC), Porton Down, Salisbury, UK. A large number of assays utilize radiolabels for detection of proliferation or catabolism; facilities for the handling and measurement of beta and gamma radiation are therefore required.

3. IN VITRO ASSAYS USING LYMPHOID CELLS

3.1 Primary cell cultures

3.1.1 *Murine thymocyte assay*

This is the original LAF assay (1) and is based on the potentiation of lectin-induced, sub-optimal thymocyte proliferation by IL-1. Biologically, the assay involves the stimulation of production of T cell growth factors and/or their receptors (e.g. interleukin 2, interleukin 4) by interleukin 1.

Thymocytes from mice of various inbred strains can be used, however, strain C3H/HeJ is most frequently used as the cells are hyporesponsive to endotoxin (see Appendix I). If thymocytes from other strains are used, the samples can be assayed in the presence of polymyxin B (25 μg/ml) to neutralize any endotoxin present. Polymyxin B is an antibiotic that forms inert non-pyrogenic complexes with bacterial endotoxin but not with protein pyrogens such as IL-1 (6). The mice are usually used within the age range of 6–8 weeks. A typical assay procedure is as follows

(i) Kill mice by CO_2 asphyxiation and under aseptic conditions, remove both lobes of the thymus and place in a Petri dish containing RPMI 1640 cell culture medium.

(ii) Dissociate the thymus gently between two sterile ground glass slides to obtain a single cell suspension.

(iii) Centrifuge the cell suspension at 100 *g* for 10 min and resuspend in fresh culture medium at a concentration of 1×10^7 cells/ml.

(iv) Distribute the cells into the wells of a 96-well flat-bottomed microtitre plate at 1×10^6 cells/well (100 μl). Add standard IL-1 or test samples to triplicate wells in 50 μl volumes and make up the total volume to 200 μl with a sub-optimal concentration of phytohaemagglutinin (PHA) (0.5–4 μg/ml) or concanavalin A (Con A) (0.5–2 μg/ml) in culture medium. Samples should always be assayed in multiple (5- to 10-fold) dilutions (minimum three concentrations — usually five or more) to ensure comparisons are made at an appropriate part of the dose–response relationship. Each dilution of unknown and each concentration of standard IL-1 should be assayed in triplicate and the mean value taken. As in all bioassays each culture plate should include negative controls (cells with medium alone and cells with lectin plus medium) and positive controls (lectin plus standard IL-1). Lectin preparations vary widely in their potency so the optimal concentration for an IL-1 assay should be checked beforehand. See Appendix II for production of an in-house IL-1 standard.

(v) Incubate the cultures for 72 h at 37°C. At 6–24 h before harvesting add tritiated thymidine (0.5 μCi/10 μl/well; sp. act. 5 Ci/mmol) to each well. If short pulse times are used (6 h) then tritiated thymidine of higher specific activity will be needed.

(vi) Harvest the cells on to glass fibre filter discs using an automated cell harvester. Dry the filter discs and assess the incorporated radioactivity by liquid scintillation counting.

The murine thymocyte assay is the most widely used assay for the routine measurement of IL-1 activity at present. It is sensitive to IL-1 at approximately 10–50 pg/ml and is reasonably easy to perform. The main disadvantage of the assay is its lack of specificity; IL-2 also stimulates thymocyte proliferation as a co-mitogen and positive IL-1 samples should also be checked for IL-2 content (see below). As with other assays involving the measurement of cell proliferation the procedure is subject to interference by inhibitors of T lymphocyte proliferation and these should be removed if possible before testing. As a minimum procedure samples for IL-1 testing by T cell prolifera-

Table 1. Proliferative responses of C3H/HeJ thymocytes to human recombinant IL-1 β.

Stimulus	*[^{3}H]Thymidine incorporation (c.p.m.)*[a]
None	331 ± 78
PHA	4545 ± 753
IL-1 50 ng/ml	50 358 ± 3921
5 ng/ml	30 062 ± 1968
0.5 ng/ml	18 305 ± 1228
0.05 ng/ml	9414 ± 1477

[a]Mean ± SEM of triplicate.

tion should be dialysed against sterile, pyrogen-free, phosphate-buffered saline (PBS) to remove low molecular weight inhibitors (e.g. prostaglandins). Typical results of an LAF assay are shown in *Table 1*.

3.2 T lymphocyte clones

3.2.1 *D10.G4.1 proliferation assay*

D10.G4.1 is a cloned murine helper T cell line specific for hen's egg conalbumin (7). The line proliferates minimally to Con A in the absence of IL-1 or feeder cells and is 100- to 1000-fold more sensitive to IL-1 than to IL-2.

D10.G4.1 cells (D10) are maintained in 25 cm^2 tissue culture flasks in RPMI 1640 containing 5% FCS and 60 μM 2-mercaptoethanol at a starting concentration of approximately 1 × 10^5 cells/ml. In addition, D10 propagation requires specific antigen (conalbumin), antigen-presenting cells (feeders) and T cell growth factors (IL-2).

Once a week the cells should be split 2- to 3-fold and fed with fresh medium containing 10 % rat IL-2 (see Chapter 16) and conalbumin (100 μg/ml; Sigma). At this time it is also necessary to prepare inactivated H-2K feeder cells as follows.

(i) Dissociate H-2K spleen (i.e. from C3H mice) into a single cell suspension in PBS between two sterile ground glass slides.
(ii) Lyse the red blood cells by hypotonic shock.
(iii) Wash the cells by centrifugation at 100 *g* for 10 min and resuspend them at 2 × 10^7 cells/ml in PBS containing mitomycin C (50 μg/ml; Sigma) and incubate at 37°C for 1 h.
(iv) Wash the cells by centrifugation four times in PBS to remove the mitomycin C and add to D10.G4.1 cultures at a final concentration of 5–10 × 10^5 cells/ml.

Between the weekly addition of feeder cells, culture medium should be replaced with fresh medium containing IL-2 (3–4 day intervals). D10 cells can survive for at least 2 weeks without feeder cells in medium. To assay for IL-1 the cells should be used at least 8 days after addition of feeder cells and at least 3 days after addition of IL-2. The assay procedure is as follows.

(i) Resuspend D10 cells at 2 × 10^5 cells/ml and dispense 100 μl volumes into 96-well flat-bottomed microtitre plates. IL-1 samples (triplicates) are added at appropriate dilutions in 50 μl together with 50 μl of a sub-optimal concentration of Con A. Incubate the cultures at 37°C for 48 h.

Table 2. Proliferative response of D10.G4.1 to human recombinant IL-1 β.

Stimulus	*[^{3}H]Thymidine incorporation (c.p.m.)*[a]
None	250 ± 18
Con A	336 ± 27
IL-1 100 ng/ml	37 562 ± 2652
10 ng/ml	85 186 ± 5428
1 ng/ml	80 543 ± 4826
0.1 ng/ml	46 838 ± 1957
10 pg/ml	23 931 ± 1542
1 pg/ml	5629 ± 753

[a]Mean ± SEM of triplicates.

(ii) Pulse each culture with tritiated thymidine (0.5 μCi/10 μl/well; sp. act. 5 Ci/mmol) and re-incubate for 18–20 h at 37°C.

(iii) Harvest the cells on to glass fibre filter discs and assess the incorporated thymidine uptake by liquid scintillation counting.

The D10.G4.1 proliferation assay has several advantages over the thymocyte assay; the sensitivity of the assay is greatly increased and it is relatively insensitive to IL-2. However, the maintenance of the cell line is very time-consuming and relatively expensive. Typical results are given in *Table 2*.

A subclone of D10.G4.1 has recently been described, known as D10.S (8) which, in the absence of mitogen, proliferates in response to IL-1 in the attogram/ml range. At this concentration IL-1 is present at approximately 5–10 molecules per cell.

3.2.2 *LBRM-33-1A5 conversion assay*

Several IL-1 assays have been developed using murine T cell lines that only produce substantial quantities of IL-2 in response to IL-1, the IL-2 produced is proportional to the IL-1 concentration in the stimulating sample and is assayed using an IL-2-dependent T cell line such as CTLL-2 (9).

The LBRM-33-1A5 conversion assay is based on the principle that IL-1 converts the line from an IL-2 non-producer to an IL-2 producer. LBRM-33-1A5 (available from ATCC) is a subclone of the radiation-induced splenic lymphoma LBRM-33 (10). Addition of metabolically inactivated LBRM-33-1A5 cells with IL-1 and mitogen to the CTLL-2 cell line results in a 24 h IL-1 assay which is several times more sensitive than the murine thymocyte assay.

Maintain LBRM-33-1A5 cells in RPMI 1640 supplemented with 10% FCS and 2.5×10^{-5} M 2-mercaptoethanol at 5×10^5 cells/ml. Long-term passage in culture (>4 months) leads to the emergence of a cell phenotype that produces IL-2 in response to mitogen alone. LBRM-33-1A5 dependence of IL-1 for IL-2 production can be maintained by regular subcloning every 3 months. Alternatively fresh cells can be recovered from cryopreserved stores. CTLL-2 are maintained as described in Chapter 16. Assay IL-1 as follows.

(i) Resuspend LBRM-33-1A5 cells at 5×10^6 cells/ml in RPMI 1640 culture

Table 3. Typical data from LBRM-33-1A5 conversion assay using recombinant IL-1 β.

Stimulus	*[³H]Thymidine incorporation (c.p.m.)*[a]
Medium	253 ± 105
PHA	1659 ± 164
IL-1 50 pg/ml	15 523 ± 862
5.0 pg/ml	13 797 ± 899
0.5 pg/ml	11 642 ± 531
50 fg/ml	5433 ± 269

[a]Mean ± SEM of triplicates.

medium containing mitomycin C (50 μg/ml) and incubate at 37°C for 1 h.

(ii) Wash LBRM-33-1A5 four times by centrifugation at 100 *g* for 10 min at room temperature resuspending each time in fresh medium.

(iii) Resuspend the LBRM-33-1A5 cells at 5 × 10^5 cells/ml and add 100 μl of the cell suspension to the wells of a 96-well flat-bottomed microtitre plate together with an equal volume of IL-1 sample and a suboptimal concentration of PHA.

(iv) Wash the CTLL-2 cells free of growth medium and resuspend at 8 × 10^4 cells/ml. Add 50 μl of CTLL-2 cell suspension to LBRM-33-1A5 cultures and incubate at 37°C for 20 h.

(v) Pulse the cultures with tritiated thymidine (0.5 μCi/10 μl/well; sp. act. 25 Ci/mmol) for 4 h and harvest the cultures onto glass fibre filter discs using an automated cell harvester.

(vi) Dry the filter discs and assess the incorporated radioactivity by liquid scintillation counting.

The advantage of using this assay is its increased sensitivity, (100- to 1000-fold more sensitive than the thymocyte assay) and its relatively short incubation period. However, as with the thymocyte assay, IL-2 is also active and positive samples should be screened for IL-2 activity in the CTLL assay. The assay does not involve macrophages so it is insensitive to endotoxin. Typical results from the LBRM-33-1A5 conversion assay are shown in *Table 3*.

A thioguanine-resistant subclone mutant of the LBRM-33-1A5 cell line has been isolated (11) known as LBRM TG-6. These cells do not require metabolic inactivation as they are sensitive to hypoxanthine and azaserine. Essentially, the assay is carried out as for the LBRM-33-1A5 assay but with elimination of the mitomycin C step. Four hours before the addition of the tritiated thymidine, hypoxanthine (10^{-4} M; final concentration) and azaserine (10 μg/ml; final concentration) are added to the cultures. This reduces LBRM TG-6 thymidine incorporation by 99% without affecting the proliferation of the IL-2-dependent cell line.

3.2.3 *EL-4 cell line assay*

EL-4 is a murine T cell thymoma (available from ATCC) that can also be used in conjunction with an IL-2-dependent cell line to assay IL-1 (12).

EL-4 are maintained in continuous culture in RPMI 1640 supplemented with 10% FCS at between 5 and 10 × 10^5 cells/ml. The CTLL-2 IL-2-dependent cell line is maintained as before. The assay is performed in two stages; initially EL-4 cells are stimulated with IL-1 and the calcium ionophore A23187 causing release of IL-2 into the medium. Medium is then taken from these cultures and added to CTLL-2 cells and their proliferation assessed by tritiated thymidine incorporation. A detailed assay protocol is as follows.

(i) Prepare the calcium ionophore, A23187 as a 1×10^{-3} M stock solution in dimethyl sulphoxide (DMSO) and store at -20°C.
(ii) Add 100 μl of serial 2-fold dilutions of IL-1 samples in triplicate to the wells of a 96-well flat-bottomed microtiter plate.
(iii) Dilute the stock solution of A23187 to 1×10^{-6} M in RPMI 1640 and add 50 μl to each well.
(iv) Wash EL-4 cells by centrifugation at 100 *g* for 10 min twice and resuspend at 2×10^{6} cells/ml. Add 100 μl of EL-4 cell suspension to the wells of the microtitre plate and culture overnight at 37°C.
(v) Remove the top 100 μl of culture fluid from each well and transfer it to a second 96-well microtitre plate.
(vi) Wash the CTLL-2 cells free of growth medium and resuspend them in RPMI 1640 at 1×10^{5} cells/ml. Add 100 μl of CTLL-2 cell suspension to each well of the second microtitre plate and culture for 20 h at 37°C.
(vii) Pulse the cultures with tritiated thymidine (0.5 μCi/10 μl/well; sp. act. 25 Ci/mmol) and incubate for a further 4 h before harvesting onto glass fibre filter discs. Assess the thymidine incorporation by liquid scintillation counting.

The EL-4 cell line assay has approximately the same sensitivity as the LBRM-33-1A5 cell line assay and is insensitive to endotoxin but has the advantage in that the line is phenotypically stable. IL-2 is again active in this assay and positive IL-1 samples should be checked for IL-2 activity.

A modification of the EL-4 cell line assay has recently been reported (13). The subclone EL-4 NOB.1. was derived by maintenance of EL-4 for 6 weeks at high cell density. EL-4 NOB.1. is maintained by RPMI 1640 supplemented with 5% FCS and is grown to a maximum cell density of 5×10^{5} cells/ml before subculturing. EL-4 NOB.1. produces IL-2 in response to IL-1 alone and there is no requirement for mitogen or calcium ionophore. The assay may be performed in two stages as for the EL-4 cell line or as a 24 h one-stage assay. The one stage assay is described here.

(i) Wash EL-4 NOB.1. cells twice by centrifugation at 100 *g* for 10 min and resuspend at 2×10^{6} cells/ml.
(ii) Add 100 μl of EL-4 NOB.1. cell suspension to the wells of a 96-well flat-bottomed microtitre plate, together with 50 μl of appropriate dilutions of the test sample.
(iii) Wash CTLL-2 cells free of growth media and resuspend them at 8×10^{4} cells/ml. Add 50 μl of cell suspension to each well and incubate at 37°C for 20 h.
(iv) Pulse the cultures with tritiated thymidine (0.5 μCi/10 μl/well; sp. act. 25 Ci/mmol) and incubate for 4 h at 37°C.
(v) Harvest the cultures onto glass fibre filter discs and assess the thymidine incorporation by liquid scintillation counting.

Unlike the LBRM-33-1A5 cell line it is not necessary to inactivate the EL-4 NOB.1.

line as the cells do not incorporate significant amounts of thymidine during the assay. The assay is highly sensitive to IL-1 and in this laboratory has responded to as little as 10 fg/ml of human recombinant IL-1 and maximal proliferation is attained with 50 pg/ml IL-1. The assay can also be made unresponsive to IL-2 by incorporating a pre-incubation step in the assay as follows.

(i) Follow steps (i) and (ii) for the EL-4 NOB.1. assay.
(ii) Incubate the plates at 37°C for 4 h.
(iii) Centrifuge the microtitre plates to pellet the cells and remove supernatants. Resuspend cells in 250 μl of RPMI 1640 and repeat the washing procedure four more times.
(iv) Resuspend in 150 μl of RPMI 1640 supplemented with 5% FCS.
(v) Wash the CTLL-2 cells free of growth media and resuspend at 8×10^4 cells/ml. Add 50 μl of cell suspension to each well and incubate for 20 h at 37°C.
(vi) Pulse with tritiated thymidine and harvest the cultures as described above.

Typical results from the EL-4 NOB.1/CTLL-2 assay are shown in *Figure 2*.

3.2.4 *Colorimetric assay for cell proliferation and activation*

Assays involving proliferation or activation of lymphoid cells can be modified so that the results may be read using a multiwell scanning spectrophotometer (ELISA reader) instead of by liquid scintillation counting of incorporated radiolabel. The colorimetric assay as described by Mosmann (14) is based on the tetrazolium salt 3-(4,5 dimethylthiazol-2-yl)-2,5-diphenyl tetrazolium bromide (MTT) a pale yellow substrate that is cleaved by active mitochondria to produce a dark blue formazan product. To assay for proliferation or activation, cells are incubated as previously described and then the following protocol is followed.

(i) Dissolve MTT in PBS at 5 mg/ml and sterilize by passage through a 0.22 μm membrane filter. This procedure also serves to remove any insoluble residue present in some batches of MTT.
(ii) Add 10 μl of the MTT solution per 100 μl of culture medium and incubate at 37°C for 4 h.
(iii) Add 100 μl of acid propan-2-ol (0.04 M HCl in propan-2-ol) and mix thoroughly to dissolve the formazan product. Acidification converts the phenol red in the culture medium to its yellow acidic form, this is necessary since the red form interferes at the wavelength most suitable for blue MTT formazan measurement.
(iv) Leave the microtitre plates for a few minutes at room temperature to ensure that all formazan crystals are dissolved.
(v) Read the plates on a microelisa reader, using a test wavelength of 570 nm, a reference wavelength of 630 nm and a calibration setting of 1.99. Plates should be read within 1 h of acid propan-2-ol addition.

The advantages of the colorimetric assay are that large numbers of samples can be read in a short period of time without the sensitivity of the assay being reduced (*Figure 1*). It also reduces costs and potential hazards of radioisotope handling and disposal.

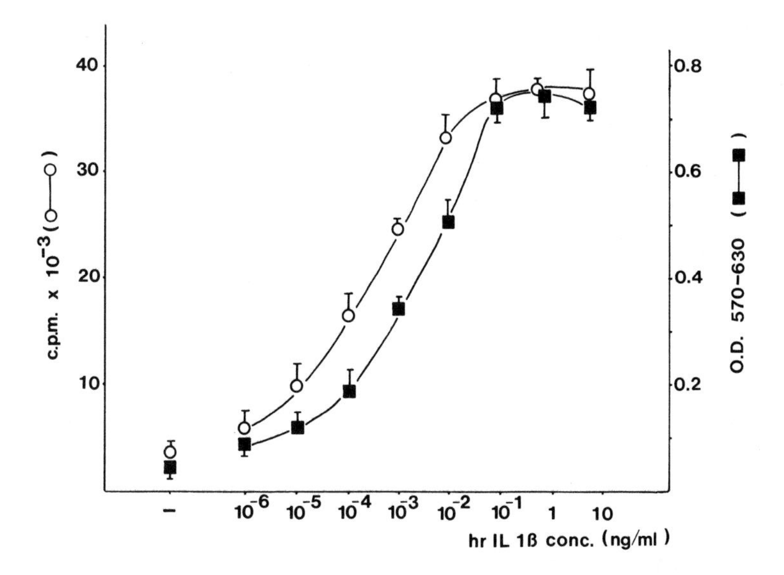

Figure 1. EL4 NOB.1/CTLL-2 assay (two stage) for IL-1. EL4 NOB.1 cells were cultured with increasing concentrations of human recombinant IL-1 β (horizontal axis). At 24 h supernatants were transferred to CTLL-2 cells for 18 h to measure IL-2 content. Cell proliferation was measured by [^{3}H]thymidine uptake during the last 4 h of culture (left vertical axis = c.p.m. $\times 10^{-3}$)or by colorimetric assay with MTT (read at OD 570−630 mM, right vertical axis).

Another colorimetric enzyme assay has been described by Landegren (15) involving the lysosomal enzyme *N*-acetyl-β-D-hexosaminidase, the assay protocol is as follows.

(i) Dissolve hexosaminidase substrate, *p*-nitrophenol, *N*-acetyl-β-D-glucosaminide at 7.5 mM in 0.1 M citrate buffer, pH 5 and mix with an equal volume of 0.5% Triton X-100 in water. Aliquot and store at −20°C.

(ii) Terminate cell cultures by centrifugation of microtitre plates to pellet the cells, remove supernatants and resuspend in PBS. Centrifuge the plates a second time and again remove the supernatant.

(iii) Add 60 μl of enzyme substrate and incubate at 37°C for 1−4 h.

(iv) Develop the colour reaction and stop the enzyme activity by adding 90 μl of 50 mM glycine buffer, pH 10.4 containing 5 mM EDTA.

(v) Measure the absorbance at 405 nm in a microelisa reader.

The enzyme is quite stable, and plates can be stored frozen for a number of days. Enzyme activity is also constant over time and incubation times can be extended for up to 16 h. The PBS washing step is essential as serum contains high levels of hexosaminidase.

4. IN VITRO ASSAYS EMPLOYING NON-LYMPHOID CELLS

As previously discussed IL-1 acts on numerous non-lymphoid cell types. These properties have been utilized to develop assays for IL-1 activity.

4.1 Fibroblast assays

4.1.1 *Fibroblast proliferation*

In many chronic inflammatory diseases, fibrosis of the diseased tissue occurs and this is often present in the same area as the infiltrating mononuclear cells. Although *in vivo* information is lacking, *in vitro* studies have shown that IL-1 is a potent stimulator of fibroblast proliferation (16). Fibroblasts from many sources can be used in IL-1 assays although most studies employ early passage human dermal fibroblasts obtained either from newborn human foreskin preparations or cell lines from the ATCC (e.g. CRL 1445).

Fibroblasts are grown in 25 cm^2 flasks in Dulbecco's modified Eagle's medium (DMEM) containing 10% FCS, antibiotics and glutamine (fibroblast culture medium). The cells are fed twice a week with fresh medium and sub-cultured 1:3 approximately every 10 days. Fibroblasts should be used for the assay in early passage, that is between passage one and five, as older cells lose their responsiveness to IL-1. A typical fibroblast proliferation assay is as follows.

(i) Remove the fibroblast culture medium from the flask and briefly rinse the cell monolayer with PBS.

(ii) Treat the fibroblasts with a solution containing trypsin (0.5 mg/ml) and EDTA (0.2 mg/ml) for 5–10 min at 37°C.

(iii) Add an equal volume of fibroblast culture medium and wash the resulting single cell suspension by centrifugation at 100 *g* for 10 min. Resuspend the fibroblasts at 1×10^5 cells/ml.

(iv) Dispense 100 μl of the fibroblast cell suspension into the wells of a 96-well flat-bottomed microtitre plate and incubate overnight at 37°C.

(v) Gently aspirate the medium from the wells and replace it with 100 μl of appropriate dilutions of the IL-1 samples and re-incubate the plates at 37°C for 48 h.

(vi) Pulse the cultures with tritiated thymidine (0.5 μCi/10 μl/well; sp. act. 5 Ci/mmol) for 18–20 h.

(vii) Before harvesting aspirate the medium from each well and replace it with 100 μl of trypsin–EDTA solution. Incubate at 37°C for 10–15 min, harvest onto glass fibre filter discs and assess the incorporated thymidine by liquid scintillation counting.

Although the fibroblast proliferation assay is insensitive to IL-2 and endotoxin, it is now known that there are many factors distinct from IL-1 that stimulate fibroblast proliferation.

4.1.2 *Production of fibroblast prostaglandin*

Fibroblasts derived from primary cell culture or established fibroblast lines may be used for this assay which measures the induction of prostaglandin E_2 (PGE_2) synthesis in fibroblasts by IL-1 (17). Maintain fibroblasts in continuous culture as previously described for the fibroblast proliferation assay. The assay is then carried out as follows.

(i) Seed the fibroblasts in the wells of a 24-well multiwell plate and grow to confluency in fibroblast culture medium supplemented with 5% FCS.

(ii) Twenty four hours before addition of IL-1, wash the cell monolayers and in-

Table 4. Production of PGE_2 by dermal fibroblasts in response to human recombinant IL-1 β.

Stimulus	*PGE_2 (pg/0.1 ml)*[a]
Medium	Not detectable
IL-1 100 ng/ml	562 ± 29
10 ng/ml	351 ± 16
1 ng/ml	129 ± 26
0.1 ng/ml	65 ± 12

[a]As determined by radioimmunoassay.

cubate overnight in fibroblast culture medium without serum.

(iii) Aspirate the medium and replace it with appropriate dilutions of IL-1 samples in serum-free media. Incubate at 37°C for 24 h.

(iv) Remove the fibroblast supernatants and store at −70°C.

(v) Assay for PGE_2 using a radioimmunoassay (RIA) kit (e.g. Seragen Inc.).

As for the fibroblast proliferation assay, this system is highly insensitive to the effects of endotoxin. It will however, yield positive results with supernatants containing other fibroblast-activating factors. The assay is relatively insensitive and will only detect 0.5 ng/ml of IL-1; typical results are shown in *Table 4*.

4.2 Assays involving bone, cartilage and synovium

The skeletal tissue activities of IL-1 suggest that it could be a common mediator of many pathological changes that occur in rheumatoid arthritis. These include fibroblast proliferation, prostaglandin release and resorption of cartilage and bone. A number of assays have therefore arisen which measure the effects of IL-1 on synovium, cartilage and bone.

4.2.1 *Proteoglycan release from cartilage*

Cartilage explants grown in suitable culture conditions synthesize extracellular matrix components representative of those found in normal tissue. During culture a stable equilibrium is established where proteoglycan breakdown and chondrocyte proteoglycan synthesis are in balance. IL-1 has been shown to induce chondrocytes to break down cartilage which is measured by chondroitin sulphate release (18). A brief protocol for this, the 'catabolin' assay, is as follows.

(i) Obtain nasal septum cartilage from recently slaughtered 1 year old cows.

(ii) Cut cartilage discs measuring 2 mm diameter up to 1−2 mm thick and incubate in DMEM containing 5% FCS and polymyxin B (20 μg/ml) in 24-well multiwell plates at 37°C for 5 days.

(iii) Transfer the discs to a 96-well flat-bottomed microtitre plate (one disc per well) and incubate in 150 μl of medium containing appropriate dilutions of the sample under test (in quadruplicate) for 16−40 h at 37°C.

(iv) Remove the media and measure the chondroitin sulphate concentration by adding 100 μl of culture supernatant to 4 ml of 1,9-dimethyl-methylene blue (Serva) solution (16 mg of dye plus 5.0 ml ethanol in 1 litre of sodium formate buffer, 0.1 M, pH 3.5).

(v) Read the absorbance at 535 nm immediately and compare with a standard curve of shark fin chondroitin sulphate (range 0–50 μg/25 μl).

Human articular cartilage is resistant to IL-1-induced proteoglycan loss, but IL-1-induced proteoglycan breakdown has been demonstrated *in vivo* in rabbits following intra-articular IL-1 injection (19). In addition to accelerating proteoglycan breakdown, IL-1 has modulating effects on new proteoglycan synthesis.

4.2.2 *Bone resorption*

Bone resorption is regulated by a number of humoral factors, such as parathyroid hormone, 1,25-dihydroxy vitamin D_3, prostaglandins of the E series and, more recently, IL-1. The cell type thought to be primarily responsible for bone resorption is the osteoclast but studies with monolayer cultures of osteoblast-enriched populations of bone cells have suggested that these cells may also make a significant contribution to resorption by providing signals to osteoclasts. The assay used to measure bone resorption is the $^{45}Ca^{2+}$-labelled mouse calvarial system (20) which is described below.

(i) Label the skeletons of mice aged 1–2 days by subcutaneous injection of ^{45}Ca (1 μCi [^{45}Ca]$CaCl_2$; sp. act. 16.8 μCi/mg Ca^{2+}) 4 days before experiments.
(ii) Obtain paired explants of half calvariae by micro-dissection and pre-culture for 24 h in DMEM to remove non-incorporated ^{45}Ca.
(iii) Transfer the explants to 0.5 ml of DMEM/0.2% bovine serum albumin (BSA) containing IL-1 or control medium and culture for 3 days at 37°C.
(iv) Remove the calvaria and dry in a desiccator. Dissolve the bone in 0.5 ml of 90% formic acid at 60°C and add to the scintillation cocktail. Assess the retained ^{45}Ca by scintillation counting.
(v) Add 0.5 ml of supernatant to the scintillant and assess the released ^{45}Ca as above. The percentage release of ^{45}Ca is determined using the following equation.

$$\% \text{ release} = \text{medium c.p.m.}/(\text{medium c.p.m.} + \text{bone c.p.m.}) \times 100$$

All assays should be performed in quadruplicate using bones from a number of different mice. Some release of ^{45}Ca into the medium is due to passive exchange with the cold Ca^{2+} in the medium. The cell-mediated resorption can be calculated by subtracting the ^{45}Ca released from dead bones, killed by three cycles of freezing and thawing. The assay can detect approximately 1 ng/ml of IL-1 and typical results are shown in *Table 5*.

Table 5. Bone resorption induced by human recombinant IL-1.

Stimulus	*% Release of* ^{45}Ca
Medium	20
IL-1 100 ng/ml	80
10 ng/ml	55
1 ng/ml	25
0.1 ng/ml	20

4.2.3 *Induction of collagenase release from synovium*

Isolated human adherent synovial cells obtained from rheumatoid synovectomy samples produce large amounts of collagenase and prostaglandins in primary culture. The phenotype of these cells appears to be dendritic/fibroblastoid (synoviocytes). They do not possess markers of monocytic or lymphoid cells and are different from normal dermal fibroblasts in that they produce less collagen per cell. The ability of synoviocytes to produce collagenase and prostaglandins *in vitro* declines with time in culture but can be restored with the addition of IL-1 (21). Preparation of the synoviocytes is as follows.

(i) Obtain human synovial tissue from rheumatoid arthritis patients at surgery and within 2 h wash the sample three times in PBS.
(ii) Dissect 2 mm^3 pieces of the superficial layer of synovium and place them in serum-free DMEM. Prepare a cell suspension by sequential treatment with proteolytic enzymes.
(iii) Incubate the synovium for 3–4 h at 37°C with 1 mg/ml bacterial collagenase (Sigma) and 10 μg/ml testicular hyaluronidase with gentle shaking. Mix the digest well by aspiration with a Pasteur pipette.
(iv) Add an equal volume of trypsin/EDTA solution [0.05%/0.02% (v/v)] and carry out incubation for a further 1 h.
(v) Centrifuge the suspension at 400 *g* for 10 min and wash three times in PBS. Resuspend the pellet in 20–40 ml of DMEM/10% FCS and dispense 2 ml aliquots in 60 mm Petri dishes. Culture for 24 h at 37°C.
(vi) Wash the cell monolayers vigorously with PBS three times to remove non-adherent cells. Replace with 2 ml of fresh medium and maintain as described for dermal fibroblasts (Section 4.1.1).

To assay for IL-induced collagenase/PGE production, second passage synoviocytes are used. At the second passage basal collagenase/PGE production is usually lower than in the primary cultures and allows the measurement of an IL-1-induced response.

(i) Obtain second passage synoviocytes by trypsin/EDTA treatment, centrifuge the cells and resuspend them in medium containing 1% FCS.
(ii) Dispense the synoviocytes into 16 mm diameter wells at 1×10^5 cells/well and incubate for 3–6 days at 37°C.
(iii) Remove the medium and replace it with aliquots of test material. Incubate for a further 3 days and obtain cell-free supernatant for collagenase/PGE assay.

Prostaglandin E_2 is assayed by RIA as described previously (Section 4.1.2). Collagenase activity is measured by following the release of ^{14}C-labelled peptides from reconstituted trypsin-resistant fibrils of ^{14}C-labelled collagen as described below.

Collagen is extracted and purified from rat skins as described by Cawston and Barrett (22). The collagen is then acetylated with [^{14}C]acetic anhydride as follows.

(i) Dissolve collagen (250 mg) in 50 ml of 0.2 M acetic acid at 4°C and dialyse against 10 mM disodium tetraborate adjusted to pH 9.0 with NaOH and containing 200 mM $CaCl_2$.

(ii) Place the collagen in a conical flask at 4°C and slowly stir with a large magnetic bar.
(iii) Condense [1-^{14}C]acetic anhydride (250 μCi) by placing in dry ice, add 1 ml of dry dioxane to the vial and add the mixture to the collagen.
(iv) Stir the collagen mixture for 30 min at 4°C and then dialyse at 4°C against 50 mM Tris-HCl, pH 7.6, containing 200 mM NaCl, 5 mM calcium acetate and 0.03% toluene until the radioactivity in the diffusate has fallen to background levels.
(v) Dilute the labelled collagen to 1 mg/ml with 0.2 M acetic acid and then further dilute with unlabelled collagen (1 mg/ml) until the collagen contains approximately 100 000 d.p.m./mg. Store at −20°C.
(vi) Before use in the collagenase assay thaw and dialyse against 50 mM Tris-HCl, pH 7.6 containing 200 mM NaCl and 0.03% toluene.

The resulting labelled collagen can then be used to asssess collagenase activity utilizing the assay procedure of Cawston and Barrett (22) as described below.

(i) Add 100 μl of collagenase solution to 100 μl of 100 mM Tris-HCl buffer, pH 7.6 containing 15 mM $CaCl_2$ in a microcentrifuge tube. In each experiment include buffer and trypsin (10 μg) controls.
(ii) Determine the total lysis of collagen by adding 10 μg of clostridial collagenase to the control tubes.
(iii) Add 100 μl of acetylated collagen, prepared as above, to each tube and incubate in a shaking water bath at 37°C for 20 h.
(iv) Centrifuge the tubes at 10 000 *g* for 15 min to remove undigested collagen fibrils. Remove 200 μl of each supernatant and assess ^{14}C levels by liquid scintillation counting.

Alternatively an RIA for collagenase is now commercially available.

4.3 Induction of PGE_2 synthesis in brain cells

IL-1 exerts its pyrogenic activity by an action on the hypothalamus associated with the induction of PGE_2 synthesis. This property has been developed into an assay system using rabbit hypothalamus (23).

(i) Kill rabbits by rapid CO_2 asphyxiation, promptly dissect the brain and place in cold (4°C) minimal essential medium (MEM). Dissect the optic chiasm away and remove the anterior hypothalamus.
(ii) Mince the hypothalamus into 0.5 mm^3 fragments and suspend in cold MEM containing polymyxin B (10 μg/ml) at a concentration of 10 mg tissue /ml.
(iii) Aliquot 1 ml portions into polypropylene test tubes and incubate at 37°C for 30 min.
(iv) Wash the tissue with 3 ml of warm MEM and centrifuge at 100 *g* for 5 min to pellet the tissue.
(v) Resuspend the tissue in duplicate 1 ml volumes of fresh MEM containing an appropriate concentration of the test sample previously incubated at 37°C for 2 h with polymyxin B (100 μg/ml) prior to dilution. Incubate the tubes for 1 h at 37°C in a shaking water bath.

(vi) Centrifuge at 100 *g* for 5 min, collect the supernatants and store them at −70°C.
(vii) Assay the supernatants for PGE_2 using an RIA kit (Seragen Inc.).

This assay is highly sensitive to the effects of endotoxin, hence the use of polymyxin B throughout the assay (Appendix I). This system will obviously not be used for routine testing of IL-1 but rather in specific studies on central nervous system effects of IL-1. *In vivo* assays for pyrogenic activity of IL-1 are described later.

5. SOLID PHASE IL-1 ASSAYS

5.1 **Radioimmunoassay (RIA)**

At the time of writing the only solid phase assay for IL-1 available is an RIA initially described by Dinarello *et al.* (24).

The assay is based on competition between the test sample and iodinated IL-1 for a limited number of binding sites on a polyclonal anti-IL-1 immobilized on Sepharose 4B (Pharmacia). A brief summary of the assay procedure is presented below.

(i) Attach purified anti-IL-1 to Sepharose 4B and suspend in PBS.
(ii) Add the test material and incubate at room temperature for 24 h.
(iii) Add ^{125}I-labelled IL-1 to each tube and incubate for a further 24 h.
(iv) Centrifuge the tubes at 1000 *g* for 10 min, remove the supernatant and wash once with water.
(v) Determine the radioactivity in the supernatant, wash and pellet by gamma counting.

An RIA for IL-1 β based on the above technique is now commercially available from Cistron Biotechnology. The sensitivity of the assay is 250 pg/ml which, in comparison with bioassay systems, is greatly reduced. Typical results are shown in *Table 6*. However, one advantage of the immunoassay is that it is possible to measure concentrations in samples containing interfering factors that would otherwise mask IL-1 bioactivities. Another potential advantage of immunoassay is that it is possible to measure one form of IL-1 (i.e. either α or β gene products) a specificity that is, at present, not possible with any bioassay. A possible disadvantage of immunoassays is that the antibody may recognize epitopes of the molecule that are not necessarily relevant to

Table 6. Typical standard curve from IL-1 β RIA.

IL-1 Concentration	*c.p.m.*
Total counts	15 822
NSB	1258
5 ng/ml	2253
2 ng/ml	3552
1 ng/ml	4656
500 pg/ml	5522
250 pg/ml	5711
Zero standard	6059

biological activity. This problem may be overcome in the future by the use of monoclonal antibodies directed at the active site of IL-1.

6. IN VIVO ASSAYS OF IL-1 ACTIVITY

In vivo assays are based on the systemic effects of IL-1 and, in particular, the induction of fever, hepatocyte acute phase protein synthesis and changes in blood iron and zinc levels following IL-1 injection in animals.

6.1 **Pyrogen assays**

Fever is recognized in higher animals as a universal response to infection and inflammation and is a regulated hyperthermia following elevated heat production and reduced heat loss. IL-1 increases the 'reference point' of core temperature regulation by an action on thermoregulatory structures associated with the hypothalamus. The adaptive value of fever (which is highly conserved in evolution) is likely to be related to large increases in T cell activation and proliferation that occur at temperatures up to 39°C. This may potentiate cell-mediated immunity and lead to rapid clearance of infectious agents (24).

6.1.1 *Mouse pyrogen assay*

For this assay 8−10 week old female mice, preferably an endotoxin-insensitive strain such as C3H/HeJ, should be used (26,25). Since mice are small and have large surface area to mass ratios they maintain their core temperature at normal laboratory temperatures by activating heat conservation and heat production mechanisms. Thus it may not be possible to generate significant fever at prevailing ambient temperatures if thermoeffector mechanisms are already employed.

To overcome this, put mice in small individual cages in an incubator maintained at 32−34°C and remove mice briefly every 10 min for rectal temperature recordings using a 1−2 mm diameter thermistor probe. Once a stable base line temperature is established (after 30−60 min) inject each mouse with 0.1−0.3 ml of test substance via a tail vein and monitor its temperature for 1 h at 10 min intervals. An occasional mouse will respond to injection with a fall in temperature. Such a response is usually regarded as an abnormal reaction to the injection and data from the individual mouse is excluded. In general, peak fever is obtained 20−30 min after injection of pyrogen. Typical results are illustrated in *Figure 2*. This is a sensitive assay for IL-1 compared with the rabbit pyrogen test, but it requires considerable operator experience of mouse-handling and is both laborious and time-consuming. Modern techniques of chronic implantation of temperature-sensitive radioemitters that can be monitored at a distance allow much more temperature data to be collected from unrestrained animals.

6.1.2 *Rabbit pyrogen assay*

Usually outbred female New Zealand White rabbits are used though other breeds are suitable. Rabbits (2.5 kg) should be trained to sit quietly in restraining stocks with an in-dwelling rectal thermistor continuously recording deep body temperatures. Daily sitting times in the restraining stocks can be increased up to 4−5 h after 5 days. In-

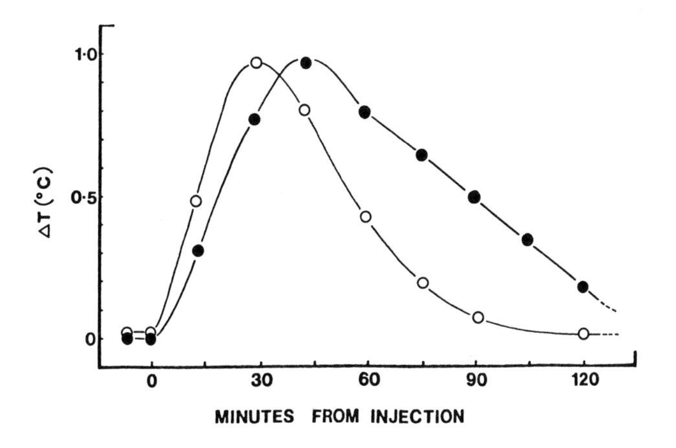

Figure 2. Rabbit and mouse pyrogen assay for IL-1. Groups of 3–6 animals are injected i.v. (mice will also respond to higher doses given i.p.) while rectal temperature is continuously recorded or recorded at regular intervals. The vertical axis shows change in temperature from baseline. The response can be assessed either as peak fever height or as area under the curve of temperature displacement ('fever index'). Mice (○-○) should be kept at high ambient temperature and rabbits (●-●) give best results when the pyrogen test is performed in a temperature controlled room (18–21°C).

dividual rabbits that are intolerant of this procedure or whose body temperature is unstable should not be used for pyrogen assays. Material to be tested for pyrogenicity is injected into the marginal vein of the ear in a volume of 0.5–2.5 ml (27).

Before injection preparations can be incubated for 2 h at 37°C with polymyxin B (100 μg/ml) to neutralize any contaminating endotoxin. Determine baseline temperatures by computing the mean value of body temperature 30 min before the intravenous injection and 10 min after the temperature response. Rabbits whose baseline temperature varies more than 0.3°C should not be used in the assay. Following injection of sample, IL-1 fever is characterized by a latency of about 10–15 min, a monophasic peak about 45 min and a return to baseline temperature by 120 min.

The mouse pyrogen assay has several advantages over the rabbit assay. First, use of endotoxin-resistant mice helps ensure that fever responses are not due to contaminating endotoxin (Appendix I). Second, the volume of test material used is greatly reduced and mice appear to be about 100-fold more sensitive to the pyrogenic effects of IL-1 than rabbits. The use of mice also ensures the availability of large numbers of replicates and unlike rabbits they do not have to be trained for the assay. However, the mouse assay is not very quantitative except within a narrow range of pyrogen concentration. The rabbit assay, on the other hand, has a range over about a 10-fold increase in pyrogen concentration (induced temperature changes from 0.2 to 0.9°C are linearly related to amount of pyrogen injected). Both assays lack specificity for IL-1 since it is now well established that interferon α (28) and tumor necrosis factor α (29) are both pyrogenic *in vivo*.

6.2 Induction of acute phase proteins

Acute phase proteins are synthesized by hepatocytes in response to infectious or inflammatory stimuli. They can be elevated in both short-lived as well as chronic infectious or inflammatory diseases. Two general categories of acute phase reactants have been described:

(1) those normally present at easily detectable levels but increased in concentration several fold following a febrile response;
(2) those normally present in minute quantities only but which increase up to 100-fold during infection or inflammation.

Examples of the first category include serum amyloid A (SAA), ceruloplasmin and fibrinogen, while C-reactive protein is an example of the second category. Serum amyloid P (SAP) is the major acute-phase reactant of mice and is the molecular homologue of CRP.

6.2.1 *Murine SAA and SAP*

Serum is obtained from mice by retroorbital or tail vein bleeding and test material is then assayed for pyrogenic activity as previously described or simply injected without temperature monitoring. Blood samples are taken again at 24 h, 48 h and 72 h after the injection and serum stored at 4°C. Serum may be tested for SAA using an RIA as described by Sipe *et al.* (30) or for SAP by rocket immunoelectrophoresis (31).

6.2.2 *Rabbit CRP*

Serum is obtained by ear arterial puncture and the animals then injected with test material. Twenty four hours later the serum is again obtained and both pre- and post-injection CRP levels determined by a radial immunodiffusion assay (32) as follows.

(i) Mix cross-reactive anti-human CRP antibody (Atlantic Antibodies) with sufficient 1.2% agarose (previously warmed to 56°C) to achieve a final dilution of 1:40.
(ii) Add 10 ml of agarose/antibody solution to plastic Petri dishes and allow this to solidify on a flat surface.
(iii) Punch wells into the agarose and add 5 μl of sera or a dilution of human CRP standard. Assay all sera in triplicate.
(iv) Allow precipitin rings to develop for 48 h at 4°C.
(v) Relate the diameter of the precipitin ring to the known CRP concentration.

7. CONCLUSIONS

IL-1 acts on a wide range of cell types and this is reflected in the variety of assays that have been developed to measure its activity. Many of the assay systems described could not be used routinely to screen a large number of samples. However, they are useful for demonstrating a relevant activity of the sample, for example, synovial fluid-mediated cartilage breakdown in rheumatoid arthritis. Many of the assays involving non-lymphoid cells are also stimulated by tumour necrosis factor (TNF) which is probably present in a large number of crude IL-1-containing supernatants, therefore the specificity of these assays is poor unless advantage is taken of the increasing availability of neutralizing antibodies specific for individual cytokines. Also as the action of IL-1 in some systems is often direct rather than inducing a lymphokine cascade, some assays are significantly less sensitive than those based on lymphocyte proliferation. IL-1 assays based on T cell lines are insensitive to TNF (33) but thymocyte assays give false positive readings because of TNF induction of IL-1 from macrophages.

T cell proliferation assays remain the most sensitive and easiest to perform routinely. For this reason the murine thymocyte assay is still the most frequently used system and is something of a 'gold standard' in IL-1 measurement. Contamination of samples with IL-2 is a significant problem with these assays, but techniques are available to overcome this (Section 3.2.3). Problems also occur due to the presence of inhibitors both in cell supernatants and in clinical samples. As yet these are largely undefined factors but their removal is essential if meaningful results are to be obtained.

This difficulty can largely be overcome by the use of an immunoassay such as the RIA currently available (Section 5.1) but the RIAs are significantly less sensitive than other assays. The sensitivity and convenience of these assays will no doubt increase with the development of better antibodies and the use of ELISA techniques. Double antibody immunoradiometric assays are being developed now and it is likely that assays based on receptor protein binding of IL-1 will be available in future. However, with the diverse nature of IL-1 effects and the apparent molecular heterogeneity of IL-1 gene products (to make no mention of multiple genes) (34) it is certain that evaluation of the role of IL-1 in biological processes will require both antibody (or receptor) binding assays and the continued use of bioassays.

8. ACKNOWLEDGEMENTS

We are very grateful to Miss Anne Mitchell for her expert help in the preparation of this manuscript.

9. REFERENCES

1. Gery,L., Gershon,R.K. and Waksmann,B.H. (1972) *J. Exp. Med.*, **136**, 128.
2. Dinarello,C.A. (1984) *Rev. Infect. Dis.*, **6**, 51.
3. Dinarello,C.A. (1986) *The Year in Immunology*. Vol. 2, Karger, Basel.
4. Lomedico,P.T., Kilian,P.L., Gubler,U., Stern,A.S. and Chizzonite,R. (1986). *Cold Spring Harbor Symp. Quant. Biol.*, **51**, 631.
5. Dower,S.K. and Urdal,D.L. (1987) *Immunol. Today*, **8**, 46.
6. Duff,G.W. and Atkins,E., (1982) *J. Immunol. Methods*, **52**, 333.
7. Kaye,J., Gillis,S., Mizel,S.B., Shevach,E.M., Malek,T.R., Dinarello,C.A., Lachmann,L.B. and Janeway,C.A. (1984) *J. Immunol.*, **133**, 1339.
8. Orencole,S.F., Ikejima,T., Cannon,J.G., Lonnemann,G., Saijo,T. and Dinarello,C.A. (1987) *Lymphokine Res.*, **6**, 1210A.
9. Gillis,S., Fern,M.M., Ou,W. and Smith,K.A. (1978) *J. Immunol.*, **120**, 2027.
10. Conlon,P.J. (1983) *J. Immunol.*, **131**, 1280.
11. Larrick,J.W., Brindley,L. and Dole,M.V. (1985) *J. Immunol. Methods*, **79**, 39.
12. Simon,P.L., Laydon,J.T. and Lee,J.C. (1985) *J. Immunol. Methods*, **84**, 85.
13. Gearing,A.J.H., Bird,C.R., Bristow,A., Poole,S. and Thorpe,R. (1987) *J. Immunol. Methods*, **99**, 7.
14. Mosmann,T. (1983) *J. Immunol. Methods*, **65**, 55.
15. Landegren,V. (1984) *J. Immunol. Methods*, **67**, 379.
16. Schmidt,J.A., Mizel,S.B., Cohen,D. and Green,L. (1982) *J. Immunol.*, **128**, 2177.
17. Bernheim,H.A. and Dinarello,C.A. (1985) *Br. J. Rheumatol.*, **24**, (Suppl. 1), 122.
18. Saklatvala,J., Curry,V.A. and Sarsfield,S.J. (1983) *Biochem. J.*, **215**, 385.
19. Pettipher,B.R., Higgs,G.A., Henderson,B. (1986) *Proc. Natl. Acad. Sci. USA*, **83**, 8749.
20. Gowen,M., Wood,D.D., Ihrie,E.J., McGuire,M.K.B. and Russell, R.G.G. (1983) *Nature*, **306**, 378.
21. Mizel,S.B., Dayer,J.M., Krane,S.M. and Mergenhagen,S.E. (1981) *Proc. Natl. Acad. Sci. USA*, **78**, 2474.

22. Cawston,T.E. and Barrett,A.J. (1979) *Anal. Biochem.*, **99**, 340.
23. Dinarello,C.A. and Bernheim,H. (1981) *J. Neurochem*, **37**, 702.
24. Dinarello,C.A., Renfer,L. and Wolff,S.M., (1977) *Proc. Natl. Acad. Sci. USA* **74**, 4624.
25. Duff,G.W. and Durum,S.K. (1983) *Nature*, **304**, 449.
26. Bodel,P.T. and Hollingsworth,J.W. (1968) *Br. J. Exp. Pathol.*, **49**, 11.
27. Duff,G.W. and Atkins,E. (1982) *J. Immunol. Methods*, **52**, 323.
28. Dinarello,C.A., Bernheim,H.A., Duff,G.W., Lee,H.V., Nagabhushan,T.L. and Coceani,F. (1984) *J. Clin. Invest.*, **74**, 906.
29. Dinarello,C.A., Cannon,J.G., Wolff,S.M., Bernheim,H.A., Beutler,B., Cerami,A., Figari,I.S., Palladino,M.A. and O'Connor,J.V. (1986) *J. Exp. Med.*, **163**, 1433.
30. Sipe,J.D., Ignarczak,T.F., Pollock,P.S. and Glenner,G.G. (1976) *J. Immunol.*, **116**, 1151.
31. Pepys,M.B., Dash,A,.C., Munn,E.A., Genurz,H., Osmond,A.P. and Painter,P.H. (1977) *Lancet*, **1**, 1029.
32. Mancini,G., Carbonara,A. and Heremans,J. (1965) *Immunochemistry*, **2**, 235.
33. Di Giovine,F.S., Malawista,S.E., Nuki,G. and Duff,G.W. (1987) *J. Immunol.*, in press.
34. Durum,S.K., Schmidt,J.A. and Oppenheim,J.J. (1985). *Annu. Rev. Immunol.*, **3**, 263.

10. APPENDIX I

10.1 Endotoxin and IL-1 measurement

IL-1 production can be activated in macrophages by extremely low (0.5−50 pg/ml) concentrations of bacterial endotoxin (lipopolysaccharide — LPS) (27). This amount of LPS frequently is found as a contaminant of culture media, sera, tissue culture ware and laboratory reagents. Clearly in IL-1 measurement systems that include macrophage-like cells (e.g. mouse thymic cells) the presence of LPS in the test sample could, by stimulating IL-1 production, produce very misleading results. The same problem arises if LPS is inadvertantly injected into animals during *in vivo* testing of IL-1. Several steps can be taken to minimize the risk of false-positive results due to LPS.

(i) Remember that only baking at 180°C for several hours will destroy LPS. Autoclaving and other sterilizing procedures may not be effective.
(ii) LPS occurs in a wide range of molecular sizes (from several thousand to several million Mr) so molecular sieving techniques cannot be relied on to remove LPS activity.
(iii) LPS levels in solutions can be assessed relatively easily by the simple limulus test (see below) and reagents or samples that score positive for LPS at working dilutions should not be used in monokine work.
(iv) Even when samples and reagents score negative in the limulus test they can often contain enough LPS to activate macrophage cytokine production. For this reason it is wise to check the effects of polymyxin B in an endotoxin-sensitive IL-1 production or IL-1 measurement system even (or especially) when all reagents are limulus negative. If readings are reduced by the presence of polymyxin B then it must be recognized that the results could be produced by the presence of LPS in the system. Polymyxin B is not a panacea for LPS contamination but, in the absence of specific neutralizing reagents, it is a valuable tool (6).

10.2 The limulus test

There are highly sensitive and specific methods for LPS identification (e.g. gas chromatography) but these are not generally available.

Fortunately a simple and rapid bioassay for LPS screening is available based on lysates of circulating blood cells (amoebocytes) from the horseshoe crab *Limulus polyphemus*. *Limulus* amoebocyte lysate (LAL) contains a pro-enzyme system that is activated on contact with LPS. Activation results in the production of a protein gel and this reaction is the basis of several tests for endotoxins. With instrumentation the process of gel formation can be sensitively and quantitatively monitored but for most purposes the production of an adherent gel assessed by eye is sufficiently good (27).

(i) Ensure all materials and reagents are pyrogen (LPS)-free.
(ii) Reconstitute lyophylized LAL on ice according to manufacturers instructions.
(iii) Add 100 μl of LAL solution and 100 μl of standard LPS or test solution to the bottom of 10 $\times$ 75 mm stoppered plastic tubes (pyrogen-free) and mix gently (include negative controls with pyrogen-free PBS).
(iv) Transfer the tubes to a motionless water bath at 37°C.
(v) Read the results at 1 h. The presence of an adherent gel (which remains adherent on gentle tube-inversion) is scored as a positive. Several dilutions of standard LPS solution should be assayed to check the sensitivity of the lysate. if the sample is tested at different dilutions quantitative data can be produced. It is also necessary to check that negative samples do not contain inhibitors of LAL gelatin (by addition of a known amount of LPS to the sample).

11. APPENDIX II

11.1 **Interleukin 1 assay standards**

A simple procedure for making a positive control for IL-1 assays involves stimulating monocytes with endotoxin and collecting the supernatant. This in-house standard preparation can be calibrated against an interim reference reagent for IL-1α or IL-1β available from NIBSC. (The procedure can also be used to prepare supernatant which contains TNF. An interim reference reagent for TNF is also available from NIBSC.)

(i) Harvest monocytes or monocytic cell lines, e.g. P388D1 (mouse), THP-1 (human), and prepare monolayers (if adherent see page 265).
(ii) Adjust to 1–2 $\times$ 10^6 cells/ml in RPMI 1640 containing 2–5% FCS.
(iii) Add LPS at 0.1–10 μg/ml.
(iv) Culture for 24–48 h.
(v) Harvest supernatant by centrifugation, filter, sterilize, store as aliquots at −70°C. (Supernatants may be dialysed to remove prostaglandins.)

CHAPTER 16

Production and assay of interleukin 2

A.J.H.GEARING AND C.R.BIRD

1. INTRODUCTION

Interleukin 2 (IL-2) is a glycoprotein of 15 500 daltons molecular weight produced by T lymphocytes, which stimulates cells from haemopoeitic lineages which express specific IL-2 receptors (1). The effects of IL-2 on its target cells can include increasing cellular proliferation, stimulating production of other lymphokines, upgrading IL-2 receptor expression and enhancing functions such as phagocytosis, cell killing or antibody production (2–4).

Of all the lymphokines, IL-2 is probably the easiest to work with. It was the first lymphokine for which a specific bioassay, employing a dependent cell line, was developed (5). This allowed reliable screening for monoclonal antibody production (6) and subsequent cloning of cDNA in *Escherichia coli* (7).

The cell lines which respond to IL-2 and which are the basis of the bioassay, and also cell lines which produce high titres of IL-2, are now freely available from tissue culture collections and numerous laboratories. Monoclonal antibodies recognizing IL-2 or its receptor are also easy to obtain. Crude, biochemically purified or affinity purified lymphocyte-derived IL-2, and also pure recombinant IL-2, can be obtained from a number of manufacturers. Specific immunoassays are also becoming commercially available. An interim reference reagent for IL-2 is available from the Biological Response Modifiers Program (BRMP) of the National Cancer Institute which enables standardization of IL-2 unitage between different laboratories. This interim reagent will be replaced late in 1987 by an international standard which will define the unit of IL-2. This will be available from the National Institute for Biological Standards and Control (NIBSC). This chapter will outline procedures involved in producing and assaying IL-2, the problems which can be encountered and how to avoid them. The cell lines and reagents used are given in *Table 1*, together with their commercial sources.

2. PRODUCTION OF IL-2: GENERAL

IL-2 can be obtained either from fresh tissues or from cell lines.

2.1 **Fresh tissue**

(i) Human peripheral blood leucocytes can be obtained from blood of healthy volunteers or from buffy coat residues which can be obtained from blood banks. Lymphocytes can be enriched on a Ficoll Hypaque gradient.

(ii) Leucocytes can be recovered from various lymphoid organs, for example rat, mouse or human spleen, lymph nodes or tonsil. Cell suspensions are obtained

Table 1. Interleukin 2: cell lines, reagents and some commercial sources.

Cell lines	*Sources*
Jurkat—produces human IL-2 EL-4—produces mouse IL-2 MLA-144—produces gibbon IL-2 CTLL—responds to mammalian IL-2	European Collection of Animal Cell Cultures, Porton Down, SP4 OJG,UK or American Type Culture Collection, 12301 Park Lawn Drive, Rockville, MD 20852–1776, USA
Reagents	*Sources*
Cell-derived IL-2	Accurate Chemical & Scientific Corp. (mouse and rat), Biotest (human), Collaborative Research Inc. (human), Electro-Nucleonics (human), Jannsen (human), Sigma (rat)
Recombinant IL-2	Amersham International, Amgen, Boehringer-Mannheim, Biogen, Cetus, Genzyme (also mouse), Sigma
Radiolabelled IL-2	Amersham International, Dupont-NEN
Interim reference reagent (currently available)	Biological Response Modifiers Program of the National Cancer Institutes, P O Box B Frederick, MD 21701, USA
International standard (will replace interim reference reagent late 1987)	National Institute of Biological Standards and Control, Blanche Lane, South Mimms, Potters Bar, Herts EN6 3QG, UK
Anti-IL-2 monoclonal antibodies	Genzyme
Anti-IL-2 receptor monoclonal antibodies	Becton Dickinson (human), Boehringer-Mannheim (human, mouse and rat), Coulter (human), Immunotech (human), Cistron (human)
IL-2 immunoassay	Genzyme
IL-2 receptor immunoassay	Boehringer-Mannheim, T cell Sciences

by mechanical dissociation followed by enrichment on a Ficoll Hypaque gradient.

Leucocytes are resuspended at 5×10^6/ml in culture medium such as RPMI 1640, containing 5% fetal calf serum (FCS) and 2×10^{-5} M 2-mercaptoethanol. Rodent cells are stimulated with 5 μg/ml of concanavalin A (Con A; Sigma) and human cells with 1% phytohaemagglutinin (PHA; Wellcome reagent grade). Phorbol myristic acetate (PMA) at 10 ng/ml may increase yields of IL-2, but is carcinogenic and can interfere in some assay systems (see Section 4). Calcium ionophores such as A23187 or ionomycin have also been used in combination with PMA to cause IL-2 production (8,9).

After 40–48 h culture, supernatants are harvested by centrifugation and stored at −20°C.

If a serum- and lectin-free product is required, the leucocytes are resuspended at 10^7 cells/ml in medium containing 5% FCS and lectin, as above, and incubated for 4–6 h. The cells are then washed four times with medium by centrifugation to remove any lectin and serum. The cells are then resuspended at 5×10^6 cells/ml in serum-free medium for a further 40 h and supernatants harvested as above. N.B. If the protein content of the supernatant is too low then IL-2, which is very hydrophobic, can be lost by adherence to the walls of culture or storage vessels. The addition of 100 μg/ml of protein such as bovine serum albumin (BSA) can overcome this problem.

2.2 **Cell lines**

2.2.1 *Jurkat*

This human T cell line produces high titres of IL-2 when stimulated with PHA and PMA.

Cells are maintained in medium containing 5–10% heat-inactivated FCS. For production of IL-2, cells are washed once in medium and resuspended at 10^7 cells/ml in medium containing 5% FCS, 1% PHA and 10 ng/ml PMA. After 4–6 h incubation the cells are washed four times and resuspended in medium at 5×10^6 cells/ml. Supernatant is harvested after 20 h in culture.

2.2.2 *EL-4*

This mouse thymoma cell line produces IL-2 in response to stimulation with PMA; interleukin 1 and lectins such as Con A can increase yields.

Cells are maintained in medium containing 5% FCS, they are washed once and resuspended at $1-2 \times 10^6$ cells/ml in medium containing 5% FCS and 10 ng/ml PMA. Supernatants are harvested after 24 h in culture.

2.2.3 *MLA 144*

This gibbon T cell line spontaneously secretes IL-2 without any requirement for lectin or phorbol ester. The supernatant is collected from cultures maintained in medium containing 5% heat-inactivated FCS.

3. PRODUCTION OF CONDITIONED MEDIUM FROM RAT SPLENOCYTES FOR MAINTENANCE OF ASSAY LINES

(i) Kill approximately 25 rats (e.g. by placing in a CO_2 drawer), remove spleens aseptically and collect in 100 ml of RPMI 1640.

(ii) Prepare a spleen cell suspension: a convenient method for large amounts of tissue is to use a fine mesh wire gauze through which the spleens are gently rubbed using blunt forceps. Once all the spleens have been teased leave the suspension to stand for 5 min.

(iii) Decant the cell suspension, discarding any clumps of tissue, and wash the cells once with RPMI 1640, then count the number of cells.

(iv) Adjust the concentration of cells to 4×10^6/ml in RPMI 1640 containing 5% FCS, 5 μg/ml Con A and 5×10^{-5} M 2-mercaptoethanol; transfer the cell suspension to 500 ml spinner flasks. Gas the suspension by flushing with CO_2 for approximately 1 min, then place on a magnetic stirrer in a dry incubator at 37°C and culture for 44 h with continuous gentle stirring.

(v) Centrifuge the cell suspension at 3000 r.p.m. for 15 min. Discard and clarify the supernatant by further centrifugation at 9000 r.p.m. for 20 min.

(vi) Cool the supernatant to 4°C and add solid ammonium sulphate with stirring to 85% saturation (55.9 g/100 ml). Stir for 1 h at 4°C.

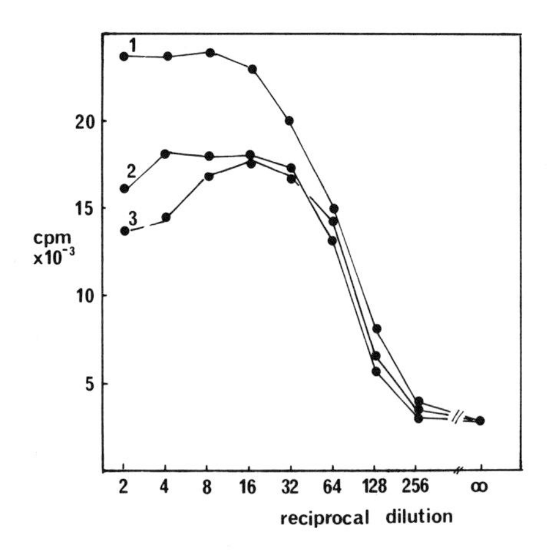

Figure 1. Titration on CTLL cells of three different batches of rat conditioned medium prepared by ammonium sulphate precipitation and Sephadex G100 column fractionation as described. Preparations 2 and 3 contain inhibitory materials which result in lower proliferation of the CTLL cells at high concentrations.

(vii) Collect the precipitate by centrifuging at 12 000 r.p.m. for 30 min. Discard the supernatant and re-dissolve the precipitate in a minimal volume ($\sim$50 ml) of 0.01 M Hepes-buffered saline (pH 7.3).

(viii) Dialyse for 1.5–2 h against 2 l of 0.01 M Hepes-buffered saline at 4°C. During dialysis a brown precipitate is formed, this is removed by centrifugation at 12 000 r.p.m. for 30 min.

(ix) Apply the clarified supernatant to a Sephadex G-100 column ($\sim$ 5 $\times$ 95 cm) equilibrated with 0.01 M Hepes-buffered saline at 4°C in approximately 10 ml aliquots.

(x) Elute the column at 1 ml/min and collect 5 ml fractions. Determine the IL-2 activity in the fractions by bioassay (see Section 4). Pool the active fractions, filter sterilize, aliquot and store at -70°C. Determine the IL-2 activity of the pool by bioassay.

This method has produced IL-2 preparations of remarkably consistent half maximal titres of between 0.5 and 2%. Different preparations may however contain inhibitory material which causes inhibition at high concentrations (see *Figure 1*).

Con A present in the original conditioned medium is retained on the Sephadex matrix and can be displaced by washing the column with 0.5 M α-methylmannoside after the fractionation process has been completed.

Further purification can be achieved using reverse phase h.p.l.c. or by affinity chromatography using immobilized polyclonal or monoclonal antibodies directed against IL-2. (Suitable procedures are described in other volumes of the *Practical Approach Series*.)

3.1 **Problems in preparation of IL-2**

IL-2 production is usually straightforward but problems can occur, resulting in low activity titres. When freshly isolated leucocytes are used, individual donor variation

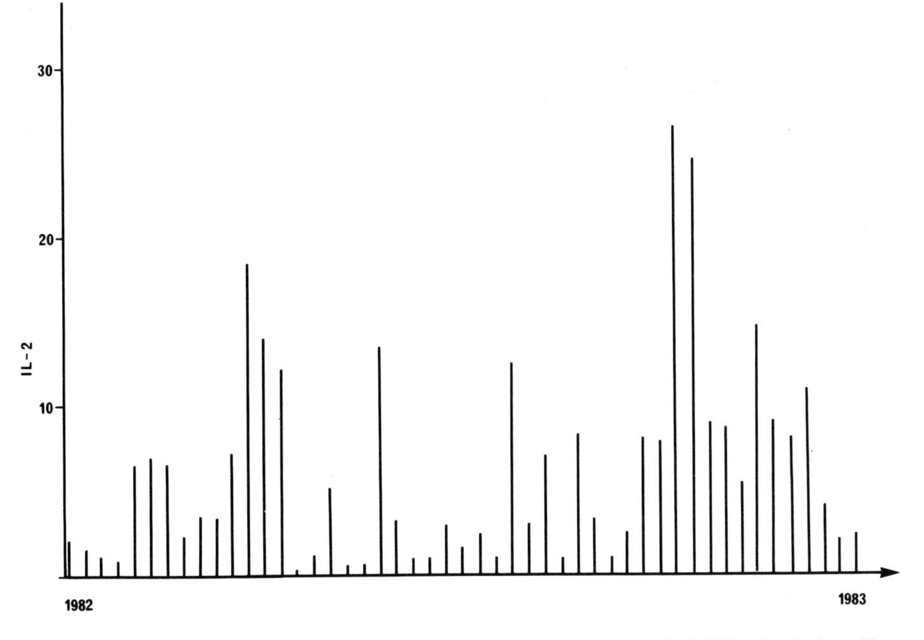

Figure 2. Variation in IL-2 levels of conditioned medium prepared by identical PHA stimulation of human buffy coat leucocytes over the period 1982–1983. Each preparation was assayed at 25% dilution on CTLL cells.

is a major factor affecting the titre of IL-2. *Figure 2* illustrates the variation in IL-2 titres of consecutive preparations of conditioned medium from human buffy coats collected over a 1 year period. Even leucocytes prepared from the same donor on different days may vary in their ability to produce IL-2.

The use of cell lines such as those already described will overcome this variation, however lines occasionally will lose the ability to produce IL-2 or their titres will drop. Re-cloning or returning to frozen stocks will restore production.

Leucocyte preparations or cell lines can produce inhibitors of IL-2 assays, these can include prostaglandins or proteins which interfere with bioassays (see Section 4). Certain sera also contain proteins which are cytotoxic for the assay lines. These inhibitors can be removed by gel filtration or dialysis or, alternatively, an immunoassay can be used.

A frequent problem with serum-free preparations of IL-2 is adsorbtion of the hydrophobic IL-2 onto the surface of containers or sterilizing filters. This can be avoided by the addition of other proteins such as albumin or by keeping the concentration of IL-2 in stock solutions above 100 μg/ml. IL-2 is very stable under these conditions and can even be kept at 4°C for several months. With the addition of protein such as human serum albumin and non-reducing sugars such as trehalose, as stabilizing agents, IL-2 can be readily freeze-dried. In the author's experience freeze-dried preparations have shown no loss in activity after 1 year of storage at 56°C.

A less common problem, but one that should be borne in mind, is that certain culture additives such as plant lectins or phorbol esters used in the generation of IL-2 can cause

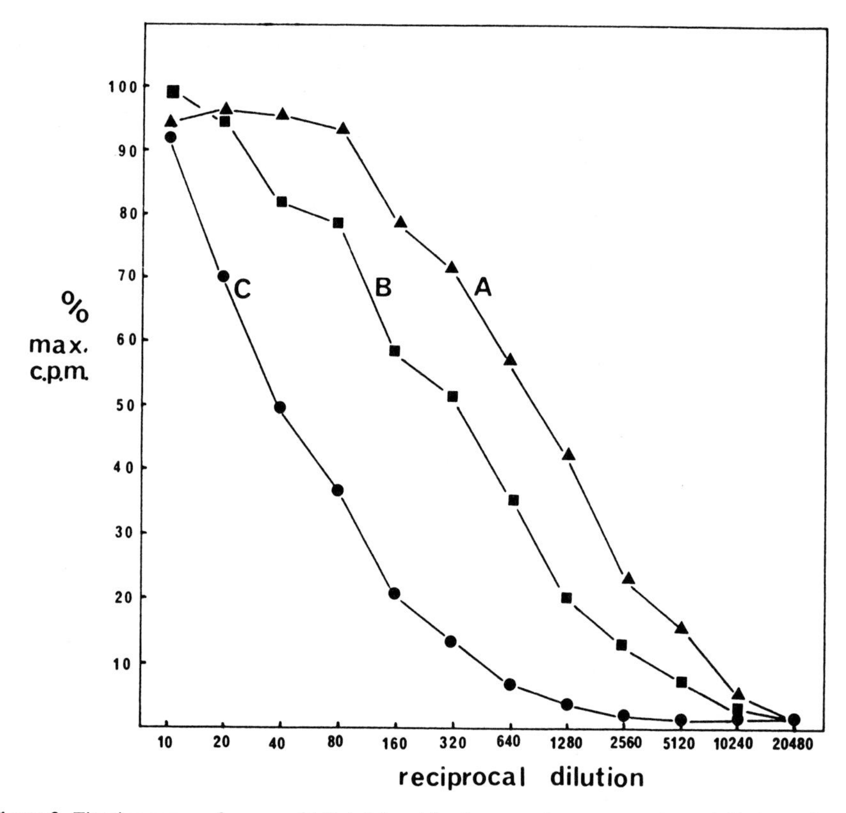

Figure 3. Titration curves of c.p.m. of tritiated thymidine incorporation versus reciprocal dilutions of samples A, B and C described in *Table 2*.

stimulation of cells used in its assay, thus giving a falsely elevated IL-2 titre. It is important that the response of assay cells to these compounds is checked (see Section 4).

4. BIOASSAY OF IL-2

Most bioassays for IL-2 rely on measuring the increase in proliferation of a dependent cell line or short-term lymphoblasts. Cell lines in common use include CTLL, MT-1, FDC-2 or HT-2. We have found CTLL to be a satisfactory assay line which is easy to maintain and which can be cryopreserved and thawed.

4.1 **Maintenance of CTLL line**

CTLL cells are maintained in RPMI 1640 containing 5% FCS which is supplemented with partially purified rat conditioned medium as described in Section 3.

Cell cultures are fed every 3 days with 5% conditioned medium, this corresponds to a concentration of approximately 50 BRMP units of IL-2 per ml. Cell density at this stage is approximately 3×10^5 cells/ml, cultures are split to 3×10^4 cells/ml for feeding.

Table 2. Transformation of data from IL-2 bioassay.

Sample	*Reciprocal dilution*	*Response (c.p.m.)*	*% maximum response*
A	10	50 215	91
	20	52 650	96
	40	52 575	96
	80	51 085	93
	160	43 048	78
	320	38 953	71
	640	30 825	56
	1280	23 133	42
	2560	12 606	23
	5120	8806	16
	10 240	3320	6
	20 480	1822	3
B	10	54 620	100
	20	52 670	96
	40	44 324	81
	80	43 416	79
	160	32 531	59
	320	28 062	51
	640	19 160	35
	1280	11 058	20
	2560	6840	12
	5120	3606	7
	10 240	1694	3
	20 480	1058	2
C	10	50 095	91
	20	38 280	70
	40	27 340	50
	80	20 558	37
	160	11 278	21
	320	7058	13
	640	4192	8
	1280	2324	4
	2560	1066	2
	5120	784	1
	10 240	822	1
	20 480	594	1

Three preparations A, B and C of IL-2 were titrated against a CTLL cell line. Response measured was tritiated thymidine uptake over a 4-h period after 18 h in culture. All values are means of triplicates. Preparation A was a recombinant IL-2 preparation from Biogen, preparation B was affinity-purified Jurkat IL-2 supplied by Dr R.Robb and preparation C was conditioned medium from PHA-stimulated human peripheral blood leucocytes.

4.2 IL-2 assay by a dependent cell line

(i) Harvest the CTLL cells 3 days after feeding with IL-2. Wash the cells twice by centrifugation in RPMI 1640 and resuspend at 1×10^5 cells/ml in RPMI 1640 containing 10% FCS.

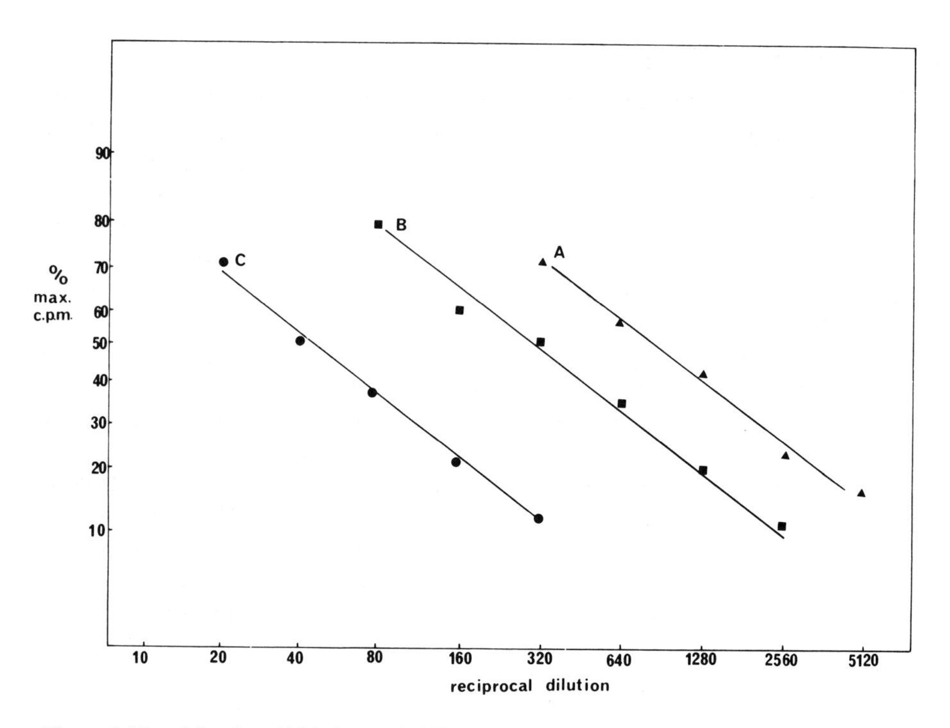

Figure 4. Plot of data from *Table 2* on probability *X* log axes. Sample A has an arbitrary titre of 100 Units of IL-2/ml. By measuring the horizontal distance between each line this gives titres of 33 Units/ml for B and 6 Units/ml for C.

(ii) Distribute titrations of a standard preparation and the samples to be measured in triplicate in 96-well microtitration plates (Nunc). The negative control is culture medium alone. Each well should contain a final volume of 50 μl.

(iii) Add 50 μl of cell suspension to each well and incubate the plates for approximately 18 h at 37°C in a CO_2 incubator.

(iv) Add 0.5 μCi of tritiated thymidine to each well and return the plates to the incubator for 3–4 h.

(v) Harvest the contents of each well onto glass fibre filters using a cell harvester (e.g. Skatron, or Dynatech) and determine the radioactivity incorporated into DNA by liquid scintillation counting.

As an alternative to tritiated thymidine for determining proliferation some laboratories use radiolabelled iodouridine, or the dye MTT which can be assessed using an enzyme-linked immunosorbent assay reader (10). See Chapter 15, this volume.

The results of a typical experiment in which three different preparations of IL-2 are titrated is shown in *Figure 3*.

Analysis of results obtained is straightforward if a standard of known unitage is included in the assay. Data can be subjected to probit analysis (11) which is concerned with the relationship between the frequency of occurrence of specific events and the dose that is responsible for induction of those events.

4.3 **Calculation of unitage**

(i) The raw data for proliferation, for example c.p.m. of [^{3}H]thymidine incorporated into the DNA, is first transformed to a percentage of maximum counts (*Table 2*). A rough estimate of relative potency can be obtained by comparing the dilution at which 50% maximum counts are obtained for each sample.

(ii) Transformed counts are then converted to log values, as are the dilutions for each sample, and plotted graphically. An alternative is to plot transformed counts and dilution on log–log graph paper or, as shown in *Figure 4*, on probability *X* log graph paper.

(iii) The unitage of each sample, relative to the standard, is calculated from the distance between the straight line portions for each sample.

In the example given in *Figure 3*, sample A is the arbitrary reference at 100 U/ml, this gives unitage of 33 for B and 6 for C.

(iv) An interim reference reagent for IL-2 is currently available from the BRMP of NCI. This preparation has an arbitrarily assigned unitage of 500 reference units/ml and can be used to calibrate suitable in-house standards for use in each assay.

In the near future this interim reagent will be replaced by an international standard which has been subjected to a full-scale collaborative study for its selection. This will be available late in 1987 from NIBSC (see *Table 1*).

4.4 **IL-2 assay by short-term lectin-induced blasts**

If the maintenance of long-term lines is not possible then a simple but less specific alternative is to use plant lectins such as Con A or PHA to stimulate the production of T cell blasts from freshly isolated leucocytes. These blasts will proliferate in response to IL-2. Human peripheral blood leucocytes or mouse spleen cells are the most commonly used sources. A mouse Con A blast assay is described below.

(i) Prepare a spleen cell suspension from a single animal as described for the rat conditioned medium.

(ii) Suspend the cells at 2×10^6 cells/ml in RPMI 1640 containing 5% FCS, 5×10^{-5} M 2-mercaptoethanol and 2.5 μg/ml Con A.

(iii) Culture the cells in 50 ml flasks for 72 h at 37°C in a CO_2 incubator.

(iv) Examine the cultures for the presence of large lymphoblasts.

(v) Wash the blasts three times with RPMI 1640 and resuspend at 1×10^6 cells/ml in RPMI 1640 containing 10% FCS, 5×10^{-5} M 2-mercaptoethanol and 25 mM α-methylmannoside.

(vi) Distribute the samples and standard preparations of IL-2 in 96-well plates as above.

(vii) Add 50 μl of the Con A blast suspension to each well and continue the assay exactly as for CTLL cells, including 18-hour culture, with the addition of 0.5 μCi of tritiated thymidine for a 4 h period prior to harvesting.

4.5 **Problems with assay of IL-2**

IL-2 titres determined by bioassay can sometimes overestimate or underestimate the amount of IL-2 present.

4.5.1 *Overestimation*

Overestimation of IL-2 levels is normally due to the presence of contaminants in the sample which stimulate the assay line. Short-term blasts in particular are sensitive to plant lectins and phorbol esters; however, even when using long-term IL-2-dependent lines such as CTLL, it must not be assumed that these substances have no effect. The line should be checked for such responses on receipt and every 6 months if grown continually. It has recently been shown that freshly isolated T cells and some long-term IL-2-dependent lines are also responsive to IL-4 (BSF-1) (12,13). Again assay lines should be checked for response to this and other cytokines. Our murine CTLL cells do not respond to human IL-4, but may respond to murine IL-4.

4.5.2 *Underestimation*

Underestimation of IL-2 levels can be caused either by the assay line becoming unresponsive or, more commonly, by the presence of inhibitory molecules in the samples.

If long-term assay lines are grown in conditioned medium containing phorbol ester or plant lectin they can become dependent on them and will not respond to IL-2 on its own. For this reason crude conditioned medium should not be used to maintain cell lines. Including a known IL-2 standard in assays should identify this problem.

Tissue fluids, serum and supernatant fluid from cultures of activated lymphoid cells all contain molecules with inhibitory effects on IL-2 measurement (14,15). These inhibitors are poorly characterized, but their effects can be seen either by giving a bell-shaped dose–response curve for samples containing sufficient IL-2, or they can be revealed by titration against a fixed amount of exogenous IL-2. In some cases where inhibitor levels are not too high, IL-2 activity can be measured by using diluted samples but this obviously limits the sensitivity of measurement. Immunoassay is an obvious solution to these problems but has limitations such as no distinction between active and inactive forms and different responses to IL-2 from different sources.

4.5.3 *Variation in titre due to plate position*

Over or underestimation of activity can also occur in large assays if samples are plated as triplicates consecutively over several (three or more) microtitration plates. In our experience if two titrations of the same preparation are placed on the first and last plates of an assay, higher levels of IL-2 are given on the last plate. The best solution to this problem is to distribute every point of each dilution series randomly, this is however impractical for large assays. A simpler solution is to assay samples not in triplicates but as a series of single dilution points, and to repeat this single point dilution series on two other microtitration plates. The position of each single point dilution series can then be randomly distributed to each plate. We have found this method to give excellent results.

5. REFERENCES

1. Robb,R.J. (1984) *Immunol. Today,* **5**, 203.
2. Smith,K.A. and Cantrell,B.A. (1985) *Proc. Natl. Acad. Sci. USA,* **82**, 864.
3. Henney,C.S., Kuribayashi,K., Kern,D.E. and Gillis,S. (1981) *Nature,* **291**, 335.

4. Kishi,H., Inui,S., Muraguchi,A., Hirano,T., Yamamura,Y. and Kishimoto,T. (1985) *J. Immunol.*, **134**, 3104.
5. Gillis,S., Ferm,M., Ou,W. and Smith,K.A. (1978) *J. Immunol.*, **120**, 2027.
6. Smith,K.A. and Favata,M.F. (1983) *J. Immunol.*, **131**, 1808.
7. Taniguchi,T.H., Matsui,T., Fujita,C., Takaoka,N., Kashima,R., Yoshimoto,R. and Hamuro,J. (1983) *Nature*, **302**, 305.
8. Farrar,J.F., Fuller-Farrar,J., Simon,P.L., Hilfiker,M.L., Stadler,B.M. and Farrar,W.L. (1980) *J. Immunol.*, **125**, 2555.
9. Simon,P. (1984) *Cell. Immunol.*, **87**, 720.
10. Mosmann,T. (1983) *J. Immunol. Methods*, **65**, 55.
11. Finney,P.J. (1952) *Probit Analysis*. Cambridge University Press, 2nd edition.
12. Lee,F., Yokoto,T., Otsuka,T., Meyerson,P., Villaret,D., Coffman,R., Mosmann,T., Rennick,D., Roehm,N., Smith,C., Zlotnik,A. and Arai,K.-I. (1986) *Proc. Natl. Acad. Sci. USA*, **83**, 2061.
13. Grabstein,K., Eiserman,J., Mochizuki,D., Sharebeck,K., Conlon,P., Hopp,T., March,C. and Gillis,S. (1986) *J. Exp. Med.*, **163**, 4405.
14. Hardt,C., Röllinghof,M., Pfizermaier,K., Mossman,H. and Wagner,H. (1981) *J. Exp. Med.*, **154**, 262.
15. Lelchuk,R., Schmidt,J.A., Hodson,K., Aston,R. and Liew,F.-Y. (1987) *Cell. Immunol.*, **104**, 126.

CHAPTER 17

Assays for interleukin 3 and other myeloid colony-stimulating factors

JOHN GARLAND

1. INTRODUCTION — CELL RENEWAL, PROGENITOR CELLS AND GROWTH FACTORS

The renewal of bone marrow-derived cells (white cells, red cells, lymphocytes, megakaryocytes) is required to replace those in peripheral blood which are removed by senescence or death; the turnover rate of many peripheral mature cells is very high, and new cells need to be continually fed into the system. This is achieved through the proliferation and differentiation of 'progenitor' cells, modelled in *Figure 1*.

A set of primitive multipotential 'stem' cells sheds daughter cells which in turn proliferate and, depending on the environment or presence of various factors, partially differentiate to become 'committed' to a particular lineage of mature cells: these are termed 'committed progenitors'. These can be induced to grow in semi-solid media producing focal colonies of growth derived from single progenitor cells, referred to as 'c.f.u.' (colony forming units). Each c.f.u. has a suffix to define the lineage; thus c.f.u.-GM refers to progenitors which produce colonies of granulocytes and macrophages, c.f.u.-E those of the erythroid series, and c.f.u.-mix those with mixed cell types and therefore nearer to the uncommitted stem cell.

The development of assay systems for haemopoietic progenitor cells followed from observations that they could be induced to proliferate and differentiate in semi-solid culture *providing* appropriate growth factors were supplied. These factors are generically termed 'colony-stimulating factors', CSFs, prefixed by the lineage they operate on. Thus, GM-CSF is a colony-stimulating factor for progenitors committed to granulocytes and/or macrophages. The colony assay thus can be used to determine progenitors, or the CSF since colony growth is broadly proportional to CSF concentration within a certain range.

It is important to understand the concept of cell self-renewal, because this determines the potential for colony growth, and differentiation from which the lineage of a c.f.u. can be deduced. If a multipotential stem cell itself differentiates, then it is lost to the pool of stem cells and can no longer function as multipotential: if all multipotential cells differentiated, then clearly there would be no capacity to regenerate committed progenitors. Therefore, there must be a balance struck between those multipotential stem cells which simply proliferate to maintain the stem cell pool (self-renew), and those which differentiate to form cohorts of committed progenitors. Cells which proliferate to maintain their original identity have complete 'self-renewal potential'; the further along a differentiation pathway a progenitor moves, so it progressively loses

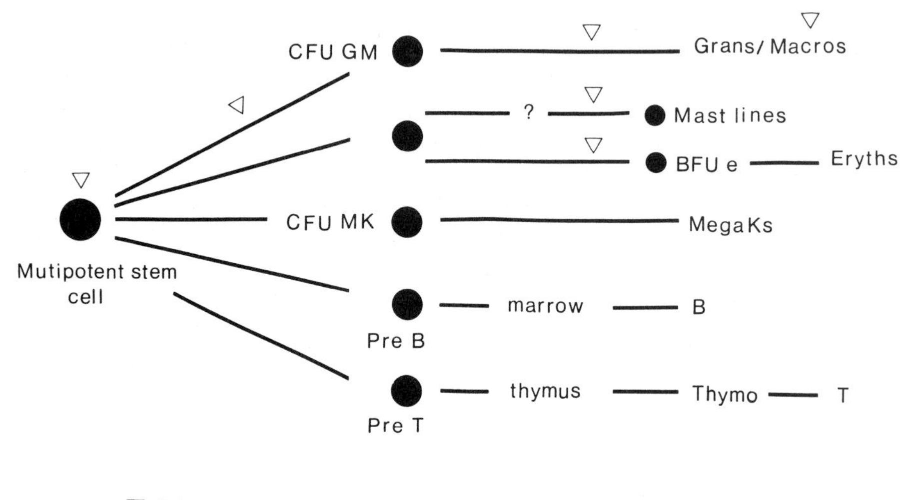

Figure 1. The simplified hierarchy in marrow cells. The pluripotent cell feeds into other lineages by a balance between self-renewal and differentiation into a progenitor committed to a particular lineage.

potential for self-renewing, thus limiting the number of progeny cells derived from it. Precisely what decides this balance is not known, but in colony assays progenitor cells usually progress towards differentiation, losing their 'self-renewal potential' and eventually finishing as a cohort of mature cells without any proliferative ability. Consequently, re-plating efficiencies of colonies from an assay are very low and most colonies have a strictly limited life-span. However, techniques do exist to grow marrow progenitor cells for longer periods, described below.

The composition and size of colonies will therefore depend upon the proliferation potential of the progenitors, and the size and relative proportion of mature to immature cells in colonies will vary markedly. Therefore it is necessary to define the criteria for a colony (e.g. >50 cells) and in some systems the relative maturity of the progenitors is also defined. Thus, erythroid colonies (which are referred to as 'burst-forming units', b.f.u.-E) grow to colonies and become haemoglobinized at different times in semi-solid culture since progenitors are at various stages along the differentiation pathway. A 'day 6 b.f.u.-E' describes a progenitor which produces a culture, scorable by the criteria defining colony size and haemoglobinization, after 6 days in culture.

Although this chapter deals exclusively with the mouse, the general concepts and techniques are readily transposed to human systems wherever cells can either be grown in semi-solid medium with a growth factor, or respond to one by proliferating.

1.1 Colony assays

Colony assays exist for all the major lineages of haemopoietically derived cells. The principle is that if marrow is plated in a semi-solid supporting medium (to reduce cell diffusion) the mature cells die; only the committed progenitors (which are at a very low concentration, ~0.1%) will proliferate, their lineage being selected by the specific growth factor included in the assay. Since lineages can only be determined on differentiated cells, it is axiomatic that some differentiation must be induced to recognize the lineage of any given colony. Generally, the growth factors used induce some differentiation which may be easily recognizable by direct microscopy, although specific stains may be needed for others. The source of growth factors is usually medium conditioned by other cells since this is relatively easy to produce. However, recombinant growth factors are now becoming available. Generally, human factors work in the mouse but not vice versa, and recombinant products may not induce identical responses as crude or purified 'native' factors. Before using recombinant or purified materials it is wise to test the sytem with crude ones.

1.2 Long-term marrow culture

In addition to the proliferation of progenitors in semi-solid media, two *in vitro* systems have been developed to grow rodent marrow progenitor cells over a long period (1,2). If marrow is seeded into medium containing serum and hydrocortisone, an adherent layer of 'stromal' cells rapidly grows into which multipotential stem cells can lodge and proliferate, and from which committed progenitors are shed (Greenberger system). The critical component is a temperature of 33°C and a high folate medium such as Fischers. Alternatively, an adherent marrow stroma is allowed to form in Fischers medium with horse serum, then a second 'recharge' of marrow added to seed the stromal layer, again keeping a temperature of 33°C (the original Dexter system). A weekly 50% medium exchange is sufficient to keep both culture systems productive for many weeks. Long-term cultures duplicate the production of both mature and progenitor cells found in *in situ* marrow with the exception of mature B and T cells; however, precursors of both are present and recent reports demonstrate that a modified culture omitting hydrocortisone and raising the temperature to 37°C will support or generate B lineage cells (3,4). Much has been learned about stem cell kinetics from these cultures and it is also from them that the phenomenon of interleukin 3 arose.

1.3 Interleukin 3-dependent cells and IL-3

Surprisingly, long-term marrow cultures do not produce significant levels of growth factors despite high production of mature granulocytes and progenitors, and the first report of 'factor-dependent' cells described *in vitro* cell lines with myeloid characteristics derived from long-term marrow cultures. These cells were absolutely dependent on medium conditioned by an *in vitro* line WEHI 3b which was routinely used as a source of GM-CSF (5). It was therefore considered that the lines were dependent on GM-CSF. Subsequently, it was shown that the factor was not GM-CSF (6), since when it has been sequenced, cloned and chemically synthesized (7–9). An interleukin 3 (IL-3) gene has now been identified in genomic DNA from the rat, monkey and human (10,11), so it is likely that most mammals have the capacity to produce it. The sequence homologies are, however, low and murine cells may not respond to primate IL-3. The importance of IL-3, as this factor is called, is that it will operate on many of the committed progenitor cells in colony assays and on the multipotential stem cell. In addition, recent studies have demonstrated that both GM-CSF and IL-3 can modulate haemopoiesis *in vivo* (12,13), thus opening up the possibility of clinical application of both growth factors, and control of factor-promoted growth with modified analogues. In addition, in the mouse the existence of dependent lines allows biological activity of IL-3 (and perhaps GM-CSF) to be determined without resorting to colony assays. However, colony assays and dependent cell assays may not measure the same parameters. The development of expression cloning (11) will undoubtedly extend the use of such assays in the isolation of genes coding for growth factors.

2. THE HAEMOPOIETIC COLONY ASSAY

This assay measures the number of progenitor cells in a given marrow sample that will respond to growth factors by forming colonies in a semi-solid support. The following refers to assaying for progenitors stimulated by GM-CSF (GM-c.f.u. assay). This assay also functions with IL-3.

2.1 **Materials required**

(i) Source of marrow.
(ii) Petri dishes.
(iii) Growth factors or conditioned medium.
(iv) Culture medium.
(v) Agar or agarose (or methyl cellulose).
(vi) 5 ml syringes and medium-bore needles.
(vii) Universals.
(viii) 30 ml bijou bottles.
(ix) White cell diluting fluid (1% acetic acid and small crystal of crystal violet enough to produce purple colour), counting chamber, white cell pipette (or, alternatively, Coulter Counter).
(x) Plastic disposable 10 ml pipettes (make sure the pipettes can be inserted into a Pi-pump or auto-pipette; a suitable adaptor can be made with polythene and rubber tubing of appropriate sizes).

The methods described below refer to mouse marrow; they may be easily adapted to other marrow samples, taking into account any special requirements.

2.2 **Preparation of mouse femoral bone marrow samples**

Note: all experiments on animals require a Home-Office licence in the UK.

The following are required: dissecting instruments sterilized in 70% alcohol; Fischers medium supplemented with 10% fetal calf serum (FCS) or horse serum; 5 ml syringes and needles, swabs and Universals. It is recommended that plastic disposable Universals are used; they have a conical end suitable for sedimenting cells.

(i) Kill the mouse by cervical dislocation.
(ii) Make a small incision over the abdomen and pull the incision apart so that the skin is removed from over the thighs in one direction and over the thorax in the other.

(iii) Snip the hind legs between knee-joint and pelvis, and remove the femur by cutting it away from the muscles (*Figure 2*).
(iv) Wash hands and swab with 70% alcohol (gloves may be worn, but are cumbersome and do not materially increase the sterility of the final assay). Using a piece of cotton swab or paper towel, clean away residual muscle from the bone (*Figures 3* and *4*). With a pair of sterile scissors, cut across the bone ends.
(v) Draw up about 5 ml of medium into a syringe, insert a medium-bore (21-gauge) needle into the marrow cavity (*Figure 5*) and squirt the marrow out into the Universal (*Figure 6*). The marrow will usually come out as a plug. Reverse the bone and similarly squirt out in the opposite direction.
(vi) To dissociate the marrow plug, draw up the contents of the Universal into the syringe and expel it quickly, repeating this several times. *Do not over-pressure*; cells may rupture if too much force is used.
(vii) When a single-cell suspension has been made, allow to settle for 2 min to sediment detritus, decant the cell suspension and count the nucleated white cells. Use either a Coulter Counter, or white-cell diluting pipette and counting chamber, e.g. improved Neubauer counting chamber.

2.3 Preparation of agar

Bacteriological agar stays liquid at 40°C but solidifies at 37°C. If agar is used, choose a purified one such as 'Noble' agar. Alternatively, low melting point agarose has the advantage that it does not solidify until 15–20°C. To make a stock solution:

(i) add 5 g of agar to 10 ml of medium *without* serum;
(ii) dissolve it by heating in a boiling water-bath.

This agar solution can be kept for many weeks at room temperature or at 4°C, and re-liquidifed in a boiling bath. If agarose is used, be sure that all the granules are dissolved.

2.4 Plating of cells

All plates should be done in triplicate. The proportion of progenitor cells in marrow is about 0.1%, therefore several dilutions of marrow cells should be set up, for example 10^{-3} through 10^{-5} nucleated marrow cells per plate (see *Table 1*).

2.5 Sources and preparation of growth factors

There are many sources of growth factors and the literature should be read for those appropriate for individual assays. Human cultures are more difficult than murine in some respects. For murine GM assays, heart or lung conditioned medium (CM) is often used. To make heart or lung CM is relatively straightforward (see *Table 2*).

2.6 Reading/staining of GM-c.f.u. assay plates

Plates should be read directly under an inverted microscope, or preferably using a colony-counting microscope with incident light. A ×30–×50 magnification is sufficient. A graticule helps break up the area into recognizable sections for counting and can be

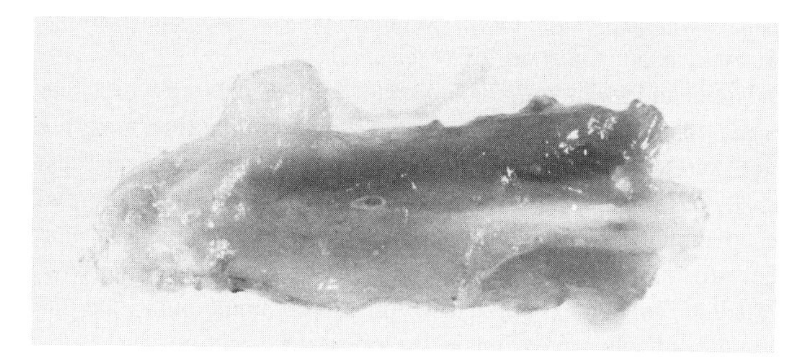

Figures 2–6 were photographed from an experiment under licence.

Figure 2. Preparation of mouse bone marrow. The excised marrow has to be cleaned to remove muscle and tendons.

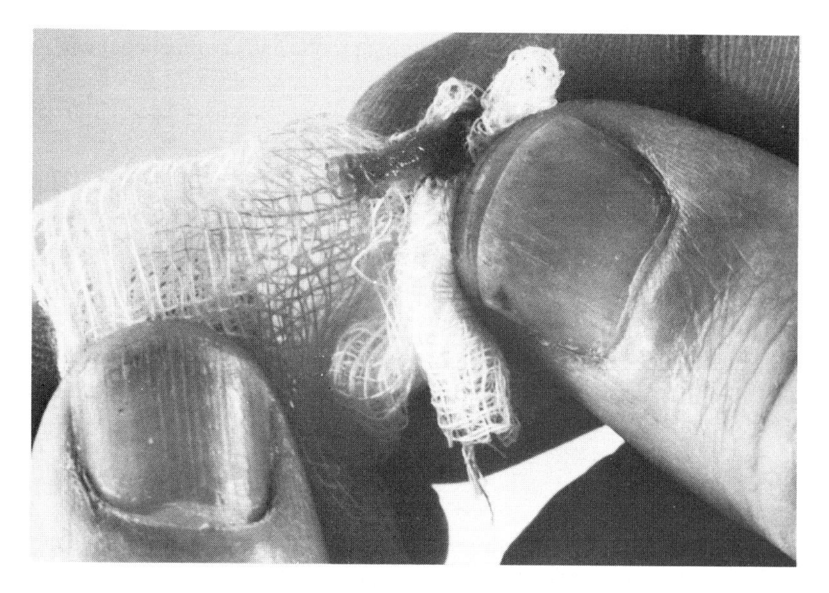

Figure 3. Cleaning the marrow shaft with a swab.

made by scratching a grid on a plastic Petri dish lid. Experience is needed in recognizing different colony types, but if a total count is sufficient, score colonies only on plates where there are between 50 and 150 colonies. If rapid *in situ* staining is needed, flood the plates with aceto-orceine (orceine in acetic acid) for 30 min, which brings out nuclear morphology, drain off the stain and view directly. Alternatively, the agar may be removed for drying and staining as below.

2.6.1 *Obtaining single cell suspensions from colony assays*

Recovery of single cell suspensions can be performed.

(i) Carefully decant the whole agar into a Universal bottle, add 5 ml of fresh medium

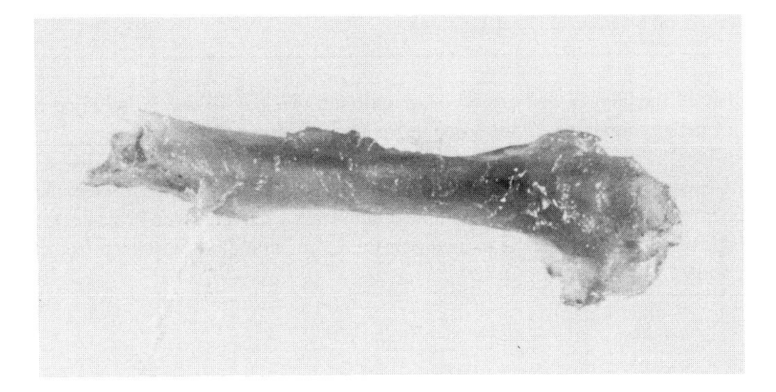

Figure 4. Cleaned femur. The ends will be cut ready to receive the needle.

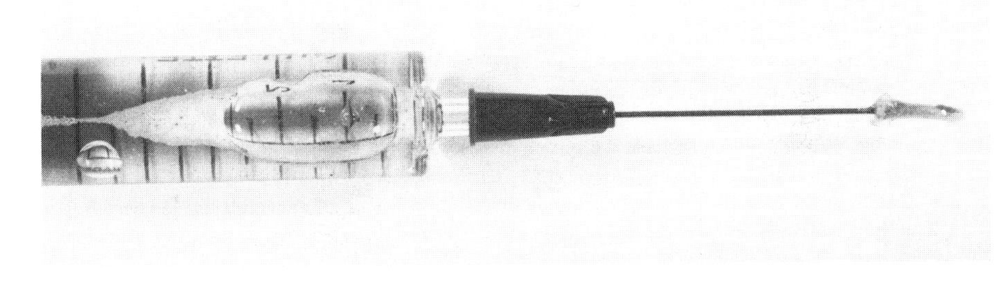

Figure 5. Inserting the marrow with a 21-gauge needle attached to syringe containing medium.

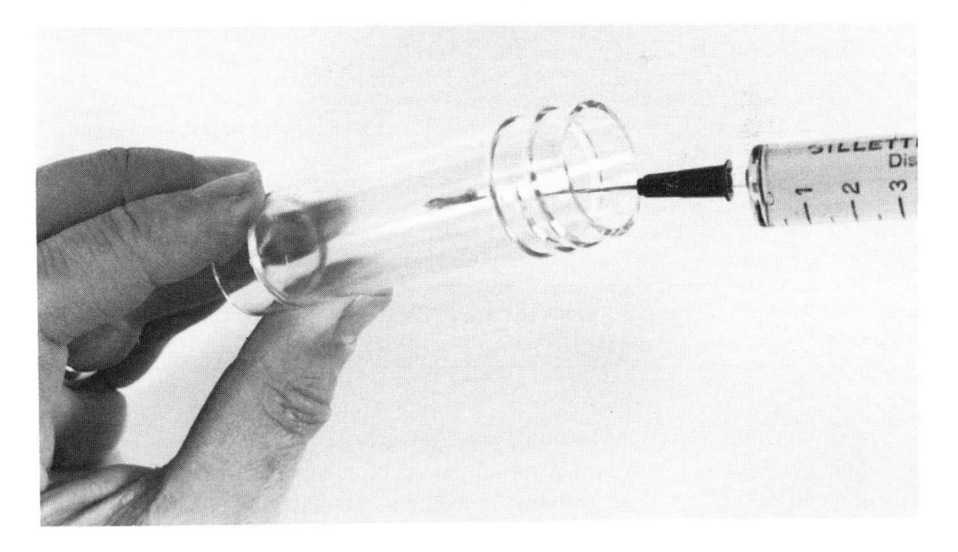

Figure 6. Squirting the marrow through with medium. The bone will be reversed and squirted through in the other direction.

Table 1. Preparation of marrow for agar colonies.

1. Ensure that all medium is well gassed or at optimum pH. Set up one 30-ml bijou bottle for each dilution of cells, containing 15 ml (enough for three plates).
2. Work out how much marrow suspension is needed for each dilution. (It is easiest to add 0.1 ml of marrow suspension from dilutions of the original.) Add to each bijou sufficient medium, growth factor and marrow suspension for the required dilution *less* 1 ml which will be made up by the addition of agar/agarose; since the agar is in medium, this 1 ml should be subtracted from the amount of medium added.

   Example: (1 bijou bottle)

	ml
growth factor (10% final concentration)	1.5
marrow cells (at appropriate dilution)	0.1 (ignore volume)
agar	1.0
total additions	2.5
total medium (15−2.5 ml)	12.5

3. Place the bijou with contents (gassed and capped) in a 37°C water bath.
4. If agarose is being used, liquify it and cool it to 37°C.
5. When both cell suspension and agarose are at 37°C, add 1 ml of agarose to the cell suspension and immediately mix thoroughly by pipetting up and down about 12 times. Use plastic pipettes for all pipetting; if glass is used, the agarose/agar will solidify inside and it is exceptionally difficult to remove. When mixed, distribute the mix to dishes, 5 ml per dish. Swirl each dish gently, and be sure not to create air bubbles through emptying the pipette vigorously.
6. Partially replace the dish lids and let cool in the hood for 5 min; this prevents undue condensation. Solidify the plates by placing in a refrigerator for 3 min, then transfer to the CO_2 incubator.
7. Read the plates at 5−7 days, checking daily after 3 days.

N.B. If agar is used, it must be kept at 40°C, added quickly to the cells, mixed rapidly and pipetted out within 1 min.

Table 2. Preparation of heart and lung conditioned medium.

1. Remove heart or lungs from a mouse.
2. Chop up into small pieces (1−2 mm) and distribute one set of lungs or one heart to 10 ml of medium (e.g. RPMI/10% FCS) in a flask. Incubate at 37°C.
3. After 5−7 days, filter the supernatant after centrifuging out debris and store frozen.
4. Assay the CM for growth activity by performing colony assays with different final concentrations of the CM, e.g. 1%, 2%, 5%, 10% and 20%, again using several marrow cell concentrations. The optimal concentration is that which produces plateau numbers of colonies consistent with each dilution of marrow cells; this should create a cell-dependent titration curve. Usually, 10% is satisfactory. Loss of colonies may be expected at high levels of CM (suppression), and linearity falls off at the lower end as well.

and whirlimix briefly to break up the agar gel.

(ii) Spin the bottle at 1500 r.p.m. for 4−5 min, decant the supernatant and resuspend the cell pellet in a small quantity of medium containing 10% FCS.

(iii) Deposit the cells onto a slide with a cytocentrifuge (600 r.p.m. for 5 min), allow to dry and stain (see *Table 3*).

This method destroys colony architecture, but the distribution of cell types may be

Table 3. Staining of colonies developed in soft agar.

1. Cut out strips of agar with a scalpel blade and gently lift onto a slide cleaned thoroughly with tissue.
2. Leave the slides to dry at room temperature for 24 h. Usually, salts in the agar will crystallize out over some part, but the rest is usable. Store at room temperature.
3. The agar should be rehydrated before staining by immersion in phosphate-buffered saline (PBS) for about 1 h. May-Grunwald/Giemsa and similar stains work well, provided the agar is not too thick.
4. If salt formation is a problem, overlay the strips with wet filter paper; as the strips dry, the salts are lifted out into the paper and it can be removed when thoroughly dry. This method can be used for whole dish blocks; lift out the agar gently and deposit onto a large glass square.
5. An alternative is to pick out colonies with a capillary tube, deposit them on slides, dry them and stain as above.

This method works well with agarose. Ordinary agar may stain as well as the cells, so be prepared to experiment. Other stains may be used (e.g. luxol fast-blue), including those developing esterase activities; however, enzyme activities do not survive very long, and they should be developed the next day.

identified, and it allows suitable staining for enzymes and special markers. This method may also be used to re-plate cells from one assay to another.

2.6.2 *Colony morphology*

In the mouse, mature granulocytes have a characteristic 'doughnut' shaped nucleus when stained; macrophages have characteristic eccentric nuclei and often a foamy or granular cytoplasm. The relative proportion of each type of mature cell varies from colony to colony, from virtually 100% mature cells of one phenotype to mixtures of macrophages and granulocytes in low proportion and colonies without any mature types at all. Without staining, colony cellular morphology is impossible to assess. Some colonies are very compact, others are very diffuse due to cell migration. Both macrophages and granulocytes are mobile cells, and the morphology depends on the agar concentration and time of development as well. No reliance can therefore be placed on the gross colony morphology to distinguish the different constituent cells. Of course, colonies from other marrow sources will have their own cell morphology as will colonies developed with other growth factors.

2.6.3 *General applicability of the technique*

The above describes a GM-c.f.u. assay. Other lineages may require different supports (e.g. methyl cellulose for erythroid progenitors and c.f.u.-mix), and will certainly need different growth factors. Particular attention should be paid to the source of growth factor and supporting medium in repeating published methods; commercially available factors and supports vary, and it is best to visit a laboratory where the colony assay needed is 'routine' so that all the nuances are explained; even so, colony assays can be temperamental and failures can be expected.

2.7 **Establishment of long-term marrow cultures**

This is included as long-term cultures are a suitable source for generating IL-3-dependent lines and may be useful as a ready-made supply of progenitor cells. The method described in *Table 4* is that devised by Dexter (1).

Table 4. Establishing a long-term marrow culture.

1. Make up Fischers medium containing 10% horse serum or 10% FCS. Add hydrocortisone succinate to a final concentration of 5×10^{-5} M.
2. Obtain femoral marrow as outlined above.
3. Use one femur's-worth of marrow cells per 25 cm^2 flask in a final volume of 10 ml of medium, and incubate at 33°C.
4. Allow the flask to grow for 5–7 days. A patchy carpet of adherent cells should develop, and the whole culture should be very cellular. Replace 50% of the medium (together with supernatant cells), or if the number of non-adherent cells is low, recharge with another femur's-worth of marrow. Thereafter, change 50% of the supernatant once a week with fresh medium. If an adherent layer does not form, no long-term culture will result regardless of changing.

The long-term culture can be checked for production of GM-c.f.u. as above. Such cultures will continue to produce progenitor cells for many months. Similar cultures have been devised for human marrow (14).

3. INTERLEUKIN 3 AND ITS PRACTICAL ASSAYS

3.1 IL-3-dependent lines

The recognition of IL-3 as a separate growth factor was initially based on the isolation of IL-3-dependent cells derived from long-term marrow cultures. The fact that they are dependent means they require more attention than other cell lines: lines must be subcultured at frequent regular intervals in the presence of the growth factor, and must be checked for dependency whenever used. Many will respond differently to other growth factors, and perhaps even to IL-3 itself. However, providing care is exercised, they may be usefully used to assay for IL-3.

3.2 Characteristics of IL-3-dependent lines

Phenotypes vary considerably. They range from basophil-like with cytoplasmic histamine to others whose lineage is uncertain. Antigenic phenotypes also are highly variable; most lines examined by the author express Fc receptors and macrophage/granulocyte antigens (15). Thy expression is variable. All require IL-3 continuously. Growth rates vary, but most grow very rapidly. At the end of the log phase (usually no more than ~48 or 72 h), decline is rapid and most semi-rich media will support their growth (RPMI, Fischers, Hams). 5% FCS is also quite sufficient; some may grow in serum-free but this depends on the serum-free medium. Iscoves modification for serum-free conditions or Northumbria Biologicals SF1 have both given good responses. Horse serum is an acceptable substitute for FCS, but it should be borne in mind that heterophile antibodies may exist (for example, horse serum contains a potent agglutinin for murine red cells) which may cloud antigenic description. The morphology also varies; some cells are small and round, others may be very pleomorphic. Most have small tufts or cytoplasmic extensions growing from one end whose function is not known (*Figures 7* and *8*). It is usually easier to acquire an established cell line than make a new one.

3.2.1 *Initiating and maintaining an established cultured line*

Usually, cultures arrive in a tube or flask. IL-3-dependent cells are very sensitive to

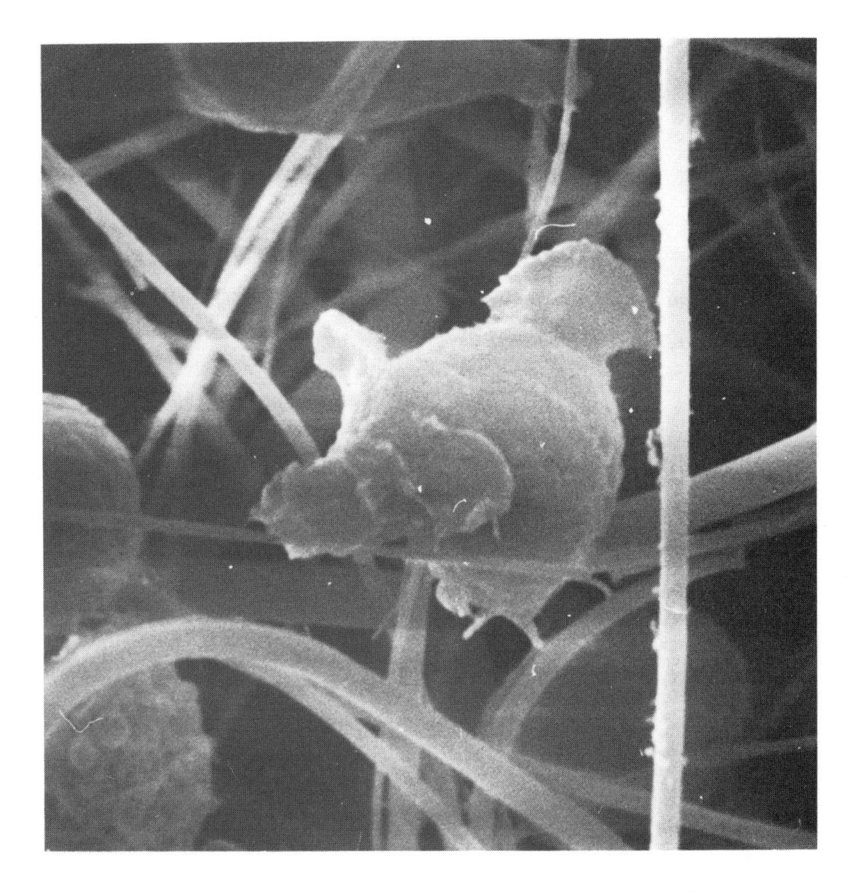

Figure 7. Scanning electron microscopy of AC2 cells. Note elongated shape and 'ears'.

prolonged shaking, therefore if the density is not high, leave the cells alone at 37°C and check for growth daily; they may be quite slow, due to being shaken up and cooled/warmed during transit. This also checks for infection. When the cells are obviously growing, subculture them at low dilution (e.g. 25%) in your own medium to adjust them gradually. If you are not already geared up, ask also for some CM from the supplier. If starting from scratch, obtain the IL-3 producer line first, and make some CM (see below). When the cells are growing, maintain them by passaging; this is easily done by 1:10 dilutions every 3–4 days depending on growth rate. Higher dilutions can be used (1:20) and temperatures reduced (e.g. to 33°C) to slow growth or increase time between passaging. If being used for experiments, it is important that not just any culture is used. Due to the short log phase, cells may vary considerably between experiments. To minimize this, subculture before the experiment and use when at a density between 1 and 5 $\times$ 10^6/ml. Keep all cultures in dry-capped flasks in a dry incubator; use a gas line for gassing.

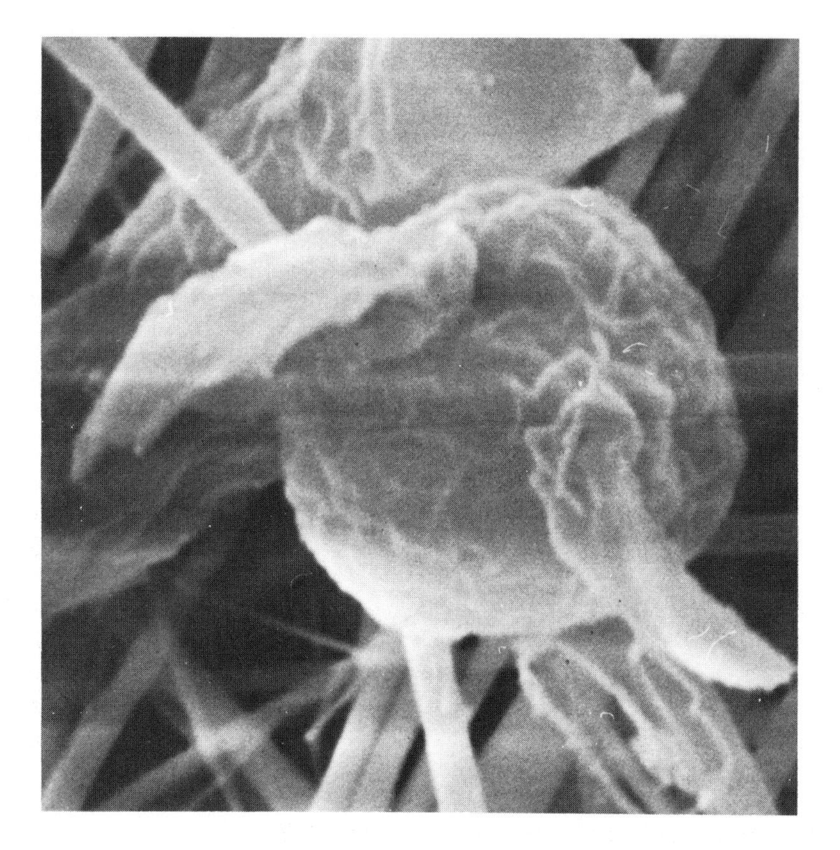

Figure 8. Scanning electron microscopy of 123 cells. This line is spherical, each cell possessing two cytoplasmic extensions.

3.2.2 *Quality control and culture inspection*

Once established, all cultures should be checked for mycoplasma contamination. To avoid cross-contamination by infection or between lines, change lines separately. Use pipettes only once, make up small amounts of medium at a time, for example 100 ml. After subbing, keep the original flask about 2 days in case something goes wrong with the subculture. Any dubious flasks should be discarded. Keep watch for cultures which grow slowly, contain dead cells and become granular; they are likely to be infected with mycoplasma. If, however, cultures are left too long before subculturing, they will contain many dead cells, will grow slowly and contain a lot of granular material from disintegrating cells. If a flask is suspect, segregate it, change it separately at very high dilution, for example 1:100, and inspect after several days.

Inspection of cultures is very important. Cultures must be incubated statically, and even shaking in experiments for 30 min will irrevocably damage the cells. Therefore, when inspecting, do not shake the flask but look at it under an inverted microscope straight from the incubator. A healthy culture will have a collection of bright cells sedimented on the bottom, with clear spaces between them. Note also the morphology

Table 5. Initiating new lines.

1.	Re-culture the 'target' cell population in medium containing differing amounts of IL-3, e.g. by varying the concentration of WEHI conditioned medium (WECM), e.g. 1%, 5%, 10%. The temperature may be 33°C or 37°C. Use 25 cm^2 flasks. Cell concentration for fresh cells should be not more than 10^6/ml.
2.	Inspect the cultures for growth daily.
3.	Maintain the cultures for 2 weeks to establish growth, changing the medium as necessary.
4.	Clone out lines by limiting dilution or agar cloning. Re-grow these clones in bulk culture, re-clone and grow up definitive lines for detailed analysis.

of cells varies with the growth phase; cultures of AC2 cells and Ea.123 are readily distinguishable in the log phase, but both are small rounded cells when entering log phase.

3.2.3 *Establishing new lines*

IL-3-dependent lines may be established from several haemopoietic sources in the mouse, such as marrow, spleen, fetal liver. Long-term cultures are highly successful in generating lines. For details of the techniques for initiating new lines see *Table 5*.

Certain points should be borne in mind. Any scheme will select out cells according to the protocol. The one given in *Table 5* will select those that grow well, will clone in agar or in single-cell culture in the concentration of IL-3 used. The age of marrow culture appears to affect the phenotypes of IL-3-dependent lines (14), and if crude lymphocyte CM is used as a source of IL-3, such lines may respond to other lymphokines from the start.

3.3 **Preparing conditioned medium as a source of IL-3**

There are a number of potential sources of IL-3; lymphocyte CM, stimulated T cell lines (e.g. EL4), but the most easily available is WEHI 3b.

3.3.1 *Preparing WEHI 3b conditioned medium (WECM)*

This hardy line will grow in low serum (1%) (but not well in serum-free medium), and rapidly produces high levels of IL-3.

(i) Inoculate a 500 ml bottle of RPMI containing 1% FCS, with cells, make sure it is gassed or not alkaline, cap it, and incubate for 2–3 weeks. Growth is easily recognizable, as a lot of lactic acid is produced and the medium goes yellow.
(ii) Harvest the CM by centrifugation (7000 r.p.m. in 250 ml bottles).
(iii) Sterilize the CM by filtration and heat treat it (60°C for 1 h) to ensure there is no carry-over of mycoplasma, and it is ready for testing and use.

Usually, 10% WECM is adequate for culture maintenance.

3.3.2 *Preparing mitogen-stimulated lymphocyte conditioned medium*

Mitogen-stimulated lymphocytes produce several lymphokines, including IL-3. Conditioned medium is therefore a useful source of IL-3, providing the presence of other factors is acceptable. Not all lymphocytes will produce IL-3 (unpublished) and as T

Table 6. Preparation of mitogen-stimulated spleen cell conditioned medium.

A simple method, using distilled water lysis to remove red cells, is as follows. Sterile distilled water and 10 times concentrated PBS are required.

1. Remove the spleen from a mouse aseptically.
2. Place the spleen in 10 ml of medium (RPMI/FCS) and disaggregate it using a pair of forceps in each hand to tear and tease it apart. Disaggregate further by syringing with a 21-gauge needle as for bone marrow.
3. Transfer to a Universal, allow to settle for 2 min and decant the cell-rich supernatant to a fresh bottle.
4. Spin the cells down, 1500 r.p.m. for 15 min, and remove the supernatant.
5. Resuspend the cell pellet in the residual medium draining to the bottom of the Universal by whirli-mixing; it should form a thick solution.
6. This requires a little practice. Have ready a 10 ml pipette containing 9 ml of distilled water and a 1 ml pipette with 1 ml 10 times concentrated PBS. Add the 9 ml of water, cap immediately, whirlimix for 10–20 sec then add the 1 ml of 10 times concentrated PBS and mix thoroughly. The red cells should all be lysed, leaving the lymphocytes intact.
7. Add 5 ml of RPMI/FCS, allow to settle for 2 min, decant and spin the supernatant at 2000 r.p.m. for 15 min. Wash the pellet once in medium by resuspension and centrifugation, and determine the cell count.
8. Weigh out sufficient Concanavalin A to produce 1 or 2 ml at 1 mg/ml in RPMI. Sterilize by filtration.
9. Plate out the cells at 1×10^6/ml and add Con A to a final concentration of 5 μg/ml. Gas and incubate at 37°C. After 3 days harvest the supernatant, spin, filter and heat treat as above.

The final stage is to assay activity. All assays are usually performed using doubling dilutions of test substance, usually in a 96-well plate.

Table 7. Titration assay for IL-3.

1. Titrate the material in question in serial doubling dilutions in a flat-bottom multiwell plate, using the medium normally used to grow the cells (but without IL-3 of course) and 100 μl volumes.
2. Include a negative control (no factor) and a positive using a known sample of IL-3 or CM.
3. Wash sufficient cells from a log phase culture by centrifuging. Use low speeds (1000 r.p.m.) and short times (2–3 min) to sediment the cells, and do not pipette vigorously to suspend. Usually, 10 ml of culture suffices with 10 ml washes. Wash three times to remove IL-3 and resuspend to a known density. The assay will function with $5 \times 10^3 - 10^5$ cells per well.
4. Add 100 μl of the washed suspension to each well and incubate in the CO_2 incubator overnight.
5. Next day, pulse each well with 1 μCi of labelled thymidine for 2–3 h, harvest and count. Before pulsing, it is always wise to visually inspect the plate, particularly if problems arise as below. The use of round-bottomed or conical well plates precludes this vital activity.

cell activation is a cascade, no lymphocyte separation should be done. Spleen cell conditioned medium is practically guaranteed to produce IL-3. The preparation of the conditioned medium is described in *Table 6*.

If lymphocyte conditioned medium is used, there may be considerable difference in the concentration needed to maintain cultures between batches, therefore always prepare and test CM well in advance.

3.4 Bioassay for IL-3 using dependent lines

This relies on the rapid death of dependent cells when IL-3 is removed. The standard assay (*Table 7*) uses incorporation of thymidine to measure residual growth after a period of incubation in the test substance.

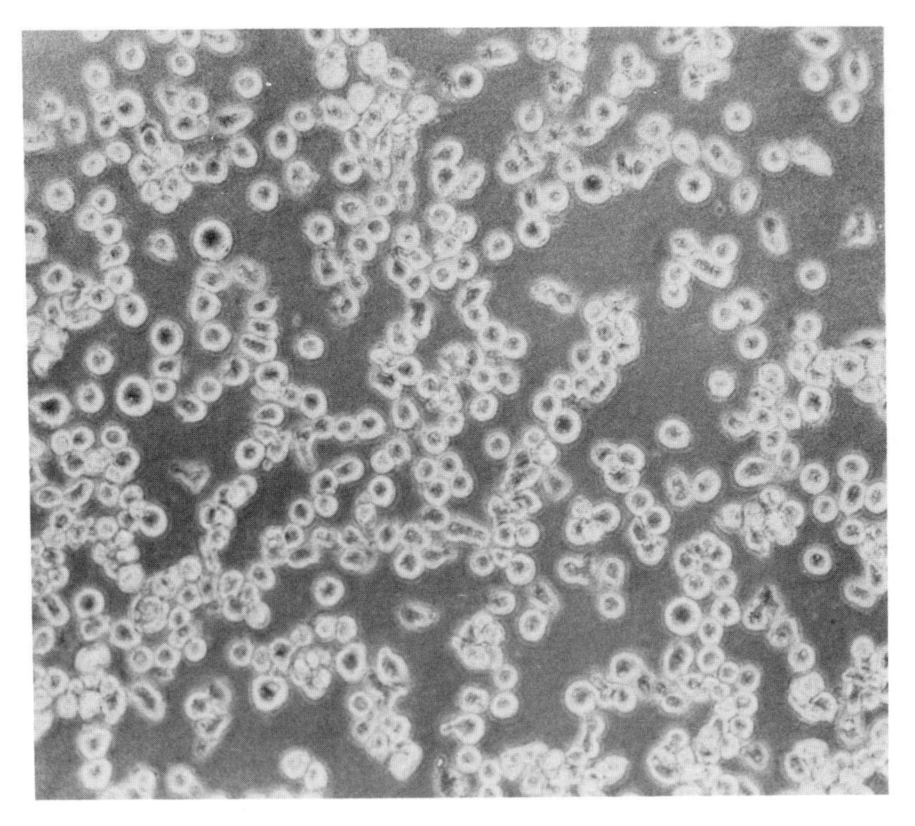

Figure 9. Viable AC2 cells. Note irregular shape and bright halos round each cell. This culture is 100% viable.

Label the plate (*Table 7*) by writing on the lid, particularly the commencing and final dilutions, remembering that the final dilutions are only 50% of the initial dilutions due to the addition of cells. The assay is read by either counting incorporation of thymidine into DNA as in *Table 7* or by viability. Viability can be determined by direct inspection. Viable cells are brightly refractile as in *Figure 9*. Dead cells lose the refractility, and become dull or ballooned (*Figure 10*). A quantitative count can be made using Trypan blue: add 50 μl of cell suspension to 50 μl of Trypan blue solution (available commercially), mix, and add to a counting chamber. Viable cells exclude the dye, therefore appear bright and white, dead cells take up the stain and appear blue.

3.4.1 *Pitfalls of the assay*

(i) IL-3-dependent cells may respond to other growth factors. Do not assume that a positive test implicitly identifies IL-3.
(ii) Values should be at least two to three times background before a 'positive' is scored.

3.4.2 *Counting problems: background responses and drift*

The biggest problem is with backgrounds. Many experimenters find with IL-3-dependent lines that 'negative control' values gradually climb over a period of time, or suddenly

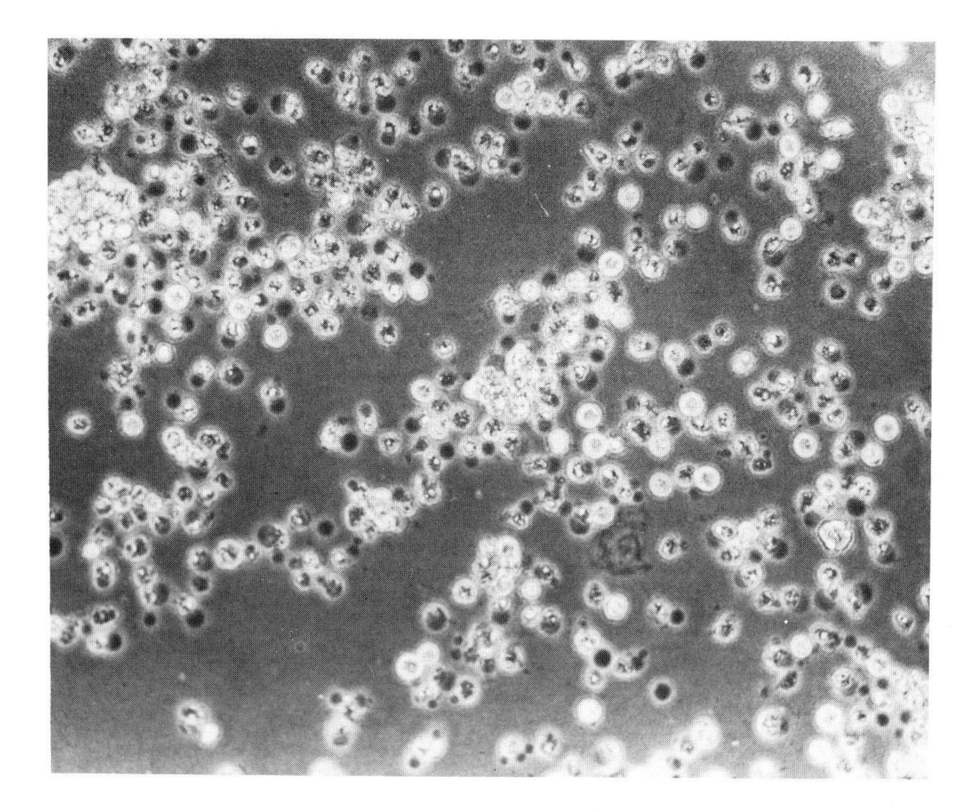

Figure 10. AC2 cells after 16 h without IL-3. Note many non-refractile cells and cytoplasmic ballooning.

rise. This may be due to the following factors.

(i) Infection with mycoplasma, which may incorporate high levels of thymidine.
(ii) Re-use of scintillation vials which collect static. This is particularly a problem with plastic vials re-used with inserts. If inserts are used, use glass vials to hold them or appropriate racks for minivials.
(iii) Contamination with other cell lines.

There may be no obvious cause. Some workers have described reversion to independence. This has happened once in the lines used by the author, although background problems have arisen more frequently.

If cultures are allowed to overgrow, this may select out cells with different characteristics. Gradually, the culture becomes significantly different from the original by 'drifting'.

The remedy for persistent high background is to re-clone the line. Because of this, it is important to keep a check on the dependency when doing other kinds of experiment; cells may have 'gone off'. Mostly, such problems right themselves within a few weeks, in the author's experience. The cause is very difficult to explain, since 'recovered' lines are just as sensitive as the original. It appears, however, that lines undergo a cyclic

Table 8. Freezing IL-3-dependent cells.

1.	The freezing mix is 30% FCS, 10% dimethyl sulphoxide (DMSO, keep frozen at −20°C for up to 6 months).
2.	Choose a culture of cell density at least 10^5/ml in log phase and in original medium; do not wash prior to freezing.
3.	Into a 30 ml bijou bottle put 3 ml of FCS and 1 ml of DMSO.
4.	Add 7 ml of cell suspension straight from the chosen culture.
5.	Swirl gently to mix and distribute 1 ml amounts to freezing ampoules.
6.	Freeze ampoules in a freezing cap or in a tray extending about 15 cm down into the neck of a liquid nitrogen Dewar.
7.	Freeze for at least 4 h, preferably overnight, then place in liquid nitrogen.

pattern of responses particularly if they are allowed to overgrow during maintenance. If no recovery occurs, it is necessary to re-clone the line or discard it and initiate a new stock from frozen (see later).

3.4.3 *Choice of assay*

Either a thymidine incorporation assay or a viability one can be used, but it is necessary to decide on the time at which the assay will be read. They may not give the same answer. Viability assays are more 'sensitive' in that cells can remain viable when not incorporating thymidine. However, the true criterion of promotion is that cells should grow. 24 and 48 h assays also may give very different responses; the 48 h one is more likely to indicate what a test substance can do to promote growth.

3.5 **Stocking IL-3-dependent cells**

IL-3-dependent cells can be easily frozen providing certain points are kept to. Use the schedule detailed in *Table 8*, which has worked well for all dependent cells examined so far. The author is greatly indebted to Dr C.Sanderson, NIMR, Mill Hill, London, for this method.

To recover cells from frozen, thaw the vial rapidly in a water bath at 37°C and immediately it thaws add to 10 ml of medium *with IL-3*. Spin gently, wash once with medium with IL-3, and culture.

3.6 **Re-cloning of cell lines**

This is straightforward. Use either the agar colony method detailed in Section 2, supplementing the agar with IL-3 CM, or use limiting/single cell dilution.

3.6.1 *Cloning by limiting cell dilution*

For this, dilute a log phase suspension to a density giving an average of 0.6−0.7 cells/100 μl, or preferably a series of dilutions from 0.1 to 3 cells/well. Add 100 μl of a dilution (in IL-3 of course) to each well in a multiplate and incubate. Inspect after 5 days; it may take a week or so for cells to grow to reasonable density. Also, wells scored as positive early on may not grow subsequently, therefore the wells should be inspected regularly between 6 and 10 days.

3.6.2 *Cloning efficiency*

The cloning efficiency varies between lines, but it is usually in the region of 25–35% in agar, and perhaps higher by limiting dilution. Despite the fact that all cells seem to be dividing or alive in bulk cultures, cloning efficiency is never 100%. When sufficient cells are growing in a well, transfer these to a flask and grow as a bulk culture. Finally check each culture for responses to and dependency on IL-3.

3.7 **20-α steroid dehydrogenase assay**

Il-3 has the ability to induce the enzyme 20-α steroid dehydrogenase in both marrow and nude mouse spleen cells. This enzyme reduces the 20-α position on progesterone to the 20-α hydroxy derivative and the enzyme activity can be estimated by the production of 20-α hydroxyprogesterone from labelled progesterone (16,17), the two compounds being separated by silica gel thin-layer chromatography (t.l.c.) (see *Table 9*). Cells are incubated with or without the putative IL-3 activity either before or during the labelling period; however, induction in spleen cells takes several hours.

Whilst the enzyme is induced by IL-3, it may also be induced by other factors, such as GM-CSF (18). Also, the enzyme is not restricted to IL-3-dependent cells (19,20), therefore care must be taken in interpretation.

3.8 **C.f.u.-mix assay**

Since IL-3 is a multipoietin, it may be used to generate mixed colonies in agar or methylcellulose. This again may not unambiguously define IL-3, but it has been very successfully used to recognize unknown multipoietin activity, from which the human IL-3 was characterized (11). The technique (*Table 10*) is similar to that used for a GM-c.f.u. assay, except methyl cellulose is used and haemoglobinization is induced by erythropoietin (21).

A couple of other assays are described in the literature:

(i) Histamine release by marrow cells (22). This assay has been used to identify IL-3 activity but the activity causing histamine release from marrow cells may not be IL-3. The literature is best consulted for the methodology.

(ii) Thy 1 induction in nude mice spleen cells (23). IL-3 induces expression of Thy 1 antigen on nu spleen cells to a variable extent, but the variation is too large for this to be of much value as an assay. Nu/nu spleen cells are obtained as above and cultured for varying lengths of time with and without the factor over 48 h. The cultured cells are harvested and stained for Thy 1 antigen using standard immunofluorescent techniques.

3.9 **Sources of IL-3-dependent cells**

IL-3-dependent cells can be grown from long-term marrow cultures and, with more difficulty, from fresh marrow. The following lines are commonly used (the responses to non-IL-3 factors should be checked with the clone received before assuming the fidelity of the responses below).

(i) *FDCP-1 and FDCP-2*. (Dexter *et al.*). Professor T.M.Dexter, Paterson Laboratories, Christie Hospital, Manchester. Both are myeloid-like. FDCP-1 may respond to IL-2 and GM-CSF, and P-2 to GM-CSF.

Table 9. Estimation of 20-α steroid dehydrogenase activity.

1. The total number of cells to be estimated should be between 2 and 5 × 10^6.
2. Wash the cells in medium and distribute equal small quantities of 1 or 2 ml to glass test tubes fitted with caps. Do not use plastic tubes.
3. Add radiolabelled progesterone ([1,2,6,7-^{3}H]progesterone) to final concentrations of $10^{-8}-10^{-10}$ M and incubate for 1–2 h. Include in the assay various concentrations of 'cold' progesterone for a given concentration of label, from 10^{-8} to 10^{-5} M. If the progesterone is in toluene, remove it by evaporation to prevent cell damage, and resuspend in a very small quantity of ethanol, e.g. 10 μl. The easiest way is to do this before adding the cells.
4. Extract the lipids by shaking with an equal volume of diethyl ether for 5 min followed by 20 min static at room temperature. Shake again, spin for 2 min at 1000 r.p.m. and freeze the aqueous layer by immersing in liquid nitrogen or by placing on solid CO_2.
5. Remove the ether layer to another tube, evaporate it to dryness with an air line and re-dissolve the lipids in 20–40 μl ether immediately prior to chromatography.
6. Spot 10 μl onto the origins of silica gel t.l.c. plates previously oven-dried and pre-run in t.l.c. solvent. Inclusion of 254 nm fluor is most helpful in identifying spots. Since the amount of lipid will not be detectable except by counting, standards (progesterone and 20-α hydroxyprogesterone) can be included in the test samples; ether containing 100 μg/ml standards can be used to re-dissolve the test samples.
7. Run the plates in ether:chloroform 3:10.
8. Dry the plates after marking the solvent front, identify the standard spots under u.v. at 254 nm and either scan for radioactivity on a strip-scanner or scrape out the standard spots and count in scintillation fluid. Progesterone runs ahead of the 20-α hydroxy derivative.

To control for the assay, thymocytes or marrow cells may be used if nude spleen cells are not available.

Table 10. C.f.u.-mix assay.

The medium commonly used is α-medium.

1. Make up a stock solution of 5% methyl cellulose in medium without serum; heat on a water bath to dissolve.
2. The final constituents should be 20–30% FCS, 0.9% methyl cellulose and the test substance. The usual source of multipoietins is 5–10% pokeweed mitogen (PWM)-stimulated lymphocyte conditioned medium; make this as for Con A spleen conditioned medium for murine studies, using 0.1% PWM. PWM usually comes as a freeze-dried powder for reconstitution in a fixed volume of water; the final concentration advised is between 5 and 10 μg/ml PWM. For human multipoietins, PWM-stimulated peripheral blood leukocyte CM is a common source.
3. Perform the c.f.u.-mix as for c.f.u.-GM, heating the methyl cellulose to liquify it. Methyl cellulose is very viscous, therefore use plastic pipettes, and make sure the solutions are not hot enough to damage the cells.
4. Incubate the plates as before.
5. To develop the erythroid colonies, erythropoietin must be added. Commercial sources are available, and the amounts are between 2 and 5 U per plate. Erythropoietin may be added at the start, or later by carefully pipetting a small amount onto the plate and rocking it to spread over the surface. Methyl cellulose plates are again viscous rather than semi-solid, and practice is required. Erythropoietin should be obtained from a source recommended by a worker doing the assay routinely. Erythroid colonies are recognized by their reddish tint. Alternatively, colonies may be stained with benzidine reagent (1 mg/ml benzidine + 1% hydrogen peroxide in PBS, made up freshly; note that benzidine is not soluble in water — dissolve in alcohol then add to buffer). Haemoglobin stains brown.

The area of mixed colonies and erythropoiesis *in vitro* is extensive; whilst the above presents an assay for c.f.u.-mix, it is strongly advised to seek the help of a specialist laboratory.

(ii) *32DCl.* (Greenberger *et al.*). Widely distributed. A myeloid line which does not respond to IL-2 but may respond to GM-CSF.

(iii) *AC2.* (Palacios and Garland). Dr J.Garland, Manchester Medical School, Oxford Road, Manchester; or European Collection of Animal Cell Cultures, PHLS, Porton Down, Wiltshire. AC2 is a basophil line, and does not respond to GM-CSF or IL-2.

(iv) *Ea.3.123.* (Palacios *et al.*). Dr R.Palacios, The Basel Institute for Immunology, Grenzacherstrasse, Basel, Switzerland.

WEHI 3b is available from ECACC above.

The ECACC provides a useful source of fully documented cell lines for lymphokine production and assay.

4. REFERENCES

1. Dexter,T.M., Allen,T. and Lajtha,L. (1977) *J. Cell Physiol.*, **91**, 335.
2. Greenberger,J. (1978) *Nature,* **275**, 752.
3. Whitlock,C., Robertson,D. and Witte,O. (1984) *J. Immunol. Methods,* **67**, 353.
4. Dorshkind,K. (1986) *J. Immunol.*, **136**, 422.
5. Greenberger,J.S., Davisson,P.R., Gans,P.J. and Moloney,W.C. (1979) *Blood,* **53**, 987.
6. Dexter,T.M., Garland,J.M., Scott,D., Scolnick,E. and Metcalf,D. (1980) *J. Exp. Med.*, **152**, 1036.
7. Fung,M.C., Hapel,A.J., Ymer,D.R., Cohen,R.M., Johnson,R.M., Campbell,H.D. and Young,I.G. (1984) *Nature,* **307**, 233.
8. Campbell,H.D., Ymer,D.R., Fung,M.C. and Young,I.G. (1985) *Eur. J. Biochem.*, **150**, 279.
9. Clark-Lewis,I., Aebersold,R., Ziltener,H., Schrader,J.W., Hood,L. and Kent,S. (1986) *Science,* **231**, 134.
10. Cohen,R.M., Hapel,A.J. and Young,I.G. (1986) *Nucleic Acids Res.*, **14**, 3641.
11. Yang,Y-C., Ciarletta,A.B., Temple,P.A., Chung,M.P., Kavacic,S., Witeck-Gianotti,J.S., Leary,A.C., Donahue,R.E., Wong,G.G. and Clark,S.C. (1986) *Cell,* **47**, 3.
12. Metcalf,D., Begley,C.G., Johnson,G.R., Nicola,N.A., Lopez,A.F. and Williamson,D.J. (1986) *Blood,* **68**, 46.
13. Kindler,V., Thorens,R., Kossodo,S., Allet,B., Eliason,J.F., Farber,N. and Vassalli,P. (1986) *Proc. Natl. Acad. Sci. USA.*, **83**, 1001.
14. Coulombel,L., Eaves,A.C. and Eaves,C.J. (1983) *Blood,* **62**, 291.
15. Garland,J.M. (1984) *Lymphokines,* **9**, 153.
16. Weinstein,Y. (1977) *J. Immunol.*, **119**, 1223.
17. Ihle,J., Rebar,C., Keller,J., Lee,J.C. and Hapel,A.J. (1982) *Immunol Rev.*, **63**, 5.
18. Hapel,A.J., Osborne,T.M., Allan,W. and Hume,D.A. (1985) *J. Immunol.*, **134**, 2492.
19. Garland,J.M., Lanotte,M. and Dexter,T.M. (1982) *Eur. J. Immunol.*, **12**, 332.
20. Fausner,A.A. and Messner,H.A. (1978) *Blood,* **52**, 1243.
21. Filho,M.A., Dy,M., Lebel,B., Luffau,G. and Hamburger,J. (1983) *Eur. J. Immunol.*, **13**, 841.
22. Ihle,J., Keller,J., Lee,J.C., Farrar,W.L. and Hapel,A.J. (1983) In *Lymphokines and Thymic Hormones: Their Potential Usefulness in Cancer Therapy.* Goldstein,A. and Chirigos,M.A. (eds), Raven Press, New York, p. 77.

CHAPTER 18

Eosinophil differentiation factor and its associated B cell growth factor activities

ANNE O'GARRA and COLIN J.SANDERSON

1. INTRODUCTION

Eosinophil differentiation factor (EDF) is a T cell derived mouse lymphokine which promotes the proliferation and differentiation of eosinophils. Although this factor is clearly a colony-stimulating factor (CSF) for the eosinophil lineage (CSF-Eo) the initial identification of the factor was carried out using liquid bone marrow cultures. Because the colony assay was too insensitive to reproducibly detect the factor it was felt the name CSF-Eo was inappropriate.

The B cell growth factor Type II (BCGF II) activity of this lymphokine was first considered when a coordinate analysis of lymphokines produced by T cell clones revealed a highly significant correlation between EDF activity and BCGF II activity (1). In addition various cell lines producing high levels of one of the factors were also high for the other (2). A T cell hybrid NIMP-TH1 has recently been described which secretes both these activities, but produces no other lymphokines (2,3). The BCGF II and EDF produced by these cells co-purify in every fractionation procedure employed: both activities are associated with a protein with an approximate molecular weight of 44 000 (2), and a pI of 5.0. Because this lymphokine showed two different biological activities the name interleukin-4 (IL-4) was proposed (2). However, because of the confusion arising from the duplication of this term (4,5), we revert to our original name of EDF. The factor has now been shown to have no effect on resting B cells but to induce both DNA synthesis and Ig secretion in naturally occurring large B cells (presumably pre-activated). Furthermore, all its bioactivities are associated with a protein band with a molecular weight of 44 000 on sodium dodecyl sulphate−polyacrylamide gel electrophoresis (SDS−PAGE) (6; and *Figure 1*). Human EDF has very recently been cloned, and found not to stimulate B cells. EDF is likely to be renamed IL-5.

2. SOURCE OF EDF

A variety of T cell sources have been shown to produce EDF (2), and a description of their production and maintenance is given below. However, most lines produce a range of lymphokine activities, which are extremely difficult to separate using standard biochemical techniques. Thus, T cell hybrids are a particularly useful source since they may be selected for the production of a limited range of activities, due to their unstable phenotype.

The medium used throughout, for growing the T cell sources of lymphokines, is RPMI 1640 growth medium (see Section 4.2), with the appropriate amounts of fetal calf serum (FCS).

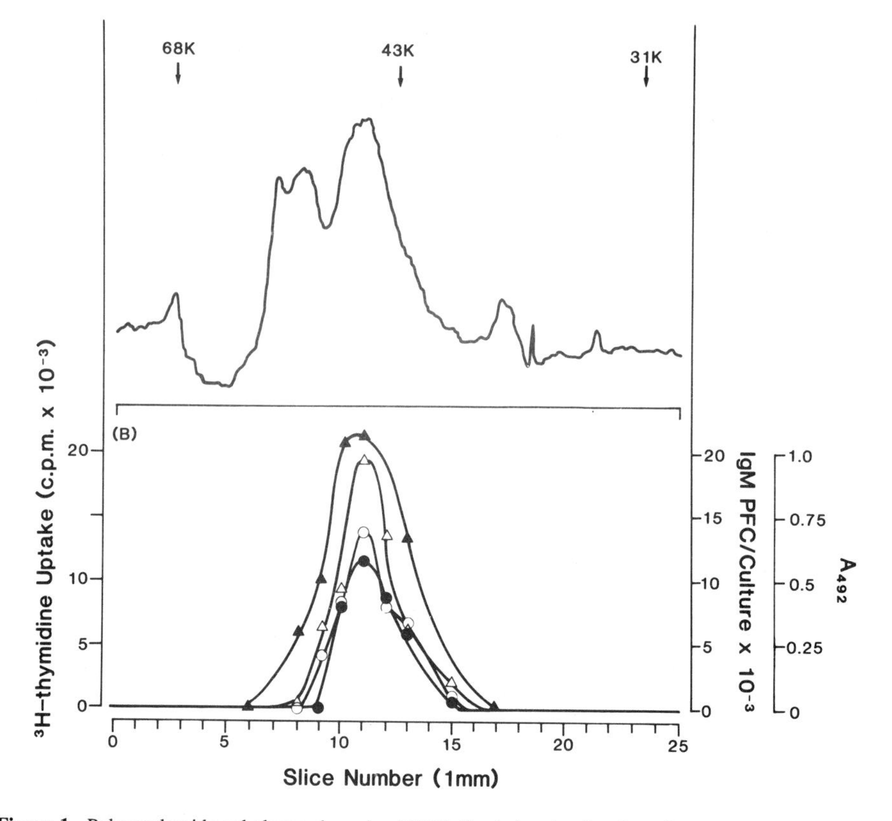

Figure 1. Polyacrylamide gel electrophoresis of EDF. Pooled active fractions from reverse-phase h.p.l.c. were separated by SDS−PAGE under non-reducing conditions. (**A**) Densitometric scan of a silver-stained portion of the gel, with positions of molecular weight markers (mol. wt $\times$ 10^{-3}) indicated. (**B**) Eluted material from each gel slice was assayed (at a final dilution of 1:40) for BCGF II activity, using the BCL_1 assay (●), for EDF activity (▲) and for its capacity to induce DNA synthesis (○) or IgM plaque-forming cells (△) in large B cells isolated from normal murine spleens. Modified from ref. 6.

2.1 EL4 thymoma

The EL4 thymoma (7,8) can be obtained from the European Collection of Animal Cell Cultures (ECACC), PHLS, Porton Down. However, it should be said that although some subclones of the EL4 cell line produce significant amounts of EDF, only sophisticated purification procedures employed to date will separate all traces of other lymphokines also produced by these cells, such as IL-2 (9). The cells may be maintained in RPMI 1640/5% FCS (see Section 4.2).

2.1.1 *Growth and stimulation of EL4 thymoma*

(i) Grow the cells to a concentration of 5×10^5/ml in RPMI 1640, containing a final concentration of 2% FCS.

(ii) Stimulate with 5 ng per ml phorbol 12-myristate 13-acetate (PMA), at 37°C. Stock solution of PMA: 1 mg/ml in dimethyl sulphoxide (DMSO), then dilute to 100 µg/ml in RPMI and store at −20°C. [Extreme care must be taken when handling PMA, since this compound is known to be a tumour promoter. Always wear gloves and rinse out glass wear thoroughly with water and detergent. Solutions of PMA should be made up in a recirculating or Class II microbiological safety cabinet, but not in a positive pressure cabinet.]

(iii) Harvest the supernatant after 24 h, by centrifugation, filter (0.22 μm) and store at −20°C.

Batches of up to 100 ml may be obtained by growing and stimulating the cells in standard plastic tissue culture flasks. Larger batches of conditioned medium may be obtained using large-scale re-useable tissue culture vessels, MCS Microcarrier stirrers (Techne, Cambridge, UK), which should be kept scrupulously clean and can be sterilized by autoclaving.

2.1.2 *Large-scale growth and stimulation of EL4 cells*

(i) *For 1 litre.*

(1) Grow up a final volume of 50 ml of cells to approximately 5×10^5/ml, in each of three plastic tissue culture flasks in supplemented RPMI/5% FCS.
(2) Transfer the total of 150 ml of cells to the 1 l flask and pour 350 ml of RPMI/5% FCS, carefully, into one of the side-arms of the vessel, without allowing any liquid to touch the neck. This may be done by tilting the flask slightly, so that the side-arm is vertical.
(3) Suck off any drops of medium at the neck with a sterile pipette, and flame the glass before sealing.
(4) Stand the vessel on the magnetic stirrer provided by Techne, at 37°C, for a few days until the concentration of cells is approximately 5×10^5/ml.
(5) Add 500 ml of RPMI, without FCS, to the vessel, in the same way as described above. (It is advisable to wipe the outside of the vessel completely with 70% alcohol, before opening, in order to avoid contamination and all procedures should be carried out in a laminar flow cabinet.)
(6) Return the vessel to the magnetic stirrer at 37°C, and leave until the cell concentration is 5×10^5/ml. At this stage, add PMA, as described above.

(ii) *For 2 litres.* The procedure is essentially as for 1 l, with the following modification.

(1) Add the initial 150 ml of cells to 850 ml of supplemented RPMI/5% FCS, in the 2 l vessel.
(2) Grow to approximately 5×10^5/ml and add 1 l of RPMI without FCS. The procedure is then essentially as above.

(iii) *For 8 litres.*

(1) Grow up 2 l of cells as above (ii).
(2) On the final split add 1 l of RPMI/5% FCS and incubate until the cell concentration is approximately 5×10^5/ml.
(3) Transfer the cells by pouring carefully as above, into the 8 l vessel, which will already contain 6 l of RPMI/2% FCS. (Prepare this in advance by directly filtering the medium into the 8 l vessel.)
(4) Grow and stimulate as in (i).

2.2 **Alloreactive T cell clones**

Although these cells usually produce very high titres of lymphokines (1) they have some disadvantages. Firstly they need to be stimulated weekly in the presence of large numbers of stimulator cells, or antigen presenting cells, thus making large-scale production difficult and expensive. Like the EL4 thymoma they invariably produce more than one lymphokine, which makes purification difficult. These lines, however, have the advantage of being antigen-specific, and this feature rules out the possibility of mitogens contaminating the conditioned medium containing the factors. We present here a description of the production, maintenance and stimulation of an alloreactive T cell as an example of one of these T cell sources.

(i) *Responder cells: preparation of spleen cell suspension.*

(1) Remove spleens aseptically from BALB/c mice (responders).
(2) Disrupt the tissue by placing on a sterile stainless steel grid (80 holes/inch), in a sterile Petri dish, and pressing through the grid with the barrel of a 5 ml plastic syringe, in 5 ml of ammonium chloride (0.83%).
(3) Transfer the resulting cell suspension to a centrifuge tube and wash out the grid and the Petri dish with a further 5 ml of ammonium chloride and centrifuge the cells at 400 *g*, for 5 min.
(4) Wash the cells twice in RPMI/5% FCS, resuspend in RPMI/10% FCS, and count.

(ii) *Stimulator cells.*

(1) One hour after irradiating CBA mice with 1200 rads, remove their spleens, and prepare a cell suspension, as above (stimulators).
(2) Deplete the irradiated CBA stimulator cells of T cells by treating spleen cells with a monoclonal anti-Thy 1 antibody, for example NIM-R1 (10; and obtainable from SeraLab), or any other commercially available equivalent, and guinea pig complement (GPC′). [GPC′ may be obtained by exsanguinating anaesthetized animals by cardiac puncture. Serum is obtained from the clotted blood and centrifuged at 1000 *g* at 4°C. The 'complement' is then diluted 1 in 3 (stock), filtered through a 0.22 μm filter, and stored at −70°C. GPC′ is also commercially available from firms such as SeraLab.]
(3) Routinely, for one mouse spleen, resuspend the pellet in 1.9 volumes of RPMI/5% FCS, 1 volume of stock GPC′ and 0.1 volume of stock NIM-R1, for 45 min, at 37°C. Wash the cells three times in RPMI/5% FCS , and then resuspend in RPMI/10% FCS and count.

(iii) *T-cell cloning by limiting dilution.*

(1) Briefly, culture the responder cells (100 cells/well) with 2×10^5 irradiated stimulator cells and IL-2, in round-bottomed microplates (Nunc, Copenhagen, Denmark), at 37°C in a humidified incubator, supplied with 5% CO_2 in air, in a total volume of 100 μl of RPMI/10% pre-selected FCS.
(2) Add a further 100 μl of medium containing IL-2 to each well. (IL-2 is produced by the EL4 thymoma, and can be obtained in conditioned medium, as described

for EDF, in Section 2.1 or in Chapter 16. Pre-select FCS using the procedure outlined in Section 2.3.)

(3) After 14 days wash the cells by removing the medium with a Pasteur pipette, and re-stimulate with irradiated T-depleted stimulator CBA cells (as described above) in a total volume of 200 μl of medium without added IL-2.

(4) After 24 h, remove the supernatants from wells showing T cell growth, for assay of the EDF or BCGF II and replace with fresh medium containing IL-2.

Only use T cell colonies from plates where less than 20% of the wells show growth. Thus, on assumption of a Poisson distribution for the number of cells seeded in each well, less than 2% of the colonies would be expected to contain more than one clone.

(5) Transfer each T cell colony, producing levels of factor which are 5-fold higher than the controls, into 24-well cluster plates (1 ml/well: Costar, Data Packaging, Cambridge, MA).

(6) Maintain these T cells by stimulating them at 2×10^5 cells/ml, with irradiated CBA spleen cells at 2×10^6/ml, at weekly intervals.

(7) Stimulate the cells in the absence of IL-2, for 24 h, and then collect the supernatants by centrifuging the cells at 400 *g*.

(8) Resuspend in RPMI/10% FCS, and IL-2, at a concentration of 2×10^5 viable T cells/ml. (Viability can be deduced using Trypan blue, at a final concentration of 0.2%. Cells which take up the dye are dead and are not counted as viable.)

(9) Transfer the cells to flasks, as they expand. In between stimulation, it is invariably necessary to split the cells into fresh RPMI/10% FCS, containing IL-2.

2.3 T hybrid cell lines

T hybrid lines appear to lose the capacity to produce lymphokines, and so repeated re-cloning is necessary to maintain active producers. Cloning efficiency depends not only on the individual hybrid, but also on the batch of FCS used. Preliminary experiments should be undertaken by cloning the parent BW5147 lymphoma in different batches of FCS, to identify a batch giving the highest cloning efficiency.

(i) Aliquot 1.5 ml of each batch of FCS into sterile tubes, add 8.5 ml of growth medium, and then add 100–200 cells in a small volume of medium.

(ii) Dispense 100 μl of this suspension into each of 48 wells of a flat-bottomed 96-well tissue culture grade microplate.

(iii) Incubate for 10–14 days at 37°C in 5% CO_2, when the clones should be easily visible and then select the batch of serum giving the greatest number of clones. Hybrid cell lines and the BW5147 may be cloned in this selected FCS in a similar way.

Spleen cells, prepared as in Section 2.2, from (CBA $\times$ BALB/c)F_1 mice infected 2 weeks previously with *Mesocestoides corti* (see Section 4.1.1) should be used for the fusion . These spleen cells may then be hybridized with an azaguanine-resistant subline of the AKR-derived thymoma BW5147 (to be made available at the ECACC) by a modification (3) of the method of Galfre *et al.* (11).

(i) Mix spleen cells (2×10^7) and 2×10^7 BW5147 in a 50 ml centrifuge tube and wash twice with serum-free RPMI 1640 (supplemented as described).

(ii) Initiate the fusion by slowly adding 1 ml of 50% (w/v) polyethylene glycol (PEG 4000; BDH Chemicals, Poole, UK) in serum-free RPMI to the cell pellet.

(iii) Dilute the suspension with 10 ml of serum-free RPMI followed by 40 ml of supplemented RPMI, containing 15% FCS, and incubate for 40 min at 37°C.

(iv) Centrifuge at 250 *g* for 5 min, and then resuspend the cells in RPMI/15% FCS, supplemented with 100 μM hypoxanthine, 100 nM aminopterin, 16 μM thymidine and 10 μM 2′-deoxycytidine (Sigma).

(v) Plate out into four 96-well flat-bottomed plates (Nunc, Roskilde, Denmark) in 200 μl volumes and maintain at 37°C in a humidified atmosphere of 5% CO_2 in air. (Hybrid growth is usually apparent by day 9, although this can sometimes take longer.)

(vi) Transfer cultures, at this stage to medium lacking aminopterin and replace the medium every 3–4 days.

(vii) Screen hybridomas for the production of EDF or BCGF II when cell colonies are 50% confluent.

2.3.1 *Production of replicate hybridoma plate*

(i) Before stimulation with PMA, split the hybridoma colonies into duplicate wells in separate flat-bottomed microplates, by resuspending the cells with a Pasteur pipette, and removing half the volume into a replica plate.

(ii) Incubate overnight, or until the wells are again 50% confluent, in supplemented RPMI/15% FCS.

(iii) Use the cells in one of the plates to produce supernatants for assay by stimulation for 24 h in RPMI/5% FCS containing 5 ng/ml PMA.

(iv) Freeze down the cells in the other microplate (master plate), and then use to recover hybridomas after the supernatants have been assayed for factor activity.

   (a) Aspirate the supernatant fluid over the hybridoma colonies and replace with 50 μl of 10% DMSO in RPMI/50% FCS.

   (b) Cover the plates with Nescofilm and then freeze in a programmable liquid nitrogen vapour freezing cabinet at 1°C/min. If this apparatus is not available insulate the plate with layers of tissue paper (3–4 cm thick), or place in a protective postal bag (Jiffy bag) and store at −70°C.

   (c) Recover the hybridomas which are positive for EDF/BCGF II, by adding 150 μl of RPMI/15% FCS (pre-heated to 56°C) and transfer to a well of a 24-well cluster plate in 1 ml of RPMI/15% FCS.

   (d) After 1–2 days when the cells have started to recover, clone by limiting dilution, by plating out the colony, after counting, into three microplates each at 0.5, 1 or 5 cells/well.

   (e) Feed cultures every 3–4 days, and treat as above.

After at least three re-clonings it will be possible to grow up large volumes of the hybrid as described for the EL4 thymoma, in Section 2.1. However it is essential that you store ampoules of freshly re-cloned cells in liquid nitrogen, and that one ampoule is only used for one culture of 2–4 l.

3. PURIFICATION OF EDF

A complete purification of EDF has not yet been published, however, because NIMP-TH1 produces undetectable amounts of other lymphokines, the following procedure gives material suitable for most biological applications. Some subclones of the EL4 cell line produce significant levels of EDF, but this purification procedure will not separate all traces of other lymphokines produced by these cells.

3.1 Production and concentration of conditioned medium

To ensure high levels of EDF a single high producing clone of (NIMP-TH1) is expanded to 8 l of RPMI/2% FCS (Section 2.3). Stimulate the cells essentially as for the EL4 thymoma (Section 2.1). Briefly, stimulate the cells with 5 ng/ml of PMA when the cell concentration is 5×10^5/ml. Remove the cells by centrifugation and discard. The supernatant is concentrated to approximately one tenth the original volume by pumping through a Pyrosart Pyrogen filter (Sartorius GmbH, Gottingen, FRG).

3.2 Ammonium sulphate precipitation

(i) Bring the concentrated supernatant to 45% saturated ammonium sulphate by adding 278 g/l of the solid salt, and stir for at least 1 h.
(ii) Centrifuge at 5000 *g* for 30 min and carefully recover the supernatant.
(iii) Discard the precipitate, and bring the supernatant to 85% saturation by adding 286 g/l of ammonium sulphate and stir for at least 2 h.
(iv) Centrifuge as before and discard the supernatant.
(v) Dissolve the precipitate in phosphate-buffered saline (PBS), pH 7.4, keeping the volume as small as possible.
(vi) Clarify by centrifugation at 5000 *g* for 30 min.

3.3 Isolation of glycoprotein fraction

(i) Pass this material (without further treatment) through a jacketed column of lentil lectin-coupled Sepharose. This can be purchased commercially (LcA-Sepharose, Pharmacia, Uppsala, Sweden) or prepared in the laboratory.
(ii) Collect the non-binding material in the flow-through peak and assay for activity. If significant activity is present the column is over-loaded.
(iii) Wash the column until the protein level returns to baseline, elute the glycoprotein fraction at 45°C with PBS containing 0.3 mol/l methyl α-mannoside and 0.2% Tween-20.
(iv) Under these conditions a discrete protein peak is eluted with the eluting buffer interface, but EDF elutes over a larger volume. Test the fractions for activity.
(v) Discard fractions containing the major eluted glycoproteins and pool the subsequent active fractions.

3.4 Gel filtration

Concentrate the pooled eluate from the lentil lectin column on a Diaflo apparatus using a YM10 membrane (Amicon Corporation, Danvers, MA), and fractionate by gel filtration on an Ultragel AcA54 column (LKB), using PBS, pH 7.4, containing 0.20% Tween-20 and 50 μg/ml PEG. Assay the eluted fractions and pool the most active.

3.5 Reverse phase h.p.l.c.

(i) Bring the pooled gel filtration fractions to 5% acetonitrile and 0.1% trifluoroacetic acid (TFA), and load on to a μ-Bondapak C_{18} RP-h.p.l.c. column (Waters Associates, Milford, MA.).
(ii) Elute the column at 1 ml/min with a gradient from 5 to 40% acetonitrile over 15 min, and then from 40 to 60% acetonitrile over 40 min.
(iii) Test the fractions for activity and pool the most active.
(iv) Freeze dry and reconstitute when required.

This material is unstable in solution and, so for biological assays, it should be stored in the presence of 1% bovine serum albumin as carrier protein.

4. EOSINOPHIL DIFFERENTIATION ACTIVITY

EDF stimulates the differentiation of eosinophils from committed eosinophil progenitor cells. It is therefore analogous to the CSF described for other myeloid cell lineages (see Chapter 17). However, at least in the mouse, very few eosinophil colonies are produced in standard semi-solid agar cultures (3), and so the term CSF-Eo was not used for this factor. The most sensitive assay for EDF is the production of mature eosinophils in liquid cultures, and EDF can be used for the production of eosinophils in bulk bone marrow cultures (12). It is most important to remember that normal mice have very few eosinophil progenitor cells, so that EDF added to normal bone marrow cultures results in only a relatively small number of mature eosinophils. A standard preparation of EDF allows the indirect assay of relative numbers of eosinophil progenitor cells, by assay for the number of mature eosinophils produced (13).

To increase the sensitivity of the assay, bone marrow from mice undergoing an eosinophilia should be used. In our hands the most convenient method for inducing an eosinophilia, is to infect the mice wth tetrathyridia (second stage larvae) of the cestode *M. corti* (14). This parasite has very low, if any, infectivity for man. The larval form is passaged by intraperitoneal injection, and under conditions of good laboratory hygiene, this form is unable to complete the life cycle, and so cannot infect other mice.

4.1 *Mesocestoides corti*

4.1.1 *Maintenance of M. corti*

(i) Inject approximately 100 μl of packed *M. corti* tetrathyridia into the peritoneum of BALB/c mice, using a 0.8 mm diameter needle. It is easiest to fill the syringe without the needle in place, but the tetrathyridia pass through this needle without being damaged. The tetrathyridia increase in numbers progressively over several months, and although the mice become distended with the volume of parasites and ascitic fluid in the peritoneum, they do not appear to suffer from the infection. Individual infected mice can be maintained for months as a source of parasite larvae.

(ii) After killing the mice collect the larvae by carefully removing the skin over the abdomen and incising the muscle layers so that the contents of the peritoneum fall into a Petri dish containing sterile PBS.
(iii) Transfer into a sterile tube and wash several times with PBS, by allowing the larvae to settle under gravity and removing the supernatant fluid.

The larvae survive for months in PBS at 4°C, providing the fluid remains sterile. Viability can be checked by placing a drop on a microscope slide, allowing to warm slightly and observing for movement under a low power microscope.

4.1.2 *M. corti-infected mice as a source of cells*

Maximum numbers of eosinophil progenitors are present in the bone marrow from 10 to 18 days after infection (13), and maximum numbers of eosinophils are present in the peritoneal exudate from 21 to 28 days (*Figure* 2).

Bone marrow cells are recovered from the femur as described in Chapter 17.

For recovery of peritoneal exudate cells, the following procedure should be used.

(i) Inject 5 ml of bench medium into the peritoneum with a 0.8 mm needle.
(ii) Withdraw the fluid using the same needle. Repeat once to ensure a good recovery of cells.
(iii) Allow any parasites to settle out under gravity and recover the cells in the supernatant. Wash the cells as soon as possible by centrifugation.

4.2 Tissue culture

Carry out all tissue culture in supplemented RPMI 1640. Three forms of this medium are used.

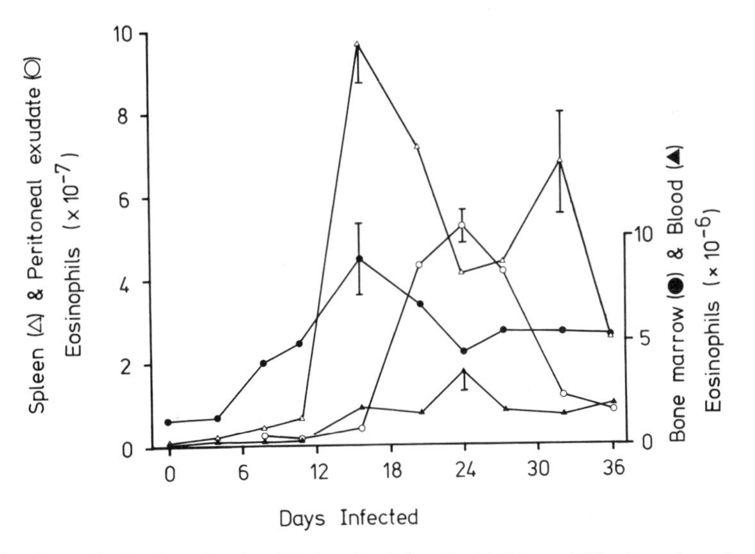

Figure 2. Tissue distribution of eosinophils in mice infected with *M. corti*. Total number of eosinophils in peritoneal exudate (○), bone marrow (●) and spleen (△). Eosinophils/ml of blood (▲). Mean of three mice in each case, but for clarity errors (±1 SD) are only shown on representative points. Ref. 13.

(i) Bench medium lacking bicarbonate buffer but containing 20 mmol/l Hepes buffer and 5% FCS. This is used for all stages of cell preparation.
(ii) Growth medium containing both bicarbonate and 10 mmol/l Hepes, with FCS. Cells are suspended in this medium at the final step before incubating in a humidified atmosphere of 5% CO_2 in air.
(iii) Bone marrow medium is growth medium supplemented with 10^{-6} mol/l of hydrocortisone sodium succinate, and 15% FCS. Add the following to RPMI 1640 (containing no buffer);

Bench medium	*Growth medium*
20 mmol/l Hepes	10 mmol/l Hepes
2 mmol/l glutamine	2 mmol/l glutamine
	24 mmol/l sodium bicarbonate
	1 mmol/l sodium pyruvate
	75 μmol/l monothioglycerol
100 U/ml penicillin	100 U/ml penicillin
100 μg/ml streptomycin	100 μg/ml streptomycin
5% FCS	5–15% FCS

4.3 Colony assay

Suspending cells in semi-solid growth medium restricts the movement of cells so that cells proliferating from a single precursor cell appear in a group or colony. The basis and background of this assay has been documented in a number of reviews (15).

Although published accounts suggest that eosinophil colonies can be identified in unstained preparations on the basis of their gross architecture, our experience is that colonies with staining characteristics typical of eosinophils occur with both 'tight' and 'loose' characteristics. It is therefore important to stain the cells to allow identification. This can be done by removing the colony from the semi-solid medium to a microscope slide and staining. However, this is a time-consuming and laborious process, and so techniques have been developed for staining the cells directly after drying the semi-solid medium.

The following assay system is our modification of the classical colony assays in agar, which is particularly suitable for staining and counting colonies in large numbers of samples. In place of Petri dishes, usually used for these cultures, this modification uses the wells of Leucocyte Migration Plates (Sterilin, Teddington, UK). Three of these cultures fit on a standard microscope slide.

It is important to note that different batches of FCS vary in their capacity to stimulate colony growth and so should be pre-selected. Also, batches of agar vary and should be pre-selected. In our hands the more purified forms of agar and agarose appear to be less satisfactory than the crude preparations.

4.3.1 *Semi-solid agar cultures*

(i) Prepare Difco Bacto-Agar (Difco Laboratories, East Molesey, Surrey, UK) by suspending 5 g in 100 ml of distilled water, and place in a boiling water bath for 5 min (do not autoclave). Store in solidified form. Before use, melt in a boiling water bath.
(ii) Prepare growth medium with 15% FCS and hold on a 37°C water bath.

(iii) Aliquot the growth factors as 40 μl volumes into the wells of the Leucocyte Migration Plate.
(iv) Add cells (to a final concentration of 2×10^5/ml) to pre-warmed (37°C) growth medium, and then add 1 volume of hot agar to 15 volumes of this cell suspension. Mix to disperse the cells and agar and immediately dispense 0.4 ml into each well. Allow the agar to solidify. If the ambient temperature is high leave the plates at 4°C for a few minutes.
(v) Place the plate in a 100 mm square Petri dish (Sterilin, Teddington, UK) containing a piece of filter paper moistened with a few drops of water. Incubate at 37°C in a humidified atmosphere of 5% CO_2 in air.
(iv) After 5–7 days colonies of cells should be visible and can be counted under a binocular dissecting microscope.

4.3.2 *Staining cells in agar cultures*

(i) Place the Leucocyte Migration Plate at an angle of about 45°. Use a stream of PBS from a wash bottle to release the agar from the well on to a microscope slide, held closely below the rim of the well. Three cultures will fit on a standard slide, and can be positioned accurately while still wet.
(ii) Cover the pieces of agar with a strip of Whatman No. 1 filter paper, and dry on a warm plate or under a hot air dryer. When dry, remove the filter paper (moisten if necessary) and fix in methanol for 5 min. The slides can be stored dry at this stage.
(iii) Stain the slices for 10 min with 0.5% Congo Red in 50% ethanol (prepare by dissolving 0.5 g of dye in 50 ml of water, then adding ethanol). Batches of Congo Red vary in their staining ability. At present we use Gurr's Congo Red (BDH, Poole, Dorset, UK).
(iv) Wash the slides in ethanol for 5 min.
(v) Stain the slides for 5 min with saturated toluidine blue (prepared by adding 1 g of stain to 100 ml of methanol, and acidifying to 0.1 mol/l HCl).
(vi) Rinse in water and air dry. Cover with a layer of immersion oil and examine first under low power, changing to high power oil immersion for detailed examination of the cells in each colony.

Eosinophils show characteristic reddish-brown granules. Mast cells show distinctive dark blue metachromatic granules. Neutrophils are polymorphonuclear, without cytoplasmic granules. Macrophages are mononuclear with extensive frothy cytoplasm.

4.4 **Long-term bone marrow cultures**

The long-term bone marrow cultures described independently by Dexter and Greenberger give rise to mature neutrophils, which are released from the adherent layer of cells on the bottom of the flasks, to remain free in the culture fluid. This apparently spontaneous differentiation appears to be due to CSF produced by adherent cells in the cultures. In our preliminary attempts to produce eosinophils in these cultures, it was found that even using bone marrow from mice undergoing eosinophilia, and containing large numbers of immature eosinophils, no mature eosinophils were produced. However, if lectin-stimulated spleen supernatants were added, a significant number

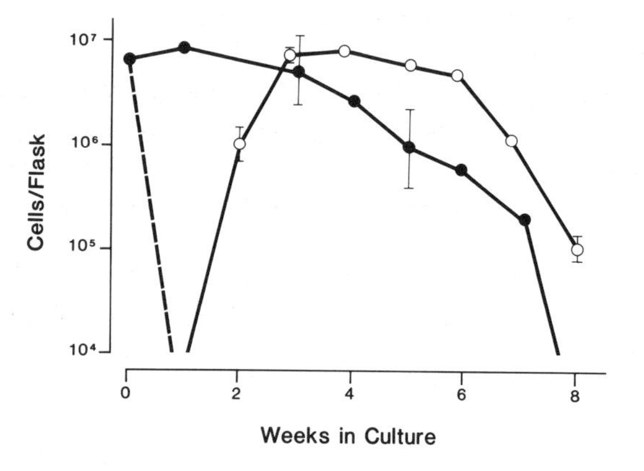

Figure 3. Number of eosinophils produced in bone marrow cultures in 25 cm^2 flasks. Control flasks without added EDF, the number of eosinophils falls within 7 days and remains below baseline (---). Cultures with EDF added at time zero, and replenished every 7 days by replacing half the medium with fresh medium containing EDF (●). Cultures maintained in the absence of EDF for 7 days, and then EDF added every 7 days (○). Mean of three flasks ± 1 SD. Modified from ref. 12.

of mature eosinophils appeared in the cultures (14). This formed the basis for the identification of EDF, by providing an assay for the lymphokine and provided a source of eosinophils for functional studies. Although in these original reports on long-term bone marrow cultures Fishers medium or alpha MEM was used, we have found that RPMI 1640 growth medium, supplemented with hydrocortisone, supports the cultures very well. As pointed out above, normal bone marrow contains very few eosinophil progenitors, and a much greater production of eosinophils is achieved with marrow from mice undergoing an eosinophilia.

4.4.1 *Maintenance of culture for eosinophil production (14)*

(i) Use 10 ml for 25 cm^2 tissue culture flasks, and proportionate volumes for larger flasks.
(ii) Add an optimal dilution of a source of EDF, this is best determined in the microplate assay (Section 4.4).
(iii) Every 7 days remove half the medium and replace with fresh medium containing more EDF.

When large numbers of eosinophils are required, all the medium can be removed, but this tends to shorten the life of the cultures. To ensure that all the eosinophils have been produced *in vitro*, the addition of EDF can be delayed for 1 week, so that all the resident bone marrow eosinophils have disappeared (*Figure 3*).

4.4.2 *Purification of eosinophils produced in vitro*

Highly active sources of EDF produce at least 80% eosinophils. The other cells are mostly neutrophils. Macrophages are only released into the non-adherent fraction in unsatisfactory or old cultures. Eosinophils in the mouse have a lower buoyant density than neutrophils (16) (the converse is true in man), and a satisfactory separation of the two cell types can be achieved by isopycnic centrifugation. We routinely use

Metrizamide [Nycomed (UK) Ltd., Sheldon, Birmingham, UK], although other high-density media produced for cell separations could be used.

(i) Prepare a 35.3% solution of Metrizamide in water, and sterilize by filtering through a 0.45 μm filter. This solution should have a refractive index of 1.3895 and is iso-osmotic.
(ii) Dilute to a Metrizamide concentration of 17% in PBS and add FCS to a final concentration of 2% (refractive index 1.3596). This concentration of Metrizamide has been determined experimentally to give the best separation of eosinophils from neutrophils, but could be varied if the separation is not satisfactory. For example, if a significant proportion of neutrophils remain at the interface, lower the Metrizamide concentration and, conversely, if too many eosinophils centrifuge to the pellet, increase the concentration (in steps of 0.5%).
(iii) Place 2 ml of Metrizamide solution in a 10 mm diameter sterile plastic centrifuge tube. Carefully layer 2–5 ml of cell suspension (up to 2×10^7/tube) in bench medium over the Metrizamide solution, and centrifuge for 15 min at 1000 *g* at 10–15°C. Recover the eosinophils from the interface, and carry out a differential count to check the purification achieved.

4.5 Microassay for the production of eosinophils (12)

(i) Take bone marrow from mice infected with *M. corti* (Section 4.1.2) and aliquot 10^6/ml cells in growth medium containing 15% FCS and 10^{-6} mol/l hydrocortisone. Aliquot 100 μl into the wells of a round-bottomed ('U') 96-well microplate.
(ii) Aliquot 10 μl volumes of samples for the assay of eosinophil differentiation activity into the wells (in practice this is most conveniently done before adding the cells).
(iii) Incubate the plate in a humidified atmosphere of 5% CO_2 in air at 37°C for 5–7 days.
(iv) Assay for eosinophils either by cell count (Section 4.4.1) or by assay for eosinophil peroxidase (Section 4.5.2).

4.5.1 *Estimation of eosinophils by cell count*

(i) Carry out a total cell count either microscopically using a haemocytometer or by means of an electronic particle counter such as a Coulter counter.
(ii) Carry out a differential count after staining with a Giemsa stain. Slides can be prepared by smearing the cell suspension on to a microscope slide and allowing to dry, by depositing the cells on the slide by means of a cytocentrifuge or by adhering the cells to the slide using the lectin concanavalin A (Con A). The cytocentrifuge gives the best cell morphology but Con A adherence is particularly adaptable to large numbers of samples.
    (a) Use 15-well multi-test slides (Flow Laboratories) in which the slide is coated with non-wettable material leaving 4 mm diameter wells of uncoated glass.
    (b) Place 2 μl of a 0.1 mg/ml solution of Con A in PBS in each well. Keep in a humidified container to prevent drying, and after 5–10 min add

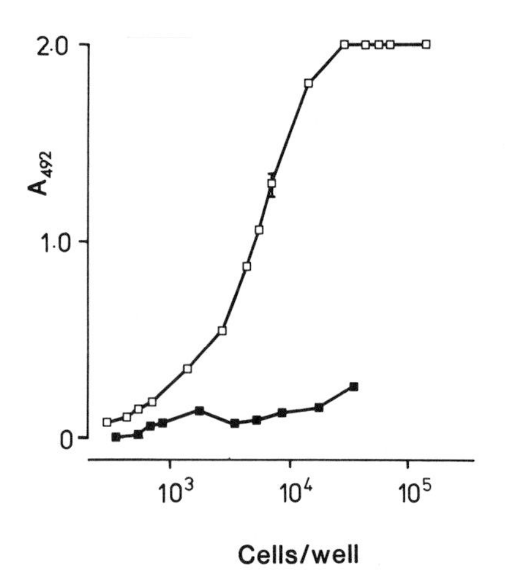

Figure 4. Peroxidase activity of eosinophils (□) and neutrophils (■) produced in bone marrow cultures. Mean of four wells ± 1 SD, most of the errors are smaller than the symbols. The eosinophil preparation contained 91% eosinophils, 2% neutrophils and 7% mononuclear cells. The neutrophil preparation contained 84% neutrophils and 16% mononuclear cells. For all practical purposes the level of peroxidase activity in neutrophils is insignificant. From ref. 17.

5 μl of cell suspension taken directly from the wells of the assay microplate.

(c) Leave for 30–40 min to allow the cells to settle and adhere to the glass. Place the slide carefully into a container of methanol and leave for 5 min to fix.

(d) Stain in dilute Giemsa. The conditions of staining must be determined, and should leave the nucleus only lightly stained, so that the eosinophil granules are prominent.

(iii) Calculate the total number of eosinophils/well and compare with control cultures carried out in the absence of growth factor.

4.5.2 *Assay for eosinophil peroxidase (17)*

(i) Prepare the following stock solutions.

(a) 0.05 mol/l Tris-HCl buffer at pH 8.0.

(b) 10% Triton X-100 (store at 4°C).

(c) 20 mg/ml *o*-phenylenediamine (OPD, Sigma No. P-1526). Aliquot and store frozen.

(d) 30% w/v hydrogen peroxide (store at 4°C).

(ii) Immediately before the use make up the peroxidase assay solution.

(a) 50 ml of Tris-HCl buffer.

(b) 0.5 ml of stock Triton X-100 solution.

(c) 0.5 ml of OPD stock solution.

(d) 6 μl of hydrogen peroxide solution.

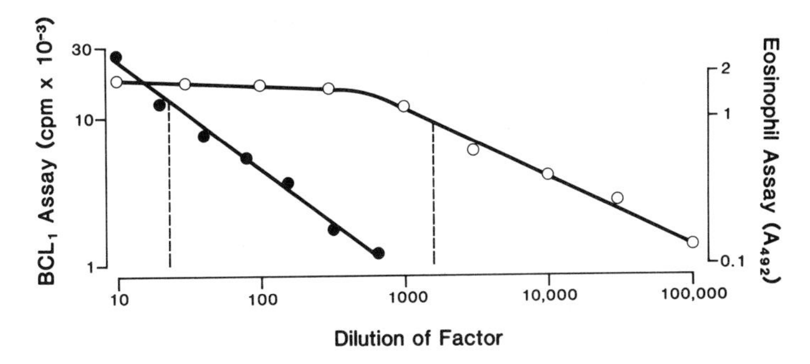

Figure 5. Titration curves for the same batch of EDF for its ability to induce eosinophil differentiation (○) and to stimulate growth of BCL$_1$ cells (●). Fifty percent end-points are shown with vertical dotted lines; for the BCGF II activity it occurs at a dilution of 1:23, and for eosinophil differentiation activity it occurs at a dilution of 1:1600. This represents a 70-fold difference in sensitivity between the two assays.

(iii) Remove most of the medium from the wells of the bone marrow culture microplate, and add 100 μl of peroxidase assay solution. Leave at room temperature for 30 min, then stop the reaction by adding 50 μl of 4 mol/l sulphuric acid. Read the absorbance at 492 nm in an automatic microplate spectrophotometer. Blank the instrument against control cultures without added growth factor.

Figure 4 shows that there is a linear relationship between eosinophil peroxidase, as measured in this assay, and eosinophil number.

4.6 Standardization of the EDF assay

To enable day to day uniformity in the assay for EDF it is necessary to establish a standard stock of material, aliquoted in small volumes and stored at −20°C.

Determination of eosinophil differentiation using the eosinophil peroxidase assay varies from day to day, and between individual mice on the same day. Two parameters vary, firstly the level of the plateau of peroxidase activity, and secondly the 50% titration end-point. This is presumably because of differences in the number of eosinophil progenitor cells in the bone marrow. It is preferable therefore to use a pool of bone marrow cells of cells from several mice and to assay the standard stock material with each assay.

The assay for BCGF II activity is less variable particularly if BCL$_1$ cells from a frozen stock are used (Section 5.1). Although this assay is approximately 50- to 100-fold less sensitive than the assay for eosinophil differentiation, it has the advantage of being a much faster assay (2 days instead of 6 days for the EDF assay). *Figure 5* shows titration curves for a EDF standard preparation in the two assays.

Neither assay is specific for the factor. IL-3 causes some eosinophil differentiation, but can be identified by assay with IL-3-dependent cell lines. The BCL$_1$ cells appear to respond also to BSF-1, though BSF-1 can be assayed in the standard co-stimulator assay with sub-mitogenic doses of anti-immunoglobulin (anti-Ig). (See Chapter 19).

(i) Assay the standard material by testing 2-fold dilutions. Estimate the 50% end-point graphically. The number of units in the starting material is the reciprocal of the dilution giving a 50% end-point. This assay is repeated on three or four occasions and the mean of the activity units calculated for the standard.

(ii) Once this is established, a correction factor (CF) for subsequent assays can be calculated:

$$\text{CF} = \frac{\text{50\% end-point for standard in assay}}{\text{No. of units attributed to standard}}$$

The number of units in other samples is then calculated:

$$\text{Relative activity for unknown} = \text{50\% end-point} \times \text{CF}$$

5. EFFECTS OF EDF ON MURINE B CELLS

BCGF II was originally detected by its capacity to induce proliferation of the murine B cell lymphoma BCL_1 *in vitro* (18,19). We have now clearly defined its effects on normal B cells, using our EDF as a source of BCGF II activity, and the detailed assays are outlined below. Throughout all B cell experiments use RPMI 1640, made up with pyrogen-free water (or specially distilled water) supplemented with 5×10^{-5} M 2-mercaptoethanol, 2 mM glutamine, 1 mM pyruvate, non-essential amino acids, penicillin, streptomycin and 5% FCS. The FCS should be pre-selected for low backgrounds in any of the B cell assays outlined in Section 5, using lipopolysaccharide (LPS): *Escherichia coli* 055:B5W (Difco, Detroit, MI) at 50 μg/ml as a positive control.

5.1 **BCL_1 lymphoma**

5.1.1 *Maintenance of the BCL_1 lymphoma*

Maintain this line (obtainable from the ECACC, Porton Down, Salisbury) by injecting 5×10^6 BCL_1 cells from the spleens of BALB/c mice previously injected with the tumour cells (4–6 weeks after inoculation). Obtain the spleen cell suspension using the method described in Section 2.2. Palpable tumours appear in these mice 4 weeks after injection. Use the tumour cells for proliferation experiments at any time between 1 week after the appearance of the tumours and the death of the animals (6–8 weeks after inoculation). Examine the mice regularly, since the mice may become very sick and die quite suddenly.

5.1.2 *Effects of EDF on BCL_1 cells*

For the BCL_1 assays, remove spleens from mice bearing the BCL_1 tumour (recoveries vary from 8×10^8 to 1.3×10^9 cells per mouse), and prepare a spleen suspension as described in Section 2.2.

Deplete T cells using a modification of the method described in Section 2.2.

(i) Resuspend one spleen in 10 ml of RPMI/5% FCS, 5 ml of stock GPC′, and 0.5 ml of NIM-R1, or an appropriate substitute at the directed dilution, for 45 min at 37°C.

(ii) Wash the cells three times in supplemented RPMI/5% FCS, and count.

For determination of DNA synthesis carry out the following procedure.

(i) Resuspend the resulting cell suspension at 2.5×10^5 cells/ml in the same medium, and plate out 100 μl volumes into 96-well microtitre plates (Nunc, Copenhagen, Denmark).

(ii) Add a further 100 μl of RPMI/5% FCS, containing the factor to be tested (see Section 4.5).

(iii) Incubate for 2 days at 37°C, in a humidified atmosphere of 5% CO_2 in air and then add 0.5 μCi/well of [^{3}H]thymidine (sp. act. 5 Ci/mmol, from Amersham International, Amersham, Bucks).

(iv) After 4 h harvest onto glass fiber paper using an automatic cell harvester (e.g. Skatron, Lier, Norway; or agents in Britain, Northumbria Biologicals Ltd, Northumberland, UK).

(v) Dry the filters down in a hot oven and place in plastic inserts, add 2 ml of non-aqueous scintillation fluid (Beckman Instruments Ltd, Geneva, Switzerland; or High Wycombe, UK) and count in a scintillation counter.

For the determination of antibody production the following steps should be carried out.

(i) Culture $2-4 \times 10^5$ BCL_1 cells in a volume of 0.5 ml in 48-well Costar plates (Data Packaging, Cambridge, MA), with the addition of appropriate factors (see Section 4.5).

(ii) Incubate the plates at 37°C, in a humidified atmosphere of 5% CO_2 in air.

(iii) After 2–5 days, collect the cells and determine the number of IgM-secreting cells [IgM plaque-forming cells (PFC)] per culture using methods previously described (20).

5.1.3 *Standardization of BCL_1 assay*

As suggested in Section 4.5 a frozen stock of BCL_1 cells may be used for the assays.

(i) Briefly, prepare the BCL_1 cell suspension and T cell deplete, using the methods described above.

(ii) Wash the cells three times, and then resuspend in cold freezing mixture (RPMI containing 20% FCS and 10% DMSO), at a final concentration of 10^7 cells/ml, in freezing ampoules.

(iii) Freeze at 1°C/min in a programmable liquid nitrogen vapour freezing cabinet at 1°C/min. If this apparatus is not available use a special container (available with most liquid nitrogen tanks), which can be held at the top of a liquid nitrogen tank, so that cells are frozen down at an appropriate rate.

(iv) Store indefinitely in a liquid nitrogen tank.

(v) For use in an assay, thaw an ampoule quickly by leaving it to stand in a water bath at 37°C.

(vi) Add the cell suspension to 9 ml of supplemented RPMI/5% FCS and centrifuge at 400 *g* for 7 min.

(vii) Resuspend in 5 ml of the RPMI/5% FCS, and count. Use in assays as above.

5.2 Effects of EDF on normal murine B cells

EDF has now been shown to act on activated B cells, which are probably at a late stage in the G_1 phase of the cell cycle but, in contrast, it has been shown to have no effect on resting B cells, which will however respond to BSF-1 plus sub-mitogenic doses of anti-mouse immunoglobulin (anti-Ig) (6,21; and *Figure 6*). The factor has been shown

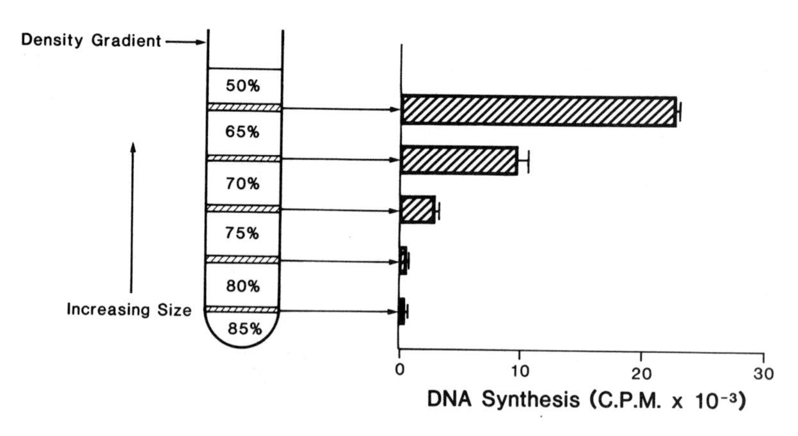

Figure 6. EDF induces large but not small B cells to synthesize DNA. B cells were fractionated on discontinuous gradients of 85%, 80%, 75%, 70% and 65% Percoll (approximate densities: 1.105, 1.099, 1.093, 1.087 and 1.082 g/ml). These fractions (83–88% Ig-positive) were cultured (at 5×10^4 cells per 200 μl) with or without partially purified EDF (2 units/ml). [^{3}H]Thymidine uptakes (mean ± SEM, $n = 3$) were measured on day 3 and are presented with backgrounds (210–3600) subtracted. Numbers in brackets are responses of identical cultures stimulated with conditioned medium, CM-T2 (5%) from an alloreactive clone (NIMP-T2), containing BSF-1, plus anti-Ig at 1 μg/ml. Modified from ref. 6.

Table 1. EDF induces large B cells to secrete IgM and IgG.

Additions	PFC/culture			
	Experiment 1		Experiment 2	
	IgM	IgG	IgM	IgG
None	1134 (39)	0	549 (63)	4 (0.8)
EDF (2 U/ml)	10 835 (374)	635 (55)	6530 (534)	710 (72)
CM-T2 (5%)	25 333 (785)	2323 (4)	50 250 (4442)	9283 (785)
LPS (10 μg/ml)	26 900 (393)	2040 (52)	59 750 (3816)	3210 (248)

Large B cells (from the 65/50% interface of Percoll gradients) were cultured at 10^5 cells per 0.2 ml, with the indicated concentrations of partially purified EDF, CM-T2, LPS, or medium. Cultures were fed with 100 μl of fresh medium containing factors on day 4. IgM and IgG plaque-forming cells (PFC) were assayed on day 7. Results are expressed as mean ± SEM of triplicate cultures. Modified from ref. 6.

to cause proliferation of B cells which have been primed with an appropriate concentration of anti-mouse immunoglobulin, for a minimum of 40 h (6). In addition, it will cause proliferation of large naturally occurring B cells (*Figure 6*), which are presumably pre-activated *in vivo*, and their maturation into IgM and IgG producing cells (*Table 1*). It will also act as a T cell replacing factor (TRF), enabling T-depleted B cells to produce an antibody response to sheep erythrocytes.

5.2.1 *Effect of EDF on large B cells*

(i) Prepare spleen cell suspensions, preferably from (× C57B1)CBA F_1 mice, and T-cell deplete as outlined in Section 2.2.

(ii) Prepare gradients of Percoll (Pharmacia, Uppsala, Sweden) as follows; add 9 parts Percoll to 1 part of 10 times concentrated saline (1.5 M). This is now 100% Percoll. From this stock make 85, 70, 75, 70, 65 and 50% solutions of Percoll by adding the 100% solution to normal strength (0.15 M) saline or Dulbecco's PBS A. Carefully layer 2 ml of 85%, then 2 ml of 80% on top of this, and so on, until the last layer of 50% is pipetted on to the top of the gradient, in a round-bottomed centrifuge tube (*Figure 6*).

(iii) Resuspend the cells obtained after T cell depletion in 1 ml of RPMI/5% FCS and carefully layer the suspension on the top of the gradient of Percoll.

(iv) Centrifuge at 1300 *g* for 15 min at 4°C in a Beckman J6-B centrifuge. Use of other centrifuges may require alteration of the speed of centrifugation in order to obtain clear bands in the gradient.

(v) Collect the large B cells from the 65%/50% interface after removing all the liquid above this interface which will contain debris and dead cells. One such gradient is sufficient in order to obtain large B cells. For initial experiments the cells obtained from other interfaces may be tested, in order to compare the proliferative effect of the factor on the different cell densities. *Never* attempt to deplete of adherent contaminating cells using Sephadex G10, since activated B cells, such as these large B cells, are also lost by this method.

(vi) Wash the B cells three times in RPMI/5% FCS in order to remove all traces of Percoll. Spin the cells the first time at 800 *g*, and then at 400 *g* for the next two washes. Resuspend in RPMI/5% FCS and count.

(vii) Re-adjust the concentration of cells in the same medium to 5×10^5 cells/ml and plate out 100 μl into flat-bottomed 96-well microtitre plates (Nunc, Copenhagen, Denmark).

(viii) Add a further 100 μl of the same medium, containing the factor under test (see Section 4.5). It is often more convenient to add this last measure to the plate before adding the cells, especially if the supernatant to be tested is to be titrated.

(ix) Incubate for 3 days under the conditions described in Section 5.1, add 0.5 μCi/well of [^{3}H]thymidine for the last 4 h, and treat as in Section 5.1.

Alternatively the large B cells can be cultured at 10^5/well, under the same conditions for 5 days for assay of Ig PFC.

In this case harvest the cells after the incubation and assay for IgM-producing cells using the method described previously (20).

In order to test for IgG PFC incubate the large B cells similarly at 10^5/well, but instead in RPMI/15% FCS for 7 days. Remove 100 μl of medium on day 4, and replace it with fresh medium containing the relevant additives and/or factors, and continue the incubation until day 7, when the cultures are harvested as above and tested for IgG-producing cells, as described previously (20).

In both cases LPS (10 μg/ml) may be used as a positive control. If the supernatants under test actually contain EDF/BCGF II they will produce very significant levels of IgM, and lower but reproducible levels of IgG (*Table 1*). As can be seen from the results in *Table 1*, the actual number of PFC may vary from one experiment to the next, although the differential values should stay the same.

5.2.2 *T cell replacing activity*

The first reference to T cell-derived factors which brought about differentiation of B cells into antibody-secreting cells, was made over 10 years ago (22,23). Since then this activity was designated as TRF, although it was not established whether this title encompassed more than one lymphokine. The TRF most well defined to date has now been shown to exert BCGF II activity, and the evidence suggests that one molecule is responsible for both activities (24). We have shown (unpublished data) that EDF acts as a TRF, enabling T-depleted B cells to produce a specific antibody response to sheep erythrocytes. Although assay systems for TRF may differ, the basis of this test is the induction of antigen-specific PFC (see also Chapter 19).

(i) Prepare enriched B cells essentially as described in Section 5.2.1.
(ii) In order to get rid of debris, load the cells onto a Percoll gradient of 85%/50% (see Section 5.2.1) and collect the cells at the interface between these two densities.
(iii) Wash as described in Section 5.2.1 and resuspend in supplemented RPMI/10% FCS at 10^7 cells/ml.
(iv) Plate out 100 μl of the cell suspension in the 96-well flat-bottomed plates described in previous sections and add a further 80 μl of the same medium, containing the supernatants under test.
(v) Finally add 20 μl of a sheep red blood cell (SRBC) suspension at 5×10^7/ml, to give a final concentration of 10^6 SRBC/well.
(vi) Incubate for 4 days, and then test for cells producing antibody against SRBC (20).

6. ACKNOWLEDGEMENTS

Anne O'Garra is supported by a Glaxo Group Research Fellowship.

7. REFERENCES

1. Sanderson,C.J., Strath,M., Warren,D.J., O'Garra,A. and Kirkwood,T.B.L. (1985) *Immunology*, **56**, 575.
2. Sanderson,C.J., O'Garra,A., Warren,D.J. and Klaus,G.G.B. (1986) *Proc. Natl. Acad. Sci. USA*, **83**, 437.
3. Warren,D.J. and Sanderson,C.J. (1985) *Immunology*, **54**, 615.
4. Noma,Y., Sideras,P., Naito,T., Bergstedt-Lindquist,S., Azuma,C., Severinson,E., Tanabe,T., Kinashi,T., Matsuda,F., Yaoita,Y. and Honjo,T. (1986) *Nature*, **319**, 640.
5. Lee,F., Yokota,T., Otsuka,T., Meyerson,P., Villaret,D., Coffman,R., Mosmann,T., Rennick,D., Roehm,N., Smith,C., Zlotnik,A. and Arai,K. (1986) *Proc. Natl. Acad. Sci. USA*, **83**, 2061.
6. O'Garra,A., Warren,D.J., Holman,M., Popham,A.M., Sanderson,C.J. and Klaus,G.G.B. (1986) *Proc. Natl. Acad. Sci. USA*, **83**, 5228.
7. Dutton,R.W., Wetzel,G.D. and Swain,S.L. (1984) *J. Immunol.*, **132**, 2451.
8. Swain,S.L. (1985) *J. Immunol.*, **134**, 3934.
9. Farrar,J.J., Fuller-Farrar,J., Simon,P.L., Hilfiker,M.L., Stadler,B.M. and Farrar,W.L. (1980) *J. Immunol.*, **125**, 2555.
10. Chayen,A. and Parkhouse,R.M.E. (1982) *J. Immunol. Methods*, **49**, 17.
11. Galfre,G., Howe,S.C., Milstein,C., Butcher,G.W. and Howard,J.C. (1977) *Nature*, **266**, 550.
12. Sanderson,C.J., Warren,D.J. and Strath,M. (1985) *J. Exp. Med.*, **162**, 60.
13. Strath,M. and Sanderson,C.J. (1986) *Exp. Haematol.*, **14**, 16.
14. Strath,M. and Sanderson,C.J. (1985) *J. Cell. Sci.*, **74**, 207.
15. Metcalfe,D. (1984) *Clonal Culture of Haemopoietic Cells: Techniques and Applications*. Elsevier, Amsterdam.
16. Sanderson,C.J. and Thomas,J. (1978) *Immunology*, **34**, 771.
17. Strath,M., Warren,D.J. and Sanderson,C.J. (1985) *J. Immunol. Methods*, **83**, 209.
18. Swain,S.L. and Dutton,R.W. (1982) *J. Exp. Med.*, **156**, 1821.

19. Swain,S.L., Howard,M., Kappler,J., Marrack,P., Watson,J., Booth,R., Watzel,G.D. and Dutton,R.W. (1983) *J. Exp. Med.*, **158**, 822.
20. Dresser,D.W. (1986) In *Handbook of Experimental Immunology. (4th edition), Vol. 2, Cellular Immunology.* Weir,D.M. (ed.), Blackwell Scientific Publications, p. 64.1.
21. Muller,W., Kuhn,R., Goldmann,W., Tesch,H., Smith,F.I., Radbruch,A. and Rajewsky,A. (1985) *J. Immunol.*, **135**, 1213.
22. Dutton,R.W., Falkoff,R., Hirst,J.A., Hoffman,M., Kappler,J.W., Kettmann,J.R., Lesley,J.F. and Vann,D. (1971) *Prog. Immunol.*, **1**, 355.
23. Schimpl,A. and Wecker,E. (1972) *Nature*, **237**, 15.
24. Harada,N., Kikuchi,Y., Tominaga,A., Takaki,S. and Takatsu,K. (1985) *J. Immunol.*, **134**, 3944.

CHAPTER 19

Assays for human B cell growth and differentiation factors

ROBIN E.CALLARD, JOHN G.SHIELDS and SUSAN H.SMITH

1. INTRODUCTION

The proliferation and differentiation of human B cells *in vitro* can be controlled by a number of soluble mediators or 'factors'. These include interleukin 1 (IL-1), interleukin 2 (IL-2), gamma interferon (IFN-γ) and several B cell growth and differentiation factors (BCGF and BCDF) of which BCGF1 (BSF1 or IL-4), BCGF2 (IL-5), low molecular weight BCGF and BCDF (BSF2) have now been purified and the genes cloned (1−3). These factors have two important properties which need to be taken into account when deciding on suitable assays, namely their lack of lineage specificity, and the range of functional activities which they display at different stages of B cell differentiation. For example, recent work with highly purified and recombinant mouse BSF1 has shown that this factor can activate resting B cells, induce cell division by activated B cells and T cells, and enhance IgG1 and IgE production by lipopolysaccharide (LPS)-stimulated B cells (4,5). Similarly, BCGF2 both induces proliferation of suitably activated B cells and B cell lines, and activates eosinophils (6) although human IL-5 has no BCGF activity (see Chapter 18). This diversity of function is further complicated by the possibility that different B cell subpopulations may not all respond in the same way. In this chapter, we have concentrated on assays which will detect the action of different factors on human B cell activation, proliferation and differentiation. In appropriate combinations, these assays can be used to distinguish between the different B cell growth and differentiation factors, and determine their action on different populations of normal and malignant B cells.

2. PREPARATIONS OF HUMAN B CELLS USED FOR ASSAYS OF BCGF AND BCDF

The way in which B cells are prepared from human lymphoid tissues can be critical for the assay of B cell growth and differentiation factors. It is important to appreciate that completely pure preparations of B cells can never be obtained from normal tissues, even by cell sorting. The degree of contamination with other cell types may interfere with the assays, and should always be determined. T cell depletion of tonsillar mononuclear cells by rosetting will routinely give preparations of greater than 95% B cells which are commonly used for assays for BCGF. Similar preparations from peripheral blood usually yield much lower proportions of B cells (20−30%), but can be used in certain circumstances, for example to test for T cell replacing factor (TRF) activity in specific antibody responses. In addition to purified normal B cells, selected B cell lines can also be used as indicators for human BCGF and BCDF. The preparation of normal B cells, and the maintenance of the indicator B cell lines is described below.

2.1 **Mononuclear cell preparation**

The first step in the preparation of human B cells is to obtain mononuclear cells free of red blood cells, platelets and polymorphonuclear cells (PMN). These are prepared from peripheral blood or tonsils as detailed below.

2.1.1 *Peripheral blood mononuclear cells (PBMC)*

(i) Take blood from volunteer donors into a syringe wetted with preservative-free heparin.
(ii) Dilute blood with an equal volume of RPMI 1640 holding medium (*Table 1*) containing 10 IU/ml of preservative-free heparin, but without fetal calf serum (FCS), and layer onto Ficoll–Hypaque (density 1.077 kg/l) (2 vols of diluted blood to 1 vol of Ficoll–Hypaque) and spin at 1000 *g* for 20 min at room temperature. The centrifuge brake should be off to prevent disturbance of the interface.
(iii) Collect the mononuclear cells from the interface, taking a portion of the Ficoll layer below, and wash ($\leq$200 *g* for 15 min) with at least an equal volume of holding medium containing 10 IU/ml of preservative-free heparin, but without FCS. The slow wash in the presence of heparin will leave most of the platelets in the supernatant and prevent clumping of the mononuclear cell pellet.
(iv) Wash the cells a second time (200 *g*) in RPMI 1640 holding medium with 5% FCS, and resuspend to between 5 and 20 $\times$ 10^6/ml.
(v) Remove a small aliquot and count on a haemocytometer. Viability determined by Trypan blue exclusion should always be in excess of 95%.

2.1.2 *Tonsillar mononuclear cells (TMC)*

(i) Surface sterilize excised tonsil with 70% alcohol for 5 sec, and then rinse in RPMI 1640 holding medium.
(ii) Put the tonsil into a Petri dish with 20 ml of RPMI 1640 holding medium containing 50 μg/ml of gentamicin and tease out the mononuclear cells by scraping from the connective tissue with a sterile scalpel.
(iii) Pipette the cell suspension into a plastic Universal tube and leave any clumps of tissue to settle for 1 min.
(iv) Remove the suspended cells and layer onto 8 ml of Ficoll–Hypaque (1.077 kg/l) in a plastic Universal tube and centrifuge at 1000 *g* for 15 min at room temperature.
(v) From this stage prepare as for PBMC cells (Section 2.1.1 steps iii–v), but use 50 μg/ml gentamicin in the media at all times as tonsil preparations usually contain some bacteria.

Table 1. Reagents used for *in vitro* experiments with human B cells.

1. Culture medium: RPMI 1640 supplemented with 25 mM Hepes, 2 mM glutamine and either FCS or horse serum as indicated. Gentamicin at 50 μg/ml can be used to inhibit bacterial growth. The RPMI 1640 medium should be purchased as either liquid medium with Hepes included (Gibco Cat No. 04I-2400) or powdered medium with Hepes included (Gibco Cat No. 079-3018). In both these formulations, the NaCl concentration is reduced to compensate for the increased osmolarity arising from the addition of Hepes buffer. In some experiments this can be crucial. For example, specific antibody responses are absent or significantly reduced if ordinary RPMI 1640 with added Hepes is used.
2. Holding medium: RPMI 1640 supplemented with 25 mM Hepes and 5% FCS where indicated. Gentamicin at 50 μg/ml is routinely used for tonsil cell preparations since these are frequently infected with bacteria. This medium contains no bicarbonate and is used exclusively for washing and holding cells before culture.
3. Phorbol ester: either phorbol 12-myristate 13-acetate (PMA or TPA) or phorbol 12,13 dibutyrate can be used. A stock solution of phorbol ester is made up to 1 mg/ml in dimethyl sulphoxide (DMSO), and stored frozen in the dark (wrapping in aluminium foil will suffice). Working dilutions are made by diluting stock solution in holding medium without FCS. Because phorbol esters in solution will eventually deteriorate, titrations should be carried out every month or so to establish potency. (Take care when handling phorbol esters which are possibly carcinogenic.)
4. Anti-IgM beads: affinity purified rabbit antibody to human IgM coupled to polyacrylamide beads (anti-IgM beads, BioRad Cat No.170-5120) is used to activate B cells. The lyophilized beads are reconstituted in water as directed, and stored in 0.1% NaN_3 at 4°C. Before use, the beads should be washed 2–3 times in holding medium, and resuspended to 40 μg/ml in culture medium. Washed beads can be stored for short periods at 4°C in holding medium containing 100 μg/ml of gentamicin.
5. Influenza virus: influenza virus used for antigenic stimulation of antibody production (Section 7) sucrose gradient purified. It was not inactivated, but inactivated virus is perfectly satisfactory. Normally, the recombinant H_3N_2 strain A/X31 was used, but other A or B strain viruses can be used equally well. As a concentrated suspension (~20 mg/ml), it can be stored in 0.01% NaN_3 at 4°C for several years without loss of antigenic activity. Diluted suspensions are made up to 10 μg/ml in holding medium with 50 μg/ml of gentamicin but without FCS, and can be stored at 4°C for several months. The virus suspension is diluted a further 50-fold in the antibody cultures which is usually sufficient to avoid inhibition by the azide present in the concentrated stock suspension. With some preparations, however, it may be necessary to dialyse out the azide. Influenza virus preparations cannot be sterilized by filtration as virus is retained by the membrane.
6. BCGF: a partially purified low molecular weight BCGF can be purchased from Cellular Products (Sera Labs in the UK, Cat No. BT-101). This is not the same as BCGFI (IL-4), and is unlikely to be human BCGFII.
7. [^{3}H]Thymidine: [^{3}H]TdR was purchased from Amersham International (Cat No. TRA 120, sp. act. 5 Ci/mmol).
8. FITC-goat anti-mIg: B cell staining with mouse monoclonal antibodies is detected with FITC-conjugated affinity purified goat antibody to mouse IgG/IgM. To avoid cross-reactions with human Ig, it is best to use antibody which is both affinity purified and absorbed against human serum proteins. Several such preparations are available. We use a TAGO antibody Cat No. 6253 (available in the UK from Tissue Culture Services).

2.2 Depletion of T cells by E-rosetting

Human T cells and some NK cells express the T11 (CD2) surface antigen which specifically binds sheep erythrocytes (E) forming E rosettes. Separation of E rosette-forming cells (E^+) from non-rosette forming (E^-) cells by density gradient centrifugation is the simplest and most effective method of depleting T cells from mononuclear cell preparations.

2.2.1 *Preparation of AET-treated sheep red cells*

E-rosettes are best formed using sheep red blood cells (SRBC) treated with S-2-aminoethylisothiouronium bromide hydrobromide (AET) to stabilize rosettes according to the method of Kaplan and Clark (7).

(i) Wash sheep blood (stored in Alsevers solution for up to 3 weeks) three times in sterile 0.14 M NaCl. Remove all the supernatant and any buffy coat on each wash.
(ii) After the last wash, remove all the supernatant to leave a packed SRBC pellet.
(iii) Incubate 1 vol of packed SRBC with 4 vols of freshly prepared AET solution (40.2 mg/ml in water, pH 9.0) at 37°C for 15 min.
(iv) Wash five times in sterile 0.14 M NaCl and resuspend in RPMI 1640 holding medium (without FCS) to give 10% AET−SRBC. These may be stored at 4°C for up to 3 weeks.

2.2.2 *E-rosette separation*

More effective T cell depletion can be obtained if Percoll (Pharmacia, Uppsala, Sweden) rather than Ficoll−Hypaque is used to separate E-rosette forming (E^+) cells from non-rosette forming (E^-) cells (8). The E^- fraction of peripheral blood obtained by this method should contain less than 2% of $CD3^+$ (T) cells compared with 2−5% when Ficoll−Hypaque is used. With two cycles of E rosetting, the percentage of residual T cells can be reduced to less than 1%. When preparing E^- cells from TMC only one cycle of E-rosetting is required to reduce the level of T cell contamination to less than 1%. T cell contamination must be reduced to this level in E^- preparations used to assay BCGFs to avoid indirect action through residual T cells. Typically, an E^- preparation from tonsil (or spleen) will contain more than 95% $B1^+$ (CD20) B cells, less than 1% $CD3^+$ (T) cells, less than 2% $UCHM1^+$ (CD14) (monocytes) and less than 2% $HNK1^+$ (NK) cells, whereas E^- cell preparations from blood after two E-rosetting cycle could contain 20% $B1^+$ (B cells), 1% $CD3^+$ (T) cells, 75% monocytes, 0−5% NK cells and 5% of null or unknown cells (*Table 2*).

E rosette formation, and separation of E^- cells on Percoll gradients is carried out as follows.

(i) Mix 10 ml of mononuclear cells at 5 × 10^6/ml with 2.5 ml of 10% AET− SRBC

Table 2. Phenotype of fractionated E^- lymphocytes.

Cell type		*% fluorescence positive cells*				
		B1	*UCHT1*	*UCHM1*	*HNK-1*	*Leu11*
Blood E^-	(2× rosetted)	21	<1	75	0	6
	(light fraction)	13	<1	78	0	10
	(heavy fraction)	92	<1	1	0	3
Tonsil E^-	(1× rosetted)	95	<1	2	2	1
	(light fraction)	98	<1	1	n.d.	n.d.
	(heavy fraction)	98	<1	< 1	n.d.	n.d.

n.d. = not done.

Table 3. Preparation of Percoll.

Percoll solutions for discontinuous density gradient centrifugation are commonly made to a predetermined percentage by mixing with buffered saline or medium. However, the density of stock Percoll varies slightly, as do the various diluting solutions, and a given percentage of Percoll does not always result in the same density. For this reason, Percoll solutions should be prepared from solutions of known densities (specific gravities). Specific gravities are easily and accurately determined by weighing in specific gravity bottles.

1. Prepare iso-osmolar stock Percoll suspension by adding 1 vol of 10 times concentrated PBS to 9 vol of Percoll. (N.B. once PBS has been added to Percoll, autoclaving will cause it to polymerize.)
2. Measure the specific gravity (SG) of the stock Percoll and the diluting medium (RPMI 1640 supplemented with 25 mM Hepes and 10% FCS) using a 10 ml specific gravity bottle. The SG of the iso-osmolar stock Percoll should be about 1.127, and the diluting medium about 1.008.

$$SG_{\text{Percoll}} = \frac{\text{weight of known volume of Percoll}}{\text{weight of an equal volume of water}}$$

3. Make up Percoll to the required specific gravity using the formula:

$$\text{percent Percoll} = \frac{(\text{SG required} - \text{SG diluting medium})}{(\text{SG Percoll} - \text{SG diluting medium})} \times 100$$

Care must be taken with liquid volumes when making up Percoll. A measuring cylinder is normally adequate, but the same cylinder should be used for both the Percoll and diluting medium to avoid inter-cylinder variation. It is worthwhile checking the diluted Percoll to be sure the specific gravity is correct.

and 1 ml of FCS in a plastic Universal tube, and centrifuge at 200 *g* for 15 min with the brake off.

(ii) Incubate for 60 min on ice.

(iii) Resuspend the rosettes by gentle rotation of the centrifuge tube. Do not resuspend by pipetting as this will disrupt some rosettes.

(iv) Layer onto 7–8 ml of Percoll (specific gravity 1.080) (*Table 3*) in a plastic Universal tube and centrifuge at 1000 *g* for 20 min at room temperature with the brake off.

(v) Remove the E^- fraction from the interface along with about 75% of the Percoll. Be careful not to disturb the red cell pellet.

(vi) Dilute with at least an equal volume of holding medium and wash twice in holding medium.

(vii) To recover the E^+ cells, remove all of the Percoll above the SRBC and resuspend the pellet with 5 ml of Gey's haemolytic solution (*Table 4*) for 1–2 min. Immediately after red cell lysis, dilute with holding medium and wash twice.

(viii) Resuspend all fractions in 2–5 ml of holding medium, and count the cells. It is most important to check each preparation for percentages of B cells, T cells and monocytes (Section 3). Preparations of E^- cells with unacceptable numbers of contaminating T cells (>2%) should be discarded.

2.3 Depletion of monocytes by adherence

Blood E^- cells contain 60–80% monocytes compared with only 1–2% in tonsillar E^- cell preparations. These can be depleted by adherence on microexudate-coated

Table 4. Gey's haemolytic balanced salt solution.

Solution A:	NH_4Cl	35.0 g
	KCl	1.85 g
	$Na_2HPO_4.12H_2O$	1.5 g
	KH_2PO_4	0.119 g
	Glucose	5.0 g
	Phenol Red	0.005 g
	Gelatine (Difco)	25.0 g
	1 litre double distilled water	
Solution B:	$MgCl_2.6H_2O$	4.2 g
	$MgSO_4.7H_2O$	1.4 g
	$CaCl_2$	3.4 g
	1 litre double distilled water	
Solution C:	$NaHCO_3$	22.5 g
	1 litre double distilled water	

All three solutions are sterilized by autoclaving and stored at 4°C. Make up Gey's solution freshly when required by mixing 7 vols of sterile water with 2 vols of solution A (warmed at 37°C for 5 min before use to melt the gelatine), 0.5 vol of solution B and 0.5 vol of solution C.

plastic Petri dishes. Contamination by monocytes in the non-adherent fraction is usually reduced to between 2 and 10% using this method. Better depletion may be obtained by passing cells through columns of Sephadex G10, or cell sorting using anti-monocyte monoclonal antibodies, but this is not usually necessary. In some assays, for example the antigen-specific antibody response to influenza virus (Section 7), small percentages of monocytes (~0.5%) are essential, presumably for antigen presentation. When required, adherent cell depletion on microexudate plates is carried out as follows.

(i) Prepare microexudate dishes by growing baby hamster kidney (BHK) cells to confluence on 90-mm tissue culture plastic Petri dishes.

(ii) Remove the BHK cell monolayer with 10 mM ethylenediaminetetraacetic acid (EDTA) in PBS.

(iii) Wash dishes vigorously several times with PBS, and sterilize with a bactericidal u.v. light source for 5 min. The treated dishes may be sealed with tape and stored at 4°C.

(iv) Incubate 2×10^7 E^- cells in 10 ml of holding medium containing 10% FCS on the microexudate dishes for 45 min at 37°C.

(v) Resuspend the non-adherent cells by gently rocking the dishes, and remove by pipette. Repeat this procedure at least twice with 10 ml of warm medium carefully added to the dish.

(vi) To remove the adherent cells, incubate for 15 min with medium containing 3 mM EDTA at 37°C, and pipette vigorously.

2.4 Preparation of heavy and light B cells

In many B cell activation and proliferation experiments, E^- preparations are fractionated on discontinuous Percoll gradients into light (<1.074 kg/l) and heavy (>1.074 kg/l) populations. The small heavy B cells are generally considered to be resting (G_0)

cells, although there is evidence to suggest that some of these cells are already activated (9). Because of this, some investigators prefer even higher density B cells to exclude those in G_0^* (10). The procedure for separating light and heavy cells on Percoll gradients is as follows.

(i) Layer up to 3 × 10^7 cells in 2−3 ml of holding medium onto 3 ml of Percoll (specific gravity 1.074) (*Table 3*) in a 10 ml conical centrifuge tube, and centrifuge at 1000 *g* for 20 min. If using blood E^- cells, monocytes should be depleted first.

(ii) Remove the low-density B cells from the interface, and discard most of the Percoll supernatant. Resuspend the loose pellet of high-density B cells in the remaining Percoll.

(iii) Wash each fraction twice, and resuspend the cells in 1−2 ml of holding medium containing 5% FCS.

(iv) Count the cells using Trypan blue.

2.5 Human B cell lines for assaying BCGF and BCDF

The advantages of using B cell lines to assay BCGF and BCDF are 3-fold.

(i) Continuous B cell lines are free of the non-B cell contamination which can cause problems even with fluorescence-activated cell sorter (FACS)-selected preparations.

(ii) B cell lines are more homogeneous than normal B cell preparations which generally consist of ill-defined B cell subpopulations and/or B cells at different stages of activation.

(iii) By judicious selection of indicator B cell lines, it is possible to distinguish between different factors. For example, the myeloma line HFB1 responds well to PHA-conditioned medium and to low molecular weight BCGF, but not to human BSF1, or BCDF.

These advantages are offset to some extent by the abnormal physiological status of B cell lines, and the uncertain relevance of their responses to normal B cell growth and differentiation. For this reason, B cell lines are best used in conjunction with other assays using normal B cells.

2.5.1 *Maintenance of B cell lines used to assay BCGF and BCDF*

Of the numerous B cell lines which respond to BCGF and/or BCDF we have selected four on the basis of their differential responses to BCGF and BCDF (*Table 5*). The best characterized of these is CESS, a lymphoblastoid line used for detecting BCDF (11). In addition, we have recently identified three other lines (HFB1, L4 and BALM4) which respond to either BCGF, or both BCGF and BCDF.

The B cell lines used in BCGF and BCDF assays are grown in medium RPMI 1640 supplemented with 25 mM Hepes, 2 mM glutamine and 10% FCS in upright 25 cm^2 or 75 cm^2 flasks at 37°C in an atmosphere of 5% CO_2 in air. Small numbers of cells can be grown in 24-well Costar plates. Antibiotics are not normally necessary, but gentamicin at 50 μg/ml can be used without interfering with the assays. All four cell

Table 5. B cell lines used to assay human BCGF and BCDF.

Cell line	Optimal cell concentration		
	Continuous culture (cells/ml)	BCGF assay (cells/well)	BCDF assay (cells/well)
HFB1	2×10^5	5 and 15 $\times 10^3$	no response
L4	2.5×10^5	4 and 12 $\times 10^3$	4 and 12 $\times 10^3$
BALM4	2.5×10^5	10 and 20 $\times 10^3$	10 and 20 $\times 10^3$
CESS	1×10^5	no response	3×10^3

lines need to be split twice a week, to the concentration given for continuous culture in *Table 5*. Both BALM4 and L4 should not be cultured at a density lower than that shown in the table as they may die at low cell densities.

One important problem which must be closely monitored is that of mycoplasma contamination which can interfere with responses to BCGF. Interpretation of [^{3}H]TdR incorporation results may be impossible if the responding cell lines are contaminated. Under these circumstances, cell proliferation can be determined with the MTT test (12) (see Chapter 15). The Epstein–Barr virus transformed line, CESS, is contaminated with mycoplasma, but this does not seem to inhibit its ability to respond to BCDF. The other mycoplasma-free lines should be checked routinely for infection. If possible, contaminated and mycoplasma-free lines should be dealt with completely separately to reduce the possibility of cross-contamination. If separate facilities are not available, mycoplasma-free cultures should be handled only after fumigation of the laminar flow cabinet with formaldehyde, preferably overnight. This is most easily done by leaving an open dish of formaldehyde in the closed cabinet with the fan off. It is also important to use separate stocks of medium and other reagents, and to keep the incubator clean and free of mycoplasma-infected cultures. There are fewer problems with contamination, especially fungi and mycoplasmas, if the incubators are kept dry.

3. INDIRECT IMMUNOFLUORESCENCE ANALYSIS

In all human B cell preparations, it is essential to monitor the purity of the prepared fractions since unacceptable contamination can sometimes occur. This is best done by indirect immunofluorescence using well-defined monoclonal antibodies to B cells, T cells, monocytes and NK cells. For small numbers of cells, staining is carried out in round-bottomed microtitre wells, or round-bottomed flexible PVC microtitration plates which can be cut to the required number of wells and supported on a rigid microtitre tray.

(i) Dispense between 1 and 3 $\times 10^5$ cells into each well of a non-sterile round-bottomed microtitre tray. Include wells for each monoclonal antibody as well as negative and positive controls.

(ii) Centrifuge the cells into a pellet at 200 *g* for 2 min using a microtitre plate attachment available for most centrifuges.

(iii) Remove supernatants by inverting the plate with a *single* flicking motion over a sink. Do not repeat this action as the cells will become resuspended and lost. Carefully wipe any droplets of medium from the surface of the plate using a

Table 6. Fixing FITC-labelled cells.

1.	Make up 0.1 g of paraformaldehyde with two drops of 1 M NaOH and 0.2 ml of water.
2.	Heat to 80°C in a water bath. Do not boil.
3.	Cool, and add 9.8 ml of PBS.
4.	Resuspend the pellet of stained cells in 200 μl of fixative at 4°C for 20 min, then wash twice.
5.	Wash the cells in holding medium containing 5% FCS before reading on the FACS.

paper tissue, then resuspend the cells by holding the plate firmly onto a vortex whirlimixer.

(iv) Add 50 μl of monoclonal antibody diluted in holding medium containing 2% FCS and 0.01% NaN_3. When checking for B cell purity, include a CD3 monoclonal antibody for T cells, and a monocyte-specific antibody such as UCHM1. For the negative control, use medium alone or a monoclonal antibody known not to react with human leucocytes. For the positive control, use a known anti-B cell monoclonal antibody such as B1 (CD20) or B4 (CD19), or an anti-common leucocyte. Mix each well individually when adding the antibody. Do not use the whirlimixer otherwise medium will spill over the sides and into adjacent wells.

(v) Incubate on ice for 30 min.

(vi) Wash the cells three times with 200 μl of holding medium, centrifuge, and remove supernatants as described in (ii) and (iii).

(vii) To the resuspended cells, add 50 μl of pre-titrated fluorescein isothiocyanate (FITC)-conjugated goat anti-mIg absorbed against human serum proteins (*Table 1*) in medium containing 2% FCS, 0.01% NaN_3 and 2% normal goat serum (NGS) to inhibit Fc receptor binding.

(viii) Incubate on ice for 30 min.

(ix) Wash four times.

(x) Resuspend the cells and make up to 0.5 ml in LP3 tubes with holding medium containing 2% FCS and 0.01% NaN_3. The stained cells can be kept on ice for 2–3 h before analysis on a FACS. If it is necessary to delay the FACS analysis, the cells can be fixed as described in *Table 6*, and kept for several days without loss of fluorescence.

For larger numbers of cells, staining can be done in LP3 tubes as follows.

(i) Incubate 0.5×10^6 cells with 100 μl of monoclonal antibody in LP3 tubes for 30 min on ice.

(ii) Wash twice in cold medium containing 2% FCS and 0.01% NaN_3.

(iii) Incubate with 100 μl of pre-titrated FITC-conjugated goat anti-mIg for a further 30 min on ice.

(iv) Wash twice, and analyse samples on a FACS, or fluorescence microscope.

4. B CELL ACTIVATION

Activation of human B cells *in vitro* can be achieved in different ways. Probably the most common method is with antibody to surface Ig receptors. Binding to surface Ig results in the breakdown of membrane phosphatidylinositol bisphosphate releasing inositol triphosphate and diacylglycerol which in turn increase levels of cytosolic Ca^{2+} and activate protein kinase C, respectively (13). Direct activation of protein kinase C with phorbol esters by-passes the first part of this pathway. B cells can also be activated with BSF1, and monoclonal antibodies to express certain surface antigens, some of which may be receptors for B cell activation or growth factors.

Early events of B cell activation, such as membrane depolarization and increased intracellular Ca^{2+} are good indicators of transmembrane signalling, and are usually measured within seconds or minutes. Later events, including size increases, expression of cell surface activation antigens and synthesis of RNA involve more complex cell functions including activation of the nucleus, and occur within hours or days. Although both early and late events can be useful indicators of B cell activation, expression of cell surface receptors for B cell growth and differentiation factors occur after membrane depolarization and increases of cytosolic calcium. For this reason, we have chosen the increased expression of B cell surface antigens as indicators of *in vitro* activation.

4.1 Activation of B cells with anti-IgM or TPA

The most commonly used B cells for activation experiments are small heavy tonsillar E^- cells isolated on discontinuous Percoll gradients (Section 2.4). Other sources of B cells such as spleen or peripheral blood can also be used, but peripheral blood B cells are not very satisfactory because of the relatively small numbers of B cells obtained, and the difficulties in removing the high proportion of monocytes, NK and null cells (*Table 2*). The culture technique for activating B cells is as follows.

(i) Prepare T cell-depleted E^- cells as described in Section 2, and resuspend to 1×10^6 cells/ml with culture medium containing 5% heat inactivated FCS, 25 mM Hepes and 50 μg/ml of gentamicin. For activation with phorbol ester, add 10 μl/ml of an appropriate working solution of TPA (*Table 1*) to give a final concentration of 1–10 ng/ml. Normally, 10 ng/ml will both activate and induce cell division whereas 1 ng/ml will activate without significant DNA synthesis. For activation with anti-IgM beads, add anti-IgM beads (*Table 1*) to give a final concentration of 10 μg/ml. The optimal doses of both TPA and anti-IgM beads should be determined in preliminary experiments.

    Note that the appropriate controls for B cells activated with TPA or anti-IgM are E^- cells cultured under the same conditions but without the activation signal. About 20% more cells are required in control cultures as they do not survive as well as activated cells. Freshly prepared E^- cells are not suitable controls.

(ii) If small numbers of activated cells are required, use 2 ml cultures of 2×10^6 cells in 24-well Costar plates. To prevent drying of the cultures, put sterile water into the outer wells, and place the Costar plate inside an unsealed plastic box. For larger numbers of cells, set up the cultures at 10^6 cells/ml in tissue culture flasks.

(iii) Incubate at 37°C for 72 h in a dry incubator with an atmosphere of 5% CO_2 in air.

(iv) At the completion of the culture period, resuspend the cells and centrifuge at 1000 *g* for 20 min over Ficoll–Hypaque (1.077 kg/l) to remove dead cells and debris. For small numbers of cells, this can be done in 12×75 mm Falcon

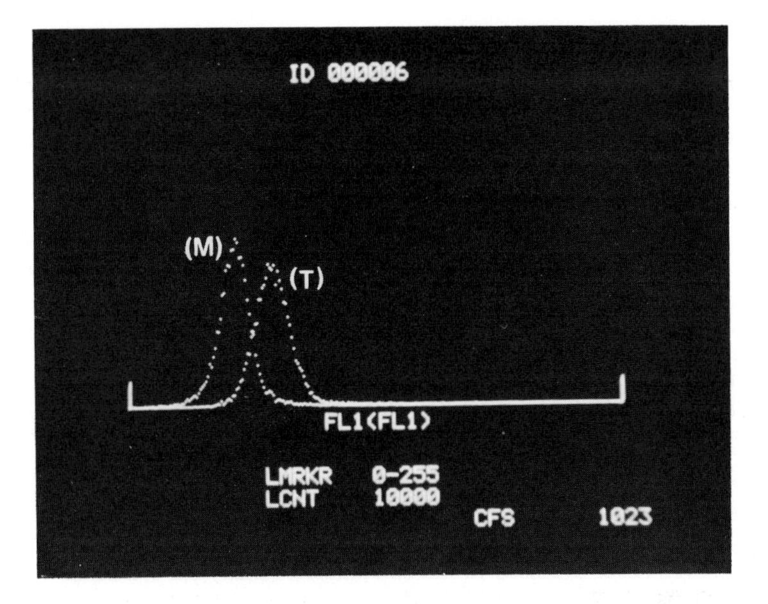

Figure 1. FACS histogram of small heavy tonsillar B cells cultured with medium alone (M) or TPA (T) for 3 days and stained with a CD23 monoclonal antibody.

tubes containing 1.5 ml of Ficoll–Hypaque, otherwise use plastic Universal containers with at least 7 ml of Ficoll–Hypaque.

(v) Carefully remove the cells from the interface with a Pasteur pipette, then wash twice and resuspend to 2×10^6 cells/ml in holding medium.

(vi) To assess the expression of activation antigens, stain the cells with the appropriate monoclonal antibody as in Section 3, and read on the FACS.

An example of a B cell activation experiment using a monoclonal antibody to CD23 (a B cell activation antigen) is given in *Figure 1*. The FACS histogram of CD23-labelled B cells activated with TPA superimposed on the histogram obtained from B cells cultured in medium alone clearly shows the enhanced expression of this antigen. The enhanced expression of surface activation antigens can be expressed in three different ways.

(i) The fluorescent channel showing the peak fluorescence (mode) can be compared.

(ii) The fluorescent channel subdividing the histogram into two equal proportions (median) can be compared.

(iii) The proportion of activated cells labelled can be given by setting the marker at the base of the peak obtained with control cells.

Changes in cell size on activation can be determined by comparing forward angle scatter histograms in the same way.

5. ASSAYS FOR BCGF

Factors which promote the growth of B cells can be assayed in different ways. The two most common methods are co-stimulation of normal (tonsillar) B lymphocytes, and promotion of cell division by selected indicator B cell lines.

Table 7. Co-stimulation of B cells with anti-IgM and BCGF.

Anti-IgM beads	*BCGF*	*Response* ([^{3}H]*TdR d.p.m.)*
−	−	4238 ± 466
+	−	2316 ± 394
−	+	1367 ± 137
+	+	21 127 ± 423

5.1 Tonsillar B cell co-stimulation assay for BCGF

This assay depends upon the co-stimulation of small heavy (resting) B cells with an activation signal (e.g. anti-IgM or TPA) and BCGF. Under the right conditions, the combined signals will result in B cell proliferation whereas each signal alone will not. Several different methods for activating resting B cells to respond to BCGF have been described. Most of these use either soluble or particulate preparations of anti-immunoglobulin, *Staphylococcus aureus* Cowan 1 (SAC), or phorbol esters. We have found the most reliable to be anti-IgM coupled to polyacrylamide beads, and the phorbol ester TPA. These are used in the co-stimulator assay for BCGF as follows.

(i) Prepare small resting tonsillar E^- as described in Section 2, and resuspend to 10^6 cells/ml in culture medium supplemented with 5% FCS and 50 μg/ml of gentamicin.

(ii) For each factor (BCGF containing supernatant) to be tested, dispense 10^5 cells (100 μl) into each of six wells of a sterile flat-bottomed microtitre tray. To three wells add 50 μl of BCGF containing supernatant and a further 50 μl of medium. To the remaining three wells add 50 μl of BCGF containing supernatant and 50 μl of anti-IgM beads at 40 μg/ml (or TPA at 4 ng/ml) (Section 4.1). For each experiment, triplicate control cultures of 10^5 cells with medium alone, and anti-IgM (or TPA) alone, should also be included. If using a dry incubator, fill the outer wells with sterile water and place the plate in an unsealed plastic box to minimize evaporation.

(iii) Incubate for 72 h at 37°C in an atmosphere of 5% CO_2 in air.

(iv) Add 1 μCi of [^{3}H]TdR in 10 μl of saline to each culture well using a 0.5 ml Hamilton Stepper syringe, and incubate at 37°C for 6−8 h.

(v) Harvest the labelled cells onto glass fibre discs with an automated Cell Harvester and count incorporated [^{3}H]TdR on a liquid scintillation counter. Results should be corrected for quenching and efficiency, and expressed as disintegrations per minute (d.p.m.).

A typical result from a BCGF co-stimulator experiment is given in *Table 7*. In this case, tonsillar E^- (B) cells were activated with anti-IgM beads, and the source of growth factor was supernatant from phytohaemagglutinin-activated peripheral blood T cells. A 10-fold stimulation index with counts up to 30 000 is typical for this sort of experiment.

It is important to note that the choice of activation signal in these experiments is not just a matter of convenience. Different B cell activators may have quite different effects and prime B cells to respond in different ways to the various B cell growth factors.

For example, TPA, but not anti-IgM, is a powerful inducer of the CD23 antigen. It may, therefore, be important to try different activation signals when screening for unknown B cell growth factors.

5.2 Assay of BCGF on B cell lines

Under conditions of low cell densities, certain B cell lines depend upon the addition of exogenous growth factors for continual proliferation. Using the cell lines listed in *Table 5* (HFB1, L4 and BALM4) this property can be exploited to assay for BCGF as follows.

(i) Harvest cells from a vigorous log-phase growth of L4, BALM4 or HFB1. The cells should be sub-cultured 24–48 h beforehand. Cultures which contain many dead cells, or are growing slowly, will not perform well in this assay. Wash the cells once and resuspend to the appropriate concentrations for the BCGF assay (see *Table 5*) in RPMI 1640 medium supplemented with 25 mM Hepes, 2 mM glutamine and 5% FCS.

(ii) Add 100 μl/well of each factor to be tested to six wells of a sterile flat-bottomed microtitre plate. Add medium only to one set of six wells as a negative control.

(iii) For each of the two recommended cell concentrations for the BCGF assay (*Table 5*), add 100 μl of cells to three wells with factor and three wells of medium control. If using a dry incubator, fill the outer wells with sterile water, and put the plate in an unsealed plastic box to minimize evaporation.

(iv) Incubate for 72 h at 37°C in an atmosphere of 5% CO_2 in air.

(v) Add 1 μCi/well of [^{3}H]TdR in 10 μl of saline using a 0.5 ml Hamilton Stepper syringe and incubate for 6–8 h.

(vi) Harvest the labelled cells onto glass fibre discs with an automatic cell harvester, and count incorporated [^{3}H]TdR on a liquid scintillation counter.

Note that in these experiments the starting cell concentration is critical. At too high a concentration, the cells proliferate well without any exogenous factor. If the concentration is too low, the cells will die even in the presence of added factor. We routinely use two cell concentrations of each cell line to test for factor activity (*Table 5*). With the normal day-to-day variation, either the lower or the higher concentration may give an optimal response. A typical set of results is shown in *Table 8*.

Table 8. Response of selected B cell lines to BCGF.

Cell line	*Cells/well* ($\times 10^{-3}$)	[^{3}H]*TdR incorporation (d.p.m.)*	
		Medium	*BCGF*
BALM4	20	38 681 ± 11 217	93 226 ± 12 119
	10	6700 ± 1474	35 614 ± 4630
HFBI	15	40 275 ± 5236	95 734 ± 13 403
	5	3191 ± 925	37 446 ± 1123
L4	12	7441 ± 1359	115 329 ± 3059
	4	1206 ± 296	38 451 ± 570

6. BCDF ACTIVITY MEASURED ON B CELL LINES

Human BCDF can be assayed in a co-stimulator assay with SAC activated B cells (14), or on certain indicator B cell lines. As these both appear to respond to the same factor, we will describe only the use of the B cell lines L4, BALM4 and CESS (*Table 5*), all of which respond to BCDF by increased immunoglobulin synthesis. With CESS (the most commonly used indicator B cell line for BCDF) it was originally reported that the minor population of surface IgG-positive cells were more responsive to BCDF than unseparated CESS cells (11), but we have not found it necessary to isolate this subpopulation. Good responses from each of the three lines are only obtained with cultures which are growing well. The following description of the assay is for CESS. In the case of L4 and BALM4, lower concentrations of immunoglobulins are secreted, and it is necessary to culture for a longer period (6–7 days) and at higher cell concentration.

(i) Harvest cells from a vigorous log-phase culture of CESS cells subcultured 24–48 h previously, wash once and resuspend to 3×10^4 cells/ml in culture medium supplemented with 5% FCS. Cultures of greater than 90% viability can be used directly. If viability is less than this (60–90%), viable cells should be separated on a Ficoll–Hypaque gradient (1.077 kg/l) and washed twice before use. Cultures containing less than 60% viable cells should not be used.

(ii) Add 100 μl of medium (negative control) or factor to be tested in triplicate wells of a sterile flat-bottomed microtitre plate.

(iii) Add 100 μl of CESS cells at 3×10^4/ml to each well. If using a dry incubator, fill the outer wells with sterile distilled water, and put the plate in an unsealed plastic box to minimize evaporation.

(iv) Incubate for 5 days at 37°C in an atmosphere of 5% CO_2 in air.

(v) Remove 100 μl of supernatant from each well (without disturbing the cells) for estimation of IgG secretion (Section 9.1). These can be transferred directly to the enzyme immunoassay (EIA) plate, or stored at −20°C until assayed.

IgG production by CESS is increased 5–50 times by the presence of BCDF compared with cultures in medium alone (*Table 9*). When using L4 or BALM4 in a BCDF assay, it is advisable to use two different cell concentrations as the response is cell dose dependent (*Tables 5* and *9*).

Table 9. Response of selected B cell lines to BCDF.

Cell line	*Cells/well* ($\times 10^{-3}$)	*IgG production (ng/ml)*	
		Medium	*BCDF*
CESS	3	47 ± 13	505 ± 61
L4	12	14 ± 8	176 ± 30
	4	< 9	< 9
BALM4	20	6 ± 1	20 ± 4
	10	< 2	12 ± 4

7. T CELL REPLACING FACTOR

In addition to assays for BCGF and BCDF, the activity of soluble factors on B cells can be determined by their ability to replace T cells in specific antibody responses *in vitro*. In man, TRFs can be readily assayed on thoroughly T cell depleted E^- cell preparations obtained from blood, tonsil or spleen (15,16). In this assay (described below), E^- cells are cultured with influenza virus in the presence and absence of TRF, and antibody production determined by specific EIA. Antibody responses obtained in the presence of TRF are usually comparable with those obtained in the presence of T (E^+) cells.

(i) Resuspend E^- cells (Section 2) to $1-1.5 \times 10^6$/ml in culture medium supplemented with 10% horse serum (Gibco) and 50 µg/ml of gentamicin, and dispense 0.5 ml into 12 × 75 mm culture tubes (Falcon 2054). Note that the horse serum is stable for many months at −20°C, but does not last more than a few days at 4°C.

(ii) Dilute influenza virus antigen (*Table 1*) to 10 times optimum concentration (usually 2.0 µg/ml) in the same medium, and add 100 µl to each culture tube.

(iii) Add 0.5 ml of medium to three tubes for a negative control, and 10^6 E^+ (T) cells in 0.5 ml of medium to three tubes for a positive control.

(iv) Dispense 0.2−0.5 ml of factor to be assayed into the remaining tubes, and make up the total volume of each culture to 1 ml with culture medium.

(v) Loosely replace the caps and incubate for 7 days at 37°C in a dry incubator with an atmosphere of 5% CO_2 in air.

(vi) At the end of the 7-day culture period, wash the cells once in holding medium containing 5% FCS, and resuspend in 0.5 ml of culture medium containing 5% FCS. Cover the tubes with aluminium foil rather than re-cap them, and incubate for a further 12−18 h at 37°C. This step minimizes background in the EIA by removing free virus and horse serum present in the culture medium.

(vii) Collect supernatants for assay of specific antibody by solid phase EIA (Section 9.2). IgG subclass antibody can also be measured (Section 9.3). Supernatants for EIA may be stored at −20°C before assay.

A typical result from a TRF-induced antibody response to influenza virus is given in *Table 10*.

Table 10. Specific antibody response obtained with TRF.

Responding cells	*Antigen*	*TRF*	*Antibody to X31 (ng/ml)*	
			IgM	*IgG*
*PBM E^-	−	−	<1	<1
	+	−	<1	2
	−	+	<1	<1
	+	+	2	220 ± 80
+ E^+	+	−	2	241 ± 41

*Depleted of T cells by two cycles of E-rosetting with AET-SRBC (Section 2.2).

The antibody cultures are usually carried out in triplicate in 12 × 75 mm capped Falcon tubes (Cat No. 2054), but a micromethod is also available using five times fewer cells in round-bottomed microtitre plates (17). In this case, the responses obtained are more variable, and more replicates (>6) should be used. The micromethod is essentially the same except 10^5 E^- cells in 200 μl are cultured in round-bottomed microtitre wells. Washing of the cells in microtitre wells (see step vi above) is best carried out by sucking off the medium from each well using a 21 gauge needle attached to a suction pump. If the needle is slid down the side of the microtitre well as far as the top of the curvature, the cell pellet will not be disturbed. The cells can then be washed by resuspending in medium and repelleted by centrifugation at 200 *g* for 5 min on a centrifuge with a microtitre tray adaptor.

8. SOURCES OF B CELL GROWTH AND DIFFERENTIATION FACTORS

The most commonly used source of BCGF and BCDF is a three-day supernatant from phytohaemagglutinin-activated T cells. Various procedures can be used to partially purify these factors, but it is extremely difficult to obtain homogeneous preparations, most of which will be contaminated with other lymphokines such as IL-1, IL-2 and IFN. See relevant chapters in this volume. This difficulty can be partially resolved by developing continuous cell lines which secrete BCGF and/or BCDF, but not any of the other recognized lymphokines. Some success has been had with T−T hybridomas (18), and HTLV-I transformed T cell lines (19). More recently, non-lymphoid cell lines secreting BCGF and BCDF have been described (20). Factors from each of these sources are active in the assays described here, but an unequivocal characterization of the different factors active in the different assays will depend upon the availability of recombinant material. Four human B cell growth and differentiation factors have recently been cloned, namely BSF1 (IL-4), BCGFII (IL-5), BSF2 (BCDF) (1−3) and low molecular weight BCGF (24). A partially purified low molecular weight BCGF (21) is commercially available (Cellular Products). The field of human B cell growth and differentiation factors is still in a state of flux, and it is likely that other factors active on B cells will be described and cloned in the near future. It is important to note that certain strains of mycoplasma can induce B cell growth and differentiation (22,23). Because of this, only factors from mycoplasma-free sources should be used.

9. ENZYME IMMUNOASSAYS FOR IMMUNOGLOBULIN SECRETED BY HUMAN B CELLS

All the assays we have used for detection of human Ig and specific antibody are based on solid phase EIA. In each case either horseradish peroxidase (HRP) or alkaline phosphatase-conjugated antibodies can be used.

9.1 Measurement of Ig secretion by B cell lines

The effect of B cell differentiation factors is easily monitored by measuring Ig production with an EIA as follows.

(i) Dispense 75 μl of affinity purified goat anti-human IgG (Sigma Cat No. I3382) at 0.5 μg/ml in bicarbonate buffer pH 9.6 (*Table 11*) and incubate at room temperature overnight. Individual batches of antisera must be pre-titrated to determine the optimal signal-to-noise ratio with high, medium and low concentrations of IgG.

(ii) Wash twice with bicarbonate buffer.

(iii) Add 100 μl of 4% NGS diluted in bicarbonate buffer to each well and leave at room temperature for 30–90 min.

(iv) Wash twice with 0.05% Tween 20 in normal saline.

(v) Add 75 μl of test supernatant, or standards diluted in normal saline containing 0.05% Tween 20 and 4% NGS, and incubate at room temperature for 60–90 min. An 11-point standard curve using doubling dilutions of either pooled normal human serum or partially purified IgG at 1000 ng/ml is set up in duplicate on each plate along with a buffer only zero standard.

(vi) Wash five times with 0.05% Tween 20 in normal saline.

(vii) To each well, add 75 μl of HRP-conjugated affinity purified goat anti-human IgG (Sigma Cat No. A6029) diluted to 1:1000 in normal saline containing 0.05% Tween 20 and 4% NGS. (Alkaline phosphatase-conjugated goat anti-human Ig can be used just as well.) Individual batches of enzyme-coupled antisera should be titrated to determine the optimal dilution.

(viii) Wash five times with 0.05% Tween 20 in normal saline.

(ix) To each well, add 75 μl of *o*-phenylenediamine diluted to 0.5 mg/ml in phosphate/citrate buffer (*Table 11*) containing 0.015% hydrogen peroxide. This reagent is made up just prior to use. For alkaline phosphatase, use *p*-nitrophenyl phosphate substrate (Sigma Cat No. 104-0) at 1 mg/ml in bicarbonate buffer pH 9.6 (*Table 11*) plus 10^{-4} M $MgCl_2$.

(x) Incubate in the dark at room temperature or at 37°C to allow colour development. For HRP, 15 min is usually sufficient. Further development is then stopped by adding 40 μl of 2 M H_2SO_4 to each well, and the absorbance read on an automatic EIA plate reader at 492 nm. For alkaline phosphatase, colour development is slower (>1 h), and the absorbance is read at 405 nm.

All incubations take place in an humidified box. Antisera can be stored undiluted in small aliquots at −20°C, and frozen and thawed a maximum of three times. Immuno-

Table 11. Buffers for EIA.

1.	Bicarbonate buffer, pH 9.6	
	Na_2CO_3	1.59 g
	$NaHCO_3$	2.93 g
	NaN_3	0.20 g
	Distilled water	1 litre
2.	Phosphate citrate buffer, pH 5.0	
	A. Na_2HPO_4	28.4 g
	B. Citric acid	21.0 g
	Both made up to 1 litre in distilled water	
	Mix equal volumes of A and B just prior to use	

globulin standards should be diluted in assay buffer to 10 μg/ml and stored frozen in small aliquots. By substituting class-specific antisera at stages (i) and (vii) above, it is also possible to measure levels of IgA and IgM in culture supernatants.

9.2 EIA for specific antibody production

Specific antibody production *in vitro* can be readily measured by a solid phase EIA. In each assay, a standard curve is constructed to enable the results to be expressed in ng/ml. The standard is obtained by measuring the IgG concentration (Section 9.1) in pooled supernatants from influenza-stimulated PBM which contain only specific antibody. Secondary standards from normal human serum can be used after calibration. Specific antibody containing supernatants are assayed as follows.

(i) Prepare a suspension of influenza virus (same strain as used for stimulation *in vitro*) to 20–100 μg/ml in PBS plus 0.02% azide, and dispense 75 μl into each well of a flat-bottomed non-sterile microtitre tray.
(ii) Incubate for 1 h at 37°C.
(iii) Recover the virus (which may be used at least 20 times) and store at 4°C.
(iv) Wash the plates twice with PBS and once with 1% bovine serum albumin (BSA) (fraction V, Sigma) in PBS.
(v) Incubate with 100 μl/well of 1% BSA in PBS for 1 h at 37°C to block the remaining non-specific binding sites.
(vi) Wash once with PBS containing 1% BSA.
(vii) Add 75 μl of standard (serially diluted from 1:1 to 1:128) to eight consecutive wells for the standard curve, and 75 μl of medium or PBS containing 1% BSA to three wells for the negative control. Add 75 μl of test supernatant to each of the remaining wells.
(viii) Incubate for 1 h at 37°C.
(ix) Wash the plate twice with PBS and once with PBS containing 1% BSA.
(x) Add 75 μl of alkaline phosphatase-conjugated affinity purified goat anti-human IgG (Sigma Cat No. A3150), diluted in PBS containing 1% BSA. This particular antibody can normally be used at 1:1000, and stored at 4°C.
(xi) Incubate for 1 h at 37°C.
(xii) Wash twice in PBS and twice in distilled water.
(xiii) To each well, add 100 μl of *p*-nitrophenyl phosphate substrate for alkaline phosphatase (Sigma Cat No. 104-0) at 1 mg/ml in bicarbonate buffer pH 9.6 containing 10^{-4} M $MgCl_2$ (*Table 11*) and allow the colour to develop (~1 h at 37°C).
(xiv) Read absorbance on a Multiskan automatic EIA plate reader at 405 nm.

The results are expressed in ng/ml calculated from the standard curve of logit OD plotted against $\log_2$ antibody concentration where:

$$\text{logit}_{OD} = \ln \frac{P}{1 - P}$$

where $P = OD/OD_{max}$ and OD_{max} is the optical density obtained when all the NPP is hydrolysed, usually about 11.0.

9.3 Assay for IgG subclass antibody to influenza virus in tissue culture supernatants

(i) Coat non-sterile flat-bottomed microtitre wells with 75 μl of purified influenza virus, block, and add standards and test supernatants as described in Section 9.2 steps i–ix.

(ii) Add 75 μl of monoclonal antibody to human IgG subclass diluted in PBS containing 1% BSA. We use monoclonal antibodies from Unipath Ltd. to IgG1 (NL16) diluted 1:1000, IgG2 (GOM 1), IgG3 (ZG4) and IgG4 (GB7B) all diluted 1:500. Optimal dilutions should be determined for each batch of monoclonal antibody.

(iii) Incubate for 1 h at 37°C, then wash the plate twice in PBS and once in PBS containing 1% BSA.

(iv) Add 75 μl of alkaline phosphatase-conjugated goat anti-mouse IgG. This antibody should be affinity purified and absorbed against human serum proteins to remove any cross-reacting antibody. The TAGO antibody (Cat No. 6550) used in our laboratory was diluted 1:1000 in PBS containing 1% BSA and 0.01% NaN_3.

(v) Incubate for 37°C for 1 h, then wash twice in PBS and twice in distilled water.

(vi) Add 100 μl of *p*-nitrophenyl phosphate substrate for alkaline phosphatase (Cat No. 104-0) at 1 mg/ml in bicarbonate buffer pH 9.6 containing 10^{-4} M $MgCl_2$ (*Table 11*). Incubate for 1–2 h at 37°C, or overnight at 4°C for colour development. Read absorbance on a Multiskan plate reader at 405 nm.

In the absence of callibrated standard preparations of specific IgG subclass antibodies to influenza virus, it is necessary to express the results simply as OD_{405}, or as units per ml calculated from the standard curve of logit OD versus $\log_2$ dilution of a serum or purified human Ig standard. Under these conditions, it is not possible to make quantitative comparisons between IgG subclasses, but by using units/ml, it is possible to make valid intra- and inter-experimental comparisons for any one IgG subclass.

10. REFERENCES

1. Yokota,T., Otsuka,T., Mosmann,T., Banchereau,J., Defrance,T., Blanchard,D., DeVries,J., Lee,F. and Arai,K.-I. (1986) *Proc. Natl. Acad. Sci. USA,* **83**, 5894.
2. Hirano,T., Yasukawa,K., Harada,H., Taga,T., Watanabe,Y., Matsuda,T., Kashiwamura,S.-I., Nakajima,A., Koyama,K., Iwamatsu,A., Tsunasawa,S., Sakiyama,F., Matsui,H., Takahara,Y., Taniguchi,T. and Kishimoto,T. (1986) *Nature,* **324**, 73.
3. Azuma,C., Tanabe,T., Konishi,M., Kinashi,T., Noma,T., Matsuda,F., Yaoita,Y., Takatsu,K., Hammarstrom,L., Smith,C.I.E., Severinson,E. and Honjo,T. (1986) *Nucleic Acids Res.,* **14**, 9149.
4. Lee,F., Yokota,T., Otsuka,T., Meyerson,P., Villaret,D., Coffman,R., Mosmann,T., Rennick,D., Roehm,N., Smith,C., Zlotnik,A. and Arai,K.-I. (1986) *Proc. Natl. Acad. Sci. USA,* **83**, 2061.
5. Noma,Y., Sideras,P., Naito,T., Bergstedt-Lindquist,S., Azuma,C., Severinson,E., Tanabe,T., Kinashi, T., Matsuda,F., Yaoita,Y. and Honjo,T. (1986) *Nature,* **319**, 640.
6. Sanderson,C.J., O'Garra,A., Warren,D.J. and Klaus,G.G.B. (1986) *Proc. Natl. Acad. Sci. USA,* **83**, 437.
7. Kaplan,M.E. and Clark,C. (1974) *J. Immunol. Methods,* **5**, 131.
8. Callard,R.E. and Smith,C.M. (1981) *Eur. J. Immunol.,* **11**, 206.
9. Proust,J.J., Chrest,F.J., Buchholz,M.A. and Nordin,A.A. (1985) *J. Immunol.,* **135**, 3056.
10. Walker,L., Guy,G., Brown,G., Rowe,M., Milner,A.E. and Gordon,J. (1986) *Immunology,* **58**, 583.
11. Muraguchi,A., Kishimoto,T., Miki,Y., Kuritani,T., Kaieda,T., Yoshizaki,K. and Yamamura,Y. (1981) *J. Immunol.,* **127**, 412.
12. Mosmann,T. (1983) *J. Immunol. Methods,* **65**, 55.
13. Cambier,J.C., Monroe,J.G., Coggeshall,K.M. and Ransom,J.T. (1985) *Immunol. Today,* **6**, 218.
14. Saiki,O. and Ralph,P. (1981) *J. Immunol.,* **127**, 1044.

15. Callard,R.E. (1979) *Nature,* **282**, 734.
16. Callard,R.E., Booth,R.J., Brown,M.H. and McCaughan,G.W. (1985) *Eur. J. Immunol.,* **15**, 52.
17. Zanders,E.D., Smith,C.M. and Callard,R.E. (1981) *J. Immunol. Methods,* **47**, 333.
18. Butler,J.L., Falkoff,R.J.M. and Fauci,A.S. (1984) *Proc. Natl. Acad. Sci. USA,* **81**, 2475.
19. Shimizu,K., Hirano,T., Ishibashi,K., Nakano,N., Taga,T., Sugamura,K., Yamamura,Y. and Kishimoto,T. (1985) *J. Immunol.,* **134**, 1728.
20. Rawle,F.C., Shields,J.G., Smith,S.H., Iliescu,V., Merkenschlager,M., Beverley,P.C.L. and Callard, R.E. (1986) *Eur. J. Immunol.,* **16**, 1017.
21. Maizel,A., Sahasrabuddhe,C.G., Mehta,S., Morgan,J., Lachman,L. and Ford,R. (1982) *Proc. Natl. Acad. Sci. USA,* **79**, 5998.
22. Proust,J.J., Buchholz,M.A. and Nordin,A.A. (1985) *J. Immunol.,* **134**, 390.
23. Ruuth,E. and Lundgren,E. (1986) *Scand. J. Immunol.,* **23**, 575.
24. Sharma,S., Mehta,S., Morgan,J. and Maizel,A. (1987) *Science,* **235,** 1489.

APPENDIX I

Suppliers of Specialist Items

Accurate Chemicals and Scientific Corp., 300 Shames Drive, Westbury, NY 11590, USA.

American Type Culture Collection, Sales Department, 12301 Parklawn Drive, Rockville, MD 20852, USA.

Amersham International Plc, White Lion Road, Amersham, Bucks HP7 9LL, UK; Amersham Corp., 2636 South Clearbrook Drive, Arlington Heights, IL 60005, USA.

Amgen Biologicals, 1900 Oak Terrace Lane, Thousand Oaks, CA 91320, USA.

Amicon Ltd, Upper Mill, Stonehouse, Gloucestershire GL10 2BJ, UK; Amicon Corporation, Danvers, MA, USA.

BDH Chemicals Ltd, Broom Road, Poole, Dorset BH12 4NN, UK; BDH Chemicals, Gallards Schlesinger Chemicals Mfg Corporation, 584 Mineola Avenue, Carle Place, NY 11514, USA.

Beckman Ltd, Progress Road, Sands Industrial Estate, High Wycombe, Buckinghamshire, UK; Beckman Instruments, 6200 El Camino Real, Carlsbad, CA 92008, USA.

Becton Dickinson UK Ltd, Laboratory Division, Between Towns Road, Cowley, Oxford OX4 3LY, UK; Becton Dickinson, 490-B Lakeside Drive, Sunnydale, CA 94086, USA.

Bethesda Research Labs, PO Box 35, Trident House, Renfrew Road, Paisley PA3 4EF, Scotland, UK; Bethesda Research Labs, Gaithersberg, MD 20877, USA.

Biogen, 14, Cambridge Center, Cambridge, MA 02142, USA.

Bio-Rad Laboratories Ltd, Holywell Industrial Estate, Watford, Herts WD1 8RP, UK; Bio-Rad Laboratories, 2200 Wright Avenue, Richmond, CA 94804, USA.

Biotest (UK) Ltd, 171 Alcester Road, Moseley, Birmingham B13 8JR, UK; US Suppliers: Seralc Corp., NW 15960, 15 Avenue, Miami, FL, USA.

Boehringer Mannheim, Bell Lane, Lewes, E. Sussex BN7 1LG, UK; Boehringer Mannheim, 7941 Castleway Drive, Indianapolis, IN 46250, USA.

Boots/Celltech, 244–250 Bath Road, Slough, Berks SL1 4DY, UK.

Brownlee Labs Inc. (Anachem), Charles Street, Luton, Bedfordshire LU2 0EB, UK; Brownlee Labs Inc., 2045 Martin Avenue, 204 Santa Clara, CA 95050, USA.

Burroughs Wellcome, Research Triangle, NC, USA.

Calbiochem-Behring (CP Laboratories Ltd), PO Box 22, Bishops Stortford, Herts CM22 7RD, UK; Calbiochem-Behring Corp., 10933 N. Torrey Pines Road, La Jolla, CA 92037, USA.

Cambridge Biotech Labs, Uniscience Ltd, 12–14 St Anns Crescent, London SW18 2LS, UK.

Camlab Ltd, Nuffield Road, Cambridge CB4 1TH, UK.

Celltech Ltd, 224 Bath Road, Slough SL1 4DY, UK.

Cellular Products Inc., 688 Main Street, Buffalo, NY 14202, USA.

Centocor, 244 Great Valley Parkway, Malvern, PA 19355, USA.

Cetus Diagnostic, 1400 53rd Street, Emeryville, CA 94608, USA.

Chemical Dynamics Group, Hadley Road, PO Box 395, South Planfield, NJ 07080, USA.

Cistron Biotechnology (Lab Impex Ltd), Lion Road, Twickenham, Middlesex TW1 4JF, UK; Cistron Biotechnology, Box 2004, 10 Bloomfield Avenue, Pine Brook, NJ 07058, USA.

Collaborative Research, 12–14 St Anns Crescent, London SW18 2LS, UK; Collaborative Research Inc., 128 Spring Street, Lexington, MA 02173, USA.

Coulter Electronics Ltd, Northwell Drive, Luton, Beds LU3 3RH, UK.

Daiichi Pure Chemicals Co. Ltd, SF Kowa, Fumisei Building, 13-S 3-Chome, Nihonbashi, Chuo-ku, Tokyo 103, Japan.

Dextran Products Ltd, Solway House, Flanders Road, Hedge End, Southampton SO3 4QH, UK; Dextran Products Ltd, 421 Comstock Road, Scarborough, Ontario, Canada M1L 2H5.

Difco, Central Avenue, West Molesey, Surrey KT8 0SE, UK; Difco Laboratories, PO Box 1058, Detroit, MI 48232, USA.

DuPont (New England, Nuclear) UK Ltd, Wedgewood Way, Stevenage, Herts, UK; DuPont Research Products, 549 Albany Street, Boston, MA 02118, USA.

Dynatech, Daux Road, Billingshurst, Sussex RH14 9SJ, UK; Dynatech Laboratories Inc., 900 Slaters Lane, Alexandria, VA 22314, USA.

Electro Nucleonics Inc., 12050 Tech Road, Silver Spring, MD 20904, USA.

E.M.Reagents (BDH Chemicals Ltd), Broom Road, Poole, Dorset BH12 4NN, UK; E.M.Reagents, M^c/B Manufacturing Chemists Inc., 2909 Highland Avenue, Cincinnati, OH 45212, USA.

European Collection of Animal Cell Cultures, Porton Down, Salisbury SP4 0JG, UK.

Fisher Scientific Co., Howell A.R. (Reagents Ltd), 73 Maygrove Road, West Hampstead, London NW6 2BP, UK; Fisher Scientific Co., Allied Corp, 711 Forbes Avenue, Pittsburgh, PA 15219, USA.

Flow Labs, Woodcock Hill, Harefield Road, Rickmansworth, Herts WD3 1PQ, UK; Flow Labs Inc., 7655 Old Springhouse Road, McLean, VA 22102, USA.

Fluka (Fluorochem), Peakdale Road, Glossop, Derbyshire SK1 9XE, UK; Tridom Chemical Inc., 255 Oscer Avenue, Hauppauge, NY 11787, USA.

FMC Corp., Rockland, ME 04841, USA.

Genzyme (Koch-Light), Rookwood Way, Haverhill, Suffolk CP9 8PB, UK; Genzyme Corp, 75 Kneeland Street, Boston, MA 02111, USA.

Gibco-BRL, PO Box 35, Trident House, Renfrew Road, Paisley PA3 4EF, Scotland, UK; Gibco Life Technologies Inc., 421 Merrimack Street, Lawrence, MA 01843, USA.

Gillette Ltd, Great West Road, Isleworth, Middlesex, UK.

Gilson (Anachem Ltd), 15 Power Court, Luton, Beds LU2 7QE, UK; Gilson (Rainin Instruments), Woburn, MA, USA.

Hitachi Scientific Instruments, Nissei Sangyo, 4 Suttons Industrial Park, London Road, Reading, Berks RG6 1AZ, UK.

Hyclone Laboratories Inc., 1725 South State Hwy 89–91, Logan, UT 84321, USA.

Ilacon Ltd, Gilbert House, River Walk, Tonbridge, UK.

Ilford Ltd, Mobberley, Knutsford, Cheshire WH16 7NA, UK.

Immunotech, Fishpond Road, Wokingham, Berks RG11 2QA, UK.

Interferon Sciences Inc., 783 Jersey Avenue, New Brunswick, NJ 08901, USA.

Janssen, PO Box 37, Stoke Court, Stoke Poges, Slough SL2 4IY, UK; Janssen Inc., 40 Kingsbridge Road, Piscataway, NJ 08854, USA.

La Calhene, 22 Hills Road, Cambridge CB2 1JP, UK.

Lee Biomolecular Research Laboratories Inc., 11211 Sorrento Valley Road, San Diego, CA 92121, USA.

LKB, 232 Addington Road, South Croydon, Surrey CR2 8YO, UK; LKB Instruments Inc., 12221 Parklawn Drive, Rockville, MD 20852, USA.

Luckham, Burgess Hill, Sussex RH15 9QN, UK.

Miles Scientific Labs (ICN Biochemical Ltd), Free Press House, Castle Street, High Wycombe, Bucks HP13 6RN, UK; Miles Scientific Labs, 2000 North Aurora Road, Naperville, IL 60566, USA.

Millipore Co., Millipore House, 11–15 Peterborough Road, Harrow, Middlesex HA1 2YH, UK.

Nagel (Machery-Nagel Co.), PO Box 307, Düren D5160, FRG.

National Institutes of Health, Research Resources Branch, Bethesda, MD 20205, USA.

New England Biolabs, CPL Laboratories, PO Box 22, Bishops Stortford, Hertfordshire, UK; New England Biolabs, Beverly, Massachussetts, USA.

New England Nuclear (see DuPont).

Nissui Seiyaku Co., 2-5-11 Komangome, T, Toshima-ku, Tokyo 170, Japan.

Northumbria Biologicals, South Nelson Industrial Estate, Cramlington, Northumberland NE23 9HL, UK.

Nunc (see Gibco-BRL).

Nycomed (see Nyegaard).

Nyegaard Ltd, Mylen House, 11 Wagon Lane, Sheldon, Birmingham B26 3DU, UK; USA Accurate Chemical & Scientific Corp., 300 Shames Drive, Westbury, NY 11590, USA.

Oxoid Ltd, Wade Road, Basingstoke, Hampshire RG24 0PW, UK; Oxoid USA Ltd, 9017 Red Branch Road, Columbia, MD 21045, USA.

Packard Instrument Co. Ltd, 13–17 Church Road, Caversham, Bucks, UK; Packard Instrument Co. Ltd, 2200 Warrenville Road, Downers Grove, IL 60515, USA.

Perkin Elmer Ltd, Post Office Lane, Beaconsfield, Bucks HP9 1QA, UK.

Pharmacia, Pharmacia House, Midsummer Boulevard, Milton Keynes MK9 3HP, UK; Pharmacia Diagnostic, 800 Centennial Avenue, Piscataway, NJ 08854, USA.

Pierce & Warrington, 44 Upper Northgate Street, Chester, Cheshire CH1 4EF, UK; Pierce Chemical Co., Box 117, Rockford, IL 61105, USA.

Polaroid, Ashley Road, St Albans, Herts, UK.

Sartorius Instruments Ltd, 18 Avenue Road, Belmont, Surrey SN2 6JD, UK.

Schleicher & Schuell (Anderman and Co. Ltd), Central Avenue, East Molesey, Surrey KT8 0QZ, UK; Schleicher & Schuell Inc., Keene, NH, USA.

Schwarz Mann, 56 Rogers Street, Cambridge, MA 02142, USA.

Seragen Inc., 54 Clayton Street, Boston, MA 02122, USA.

Seralabs, Crawley Down, Sussex RH10 4FF, UK.

Sigma Chemicals Co. Ltd, Fancy Road, Poole, Dorset BH17 7NH, UK; Sigma, PO Box 14508, St Louis, MO 63178, USA.

Sterilin Ltd, Sterilin House, Clockhouse Lane, Feltham, Middlesex TW14 8QS, UK.

T Cell Sciences (see Boehringer Mannheim).
Techne (Cambridge Ltd), Duxford, Cambridge CB2 4PZ, UK.
Tekmar Co., Cincinnati, OH, USA.
TSK-Anachem Ltd, 20 Charles Street, Luton, Bedfordshire LU2 0EB, UK.
Unipath Ltd, Norse Road, Bedford MK41 0QG, UK.
Uniscience Labs, 12–14 St Anns Crescent, London SW18 2LS, UK.
United States Biochemicals Corporations, Cleveland, OH, USA.
Vickers Medical, Basingstoke, UK.
Waters Associates, 324 Chester Road, Hartford, Northwich, Cheshire CW8 24H, UK; Waters Associates Inc., 34 Maple Street, Milford, MA 01757, USA.
Weddel Pharmaceuticals Ltd, Wrexham, Clwyd, Wales, UK.
Whatman Chemicals, Springfield Mill, Maidstone, Kent ME14 2LE, UK; Whatman Chemicals, 9 Bridewell Place, Clifton, NJ 07014, USA.

APPENDIX II

Cloned and purified cytokines and their properties

Factor (Synonym)	*Source*	*Actions*
Interleukin 1α & β (Lymphocyte activating factor) (Epidermal cell derived thymocyte activating factor) (Haemopoietin H1)	Multiple, monocytes, lymphocytes, keratinocytes	Multiple: inflammatory, fever, cytotoxicity, interleukin 2 production haemopoietic activity
Interleukin 2 (T-cell growth factor)	T-lymphocytes	T and B cell proliferation and differentiation, macrophage activation, natural killer cell activation
Interleukin 3 (Multi CSF) (Haemopoietin H2)	T-lymphocytes	Pluripotent growth and differentiation, mast cell growth
Interleukin 4 (B-cell growth factor I) (B-cell stimulatory factor I)	T-lymphocytes	T and B cell proliferation and differentiation
Interleukin 5 (Eosinophil differentiation factor) (B-cell growth factor II)	T-lymphocytes	Eosinophil differentiation (mouse IL-5 also acts as a B-cell growth factor, human does not)
Low molecular weight B-cell growth factor	T-lymphocytes	B-cell proliferation
B-cell stimulatory factor 2 (interferon β2)	T-lymphocytes	B-cell differentiation
Interferon γ (immune interferon) (Type II interferon)	T-lymphocytes	Anti-viral, macrophage activation, MHC expression, cell growth inhibition
Interferon α (leukocyte interferon) (Type I interferon)	Multiple	Anti-viral, cell growth inhibition, induction of differentiation
Interferon β1 (fibroblast interferon) (Type I interferon)	Multiple	Anti-viral, cell growth inhibition, induction of differentiation
Granulocyte colony stimulating factor	Multiple	Growth and differentiation of granulocytes
Macrophage colony stimulating factor (colony stimulating factor 1)	Multiple	Growth and differentiation of monocytes/macrophages
Granulocyte macrophage colony stimulating factor	Multiple	Growth and differentiation of monocytes/granulocytes
Tumour necrosis factor α	Monocytes/ lymphocytes	Cytotoxicity, cacheia, haemolytic necrosis, bone resorption, fever
Tumor necrosis factor β (lymphotoxin)	T-lymphocytes	Cytotoxicity, cacheia, haemolytic necrosis, bone resorption, fever

INDEX